THE METAL *Plutonium*

Contributors

R. ABRAMSON
F. ANSELIN
H. L. BAUMBACH
W. M. CASHIN
A. S. COFFINBERRY
E. M. CRAMER
B. B. CUNNINGHAM
R. J. DUNWORTH
F. H. ELLINGER
DANA E. ELLIOTT
R. O. ELLIOTT
H. M. FINNISTON
FRANK G. FOOTE
S. FRIED
R. M. FRYAR
E. GRISON
DAVID H. GURINSKY
L. L. HAWES
A. E. KAY
L. R. KELMAN
R. M. KIEHN
P. L. KIRK
HENRY L. LAQUER
A. C. LARSON
J. A. LEE
P. G. MARDON
W. N. MINER
H. MONTI
S. F. PUGH
O. J. C. RUNNALLS
THOMAS A. SANDENAW
F. W. SCHONFELD
GLENN T. SEABORG
CYRIL STANLEY SMITH
RUSSELL E. L. STANFORD
R. E. TATE
GEROLD H. TENNEY
J. T. WABER
M. B. WALDRON
K. L. WAUCHOPE
E. F. WESTRUM, JR.
E. S. WRIGHT
W. H. ZACHARIASEN

Edited by

A. S. COFFINBERRY

and

W. N. MINER

THE METAL

Plutonium

THE UNIVERSITY OF CHICAGO PRESS

THE UNIVERSITY OF CHICAGO PRESS, CHICAGO & LONDON
The University of Toronto Press, Toronto 5, Canada

THE METAL PLUTONIUM

Composed and printed by The University of Chicago Press, Chicago, Illinois

Library of Congress Catalog Card Number: 61-17072

PREFACE

Plutonium, discovered less than twenty years ago, is the first man-made element to be produced on an industrial scale. In addition to being used as the energy source of atomic weapons and as a fuel for nuclear reactors, it has some unique and very surprising properties which make it of special scientific interest and may lead to applications other than in atomic energy. The mysteries of the remarkable behavior of pure plutonium are still very far from being solved. Their solution promises to be an exciting field for physical metallurgists and solid-state physicists for many years.

This book's carefully planned assemblage of papers by plutonium pioneers will be, the editors believe, an essential volume in the libraries of scientists and engineers interested in the properties and applications of plutonium. Introductory chapters give comprehensive background; the body of the book covers physical metallurgy and projects at various laboratories; concluding chapters describe applications in reactor fuels. The scope is international, the authors being from Canada, France, Great Britain, and the United States. Also, recognition is given Russian work to the extent that some plutonium phase diagrams published in the U.S.S.R. are reported and discussed.

The book is the outgrowth of a two-day conference held under the auspices of the American Society for Metals and the U.S. Atomic Energy Commission and planned by Ernest E. Thum, editor-in-chief of *Metal Progress* until his death in April, 1961; Edward Epremian, then of the AEC; and A. S. Coffinberry, of the Los Alamos Scientific Laboratory. In his capacity as general chairman of this conference, Coffinberry was assisted by Glenn T. Seaborg, Cyril Stanley Smith, H. M. Finniston, Edward Epremian, Eric R. Jette, John Chipman, and O. J. C. Runnalls, who served as presiding officers during the four half-day sessions.

The comprehensive character of this book arises from its being based on the first plutonium conference, which was held in November, 1957, in connection with the World Metallurgical Congress meeting in Chicago. Though publication was delayed in part by the 1958 Geneva Conference on the Peaceful Uses of Atomic Energy, we—with the help of the authors—have taken advantage of the additional time to update significantly a major portion of the material.

Very few chapters of this volume have been duplicated or superseded by papers given since the first conference. Where notable items of progress have been reported, the chapters of this book provide important background information to the new reports. We refer, primarily, to the plutonium conference at Grenoble, France, in the spring of 1960, in which many of the papers made reference to chapters in this book, and also to the plutonium symposium held in San Francisco in early 1959 by the American Institute of Metallurgical Engineers. The San Francisco and Grenoble papers in book form thus constitute, in effect, Volumes **2** and **3** of a series of which this is Volume **1**.

We call attention to the fact that this book recounts progress during the first fourteen years of plutonium work—from 1943 to 1957—plus additional more recent matters of consequence. Hence the book contains many chapters which report the results of classic investigations not published elsewhere. Examples of such reports are Smith's account of the work at Los Alamos for the Manhattan Engineering District, Laquer's studies of the elastic constants of alpha plutonium, Sandenaw's report on Comstock's (the earliest and probably the best) measurements of the magnetic susceptibilities of the plutonium allotropes, Cramer's summary of the dilatometric work done at Los Alamos over a period of several years, Elliott and Larson's chapter on delta-prime plutonium (reporting the first and almost the only work done on this phase), and Zachariasen's chapter on the crystal structures and thermal expansions of the plutonium phases. (He explains for the first time the method by which he arrives at the values of atomic valences discussed in this and other publications. This information has long been wanted by those interested in solid-state chemistry and the more theoretical aspects of metallurgy.)

The organization of the book has been carefully planned. The short historical introduction serves as a framework interrelating the various chapters. It was considered that a presentation of plutonium metallurgical research chronologically and geographically would be the most satisfactory means of communicating to metallurgists in general the highly specialized knowledge that has been the intimate concern of only a very few people. For this reason, Part I (chaps. i through ix) begins with the earliest history of plutonium as a chemical element, follows the evolution of its extractive and physical metallurgy through to its current status, and presents a conspectus of its future.

With this background establishing the milieu in which plutonium metallurgical research has been pursued, it is logical that a major portion of the book be devoted to presenting the results of the various research projects that have been undertaken at the several laboratories. Chapters x through xxvi (Part II) discuss different aspects of the physical metallurgy of plutonium and its alloys. Only in chapter xxvi is attention directed primarily to extractive metallurgy, and here only to processes for the reduction of plutonium compounds to metal.

The chapters pertaining to the various topics in physical metallurgy not only represent a widespread geographical distribution within the world of plutonium metallurgy but also cover an extensive period of time. In many instances, exemplified in parts of chapters xv, xvi, xxii, and xxv, results are reported for work done at

Los Alamos prior to 1950. When first obtained, these data were classified, and many are now published for the first time.

The concluding chapters discuss the applications of plutonium alloys and compounds as fuels in nuclear reactors. Thus chapters xxvii through xxxv (Part III), which report these discussions, explain, first, the nuclear properties of plutonium as a reactor fuel (chap. xxvii), then describe ways in which plutonium alloys have been used in the cores of nuclear reactors (chaps. xxviii and xxix), and conclude by discussing design concepts for fuels and reactors based on plutonium-containing materials.

The editors express appreciation to the planners of the Chicago meeting and to the afore-mentioned chairmen of the four sessions. We are also grateful to John Parina, chief of the book department of the American Society for Metals, for his work on the manuscripts and to the Office of Technical Information of the AEC for arranging publication.

A. S. COFFINBERRY
W. N. MINER

TABLE OF CONTENTS

PART III

PLUTONIUM IN NUCLEAR REACTORS

INTRODUCTION

A. S. Coffinberry

Plutonium was discovered early in 1941 by Glenn T. Seaborg and associates at the Radiation Laboratory of the University of California. During the remainder of that year and early 1942, its chemical properties were studied at Berkeley on the tracer scale by A. C. Wahl and others in the Department of Chemistry. Meanwhile, continuing investigations of the nuclear properties of plutonium at the Radiation Laboratory established the great value of Pu^{239} as a fissionable isotope, capable of releasing enormous amounts of nuclear energy through a chain reaction, and led to the inauguration of the wartime Plutonium Project, having as its objectives the large-scale production of Pu^{239} and the possible use of this isotope in a nuclear weapon.

The well-known Smyth Report describes the successful solution of the problem of establishing a chain reaction in natural uranium, which thereby demonstrated the possibility for large-scale production of plutonium in reactors. The solving of this problem was the primary objective of the Metallurgical Laboratory established at the University of Chicago in 1942. A further objective was the chemical separation of plutonium from the other materials present in irradiated uranium. Hence the continuation of research on the chemistry of plutonium that was undertaken at the Metallurgical Laboratory was directed toward the development of a satisfactory separation method. At the beginning of this research, tracer techniques still had to be used, but, as the result of a decision to increase appreciably the amount of plutonium then being produced in cyclotrons, B. B. Cunningham was able to perform the first of what have been called "ultramicroscale" experiments with plutonium. Moreover, as a consequence of his development of "ultramicrochemical" techniques, he and L. B. Werner were able in September, 1942, to realize the first isolation and weighing of a compound of pure plutonium (2.77 μg of PuO_2).

The quantities of cyclotron-produced plutonium ultimately made available proved to be sufficient for metallurgical investigations to be started at the Metallurgical Laboratory before the end of 1943, when reactor-produced plutonium began to be extracted from the fuel elements of the Clinton pile. During 1943 the first

Arthur S. Coffinberry is at the Los Alamos Scientific Laboratory, Los Alamos, New Mexico.

reductions of plutonium to metal were made at the Metallurgical Laboratory by H. L. Baumbach, S. Fried, and associates, who employed micrometallurgical methods that were unprecedented in the Western world. During 1944, as soon as plutonium from the Clinton works became available in gram quantities, the major effort in metallurgical research on plutonium was transferred to Los Alamos. The "large-scale" production of plutonium metal began at Los Alamos in February, 1945, following receipt there of the first shipment of the product of the Hanford reactors and their associated chemical-processing plant.

During the war years the metallurgy of plutonium was studied at Los Alamos within the Chemistry-Metallurgy Division, which was administered jointly by Joseph W. Kennedy and Cyril Stanley Smith. At that time the investigations were carried only as far as was deemed necessary for utilization of the metal as a component in nuclear weapons. Soon after the conclusion of hostilities, however, plans were made for the construction of an experimental fast reactor in which plutonium would be used as the fuel. This project led to the first plutonium-fueled reactor, "Clementine," which was brought to full power (only 25 kw, however) in March, 1949.

The metallurgical knowledge utilized in fabricating the Clementine fuel elements did not extend beyond what had been learned during the war and consequently did not provide a type of fuel suitable for commercial power reactors. It was not until 1949 that serious attention was given at Los Alamos to the study of plutonium alloys that might be used in solid-fuel elements at the high temperatures required in industrial power reactors. Intensive investigation of liquid plutonium fuel alloys did not begin until even later—in 1955.

Meanwhile, another postwar development was the evolution of the Argonne National Laboratory as the successor to the Metallurgical Laboratory at the University of Chicago. From its inception, the primary responsibility of ANL has been reactor development, but initially the types of reactors designed and constructed there and at its Arco, Idaho, extension employed only uranium as the fissile material. However, the inevitability of using plutonium as a fuel for power reactors led in 1954 to the installation at ANL of a small laboratory for plutonium metallurgical research. To this modest beginning has since been added an impressive plant for the fabrication of plutonium-alloy fuel elements, and a much larger metallurgical research laboratory for both plutonium and uranium is now being constructed there. The over-all supervision of these projects is the responsibility of Frank G. Foote; Arthur B. Schuck and L. R. Kelman have been in charge of developing the detailed plans for the fuel-element facility and the research laboratory, respectively.

At present, research in the field of plutonium metallurgy leading to publishable results is being done in the United States at only two other AEC installations—the Hanford Atomic Products Operation at Richland, Washington, and the Mound Laboratory of the Monsanto Chemical Company at Miamisburg, Ohio. Initially, the reduction of plutonium to metal was not a step included in the processing done at Hanford. In recent years, however, and especially since the development of a plutonium-fueled power-reactor concept has become a project there, both the extractive and the physical metallurgy of plutonium have become major activities at Hanford.

Since 1956, when metallurgical work with plutonium at the Mound Laboratory began, this work has consisted primarily of the production of neutron sources containing a plutonium-beryllium alloy. More recently, however, this laboratory has extended its program in plutonium metallurgy to include the investigation of several problems pertinent to the liquid-plutonium-reactor (LAMPRE) project at Los Alamos.

Although not equipped to do metallurgical research on plutonium, several other American laboratories engaged in the development of power-reactor concepts have devoted extensive study to the potentialities of plutonium as a reactor fuel. As early as 1950, T. M. Snyder, A. U. Seybolt, and others at the Knolls Atomic Power Laboratory gave consideration to ways in which plutonium could be made the fuel in a power-breeder reactor, and they requested the plutonium metallurgy group at Los Alamos to perform certain experiments pertinent to their design concept. More recently the interest at KAPL in the fuel uses of plutonium has been directed toward a cermet type of element in which plutonium is combined with uranium in an oxide phase.

In 1952 Los Alamos received a second request to undertake a metallurgical investigation in the interest of reactor development at another laboratory. This task was the determination of the solubility of plutonium in molten bismuth, information needed by D. H. Gurinsky and associates at the Brookhaven National Laboratory if they were to adapt their liquid-metal-fuel-reactor concept to operation with plutonium.

In the United States the only co-operative group of power and related industries who are actively engaged in the development of fast, rather than thermal, reactors for the generation of electric power is Atomic Power Development Associates of Detroit, Michigan. Although their Enrico Fermi Reactor will initially be fueled with enriched uranium, they are fully cognizant of the advantages of using plutonium as the logical fuel for a fast reactor. R. E. L. Stanford, A. P. Donnell, and others at APDA have given particular attention to concepts of a plutonium fuel in fluid form, either liquid or paste.

During the war the beginnings of British activities directed toward a study of the military potentialities of nuclear energy were centered mainly in the Cavendish Laboratory at Cambridge University, although production of uranium metal and hexafluoride was undertaken by Imperial Chemical Industries in the Liverpool area in 1941. However, because of the danger of enemy action and in order to associate the British and American efforts more closely, the British nuclear research program was transferred to Montreal in 1943. Growth of the Montreal project led in turn to the establishment in 1944 of a permanent reactor and chemical-processing site at Chalk River, Ontario. Construction at this site of the NRX reactor resulted in the beginning of plutonium production in October, 1947, and later led to a study of plutonium alloys by O. J. C. Runnalls that was oriented primarily toward their possible use as "spike" elements to be added to the natural uranium fuel of the NRX core.

Meanwhile, the end of the war had made possible the return to Great Britain of the United Kingdom's major program in nuclear research and resulted in the creation of several British national laboratories, such as the Atomic Energy Research Establishment (1946) at Harwell and the Atomic Weapons Research Establishment (1951) at Aldermaston. From their inception, plutonium metallurgy has played a prominent role in the research programs of these laboratories. At AERE the plutonium physical metallurgy section is supervised by M. B. Waldron, with H. M. Finniston (until recently) providing over-all metallurgical direction. At AWRE a corresponding group reports to A. E. Kay, with W. B. H. Lord in charge of metallurgical activities at the higher level.

Although the operation of plutonium-producing reactors in Great Britain is administered by the Industrial Group of the Atomic Energy Authority, research laboratories for metallurgical studies of plutonium have not existed within the widespread IG organization until quite recently. Currently coming into operation at Dounreay, Scotland, is an IG laboratory in which, along with other metallurgical investigations, those pertinent to plutonium fueling of the Dounreay fast reactor will be undertaken.

In France, plutonium metallurgy began in January, 1956, when several grams of metal extracted from the fuel slugs of the EL-2 reactor at Saclay were made available to the plutonium laboratory at Fontenay-aux-Roses. This laboratory was established by renovating and adapting to work with plutonium the venerable Fort de Chatillon, built on the outskirts of Paris soon after the war of 1870. At this laboratory, now a part of the Centre d'Étude Nucléaires de Fontenay-aux-Roses, research on the physical metallurgy of plutonium, and especially with regard to its power-reactor applications, is being pursued by a group of metallurgists and physical chemists under the direction of E. Grison.

The first account of work done in Russia on plutonium metal and its alloys was presented at a meeting held in Moscow during July, 1955, by the U.S.S.R. Academy of Sciences for the purpose of discussing peaceful uses of atomic energy. Judging from the quantity and variety of results reported at that time, it seems likely that the work was begun well in advance of the date of the meeting. Moreover, because microscale methods of metallurgical research had been developed in Moscow previous to World War II, the availability of these techniques enabled the Russian plutonium metallurgists to begin their work as soon as only microgram amounts of the man-made metal had been produced in the Soviet reactors.

Only S. T. Konobeevsky was designated as the author of this first Russian paper on plutonium metallurgy, but it was acknowledged therein that the data discussed represented the combined contributions of a group of scientists. Among the authors of subsequent Russian papers on plutonium and its alloys, the names of both Konobeevsky and A. A. Bochvar appear prominently, and from this circumstance it might be concluded that research in the field of plutonium physical metallurgy is being pursued only at the Moscow Institute of Non-ferrous Metals, an institution with which Konobeevsky and Bochvar have long been associated. However, in conversations between Soviet and Western plutonium metallurgists that took place at the Second United Nations International Conference on the Peaceful Uses of Atomic Energy (Geneva, 1958), it was learned that plutonium metallurgical research was at that time being conducted by more than one hundred graduate scientists employed at some eight different laboratories under the control of the Moscow Institute for Advanced Studies (with which, presumably, Konobeevsky, Bochvar, and their co-workers are also associated).

PART I

HISTORICAL REVIEW

CHAPTER I

THE DISCOVERY OF PLUTONIUM IN THE CYCLOTRON

Glenn T. Seaborg

Plutonium is unique among the chemical elements: it is a synthetic element, the first realization of the alchemist's dream of large-scale transmutation; it was the first synthetic element produced in visible amounts; the nuclear properties of its isotope Pu^{239} are of great importance in the affairs of man; it has unusual and interesting chemical and metallurgical properties; it is one of the most dangerous poisons that man must learn to handle; and it was discovered and methods for its production were developed in secrecy during World War II. The story of its discovery and of the research leading to its large-scale production is among the most fascinating in the history of science.

The story of the transuranium elements begins with the discovery by E. Fermi, E. Amaldi, O. D'Agostino, F. Rasetti, and E. Segrè in 1934 that the neutron irradiation of uranium produced a number of radioactive substances. On the basis of their chemical work, Fermi and his co-workers concluded that these substances were transuranium elements because they decayed by beta emission, a process that might result in elements of higher atomic number than uranium, and because they differed in chemical properties from all known elements in the region of uranium. For the following several years, these "transuranium elements" were investigated by O. Hahn, L. Meitner, and F. Strassmann, among others. At first, their experiments appeared to confirm the point of view that these were, in fact, transuranium elements. But, early in 1939, Hahn and Strassmann, in their work leading to the discovery of fission, found that at least some of the radioactive substances produced by the bombardment of uranium with neutrons were isotopes of barium and lighter elements. Most of the radioactivities previously assigned to the transuranium elements were later found to be due to fission products.

Further investigation of the fission process and products resulted in the discovery of the first transuranium element, neptunium. Several investigators, including E. M. McMillan of the University of California at Berkeley, measured the energies of the two main fission fragments by observing the distances they traveled from each other as a result of their mutual recoil from the fissioning nucleus. Mc-

Glenn T. Seaborg is now Chairman of the United States Atomic Energy Commission. He was formerly at the University of California, Radiation Laboratory and Department of Chemistry, Berkeley, California.

Millan noticed another product of the reaction of uranium with neutrons, with a half-life of 2.3 days, which did not recoil sufficiently to escape from the thin layer of fissioning uranium. He suspected that this was a product formed by simple neutron capture, a process which does not release much energy. In the spring of 1940, McMillan and P. H. Abelson were able to prove by chemical means that this product is an isotope of element 93, arising from the beta decay of the U^{239} (23-minute half-life) formed. The element was named "neptunium," in recognition of its position beyond uranium, after the planet Neptune, which lies immediately beyond Uranus.

The discovery and early investigations of plutonium were also carried out at the University of California at Berkeley. In the fall of 1940, A. C. Wahl, then a graduate student, at my suggestion, began a study of the tracer chemical properties of element 93. His work and related work on element 94 was carried out in collaboration with J. W. Kennedy, who, like myself, was an instructor in the Department of Chemistry. After McMillan's departure from Berkeley in November, 1940, and his gracious assent to our continuing the work he had begun, our group turned its major effort to the search for element 94.

These experiments began on December 14, 1940, with the bombardment of uranium oxide with 16-Mev deuterons in the 60-inch cyclotron at Crocker Laboratory. We found that an alpha radioactive substance grew into the element 93 fraction that had been separated chemically from the rest of the target material. During the following two months, we were successful in separating this alpha activity from the other elements present—in particular, elements 90–93. In these experiments, in which the positive identification of element 94 was established, we found that the element had at least two oxidation states, distinguishable by their precipitation chemistry, and that it required stronger oxidizing agents to oxidize element 94 to the upper state than was the case for element 93. The key step in the discovery of element 94 was its first successful oxidation by peroxydisulfate ion and silver-ion catalyst on the night of February 23–24, 1941. The isotope identified was found to be Pu^{238}, and it had been produced by the following reactions:

$$_{92}U^{238} + {}_{1}H^{2} \rightarrow {}_{93}Np^{238} + 2n$$

$$_{93}Np^{238} \xrightarrow[2.1\text{ days}]{\beta^-} {}_{94}Pu^{238} \text{ (90 years, } \alpha\text{)} .$$

The announcement of this significant discovery was withheld, although this work antedated the time of government support. During 1941 and early 1942, while the chemical properties of neptunium and plutonium were being studied by the tracer method, this work was carried on in self-imposed secrecy; even code names were given to elements 93 and 94.

During this time a great deal was learned about the chemical properties of plutonium. It was established that plutonium in its higher oxidation state was not carried by lanthanum fluoride or cerium fluoride, in contrast to plutonium in the lower state (or states), which was quantitatively co-precipitated with these compounds. The lower state could be oxidized to the higher state with oxidizing agents such as persulfate, argentic, dichromate, permanganate, or periodate ions, and by treatment with sulfur dioxide or bromide ion this upper state could be reduced to the lower state. The approximate oxidation-reduction potential of the couple, plutonium (reduced) → plutonium (oxidized), was estimated from these experimental results to be between −1.0 and −1.4 v. It was found that plutonium metal could not be prepared by the reduction of plutonium in aqueous solution with zinc and that plutonium did not form a volatile tetroxide. In addition, thorium iodate was found to be a carrier for a stable lower oxidation state of plutonium—probably plutonium (IV). This work resulted in the development of methods for separating plutonium from thorium, protactinium, uranium, and neptunium.

The original report on the chemical properties of plutonium, by A. C. Wahl and myself, was sent as a secret report to the Uranium Committee (the group headed by L. J. Briggs that had been co-ordinating the early work in the United States on the possible attainment of practical atomic energy from nuclear fission). In this report, element 94 was given the name "plutonium," after Pluto, the second planet beyond Uranus. The report was issued as Report No. A-135 and was published in its original form (1) after the war.

On the basis of the tracer work on plutonium, it was possible to make a number of deductions concerning its chemistry. It was thought likely that plutonium in its higher oxidation state was similar to uranium (VI) and, in a lower state, similar to thorium (IV) and uranium (IV). It was reasoned that if plutonium existed as a stable plutonium (IV) ion, it would probably form insoluble compounds or complex ions analogous to those of these similar ions.

The oxidation-reduction cycle principle, as applied to the purification of plutonium which was to become so important later, was originated. This principle was found to be applicable to any process involving the use of a substance that carried plutonium in only one of its oxidation states. For example, a carrier could be used to carry plutonium in one oxidation state and thus to separate it from uranium and fission products. The carrier and the plutonium could then be dissolved, the oxidation state of the plutonium changed, and the carrier re-precipitated, leaving the plutonium in solution. The oxidation state of the plutonium could again be changed and the cycle repeated. Only a contaminating element with a chemistry nearly identical with that of plutonium would fail to separate if a number of oxidation-reduction cycles were used. This principle applies equally well to other types of processes, such as solvent extraction, adsorption, or volatility methods.

With the added collaboration of E. Segrè, the search for the isotope of major importance, Pu^{239}, was proceeding simultaneously with the plutonium discovery experiments. The search for Pu^{239} was successful, and its value as a nuclear energy source was established during the spring of 1941. The isotope Np^{239} was prepared as follows. A sample of uranyl nitrate weighing 1.2 kg distributed in a large paraffin block, was placed directly behind the beryllium target of the 60-inch cyclotron and was bombarded for 2 days with neutrons produced by the full deuteron beam. The uranyl nitrate was then extracted with diethyl ether. Taking advantage of the then-known chemistry of neptunium and plutonium, the Np^{239} was isolated by the use of the oxidation-reduction principle, with lanthanum fluoride and cerium fluoride being used as carriers and re-precipitated six times in order to remove all uranium impurity. Measurement of the beta radiation from the Np^{239} made it possible to calculate that 0.5 μg was present to yield Pu^{239} upon decay. (The resulting alpha activity corresponded to a half-life of 30,000 years for the daughter Pu^{239}.)

Using neutrons produced in the 37-inch cyclotron in the University of California Old Radiation Laboratory and the sample containing 0.5 μg of Pu^{239}, the group found that this isotope underwent slow-neutron-induced fission with a cross-section even larger than that of U^{235}. The first demonstration that Pu^{239} underwent fission with slow neutrons occurred on March 28, 1941. The sample was placed near the screened window of an ionization chamber, which was imbedded in paraffin near the beryllium target of the 37-inch cyclotron. This gave a small but detectable fission rate when a 6-microampere beam of deuterons was used. In order to increase the accuracy of the fission cross-section measurement, the sample was purified by an oxidation-reduction type procedure which reduced the amount of carrier from about 5 mg to a few tenths of a milligram. Using this material, a fission cross-section, some 50 per cent greater than that of U^{235}, was found, which was in remarkable agreement with the accurate values determined later. Again, publication of this information was voluntarily withheld until the end of the war.

The demonstration that Pu^{239} underwent fission by thermal neutrons with a large probability, showing that all the neutrons emitted in the process could also initiate further fission, established the great value of this isotope. This discovery resulted in the establishment of the wartime Plutonium Project for the production of plutonium on a large scale for possible use in a nuclear weapon.

REFERENCE

1. SEABORG, G. T., and WAHL, A. C. "The Chemical Properties of Elements 94 and 93," in *The Transuranium Elements*, ed. G. T. SEABORG, J. J. KATZ, and W. M. MANNING, Paper 1.6, pp. 25–38. ("National Nuclear Energy Series," Division IV, Vol. **14B**.) New York: McGraw-Hill Book Co., Inc., 1949.

CHAPTER II

THE FIRST ISOLATION OF PLUTONIUM

B. B. Cunningham

In chapter i, Professor Seaborg has described the discovery of plutonium and the events leading to the first production of the isotope of mass 239, which was shown to undergo fission by thermal neutrons with a cross-section even greater than that of U^{235}. Largely on the initiative of Professor E. O. Lawrence, the authorities at Washington were quickly apprised of these events, and their attention was drawn to the potential significance of the work in terms of the eventual development of a nuclear weapon.

Early in December, 1941, a high-level decision at Washington committed the United States government to the support of a major effort to achieve the production of militarily useful amounts of Pu^{239}. Chicago was chosen as the headquarters for the project and the University of Chicago as the first site. Professor A. C. Compton was selected as project director. The goal to be achieved required the solution of two major problems: (*a*) the development of a method for the production of kilogram quantities of plutonium and (*b*) the design of chemical-processing methods for the decontamination, concentration, and purification of the plutonium so produced.

The breakthrough on the first problem came on December 2, 1942, when Fermi and his collaborators succeeded for the first time in achieving a self-sustaining chain reaction in a lattice arrangement of graphite and natural uranium. The reaction, sustained by the fission of the low-abundance isotope of mass 235, necessarily involved the production of some U^{239}, owing to the capture of a part of the fission neutrons by the high-abundance U^{238}. Two successive rapid beta decays produced the desired Pu^{239}.

More than eight months before this historic event, however, a group of chemists under the direction of Professor Seaborg was already at work on chemical-processing studies. Of necessity, this work was done at trace concentrations of plutonium, usually with Pu^{238}, which could be produced directly by cyclotron bombardment. It is noteworthy that by June, 1942, every type of processing method that has since been deemed potentially useful was under active investigation. Many of these

B. B. Cunningham is at the University of California, Radiation Laboratory and Department of Chemistry, Berkeley, California.

methods, necessarily abandoned in favor of the "least-risk" process chosen during the war, are once more receiving the attention they merit.

In all this early process research, one disturbing uncertainty remained. In no case had it been possible to conduct the experiments at anything approaching the concentrations of plutonium to be expected in real plant operation. In view of the enormous effort represented by the plutonium project, it was unthinkable that any scheme of chemical processing should be selected without having passed this crucial test.

The problem, however, was how to secure the necessary amount of plutonium, for, at that time, no chain reaction had been achieved. Small amounts of plutonium had, of course, been produced by cyclotron bombardment. Some of the desired isotope of mass 239 had been made indirectly, using the cyclotron to produce neutrons and stacking uranium near the target so that a part of the neutrons was captured by U^{238} to form U^{239}, which beta-decayed to Pu^{239}. It could be shown by calculation that even long bombardments, using existing cyclotron facilities, would produce no more than a few hundredths of a milligram of plutonium—an amount much less than that required for studies on the conventional scale of chemical experimentation. In fact, the restrictions on the quantities of plutonium obtainable in this way appeared to be so severe that the only feasible scale of operation was the microgram, an "ultramicrochemical" scale.

At that time there were only two laboratories in the United States in which ultramicrochemical methods had been explored in a systematic way: (1) the laboratory of A. A. Benedetti-Pichler at Queens College in New York and (2) the laboratory of Professor P. L. Kirk of the Division of Biochemistry at the University of California in Berkeley.

The techniques of both Kirk and Benedetti-Pichler were developed entirely for analytical purposes and relied exclusively on volumetric measurements. Nevertheless, they formed an invaluable basis of experience in ultramicro methods from which it was hoped to develop the special procedures needed for the work with plutonium. With this in mind, Seaborg in May, 1942, hired Dr. Michael Cefola, who had assisted Professor Benedetti-Pichler in developing his ultramicro methods, to work on the Chicago project.

In the meantime, the decision had been reached to attempt to expand the microchemical program to include an accurate measurement of the specific activity and half-life of Pu^{239} by direct weighing and counting. The measurement of the fission cross-section of Pu^{239} had been based on a rather uncertain value for the half-life, and it seemed very worthwhile to increase the accuracy of the measurement, if possible.

During a trip to the West Coast in June, Seaborg got in touch with me through Professor Kirk, for whom I was then working, with respect to joining his group at Chicago. Although he could not divulge the nature of the project, Dr. Seaborg stated that it was very important and mentioned that if I went to Chicago, it would be my responsibility to construct an ultramicrobalance, capable of weighing a microgram of material to 1 per cent.

I was intrigued by the problem of the balance (no one in the United States had ever constructed such an instrument) and very curious, naturally, as to what important applications it might have. (I did not, of course, suspect the nature of the Chicago project. My previous six years in biochemistry had left me unaware of the fact that physicists were getting close to the solution of the problem of atomic energy.)

The decision to join Seaborg's group was made quickly, and I started work on July 1. Dr. Cefola and I were given an abandoned darkroom for a micro-laboratory where he went ahead practicing plutonium-carrying experiments on an ultramicroscale. I started on the balance problem. A literature search had shown that ultramicrobalances of the requisite sensibility had been constructed twice before. One was built in 1914 by the Swedish physicist Hans Pettersson, and Pettersson's balance was modified from a still earlier instrument devised by two Australian physicists, B. Steele and K. Grant.

Unfortunately, both instruments were rather elaborate in design, demanding a good deal of careful machine work and considerable skill in fused-quartz manipulation. In July, 1942, the machine shops of the Chicago project were so swamped with high-priority work for the physicists that there was no hope of duplicating either the Pettersson or the Steele and Grant balances for many months. As a last resort, attention was directed toward the possibility of refining a very simple microbalance described by the Italian chemist, E. Salivoni, in 1901.

Salivoni's balance consisted merely of a glass fiber projecting horizontally from a rigid support. Weights hung from the free end produced deflections which were measured with a microscope fitted with an ocular scale. Salivoni reported that weighings accurate to 0.1 μg were possible with this simple device.

Only a few minor refinements—chiefly the substitution of silica for glass and the use of an accurate filar micrometer in place of a fixed scale—proved necessary to achieve the desired 0.01-μg sensibility. The load capacity of the balance was severely limited—to 0.5 mg, in fact—in order to retain a linear relationship between deflection and load. The pan and pan hanger had therefore to be very small. These were made of very thin platinum foil, cut and shaped under the microscope.

The balance was surrounded by a protective case, made from wood and glass plates, to protect the delicate fiber from air currents. Although only a few days were required to construct the balance and its ancillary equipment, it was then necessary to carry on a series of tests of the precision and accuracy of the instrument, to devise methods for accurate calibration, and to establish a technique for the proposed specific activity determinations.

L. B. Werner, who had worked with me at Berkeley, was invited to Chicago to aid in this program. With Werner's help, the balance tests were completed in about two weeks. Two methods were used to establish the calibration factor. The first involved the construction of two additional balances, one of which had a sensibility of about 0.1 μg, and the other a sensibility of 1 μg. The latter instrument was calibrated to 0.2 per cent with a standard 1-mg weight, and the calibrated balance was used to weigh a 0.1-mg platinum wire to 0.4 per cent.

The second balance was calibrated with the platinum wire and then used to weigh a 10-μg quartz fiber—which constituted the standard weight for calibrating the most sensitive balance. Two independent calibrations by this method agreed to 0.7 per cent.

The second method of calibration consisted of the following: A liter of thorium nitrate solution was prepared by dissolving about 2 gm of thorium nitrate in a liter of dilute nitric acid. Ten-milliliter aliquots of this solution were carefully evaporated in weighed platinum containers, and the residue dried and ignited to ThO_2. The weight of ThO_2 residue per milliliter of solution was established to 0.4 per cent by these macro experiments.

An ultramicro pipette of about 1 μl volume was calibrated for volume by weighing the 13 mg of mercury contained by the pipette. The pipette was then used to transfer 1.27 μl of the thorium nitrate solution to the previously weighed platinum weighing pan. Evaporation and ignition yielded a ThO_2 residue weighing 2.92 ± 0.02 μg. The additional deflection produced by the residue was measured, and from this an independent check of the balance calibration was obtained. Agreement with the first calibration was better than 1 per cent.

The second method of calibration duplicated the technique by which it was planned to determine the specific activity of Pu^{239}—i.e., to deliver a solution of plutonium nitrate onto the weighing pan, evaporate this, and ignite to the oxide. The oxide would then be weighed, subsequently dissolved, and aliquots taken for alpha counting.

The construction and calibration of the balance were completed before any plutonium was available for isolation. During this time, however, 5 kg of uranyl nitrate hexahydrate were being exposed to neutrons produced by a (d, n) reaction on beryllium at the Berkeley 60-inch cyclotron. After a calculated production of about 1 μg of plutonium, the bombardment was stopped. A. C. Wahl, J. W. Gofman, and others at Berkeley did a preliminary separation of uranium and fission products, and the plutonium fraction, in 15 ml. of solution containing a few milligrams of rare-earth carriers, was sent to Chicago for isolation.

A series of oxidation-reduction cycles, of the type already described by Professor Seaborg, were used. Repetition of these cycles, at smaller and smaller volumes, with less and less carrier, led ultimately to the ultramicrochemical scale of operations and to a concentration of plutonium from which the element could be expected to precipitate quantitatively as the fluoride without carrier.

This final step was actually performed on August 18, 1942. A small amount of flocculent precipitate, visible under the microscope, was observed to form after the addition of hydrofluoric acid to the carrier-free plutonium solution in a quartz microcone. The appearance of the precipitate changed quickly, however, and it was judged that it was being contaminated with impurities from the attack of hydrofluoric acid on the silica tube. Since this isolation yielded a product of doubtful purity, unsuitable for specific activity measurements, it was considered a failure. It had demonstrated, however, that the fluoride-cycle process was applicable at high plutonium concentrations.

Fortunately, arrangements had been made in the meantime for producing more Pu^{239} by cyclotron bombardment. The production site was shifted to the cyclotron at Washington University at St. Louis, to take advantage of the much higher beam current available. By early August, about 50 μg of plutonium had been produced, and half of this was allotted for isolation work.

Following the preliminary separation and decontamination from uranium and fission products and the development of a technique for applying a transparent fluoride-resistant coating to the quartz microcones, the plutonium was obtained finally as a pure, carrier-free solution of the nitrate in a few microliters of dilute nitric acid. The precipitation of the first pure compound of plutonium was performed on September 9, when an excess of ammonium hydroxide was added to the nitrate solution, producing a green flocculent precipitate of the hydroxide. This was, of course, a thrilling moment, as it represented the first time that a visible amount of a pure synthetic isotope had been seen by man.

The hydroxide was carefully washed and redissolved to form a more concentrated solution in dilute nitric acid. This solution was used on the following day in the first of the weighing experiments designed to measure the half-life.

A 1-μl aliquot of the nitrate solution was delivered onto a clean, previously weighed weighing pan, then carefully dried and ignited in a simple muffle furnace. The thin deposit of golden-yellow oxide was clearly evident under the microscope. The pan and oxide were reweighed, and the weight of the oxide found to be 2.77 μg. At Professor Seaborg's suggestion, this sample, on the original weighing pan, was preserved intact, because of its historic interest. It is still in our possession.

Two additional samples of oxide were prepared and weighed in similar fashion. These samples were subsequently dissolved in sulfuric acid, diluted to known volumes, and counted for Pu^{239} activity. The experiments gave consistent values for the specific activity and half-life of the isotope. After correcting for a previously unsuspected error in counting yield, due to multiple scattering of alpha particles from the plutonium counting plates, the half-life was computed to be 24,300 years —a figure which subsequent work has shown to be in error by only about 0.3 per cent. Fortuitously, the accuracy of the measurements was considerably better than could reasonably have been expected.

The initial isolation provided a source of pure plutonium for a wide variety of research on the properties of pure plutonium compounds and for testing process methods at plant concentrations. As more plutonium became available, many individuals contributed to its study. One of the most interesting programs in this early work was that which led to the first preparation of plutonium metal—a subject which will be discussed by Dr. Fried.

CHAPTER III

THE MICROSCALE PREPARATION AND MICROMETALLURGY OF PLUTONIUM METAL

S. Fried, E. F. Westrum, Jr., H. L. Baumbach, and P. L. Kirk

INTRODUCTION

The investigations summarized in this chapter comprise microchemical and micrometallurgical studies of plutonium made during 1943 and 1944 under the pressure of wartime conditions at the Manhattan Project Metallurgical Laboratory operated by the University of Chicago for the United States War Department. These studies led to the first successful preparation of plutonium metal and provided data essential to the macroscale technology of plutonium production that followed very soon thereafter at Los Alamos (1). Because microscale metallurgical methods are of continuing applicability in the isolation and investigation of synthetic and other rare elements, a description of these techniques is of importance as a contribution to the subject of laboratory procedures in modern chemical and metallurgical research.

A major chemical problem in the development of the release of nuclear energy was the preparation on the microgram scale of globules of plutonium metal suitable for the investigation of its physical and chemical properties. Micrometallurgy on this scale was unprecedented; an entirely original approach was necessary. The uncertainty inherent in the prediction of the physical and chemical properties of a new element (in a series the very nature of which was still under speculation), the exceedingly small amounts of plutonium available, the doubt as to the identity and composition of compounds to be reduced, and the extreme health hazard of traces of the finely divided solids aggravated the problem. The complex allotropy of plutonium (1, 2), which was not then suspected, also contributed to the difficulty of these investigations.

Preliminary experiments were made on "stand-in" compounds which it was hoped would simulate the behavior of plutonium compounds in reduction to the element. The objective was reduction on the microgram scale, with high yields of metal in a single globule. A variety of methods was tested, including reduction with atomic hydrogen, electrolytic reduction in aqueous and fused-salt systems,

S. Fried is at the Argonne National Laboratory, Argonne, Illinois. E. F. Westrum is in the Department of Chemistry, University of Michigan, Ann Arbor, Michigan. H. L. Baumbach is with the Unicorn Engineering Corporation, Los Angeles, California. P. L. Kirk is in the School of Criminology, University of California, Berkeley, California.

and thermal reduction of halides with alkali and alkaline-earth metals; only the last-mentioned proved successful and is described here. On the basis of the limited data available, uranium compounds were selected as probably representative of the chemical reactivity and physical properties of the analogous plutonium compounds. UF_4 was chosen because of its stability, non-hygroscopicity, and comparative non-volatility. It was anticipated that PuF_4 would have a similar set of desirable properties.

The first attempted reductions (3) were performed by distilling alkali metals onto uranium compounds placed in the tip of glass microcones *in vacuo* and removing excess alkali metal by heating the mixture gently with a flame. Further inconclusive attempts to reduce a few micrograms of UF_4 and PuF_4 to metal by treatment with sodium in evacuated glass capillaries (4) emphasized the need for practical criteria for identification of metal on the microscale. Methods involving the metallic characteristics of malleability, high density, luster, and especially the formation of a hydride were adapted to microscale work.

The first unequivocal production of plutonium metal was made in November, 1943 (5). About 35 μg of PuF_4 were transferred on the tip of a sharp needle to a small thoria crucible. About ten times the stoichiometric amount of barium metal freshly cut under xylene was placed in a larger crucible, the small crucible was inserted on top of the reductant, the outer crucible cap inserted, and the entire assembly placed in a microfurnace and fired to 1,400° C in a high vacuum. A number of 3-μg plutonium-metal globules were formed. The metal had a silvery luster, a density of about 16 gm/cm^3, and rapidly absorbed hydrogen at about 210° C to form a black powder subsequently identified as PuH_3 (6).

THE CERAMIC DOUBLE-CRUCIBLE SYSTEM

The reaction between condensed phases of reactants was generally violent and tended to yield dispersions of metal and slag. Techniques were therefore developed whereby only the reductant vapor came in contact with the halide to be reduced (5, 7, 8).

Although originally the small crucible with its charge of plutonium halide was placed on top of the reductant metal, after subsequent experimentation the arrangement of the double-crucible system was modified to that shown in Figure 1 (9). This reduced the tendency of the volatilizing reductant to agitate the inner crucible and disturb its contents. To this end, the top of the inner crucible was slightly castellated, the inner crucible was placed on the bottom of the outer crucible and was covered with a thin plate of ceramic to separate its contents from the solid reducing agent above it. A conical cap partially sealed the system and permitted the reductant to develop pressure and subsequently to escape gradually.

The most generally satisfactory crucibles were made of beryllia (BeO). These were fabricated by making a paste of finely ground fused beryllia with 6 M nitric acid. The stiff paste was formed into crucibles with the aid of stainless-steel mandrels. After gradual air-drying, they were slowly fired to 1,000° C in a muffle furnace. They were then shaped and bored to exact size and then fired and outgassed in high vacuum. The outer crucibles and caps were fired to 1,800° C for greater hardness and strength; the inner crucibles were partially vitrified, by heating to 1,600° C for an hour, in order to minimize penetration by the plutonium metal and yet preserve the degree of porosity useful in absorbing the slag. Thoria crucibles were similarly fabricated, except that the paste was made with ground fused thoria and 50 per cent thorium nitrate solution.

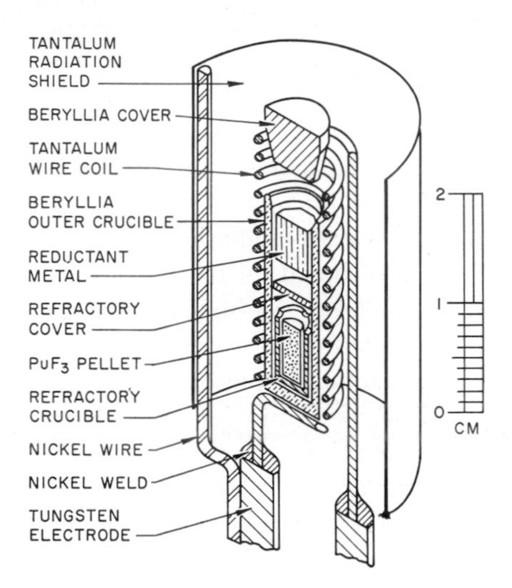

FIG. 1.—The double-crucible arrangement in the tantalum-shielded resistance furnace for the reduction of plutonium.

THE HIGH-VACUUM FURNACE

The original furnaces (5) consisted of a helix of tungsten wire imbedded in beryllia. The shielded tantalum furnace (9–11) (Fig. 1) had several advantages, in that it was more efficient, more readily outgassed, and the tantalum coil was not subject to embrittlement from repeated heating cycles. The furnace was controlled with an autotransformer in series with a step-down transformer. Crucible temperatures of about 1,600° C were obtained, using 14 amp at 12 v. These furnaces were operated *in vacuo* of 10^{-6} mm Hg in air-cooled 500-cm^3 bulbs with water-cooled, standard taper joints to permit removal of the furnace assembly. The bulbs were attached to a vacuum system, using Apiezon-W wax on large standard taper joints, permitting pumping speeds of the order of 20 l/sec at the furnace position. Because water and oxygen react with plutonium halides to produce compounds difficult to reduce, it was found necessary to outgas the furnace and crucible system for about an hour at 1,600° C at 10^{-5}–10^{-6} mm Hg prior to loading and to break the vacuum with argon purified by passage over uranium turnings at 800° C.

REACTANTS

Plutonium tetrafluoride was prepared by the hydrofluorination of PuO_2—prepared from air-dried plutonium hydroxide (12)—with hydrogen-free anhydrous HF at 600° C. Plutonium trifluoride was prepared by techniques similar to those previously described for americium trifluoride (13).

EXPERIMENTAL REDUCTION TECHNIQUE (14)

To prevent oxidation of the surface of the reductant metal and adsorption of water on the outgassed crucibles, the handling of crucibles and the entire loading procedure were performed in an anhydrous nitrogen atmosphere in a dry box provided with an evacuable air lock and a horizontal plate-glass top to facilitate operation under a stereoscopic dissecting microscope. Small amounts of anhydrous plutonium compounds (5–200 μg) were placed on the tip of a tungsten needle and loaded into the inner crucible. In the larger-scale reductions (e.g., 1–50 mg), it was found effective to compress (to about 10^4 kg cm^{-2}) the plutonium halide into small cylindrical pellets 0.5–1.5 mm in diameter with hardened steel dies in a hand press. This procedure resulted in better agglomeration and higher useful yield of the product metal. Intermediate amounts were handled with a small spatula. The loaded inner crucible was dropped inside the larger one, and the small refractory disk or cover dropped in on top. A freshly cut piece of pure barium (in large stoichiometric excess) was dropped in and then the previously fitted refractory cap plugged tightly in place. The entire crucible assembly was rapidly transferred to the furnace, which was evacuated on the high-vacuum line until the pressure was 10^{-5} mm Hg or better. This evacuation required 15–20 minutes.

Firing of the furnace system was achieved by applying the full voltage to the coil. In about 10 seconds the coil reached its maximum temperature; the crucible system required much longer. Very soon after the crucible reached red heat, the barium vaporized, gradually escaped from the crucible, and condensed as a mirror on the bulb walls. Under certain conditions a luminous green discharge existed as long as there was an appreciable vapor pressure of barium. After approximately 1 minute, or longer, if the continuing escape of barium was indicated by the discharge, the current was cut off, and the system was allowed to cool under vacuum. The last stages of the cooling process were aided with admission of highly purified argon. After removal from the furnace, the crucibles were opened and the metallic globule(s) of metal removed. The slag was generally absorbed within the pores of the inner crucible. Occasionally, the metal would partially wet the crucible and was removed only with difficulty. Fragments of refractory adhering to the metal could be removed under xylene with a sharp tungsten needle.

RESULTS OF REDUCTIONS

A series of studies on the reduction of PuF_4 with Ba (9) in BeO crucibles indicated that at least partial reduction took place at furnace temperatures in excess of 900° C. Optimum results were obtained at about 1,150° C. Although a single globule was desired, the results were considered acceptable if the over-all yield of compact globules was high and each globule represented at least 20 per cent yield. Reduction of $PuCl_3$ to metallic plutonium was achieved (10, 15) in fair yield by the action of Li, Na, K, Ba, and Ca. However, the reductions of the chloride with reductant vapor on the 5–8-mg scale did not produce large particles (i.e., larger than 0.5–1.0 mg) of metal, even with pelleted chloride. This is attributed to the higher volatility of the $PuCl_3$ as compared with PuF_4. An attempt was made to increase the yield by bringing the halide into direct contact with the molten reductant in the outer crucible at lower temperature. This resulted in less sublimation of $PuCl_3$, formation of larger particles, and increased yield. $PuBr_3$ has also been reduced to Pu metal with calcium (9). Two attempted reductions of PuOCl with potassium yielded only the phase identified both by its characteristic semimetallic appearance and by X-ray diffraction analysis as PuO (9). Attempted reductions of PuO_2 with Na and Mg vapor at about 1,200° C did not show evidence of reduction; Ca and Li vapor at 1,200° C yielded only partial reduction to a phase tentatively indentified as Pu_2O_3 (16). Plutonium fluorides were reduced with Ba in La_2O_3, tantalum, and tantalum nitride (10, 15). However, the product metal wetted both Ta and TaN and could not be separated from either.

Plutonium trifluoride is easily prepared, has desirable physical characteristics, and has been more thoroughly investigated from the standpoint of metal production on the milligram scale in this laboratory than have the other halides. Table 1

TABLE 1

REDUCTION OF PLUTONIUM TRIFLUORIDE

Reductant	Inner Crucible	Furnace Temperature (° C)	Results
Ba	BeO	1,250	Single piece of metal, good yield
Ba	La_2O_3	1,250	Single piece of metal, fair yield
Ba	CaO	1,250	Poor results, metal surface pitted
Li	BeO	1,350	Good results, metal in one piece
Ca	BeO	1,250	Obtained 60.4-mg metal from 85.3-mg PuF_3; 88 per cent yield
Ca	BeO	1,450	One piece of metal, 33.2 mg; 92 per cent yield; density, 18.7
Li	Graphite	1,000	Formation of PuC (and possibly Pu_2C_3) only

summarizes the results of the influence of various conditions on the reduction of PuF_3. Yields were excellent, and the few failures could generally be traced to faulty technique or secondary reactions with crucible material. Almost 200 microscale reductions of PuF_3 were made in beryllia.

Metal produced in thoria crucibles regularly revealed appreciable contamination by thorium. Beryllium oxide, however, yielded very pure plutonium metal. Spectrographic analyses of plutonium made in beryllia crucibles by barium reduction on the 1-mg scale indicated (in p.p.m.) from 1 to 4 of Be, 1,000 of Ca, 250 of Mg, and less than 250 of Ba, with no other metals detectable (10). Atmospheric contamination from building construction materials is believed to be in part responsible for the high calcium content. Other crucible materials (cf. Table 1) were also studied with respect to their practicability for metal production. Crucibles prepared from vacuum-fired moist La_2O_3 showed considerable promise, but the hygroscopicity complicates the handling in air. Crucibles of CaO prepared both of reagent oxide and of dense St. Genevieve limestone (some silica present) were less satisfactory.

TESTS OF SULFIDE REFRACTORIES FOR METAL PRODUCTION

Sulfide refractories suitable for the fabrication of crucibles tend to be resistant to reduction and have high melting points and relatively low volatilities. Tests were, therefore, made on the practicability of utilizing a series of sulfide crucibles prepared by a group of scientists at the University of California at Berkeley under the direction of Eastman *et al.* (17). The tests were made on the milligram scale, using high-purity calcium metal as reductant. Barium was also tried on barium sulfide refractories. A tantalum liner of 2 mil thickness was interposed between the sulfide inner crucible and the beryllia outer crucible to avert the possible formation of low-melting eutectics. Without the use of such a liner, a CeS crucible completely fused into the beryllia during outgassing below 1,700° C.

The crucible systems, furnace, and bulb were degassed for approximately 1 hour at crucible temperatures of 1,600°–1,700° C at 2×10^{-6} mm of Hg. After outgassing, the crucibles were cooled to room temperature, and argon was admitted to break the vacuum. The crucible system was then rapidly transferred to a nitrogen-filled dry box, loaded with pelleted PuF_3 and calcium, and replaced in the furnace. After evacuating to about 2×10^{-6} mm of Hg, the actual firing was done by raising the temperature of the crucible to 1,250°–1,350° C within 30 seconds. It was noted that reduction occurred within an additional 15 seconds and that the excess calcium distilled from the crucible formed a sharply defined mirror on the top and bottom of the glass bulb. The crucible system was maintained at this temperature for a total time of 4.5 minutes and was then allowed to cool to room temperature before breaking the vacuum.

Upon completion of the reduction, the products of the reaction were examined microscopically *in situ* to note any evidence of reaction, wetting, pitting, etc. Specimens of the plutonium metal were also examined metallographically and their hardness determined with an Eberbach microhardness tester. Other samples were submitted to the analytical section for spectrographic and chemical analyses. The over-all judgment on the relative suitability of these refractories is as follows: $BaS_{0.98}$—unsatisfactory; $ThS_{1.77}$ and $ThS_{0.97}$ (5 per cent ThO_2, 0.3 per cent MoO_2) —poor; $ThS_{1.50}$, $ThS_{1.00}$, and $ThUS_2$—satisfactory; $Th_{0.32}Ce_{0.65}S$, $ThCeS_2$—excellent.

Spectrochemical analyses on the metal produced in any of the cerium and/or thorium refractories revealed that the plutonium contained 0.0 ± 0.2 per cent of cerium and 0.0 ± 0.04 per cent of thorium. A few of the metal samples showed traces (less than 0.01 per cent) of beryllium and calcium. In the reductions in $BaS_{0.98}$ crucibles, barium was present to more than 5 per cent in some samples of the calcium-reduced material. Less than 0.1 per cent of barium was present in the product obtained by barium reductions in $BaS_{0.98}$ crucibles. No other elements were found to be present in concentrations greater than in the blank determinations on the reagents and electrodes.

Colorimetric analysis for sulfur showed less than 200 p.p.m. of sulfur (the practical limit of sensitivity of the method) for all metal produced in the various sulfides, with the exception of the two reductions in barium sulfide crucibles. Analysis (15) by chemical and X-ray diffraction methods indicated that the product from calcium reduction of plutonium in a BaS crucible was an impure and inhomogeneous PuS. The plutonium reduced with barium in the other barium sulfide crucible showed 0.0 ± 0.08 per cent sulfur.

THE DENSITY OF PLUTONIUM METAL

The density of plutonium metal was determined, when only milligram globules of metal were available, by using the micropycnometric method developed by Kirk and his associates (5). A series of uniform cylindrical capillaries having internal diameters of 100–200 μ were drawn. These were sealed shut at intervals of about 5 cm and cut within 5 mm of the seal, forming a short cup with a long stem. The diameter of each cup was accurately calibrated with an ocular micrometer with about 100 $\times$ magnification.

A capillary was selected with an internal diameter slightly larger than the metal globule. The metal was weighed on a quartz-fiber torsion microbalance (18) to an accuracy of about 0.01 μg. The capillary was loaded with a small droplet of dibutyl phthalate (chosen for its low volatility), which was centrifuged to the bottom of the cup. The position of the meniscus was determined relative to a reference mark to an accuracy of about 0.001 mm. The metal particle was then inserted with a micromanipulator and centrifuged to the bottom of the cup. The new meniscus position was determined with respect to the reference mark, and the movement used in conjunction with the capillary calibration to determine the volume displaced by the metal. This volume and the weight were then used to calculate the density. The method was tested by determining the specific gravity of Pt spheres (21.5 ± 0.3) and of milligram quantities of U (18.3 ± 0.4) (literature values: Pt, 21.45; U, 19.0). Determinations (19) on several samples of Pu, weighing from 4 to 28 μg, produced by reduction of PuF_4 with Ba or Mg in ThO_2 crucibles, indicated a density of 15.4 ± 0.2 gm cm^{-3}. Determination on plutonium metal similarly produced in BeO crucibles (5, 20) yielded consistently higher densities; these ranged from 17.2 to 21.0 gm cm^{-3}. With the production of larger globules of metal, the capillary pycnometers were made from regular Pyrex capillary tubing (0.2–1-mm bore) with two optically plane surfaces parallel to each other and the axis of the tube (21) to permit precise measurement. Measurements on the milligram scale confirmed the spread of densities observed earlier on the microgram scale, now known to be occasioned by a spectrum of phase transformations (2, 22).

PHASE TRANSITIONS

The possibility of the existence of one or more allotropic modifications of plutonium metal was first suggested by Zachariasen (23), who interpreted X-ray diffraction patterns of metal samples made under various conditions. This hypothesis was strengthened by the range of densities observed and by simultaneous dilatometric studies in the Metallurgical Laboratory (24) and at the Los Alamos Scientific Laboratory. Although the milligram-scale dilatometry represented a considerable achievement in view of the reactivity of the metal, the results obtained here, as in the case of hardness and the melting point, have been superseded by subsequent extensive measurements on the gram scale made at the Los Alamos Scientific Laboratory (1, 2, 22) and the Atomic Energy Research Establishment at Harwell (25).

The authors gratefully acknowledge the assistance, advice, and co-operation given by A. E. Florin, Z. V. Jasaitis, S. Katz, F. Meyer, H. P. Robinson, and R. S. Rosenfels.

REFERENCES

1. SMITH, C. S. *Metal Progress*, **65**:81–89, 1954.
2. ———. *Physical Review*, **94**:1068, 1954.
3. KIRK, P. L., and ROSENFELS, R. S. Manhattan Project Rept. CK-514 (March, 1943).
4. ———. Manhattan Project Rept. CK-591 (April, 1943).
5. BAUMBACH, H. L., FRIED, S., KIRK, P. L., and ROSENFELS, R. S. Manhattan Project Rept. CK-1145 (December, 1943).
6. FRIED, S., and DIRKSEN, A. Manhattan Project Rept. CK-1372 (March, 1944).
7. BAUMBACH, H. L., ROSENFELS, R. S., and KIRK, P. L. Manhattan Project Rept. CK-736 (June, 1943).
8. ———. Manhattan Project Rept. CK-987 (October, 1943).
9. FRIED, S., and WESTRUM, E. F., JR. Manhattan Project Rept. CK-1512 (April, 1944).
10. WESTRUM, E. F., JR. Manhattan Project Rept. CK-1586 (May, 1944).
11. ———. Manhattan Project Rept. CK-2495 (January, 1945).
12. WESTRUM, E. F. "The Preparation and Properties of Plutonium Oxides," in *The Transuranium Elements*, ed. G. T. SEABORG, J. J. KATZ, and W. M. MANNING, Paper 6.57, pp. 936–44. ("National Nuclear Energy Series," Division IV, Vol. **14B**.) New York: McGraw-Hill Book Co., Inc., 1949.
13. WESTRUM, E. F., JR., and EYRING, L. *Journal of the American Chemical Society*, **73**: 3396, 1951.
14. A series of photographs of microscale techniques for plutonium metal production appears in *Life*, **21**:68, 1946; also in *Life International*, **1**:30, 1946.
15. WESTRUM, E. F., JR. Manhattan Project Rept. CK-1701 (June, 1944).
16. ———. Manhattan Project Rept. CN-2159 (October, 1944); also W. H. ZACHARIASEN, private communication.
17. EASTMAN, E. D., BREWER, L., BROMLEY, L. A., GILLES, P. W., and LOFGREN, N. L. *Journal of the American Chemical Society*, **72**:4019, 1950.
18. KIRK, P. L., CRAIG, R., GULLBERG, J. E., and BOYER, R. Q. *Analytical Chemistry*, **19**:427, 1947.
19. FRIED, S., and KATZ, S. Manhattan Project Rept. CK-1512 (April, 1944).
20. KATZ, S. Manhattan Project Rept. CK-1586 (May, 1944).
21. ———. Manhattan Project Repts. CK-1367 and CK-1377 (February, 1944).
22. JETTE, E. R. *Journal of Chemical Physics*, **23**:365–68, 1955.
23. ZACHARIASEN, W. H. Manhattan Project Repts. CK-1367 and CK-1377 (February, 1944).
24. KATZ, S. Manhattan Project Rept. CK-1763 (July, 1944).
25. LORD, W. B. H. *Nature*, **173**:534–35, 1954; also BALL, J. G., ROBERTSON, J. A. L., MARDON, P., LEE, J. A., and ADAMS, E. T. *Nature*, **173**:535, 1954.

CHAPTER IV

PLUTONIUM METALLURGY AT LOS ALAMOS DURING 1943-45

Cyril Stanley Smith

INTRODUCTION

In this chapter is described, in a non-technical and reminiscent way, some of the work on plutonium metallurgy that was carried out at the Los Alamos Laboratory during World War II. On August 6, 1945, the purpose of the laboratory, whose very existence had been previously kept as quiet as possible, was dramatically revealed to the world by the Hiroshima explosion. The work of the physicists subsequently received much publicity. The achievements of the metallurgists were perhaps less spectacular and certainly of less intellectual significance, but they were no less essential to the success of the project. The famous Smyth Report, which has many pages devoted to the work of the nuclear physicists, says, on metallurgy, just the following:

> The Chemistry and Metallurgy Division of the Los Alamos Laboratory was under the joint direction of J. W. Kennedy and C. S. Smith. It was responsible for final purification of the enriched fissionable materials, for fabrication of the bomb core, tamper, etc., and for various other matters. In all this Division's work on enriched fissionable materials especial care had to be taken not to lose any appreciable amount of the materials (which are worth much more than gold). Thus the procedures already well established at Chicago and elsewhere for purifying and fabricating natural uranium were often *not* satisfactory for handling highly enriched samples of U^{235}.

Dr. Smyth's statement implies nothing about the most exciting, important, and difficult job of developing plutonium metallurgy; furthermore, the "other matters," so lightly dismissed, were of such diverse nature as to provide the ingredients of a first-class nightmare for any industrial metallurgist with production responsibilities. A high percentage of the elements in the periodic table was involved, and we soon learned that materials selected for their nuclear constants are rarely those of greatest ease of fabrication. The Los Alamos metallurgists got used to astounding requests (indeed, enjoyed them) and also became accustomed to having the physicists say that further calculations showed a certain bizarre material to be no longer of interest after we had devoted many weeks to suitable methods for shaping it.

The writer joined the project in March, 1943, at the very beginning of the New

Cyril Stanley Smith is at the Institute for the Study of Metals, University of Chicago, Chicago, Illinois.

Mexico laboratory, but four months after the first pile (nuclear reactor) had been operated in Chicago and after the broad outlines of the physics and chemistry of fissionable material and its production and use had been sketched. By August, 1945, there were 115 people in the metallurgical groups.

The first month was mostly devoted to recruitment. This was not an easy matter when all good metallurgists were already engaged in war work and security prevented us from saying what the project was about. Indeed, we hardly knew ourselves what we had to do, except that at some time, approximately two years in the future, we would have to fabricate some excessively precious, highly radioactive metal of unknown metallurgical characteristics into completely unknown shapes. As time went on, we collected an excellent staff, including R. D. Baker, J. E. Burke, Eric Jette, Alan Seybolt, and J. M. Taub, all of whom served as leaders of groups working on some aspect of plutonium metallurgy during the period under review.

For administrative purposes, metallurgy was grouped with chemistry under J. W. Kennedy—one of the discoverers of plutonium—an imaginative chemist and, despite his youth, an understanding and excellent administrator. His untimely death in 1957 was a great loss to science and to his friends, for his scientific insight was combined with a rare broad wisdom and a rich sense of humor. The entire "final chemistry and metallurgy" program was co-ordinated and directed by Charles Allen Thomas (assisted by J. C. Warner), who established the balance of effort between Los Alamos and the other Manhattan Project laboratories whose work was integrated with ours. Continued help came particularly from the groups working under Glenn Seaborg and John Chipman at Chicago, F. H. Spedding at Ames, and the refractory laboratory under F. H. Norton at Massachusetts Institute of Technology. A. B. Kinzel visited Los Alamos at intervals and gave invaluable advice. Karl T. Schwartzwalder was consulted on ceramic problems and was most helpful.

Finally, but first in importance, the writer must mention J. Robert Oppenheimer, director of the laboratory. He followed the work on plutonium closely and often showed a more realistic appreciation of metallurgical difficulties than did some of the metallurgists themselves. He personally participated in the metallurgical decision as to the type of plutonium to use in the first bomb cores. The metallurgists, like all others in the laboratory, owed much to his wise and inspiring leadership.

PLUTONIUM METALLURGY[1]

Plutonium metallurgy was the most important, the most interesting, and the most dangerous of the responsibilities of the Los Alamos metallurgists. Its toxicity required all operations to be carried out with extreme precautions against breathing even traces of dust, because the radiation from as little as a microgram, if ingested, is highly dangerous. Since it was a brand-new element—artificial, in fact—we

[1] A note on the prehistory of plutonium may be added here. The hot springs in the Phrygian city of Hierapolis once included a cavity running deep into the earth, from which belched a hot and stinking sulfurous vapor. Since the hole was thought to connect with the nether regions, the vapor was called "plutonium," and a temple was built at the site, housing unmentionable rites until its destruction by Christians in the fourth century. The name, moreover, has been pre-

knew nothing whatever about its metallurgy; yet we had to be ready to go into full-scale production the moment it was available in sufficient quantities to make one bomb. The purity specifications were extreme, for undesirable neutrons would be formed by the interaction of alpha particles resulting from its radioactive decay with most of the light elements, which might exist as impurities.

Initially, we made the incorrect assumption that plutonium would be something like uranium, and we studied metallothermic and electrolytic processes of reduction on uranium in order to gain metallurgical and analytical experience. Hot-wire and atomic-hydrogen reduction methods also figured in our early plans. It was agreed that Los Alamos would not work with batches of less than 1 gm, because the Metallurgy Laboratory at Chicago was doing excellent work in chemistry and metallurgy on a truly microscopic scale.

The first reductions at Chicago were made on about a 50-μg scale, producing tiny spheres which were sometimes malleable, had a density of about 16, and melted below 800° C. This micrometallurgy was an unsung, but truly magnificent, performance and served to give the macrometallurgists at Los Alamos an inkling of

---

viously applied to a metal. Early in the nineteenth century, C. D. Clarke, a pioneer in the use of the oxyhydrogen blowpipe, found metallic beads in the residue obtained on fusing barytes ($BaSO_4$). In a paper published in 1817 (*Jour. Science and Arts,* **2**:104–23) he named his new element plutonium "because we owe it entirely to the dominion of fire." However, Michael Faraday repeated

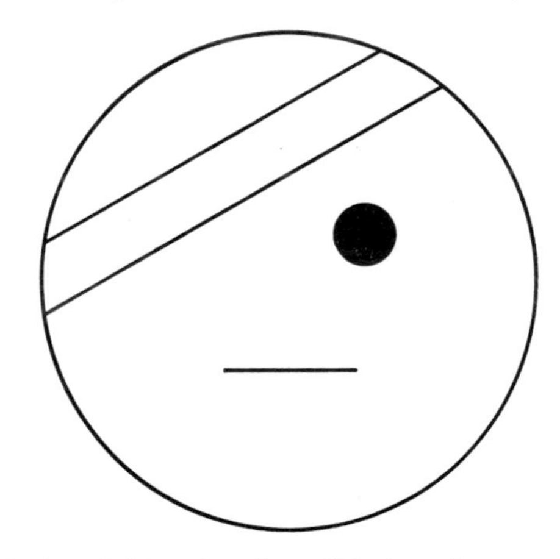

FIG. 1.—The alchemical symbol for plutonium. This figure is a reproduction of the colophon placed at the end of a book (still classified) entitled *The Chemistry, Purification and Metallurgy of Plutonium,* edited by Charles Allen Thomas and John C. Warner, and published in December, 1944, by the Army Service Forces, United States Engineer Office, Manhattan District, Oak Ridge, Tennessee.

the experiment with negative results (same volume, pp. 461–62), and the name was released to be later used for an element to whose diabolic attributes it is more appropriate.

Here also we add the hitherto-secret alchemical symbol for our metal. This was devised by Charlie Thomas in 1944, being first drawn by him to complete a set of alchemical symbols for the metals that the writer had drawn while doodling during a rather dull meeting. The reproduction (Fig. 1) is of the colophon of the still classified book on the chemistry and metallurgy of plutonium, which was edited by C. A. Thomas and J. C. Warner and issued in a very limited edition by the Manhattan District, U.S. Army Engineers, at the end of 1944.

what they were to be up against a critical few weeks before gram lots became available (see chap. iii).

VARIABLE DENSITY

The first portent of really severe problems ahead came from reports from Chicago of widely varying densities of these almost invisible beads of metal. There was equally good evidence for a density of about 16 and for about 20! The first determination of the crystal structure supported the 16 figure. It was subsequently found that *both* were correct, for the metal has allotropic forms with a very wide spread in density. An argument between John Chipman and Joe Kennedy on the true density of the metal resulted in a bet, which had to be called off when it was found that both were right. By the middle of 1944 it was known that no less than

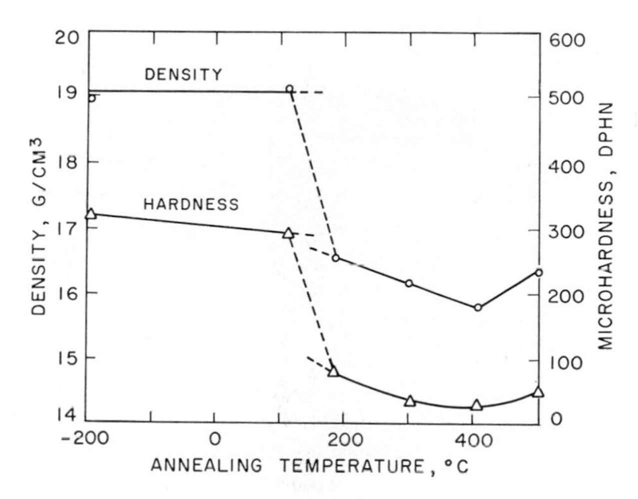

FIG. 2. The density and hardness of plutonium after annealing from 30 to 45 minutes at the temperatures indicated. The sample studied contained sufficient impurities to cause retention to room temperature of low-density phases stable at higher temperatures.

five phases existed in the metal between room temperature and the melting point, with densities between 15.9 and 19.8 (Fig. 2). This discovery greatly alarmed us, for we anticipated that great difficulty in fabricating the metal would result from the large volume change, although it was likely that at least one of the many phases would be malleable and easily shaped.

The small amount of plutonium in Los Alamos at the end of 1944 had to serve for all the chemical and metallurgical development work. It was remelted, reshaped, repurified, and re-reduced many times as we tried out, on gradually increasing scales, the processes that had been developed with 1-gm lots (see Table 1). When at last there was enough material for one bomb core, all experimental material was called back, and our entire efforts were directed at shaping the whole available supply into the final form as quickly as possible. The first phase of metallurgy at Los Alamos was completed when, at about noon on July 15, 1945, the writer placed the plutonium core in the subassembly in McDonald's farmhouse located on the test site at Alamogordo. Seventeen hours later it became apparent that explosives were no longer the exclusive province of the organic chemist.

Production of Plutonium

At the time that the first production of plutonium was attempted on the 1-gm scale at Los Alamos, the microchemical studies at Chicago had indicated that the melting point was perhaps as low as 800° C, although there was much unwarranted skepticism about this figure. The extreme reactivity of the metal resulted in a coating of oxide or nitride, which, being of very high melting point, prevented clear observation of the metal. Since some of the earlier successful microreduction at Chicago had been at 1,300° C, our initial trials at Los Alamos were carried out at a

Fig. 3.—A 10-gm regulus of plutonium as obtained from the reduction process in 1944. Magnification: 6×.

high temperature. The first attempted reduction, using a metallothermic reaction in a closed vessel or stationary "bomb," was a failure; we got a partially fused coky mass. The second, carried out by T. Magel and N. E. Dalles at a high temperature in a 900-rpm centrifuge, produced a fine button of clean metal. Several gram-scale reductions were made in the centrifuge before it was displaced by stationary bomb techniques, which in the meantime had been brought to perfection by R. D. Baker. Figure 3 shows the appearance of one of the first 10-gm reductions made by Baker. Shortly thereafter other methods also became workable, including electrolysis from a fused-salt bath, a scheme devised by Morris Kolodney. The drops of molten plutonium accumulating at the cathode provided the first irrefutable proof that the melting point of the metal was relatively low.

REFRACTORIES

Despite its fusibility, however, the handling of molten plutonium is rendered difficult by its extreme reactivity. No ordinary refractory was found to be satisfactory for crucibles in which to melt and cast very small pieces of the metal. Fortunately, an exotic one—cerium sulfide—had been studied by the chemistry group at Berkeley for use in vacuum at high temperatures, and they willingly gave us a supply of small crucibles. For several critical weeks these were, indeed, the only available containers in which the metal could be melted without serious contamination. As larger amounts of metal were melted at one time with relatively smaller areas of contact with the crucible, the refractory problem became somewhat simpler. Nevertheless, without the production of special crucibles, first in the Chemistry Department of the University of California at Berkeley and somewhat later at the Massachusetts Institute of Technology, work at Los Alamos would have been seriously delayed.

MALLEABILITY

The first button of plutonium we made was of low density and was malleable. As the purity of the metal increased in later reductions, both density and brittleness increased. At once it was suspected that the behavior of the metal was complicated by the existence of polymorphic transformations. The discovery of the first transition point at about 120° C followed a few days later from an experiment in an oil-filled volumetric dilatometer, which revealed a volume change on transformation greater than that of any known metal except tin.

During our first trials of shaping methods we encountered many manifestations of the enormous change in volume on transforming. A beautifully flat sheet of metal, hot-rolled after heating in an oil bath, would curl up like a saucer as it cooled and transformed, and cylinders would develop strongly concave ends. Fortunately, however, the metal stayed more or less in one piece and did not fall to a powder as tin does, which suffers a comparable volume change when it transforms to the gray modification at low temperatures.

Since plutonium does not shrink on solidification, solid castings with smooth tops can be made. Though machining is possible, it was generally avoided in the 1944–45 years, partly because of the health hazard and partly because we could afford to produce no scrap whatever.

CORRODIBILITY

Both to prevent the rapid corrosion of plutonium in moist air and to give some measure of health protection, it was necessary to apply an impervious coating of some kind to the finished pieces. When the plutonium core for the Alamogordo test was completed, it was beautiful to gaze upon. However, a few blisters resulting from corrosion began to be evident 3 or 4 days before the scheduled date. For a time, postponement of the whole event was threatened, but fortunately a simple metallurgical trick was devised in time to salvage both the core and the metallurgists' reputations.

TRANSFORMATIONS

Parallel with the study of fabrication methods went some work on the nature of the allotropic transformations and the kinetics of transformation. It was soon found that the unprecedented number of five modifications existed between room temperature and the melting point, producing great metallurgical complexity and opening a fascinating field for fundamental research. Perhaps unfortunately, we were able to solve the fabrication problems empirically with very little fundamental knowledge, and a well-planned program of basic research had to be postponed until the postwar years.

Table 1 lists some of the physical properties of plutonium as they were believed to be at the end of 1945. When one considers the completely unorthodox nature of

TABLE 1

PROPERTIES OF PLUTONIUM METAL AS THEY WERE BELIEVED TO BE IN 1945*

Phase	Temperature Range of Stability† (° C)	Crystal Structure	Density at 25° C.	Linear Expansion Coefficient (Average)	Electrical Resistivity Microhm-cm	Temperature Coefficient of Resistivity
Alpha..	Below 117	Orthorhombic, like U (doubtful)	19.8	55×10^{-6} per °C	150 at 25°C	-29.7×10^{-4}
Beta...	117–200	Unknown (complex)	17.8	35	110 at 200°	Zero (approx.)
Gamma	200–300	Unnkown (complex)	....	36	110 at 300°	Zero (approx.)
Delta..	300–475	Face-centered cubic	16.0	−21	102 at 400°	$+1.5\times10^{-4}$
Epsilon.	475–637	Body-centered cubic	16.4	4	120 at 500°	Zero (approx.)

* See chaps. x and xv for more recent values.
† Melting point 637° ± 5° C.

the metal and the environmental pressures under which its properties were first determined, it is indeed remarkable that these wartime values are so close to the later ones reported elsewhere in this volume (chaps. x and xv).

By the end of 1945 the existence of five phases above room temperature had been confirmed by dilatometric and electrical resistance measurements, with additional confirmation in some cases from crystal-structure determination and the mechanical properties. The two modifications existing at the highest temperatures below melting are face-centered cubic and body-centered cubic in structure, while the three lower phases are complex, and the diffraction patterns had not been satisfactorily analyzed by the end of 1945.

There is considerable hysteresis on heating and cooling, as can be seen in the typical dilatometric curves in Figure 4. Figure 5 shows the progress of the alpha-beta transformation isothermally at various temperatures. The rate of transformation is very sensitive to prior treatment and to the presense of impurities.

Not only is plutonium anomalous in the large number of allotropic modifications in which it exists, but also the electrical resistance of all the phases (Fig. 6) is ex-

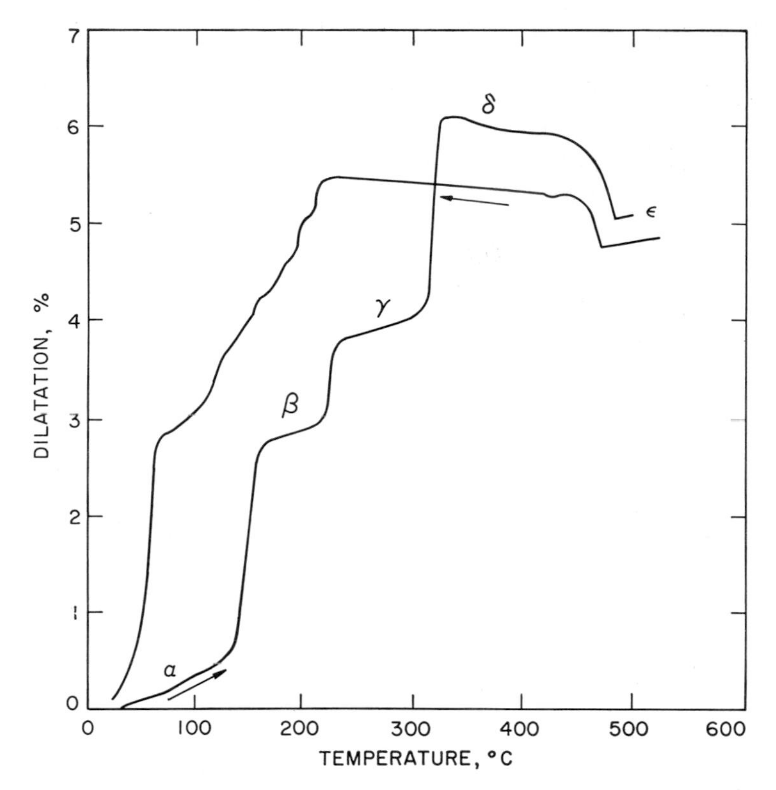

Fig. 4.—An early dilatometer curve for plutonium, showing hysteresis between the temperatures of phase transformations on heating and on cooling.

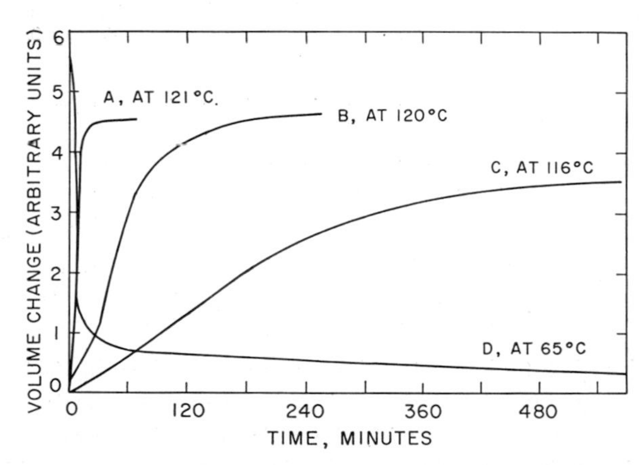

Fig. 5.—Curves showing the rate of volume change for the alpha-beta transformation at various temperatures. Curves A, B, and C are for the transformation of alpha to beta at the indicated temperatures. Curve D is for the transformation of beta to alpha at 65° C.

tremely high—that of alpha exceeding all other elements generally considered to be metallic. Its temperature coefficient of resistivity is negative, generally a characteristic of a semiconductor. The high positive thermal expansion coefficient of alpha is noteworthy, and the high *negative* expansion coefficient of the face-centered cubic delta phase is without parallel among metals.

In 1945 the theory of metals was (as today it still is) unable to predict any of these structures or properties. Since plutonium is a member of a new rare-earth series (commencing with actinium or thorium, depending on valence), there are many electron-energy levels of nearly identical energy, and the electron state is

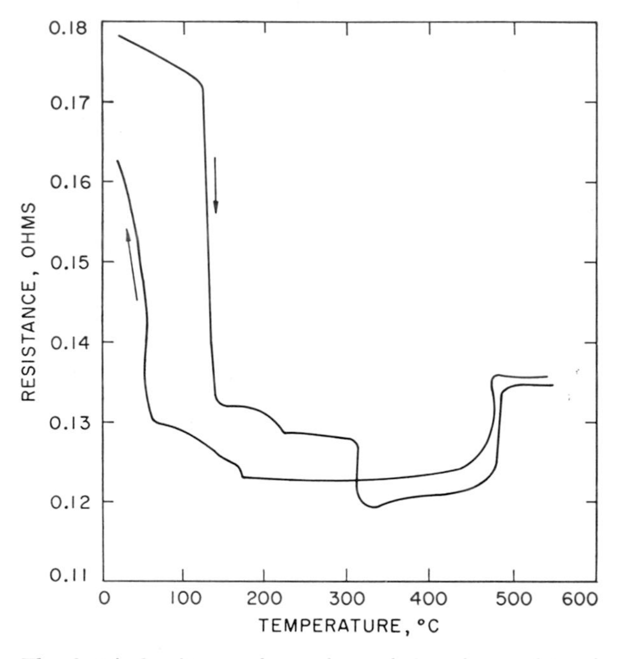

FIG. 6.—The electrical resistance of a specimen of plutonium as determined in 1945

easily affected by external factors. The large volume difference between delta and alpha is probably due to the forcing of a "valence" electron into the 6d shell.

The mechanical properties of the metal vary greatly with the amount of each phase that may be present. Early in 1944, tensile and compression tests were made on 1-gm samples by means of a Chevenard microtesting machine borrowed from Canada.

The properties of plutonium listed above were determined in 1944 and 1945 by a group working under the direction of Eric Jette; Frank Schnettler made many of the more difficult measurements with the assistance of Messrs. Martin, Wilson, Davis, Loeb, Litz, Spindler, Goldman, and Simmons. E. F. Hammel had charge of remelting and casting, while reduction was carried out by R. D. Baker and Morris Kolodney and their assistants. It is unfortunate that the pressure of wartime activities and the confusion of the immediate postwar period prevented the preparation of complete documents by these individuals. The data given in this paper have been gleaned from progress reports from which it is often impossible to credit individuals for any given experiment. The entire Chemistry-Metallurgy Division shares credit for each achievement.

Working at Los Alamos in wartime was an exciting and moving experience. The knowledge of the broad problems that would face mankind if the bomb were successful, the excitement of discovering the properties of a brand-new element, the pressure and the isolation of the work, the personal association with many of the leading scientists of the day, the background of the fantastically beautiful New Mexico landscape—all combined to make the period unforgettable. In retrospect, the most valuable aspect from the point of view of a metallurgist is the fact that, throughout the project, physicists and metallurgists, through enforced close association, learned to appreciate the value of each other's special knowledge and viewpoint. This is in some degree responsible for the increased activity in the physics of metals that has been evident in the postwar years.

CHAPTER V

LATER PLUTONIUM METALLURGICAL RESEARCH AT LOS ALAMOS

A. S. Coffinberry

ORGANIZATIONAL CHANGES AND PLUTONIUM METALLURGY AT LOS ALAMOS DURING EARLIER POSTWAR YEARS

With the end of World War II on August 15, 1945, the original *raison d'être* for Project Y at Los Alamos, New Mexico, ceased to exist. The impressive array of scientific talent, largely recruited from academic circles, that had been assembled there to do a specific job had hit the bull's-eye of their immediate target, and at once a majority of this personnel, after many months of living a most circumscribed existence for both themselves and their families, became imbued with a restless urge to return to their former pursuits at universities, in industry, etc. It is interesting, however, that a not inconsiderable number of those who had looked forward so longingly to the day when they could escape from the restrictions of a highly secret military project hurried back to Los Alamos after only a brief encounter with postwar conditions in the outside world. These people, together with others who elected to remain at Los Alamos because they had acquired an attachment for the Land of Enchantment or had become intrigued with some aspect of the new scientific developments that were sure to evolve from the feverish research of the war years, formed a nucleus of those who hoped and planned that from the specifically oriented activities of the wartime project there would emerge a well-ordered and broadly balanced research establishment, justified by the need for a continuing development of the military uses of atomic energy but providing, as well, for fundamental investigations in all branches of nuclear and related science.

The hopes of these planners were launched on a course toward fulfilment when, on October 1, 1945, Norris Bradbury, the laboratory director newly appointed to succeed J. R. Oppenheimer, presented to the Coordinating Council of Project Y the statement of an "interim philosophy" designed to guide the laboratory until such time as regulation of its policies would be taken over by an as yet uncreated agency of the federal government. In late October, 1945, Eric Jette became the division leader of the Chemistry and Metallurgy Research (CMR) Division, a reorganization of the former Chemistry and Metallurgy (CM) Division that had been admin-

A. S. Coffinberry is at the University of California, Los Alamos Scientific Laboratory, Los Alamos, New Mexico.

istered jointly during the war years by Joseph W. Kennedy and Cyril Stanley Smith. By this time Project Y had become quite generally, but unofficially, known as the "Los Alamos Laboratory." It was not until early May, 1946, however, that the official title became "The Los Alamos Scientific Laboratory," when Jette, as acting laboratory director during an absence of Bradbury, had this name placed on a new letterhead.

By no means the least intriguing of the fleeting glimpses of new scientific vistas that were revealed during the rush of wartime activities was the rather sketchy insight that had been obtained into the idiosyncrasies of plutonium metal. Although several different groups in CM Division had contributed importantly to the development of plutonium metallurgical technology, Jette and his associates—the Plutonium Physical Metallurgy Group—had been the ones most intimately introduced to the fantastic behavior of this weird new element. It was to be expected, therefore, that, in his reorganization of the Chemistry-Metallurgy Division, Jette should have placed plutonium metallurgical research high on the agenda, and this he did. Thus it came about that, although a number of organizational changes have occurred during CMR Division's existence, throughout most of the twelve-year period since this division was created, two groups within it have been concerned almost exclusively with plutonium metallurgy, and important contributions to the metal physics of the element have been made by personnel in two other groups.

The successor to Jette's wartime CM-8, the new Plutonium Physical Metallurgy Group, was CMR-5 (now CMF-5), and three other groups of the earlier CM Division eventually became merged into CMR-11 (now CMB-11), the Plutonium Production Group. As originally constituted, CMR-9 (now CMF-9) was a Metal Physics Group, and, before it became the Cryogenics Group in 1953, its program called for precision measurements to be made on many physical properties of pure plutonium and a few of its alloys. The shift of emphasis within CMR-9 from metal physics to cryogenics was accompanied, however, by the eventual transfer into a newly formed Non-Plutonium Physical Metallurgy Group, CMR-13 (now CMF-13), of the four members of CMR-9 who had done the most work on plutonium. On January 1, 1953, R. B. Gibney became the group leader of CMR-13, and T. A. Sandenaw, H. L. Laquer, and A. A. Comstock, as members of his group, continued to study plutonium in addition to other metals. Laquer has since returned to CMF-9, and Comstock had left Los Alamos before his untimely death in 1956. More recently, Sandenaw has been joined by C. E. Olsen and C. C. Herrick, and these three comprise a section within the otherwise non-plutonous CMF-13 who are currently concentrating their attention on plutonium metal physics.

With regard to the histories of the two groups—CMF-5 and CMB-11—that have had the most to do with plutonium metallurgy at Los Alamos, it is appropriate to refer to two important policy recommendations made by Norris Bradbury at two different turning points in the laboratory's history. The first of these was contained in the "interim philosophy" referred to above. Specifically, Bradbury stated that "we have an obligation to the nation" to continue the production of nuclear weapons during the interim period.

Bradbury's conviction as to what the laboratory's longer-range production responsibilities should be was expressed to the first Atomic Energy Commissioners a few weeks before a second turning point in the laboratory's history, namely, when the AEC assumed control of the laboratory's future on January 1, 1947. In a letter dated November 14, 1946, he stated:

> The problem of the "production" of atomic weapons has been considered above. It is believed that no immediate change can be made in the extent of the "production" now being carried out at Los Alamos. However, if the philosophy of maintaining Los Alamos as an atomic weapon research center is carried out, it is suggested that plans be made to remove as much as possible of this routine activity from this site. This has the additional advantage of disseminating the knowledge of the necessary techniques, as well as decreasing the seriousness to the nation of a major accident or catastrophe at Los Alamos.

In other words, even though Los Alamos was required to be the sole atomic-weapon production center for some undetermined period of time, it was desirable not only that new production sites be set up elsewhere as rapidly as possible but that ultimately all production activities should be removed from Los Alamos, leaving there a program concerned only with weapon development, fundamental research in nuclear science, and (as entered the program later) development of certain non-military applications of nuclear energy.

The histories of CMF-5 and CMB-11 reflect the changes in laboratory activities that resulted from the implementation of these policies. Although CMR-5 was always a research group and never assumed any production function, during the first few years of its existence its research problems pertained mostly to the requirements of weapon development and plutonium production. In addition to its program for fundamental plutonium metallurgical research, this group has seldom been free from high-priority responsibilities in the weapon program.

From the standpoint of unmitigated commitment to the continuing demands of metal production and weapon development, however, the members of Group CMB-11 must be recognized as the major contributors to practical plutonium metallurgy, both extractive and adaptive. Although security regulations do not permit a detailed account of the activities of this very efficient group to be presented in the open literature, it should be emphasized that the personnel of CMR-11 were the pioneers in developing the reduction, casting, forming, and machining processes for plutonium materials. Hence most of what is now being done along these lines at other AEC installations and much of what has been published elsewhere regarding these techniques originated at Los Alamos.

When the war ended, chemical methods for the production processing of plutonium were, by present-day standards, primitive; recovery methods for scrap and residues were almost totally undeveloped; and, despite the operational success of the Trinity and Nagasaki bombs, the exotic details of plutonium fabrication had been very little explored. This was the situation in late 1945 when CMR-11 began its production and development program, which was eventually to be directed by R. D. Baker, who became group leader in 1948.

Initially, all the product output of both Oak Ridge and Hanford came to Los Alamos for final processing, and production was one of CMR-11's earlier responsibilities. Later, however, as production activities were gradually transferred elsewhere, an opportunity was provided for the group to attack much more effectively both the chemical and the metallurgical problems of plutonium purification, reduction, and fabrication. Mechanized and remotely operated methods of chemical processing were developed, the production of high-purity metal for research requirements was accomplished, and quality control in the fabrication of weapon components was brought to a high state of development. The designing of special equipment and methods for handling plutonium and for dealing successfully with its health hazard were most important features of this program.

Concurrently with these activities, CMR-11 shared with CMR-5 the responsibility for producing various special shapes and compositions of plutonium metal and alloys as a service to other groups and divisions at Los Alamos, as well as to other laboratories throughout the country. Most of these pieces were requested by physicists who wanted them for cross-section measurements, irradiation studies, or some type of criticality determination. The smaller pieces, such as disks or foils, were made by CMR-5, and the requests for larger sizes or larger numbers of pieces were usually filled by CMR-11. Although the present CMB-11 retains its responsibility to the weapon-testing program, it has recently come to play an important role in the LAMPRE, or liquid-plutonium reactor, activities at Los Alamos. In addition to conducting tests on fuel systems, a section of the group is studying fuel-reprocessing methods that will eventually be used in the power-producing operation of a LAMPRE-type reactor.

The first group leader of CMR-5 was George L. Kehl. When Kehl left Los Alamos in February, 1946, R. D. Baker became the acting group leader, in addition to his responsibilities as group leader of CMR-8. (Baker was group leader of both CMR-8 and CMR-11 from May, 1948, to September, 1956, when he became CMB division leader.) Baker remained in charge of CMR-5 until August 1, 1946, when F. M. Walters assumed the acting group leadership in addition to his duties as associate CMR division leader. A. S. Coffinberry succeeded Walters as group leader of CMR-5 in July, 1947. F. W. Schonfeld was made alternate group leader a few weeks later. Since February, 1958, Schonfeld has been the CMF–5 group leader.

During most of its existence, CMR-5 has functioned as a group within Jette's CMR Division. However, when Jette left Los Alamos in September, 1956, to become director of the Union Carbide Research Institute, CMR Division was subdivided to form two new divisions—CMB under Baker and CMF under R. D. Fowler. At that time CMR-5 became CMF-5, a group within the newly created CMF Division.

THE HISTORY OF THE PLUTONIUM PHYSICAL METALLURGY GROUP

It was during Baker's stewardship of CMR-5 that fuel elements for "Clementine," the Los Alamos fast reactor, were required. Although production of these plutonium elements was completed by CMR-11, their development was initiated in CMR-5. They were made by extruding delta-phase alloy, which was then machined, nickel-coated, and finally jacketed in mild steel. Although this work did not represent the first experience gained in extruding plutonium, it resulted in a considerable increase in the knowledge of delta-phase fabrication.

Throughout an extensive career in metallurgical research, one of F. M. Walters' special interests had been dilatometry as an experimental tool. It is not surprising, therefore, that during his regime in CMR-5 particular attention was given to dilatometric studies of pure plutonium and a number of its alloys. F. H. Ellinger, who joined the group during this period, greatly improved the application of X-ray diffraction to studies of plutonium and its compounds. F. W. Schonfeld and W. C. Spindler learned how to fabricate the alpha phase successfully and made measurements of certain mechanical properties of plutonium that have not since been repeated at Los Alamos.

Space limitations do not permit a detailed discussion of all the interests and activities that have been pursued by CMR-5 since July, 1947, nor can as many names be mentioned as would be required to give proper credit to all who contributed importantly to the group's program throughout the period since then. A few of the major trends and special developments that have characterized that program are worthy of note, however.

Mentioned above are the group's service activities as they related to the fabrication of special shapes and alloys to be used in nuclear physics investigations. A further service of this type, which may be said to have brought the Plutonium Physical Metallurgy Group its first recognition outside Los Alamos, was the production of the intermetallic compound $PuBe_{13}$ for use as a neutron source. This compound was first prepared in 1950 by C. R. Tipton, R. D. Moeller, and F. W. Schonfeld, and since then its popularity as a stable and predictable calibration standard has increased to the point where it became necessary to transfer manufacture of the sources to the Mound Laboratory at Miamisburg, Ohio.

Regarding the organization of CMR-5, it is appropriate to quote a few sentences from a November, 1953, memorandum on "The Organization and Activities of Group CMR-5":

> Organizationally CMR-5 is subdivided into six sections: The Fabrication, Thermal-Analysis, Diffraction, Metallographic, Corrosion, and Gas-Metal Sections. . . .
>
> It will become apparent that the study of plutonium-alloy phase diagrams is a primary concern of CMR-5. . . . Because the peculiar nuclear properties of plutonium give it special significance as a fuel for breeder-type reactors, this alloy program has its widest applicability, and is of greatest interest, in relation to the reactor uses of plutonium. Although (in 1953) interest in the reactor uses is probably greater at other AEC laboratories than at Los Alamos, these other laboratories have not in the past possessed facilities for undertaking metallurgical research on plutonium, and even now (in 1953) are equipped to perform such research in only a very limited way. Hence CMR-5 continues to be called upon to supply information that is needed for the designing of reactors of a variety of types.

One of the earlier bits of evidence that the Los Alamos Laboratory felt a sense of responsibility for the plutonium fuel needs of the nation's over-all reactor program consists of a memorandum dated August 22, 1946, and entitled: "Conference on Alloys for Breeders; Present: Hilberry, Nordheim, Peterson, Jette, Baker, Walters." In the light of current reactor concepts, this memorandum is most interesting. Heat transfer and corrosion resistance were recognized as "the important considerations for breeders, but especially the former." The volume of permanent gases produced by the fission processes was regarded as "trivial"!

> As far as alloying elements are concerned, the high energy neutrons are not bothered by impurities except those which slow them down. The lighter elements are more easily tolerated than the heavier ones. No particular objection to indium, tin, lead, mercury, or even cadmium; bismuth would give polonium which might or might not be desirable.
> . . . An alloy 80% U, 10% Pu, and 10% of anything else would be satisfactory atomic percentages.

The final paragraph reads:

> Present order of interest in the alloys discussed is:
>
> 1. Solid Pu Alloys
> 2. Al or Be Alloys of Pu
> 3. Liquid Alloys
> 4. Thorium-plutonium Alloys.

Because of the pressure of other problems, however, no action to implement these suggestions was undertaken for almost a year. During the summer of 1947 Walter Zinn and Edward Teller were in residence at Los Alamos as consultants, and conferences with each of these scientists were arranged by Jette, Walters, and Coffinberry. The suggestions of both Teller and Zinn regarding the reactor uses of plutonium were similar to the recommendations that had been made at the earlier conference, but the discussions had a considerably more quantitative character. Zinn expressed the opinion that the metallurgical problem of greatest immediate concern to reactor physicists was the reason for blister formation on the surface of the cans that contained the active materials. He called attention to a formula for cycling time in which important variables were the specific power of the reactor and the per cent burnup as limited by blister formation. The most easily controlled variable in this formula was the dilution of fissionable atoms in the fuel.

Both Teller and Zinn expressed interest in the idea of using low-melting metals as liquid fuels in which plutonium might be dissolved. This interest led to the following suggestion made in a memorandum of September 23, 1947, Coffinberry to Jette:

> As a result of conferences with E. Teller and W. H. Zinn, it seems certain that lead and bismuth are destined to be of considerable interest to physicists as alloying elements with plutonium in reactors. It is a coincidence that these two elements, together with tin [all three of them low neutron absorbers], are the metals most extensively represented in the compositions of all common low-melting alloys (excluding amalgams). Since, from the standpoint of present interest in applications, the probable relative importance of plutonium alloys lies in the order (1) low temperature alloys, (2) high temperature alloys, (3) corrosion resistant alloys, it is suggested that an investigation of the binary, ternary, and quaternary alloys of plutonium with lead, bismuth, and tin be made the beginning of a systematic and rather thorough study of the alloys of plutonium with B subgroup elements.

The inevitable fate of this program is now clearly recognized by all plutonium metallurgists. The B subgroup metals were found to be the very elements which form the highest-melting compounds with plutonium and which are most quickly "dried up" by the addition of plutonium to their melts. Eventually, however, this program was carried out in part by R. O. Elliott and E. M. Cramer, to the extent of determining the liquidus curve at the high-bismuth end of the plutonium-bismuth system.

But the question of exactly what metallurgical program might be expected to do the reactor physicists the most good was still felt to be largely unanswered, and Jette insisted that the physicists should make up their minds quite definitely, in order to avoid the expenditure of wasted effort in experimentally difficult investigations of plutonium alloys. It was highly opportune, therefore, that Jette was able one day in 1949 to engage Samuel Untermeyer and Harvey Brooks at lunch simultaneously. Proceeding to pin them down to definite commitments, he shortly thereafter received letters from both these power-reactor enthusiasts in which two lists of alloying elements—first and second choices—were recommended for study. The first choices were aluminum, magnesium, titanium, vanadium, nickel, and zirconium. Somewhat less desirable, but reasonable, second choices were indicated to be manganese, iron, cobalt, and molybdenum.

Throughout the following year the binary phase diagrams based on the first choices were made the subjects of six projects in the CMR-5 research program, and to these was added, because of its possibilities with regard to internal breeding, a seventh project on the plutonium-uranium system. But the second choices suggested by Untermeyer and Brooks were overlooked until long afterward.

The principal features of the plutonium-aluminum diagram were determined rather promptly, and throughout a period from 1950 to 1952 the results of this investigation and much of the rest of the over-all alloy program appeared in a series of classified reports, such as LA-1000 on the plutonium-aluminum system by Schonfeld and Moeller, LA-1304 on the plutonium-nickel system by G. W. Wensch and D. D. Whyte, and CMR-5-2065 on plutonium-uranium by F. H. Ellinger. Only the rough outlines of the phase diagrams of plutonium with titanium, vanadium, and zirconium were determined, however.

When, in the fall of 1955, K Division, the power-reactor division at Los Alamos, became interested in low-melting alloys of plutonium that could be used as fuel in a liquid-metal homogeneous reactor, CMR-5 knew of only one possibility with a high enough plutonium content to meet their requirements—a eutectic in the plutonium-nickel system. Within a very few weeks, however, several other binary systems were investigated briefly, such as those of plutonium with manganese, iron, cobalt, and copper, metals suggested by the form of their phase diagrams with uranium and the rare earths. From this work it was learned that the two most promising possibilities among binary alloys were plutonium-rich eutectics in the plutonium-iron and plutonium-cobalt systems.

Since that time, the present Group CMF-5 has been much concerned with the study of liquid-fuel alloys and of materials to contain them. More recently the program has been expanded to include ternary alloys involving cerium as a diluent for the plutonium-rich eutectic compositions of the binary systems. Cerium has been found to be the most satisfactory metal that can be added to the plutonium-cobalt and plutonium-nickel alloys as a third component without appreciably increasing the low-melting temperatures of the binary eutectics.

For many years following the first reduction of plutonium to metal, metallographers at Los Alamos were unable, under either bright-field or polarized illumination, to reveal any semblance of grains or grain boundaries in the room-temperature (alpha) phase. All techniques of polishing and etching that were tried failed to reveal anything but either a uniformly bright field (except for impurity inclusions) or a pattern of stains bearing no relation to any possible grain structure. Finally, one day Schonfeld quite accidentally treated a specimen in such a way that an appearance suggesting grains was revealed under polarized light. He failed to obtain a photograph of this, however, and was never able to reproduce the same

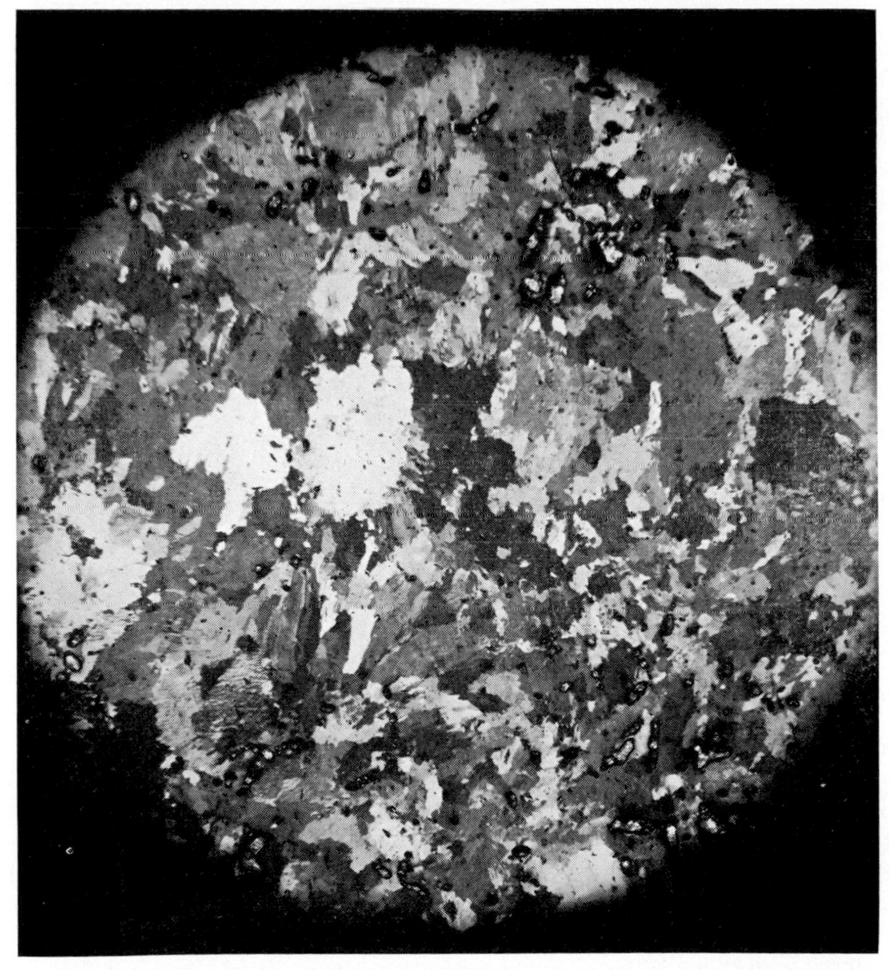

FIG. 1.—The first photomicrograph (125×) of alpha plutonium obtained with polarized light in such a way as to reveal differences in grain orientation. This specimen contained many small voids.

condition. Later, E. M. Cramer also accidentally obtained a similar result, but he also could not reproduce it immediately. Ultimately, it was C. O. Matthews who, in early 1954, first developed a metallographic procedure that would reproducibly reveal differences in grain orientation in alpha plutonium. Figure 1 is the first photomicrograph of alpha plutonium obtained by Matthews in this way.

The specimen shown in Figure 1 is high-purity metal that had been used in dilatometer experiments that led to the discovery of the delta-prime phase of plutonium in 1953. Before his death in April, 1953, F. M. Walters had suggested an

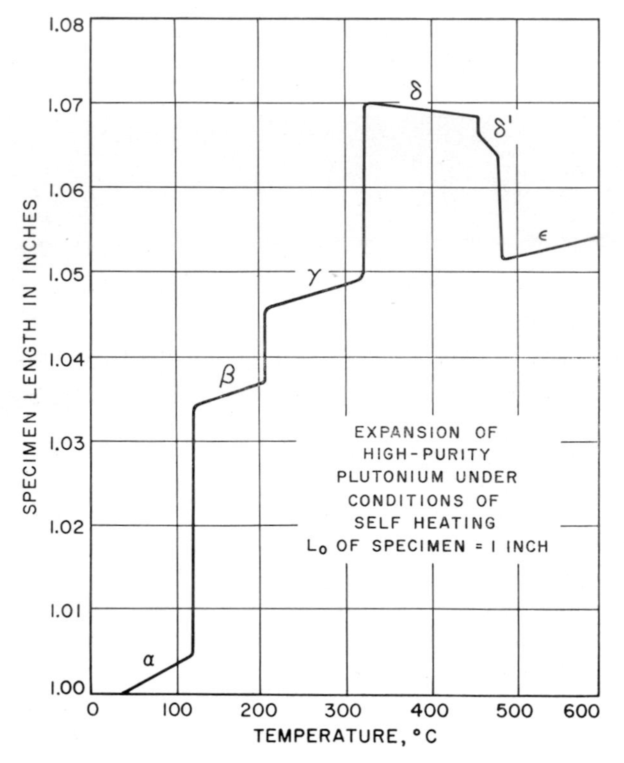

FIG. 2.—A typical dilatometer curve of high-purity plutonium obtained by the self-heating method.

explanation for anomalous behavior observed in the melting and freezing of plutonium that took account of its radioactive self-heating and its very low thermal conductivity. Analysis of the melting-point problem on this basis served to call attention to the fact that, if one undertook to obtain either thermal-analysis or dilatometric curves of plutonium in the manner usual for metals having good thermal conductivity, the results obtained must necessarily be erroneous because of the existence of an appreciable temperature gradient within the metal specimen at all times and especially during phase transformations. It was apparent that the only way in which such a gradient could be avoided (and it could be avoided only during heating) was to allow the specimen to heat itself through its own radioactivity and to maintain the temperature of a surrounding heating coil at the same temperature as that of the specimen, so that heat would flow neither into nor out of it.

When this was tried as nearly as could be realized experimentally, it was found that the dilatometer trace had changed from its previous alloy-like form to a curve quite characteristic of a pure metal, and it was this sharpening of the dilation curve that clearly indicated the existence of a sixth allotropic modification of plutonium —the delta-prime phase. The form of dilatometer curve obtained by the self-heating method is shown in Figure 2. Its improvement with respect to sharpness of the transformations is obvious if one compares this figure with the early dilatometer curve shown in Figure 4, chapter iv. Dr. Walters did not live long enough ever to learn about delta-prime plutonium, but it must be recognized that his analysis of the melting-point anomaly was a development very instrumental in leading to the discovery of the new phase.

An account of the more noteworthy accomplishments of CMF-5 would not be complete without mention of the contributions of W. H. Zachariasen in solving the crystal structures of the alpha and gamma phases of plutonium and in determining the unit cell of the beta phase. Since 1950, Dr. Zachariasen has been a consultant to the group, and his summers spent in Los Alamos have been exceedingly fruitful in their yield of important crystal-structure results relating to the transuranium elements and their compounds. Because no single crystals of any plutonium allotrope have ever been produced, the results for the alpha, beta, and gamma structures had to be obtained from powder patterns. It is highly improbable that any scientist other than Zachariasen could have solved three structures as complex as those of alpha, beta, and gamma plutonium from powder patterns alone, at least not by means of the direct approach that he employed. Also essential to his success, however, was the experimental work of Ellinger in preparing the excellent powder patterns with which he worked.

THE PLUTONIUM HEALTH HAZARD

A subject of constant concern to any group working with plutonium has been mentioned only in connection with the developmental program of CMB-11. This is the matter of the plutonium health hazard, to which a great amount of attention was necessarily given as an inescapable adjunct to all plutonium research activities.

At Los Alamos the history of the health problem associated with plutonium began with the first small quantities of plutonium-containing material that were received at Site Y in 1944. The subject of the health and safety practices followed at Los Alamos during the war years is discussed by David Hawkins in his fascinating history (classified) of Project Y from its inception to the end of the war. To quote Dr. Hawkins:

> A Health Group reporting to the Director was part of the Laboratory administration from the beginning. . . .
>
> In the original plan of Laboratory activities it was assumed that biological and physical research related to health problems would be entirely the responsibility of other laboratories within the Manhattan District. Reliance on the work of others did not, however,

always provide necessary information at the time it was needed. Research sections were set up as required within the Health Group or by its request in other groups. Thus the development of apparatus needed for monitoring was undertaken at Los Alamos in the spring of 1944, and a large share of the instruments were built in the Electronics Group. In August, 1944, it became necessary to investigate biological methods of testing for overexposure to radioactive poisons, and this work was undertaken by a section of the Health Group.

Its activities up to the end of 1943 were confined largely to the hazards of external radiation from accelerating equipment and radioactive sources. The danger of heavy-metal poisoning from uranium had to be guarded against, as did other chemical hazards, but these problems were not serious.

The really serious problems of the Health Group appeared in February, 1944, with the arrival at Los Alamos of the first quantities of plutonium.

Although the exact nature of the plutonium health hazard was certainly not understood at that time, its general similarity to the radium hazard had just been discovered, and, as a result, Dr. Louis Hempelmann, accompanied by representatives from Oak Ridge and the Chicago Metallurgical Laboratory,

visited a luminous paint company in Boston to learn how the radium hazard was handled in that industry. On his return three committees were established in the Chemistry and Metallurgy Division to develop methods for control of the plutonium hazard. An instrumentation committee was appointed to design counters suitable for measuring the radioactive contamination of laboratories and personnel. A second committee was responsible for the design of apparatus and equipment for handling plutonium. Apparatus was designed by this committee in consultation with the chemists concerned, and was built or procured by the Chemistry and Metallurgy Service Group. A third committee drew up rules and recommendations for the safe handling of radioactive materials. The procedures recommended were put into effect in March, 1944, with the understanding that willful non-cooperation would result in immediate dismissal from the Laboratory. A section of the Service Group was established under W. H. Popham to enforce these procedures. It had the positive functions of providing personnel with proper protective equipment, laundering this equipment, monitoring the laboratories and decontaminating them when necessary, and of keeping complete records. This group [in CM Division] worked very closely with the Health Group.

In addition to organizing the safety measures described above, the Health Group carried on an extensive educational campaign among those working with plutonium. Lectures were given on the toxicology of plutonium, and numerous conferences were held with operating groups to work out the application of general recommendations. The Health Safety Handbook was given to new members of the Division.

Despite these precautions the members of the Health Group and of the Chemistry and Metallurgy Division were not satisfied with the progress of biological studies on plutonium made by the other projects responsible for this work. This dissatisfaction was crystallized by an accident which occurred in August, 1944, when in a minor chemical explosion a number of milligrams of plutonium were thrown into the face of one of the chemists. A research program was undertaken, aimed primarily at developing tests for detecting overdosage of plutonium.

Another continuing difficulty was the lack of adequate monitoring equipment. Alpha-ray counters lacked either sensitivity or portability, and were not received in adequate numbers. The lack of sensitive portable meters made it necessary to swipe surfaces suspected of contamination with oiled filter paper and measure the activity collected with stationary counters. Contamination of hands and nostrils was measured in the same fashion. Because instruments received from Chicago did not meet the local monitoring requirements, development of such equipment was begun in the Electronics Group of the Physics Division in May, 1944.

By August, 1944, methods for the purification of plutonium were well advanced, but one of the most serious difficulties, from a technical point of view, was the prevention of contamination of the plutonium from dust, etc., that would undo the work of purification. It was this factor that made necessary the construction of an air-conditioned laboratory building, designated "D Building," and shown in Figure 3. Thus the first dust-tight structure built to house plutonium research and production operations was designed to keep dust out, not to keep plutonium in—although ultimately the latter function came to be the more important of the two.

At about this same time, increased knowledge of the seriousness of plutonium poisoning resulted in the development of inclosed apparatus wherever possible, so

FIG. 3.—D Building, an air-conditioned laboratory for plutonium chemistry and metallurgy built at Los Alamos during World War II.

that D Building was equipped throughout with glove boxes, or "dry boxes," as they were then called.

It will be noted that from the very beginning of work with plutonium the Chemistry-Metallurgy Division seemed to be even more concerned about the health hazard than was the Health Group—and for good reason, they were the ones who had to work with this highly toxic material. With the war's end and the assumption by Jette of the leadership of CMR Division, the time had arrived when attention could be given to the health problems of plutonium in a more thorough fashion. It was one of Jette's traits as an administrator that matters relating to the health and safety of personnel were assigned an importance that overrode every other consideration. This philosophy was relentlessly impressed upon all who worked under his supervision during the eleven years that he was a division leader. It resulted in a health and safety record for the division that outshone by an order of magnitude even the very outstanding record of the laboratory as a whole.

As Jette saw the role of CMR Division in the over-all scheme of AEC laboratories, its special function was to work with large quantities of radioactive and fissionable materials, in contrast to the microgram and milligram scales of operation that had been developed at the Clinton, Berkeley, and University of Chicago laboratories. He gave expression to this view and its consequences in a letter to Bradbury in which he outlined the division's research program as of April 10, 1946:

1. Our problems are concerned with the production, isolation and utilization of large quantities of fissionable and radioactive materials. And the results of very small scale investigations have generally proved inadequate and unsatisfactory for our purposes.

2. The protection of the men working on the larger amounts of such materials involves elaborate equipment, special techniques, and medical inspection and auxiliary services which we have developed to a rather high degree during the past few years. The health hazards involved in dealing with large quantities, whether expressed in mass or radiation energy, are known to and appreciated by few persons who have not been in direct contact with work on these scales. The spreading of contamination from the laboratories to the surrounding community involves problems which become quite serious as the scale of operations increases.

The Health Instruments Group, CMR-12, was established in 1945 under W. H. Hinch as a part of CMR Division. Since 1951, however, this group has been incorporated into the Health Division, with D. D. Meyer as group leader. Its "watchdog" functions included monitoring and decontamination activities in the technical areas, responsibility for the care and use of counters and meters for detecting radioactivity, and laundry functions relating to protective clothing and respirators. Besides this "police" work, much effort was put into educating personnel in the importance of health-safety rules and regulations. CMR-12 also carried on investigations directed toward the development of radiation-detection equipment that would be more sensitive and more stable and rugged than previously used types. Finally, this group and its successor, H-1, played an important role in the development of a philosophy for dealing with the plutonium health hazard that was basic in designing a new building to house research activities in the fields of plutonium chemistry and metallurgy.

THE NEW CHEMISTRY-METALLURGY BUILDING

During the war years, partly because of ignorance and partly because of the stress of wartime conditions, operations with plutonium in D Building were conducted with greater laxity than has ever been tolerated since. As a consequence, by the time the war ended, D Building was known to be "hotter than a firecracker." Only by heavy and frequent applications of paint on walls, floors, ceilings, and other surfaces could the contamination on them be safely prevented from getting into the laboratory air. For this reason and because the old D Building was seriously overcrowded, plans for a new D Building began to be formulated very early in Jette's regime.

Shortly after his arrival in Los Alamos, the author, along with the other CMR group leaders, was asked to make suggestions regarding how the new building should be designed. At about that time F. M. Walters brought to his attention the fact that, because plutonium and its compounds are heavy, the conventional practice of exhausting air upward from fume hoods and other inclosures was not logical for the removal of plutonium-contaminated air, and he suggested that the exhaust ductwork be made to carry its burden downward and out of the building through ducts beneath the floor. This system was tried in the CMR-5 laboratories of the old D Building, and it proved to be so successful and was such an improvement over the former ventilating arrangements that it was adopted as standard for the new building, designated the "CMR Building" and shown in Figures 4 and 5.

Figure 4 is an architect's drawing of the CMR Building showing its over-all layout. It is seen that the building consists primarily of five large "plutonium" wings interconnected by a narrow, windowless "spinal corridor" perpendicular to these wings (see also Fig. 6; the length of the spinal corridor is 650 feet, approximately $\frac{1}{8}$ mile). In addition, there are a smaller Administration Wing (containing the main entrance to the building, at the left in Fig. 5) and, directly behind this, a "non-plutonous" laboratory wing, shorter than the plutonium wings and having

FIG. 4.—Architect's drawing of the CMR Building at Los Alamos

FIG. 5.—Wing 2 and the front entrance of the CMR Building

laboratories on three floor levels. Although the entire CMR Building is air-conditioned, only the five plutonium wings contain the elaborate and extensive ventilating equipment required to deal adequately with the health hazard of plutonium.

Figure 6 is a plan view of the CMR Building in which is indicated the general type of activity pursued in each of the five plutonium wings. It is seen that Wing 4, although equipped for work with plutonium, is used primarily for work pertaining to U^{233} and U^{235}. Also performed in Wing 4, however, is the research of the plutonium metal-physics section of Group CMF-13. Wing 1 is the "non-plutonous"

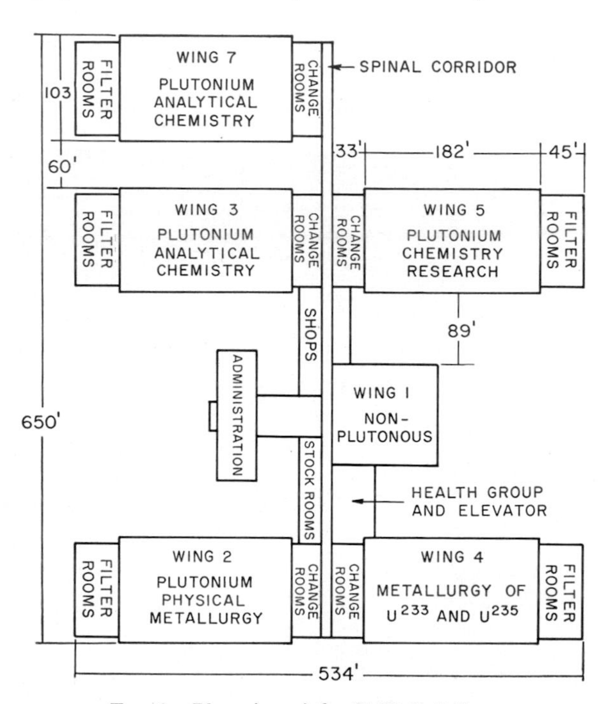

FIG. 6.—Plan view of the CMR Building

laboratory wing, and in it are housed the research activities of CMF-13 in the field of non-plutonous physical metallurgy, together with the laboratories of an instrument-development and servicing group. Wing 2 is occupied by Group CMF-5 and thus contains the laboratories for research on the physical metallurgy of plutonium. Wings 3, 5, and 7 are devoted to research and service work in plutonium chemistry, as indicated in Figure 6. The facilities of Group CMB-11 for the extractive and fabrication metallurgy of plutonium are housed in a different building.

Figure 5 is a photograph of Wing 2, with a portion of the Administration Wing shown at the left. Wing 2 is typical of all the plutonium wings, since all five of them are identical in general plan and content of service equipment. Each of these wings is, in fact, a separate building, insofar as it contains a ventilating system, clothing-change rooms, and other service facilities that are separate and complete in every wing. A typical plan view of a plutonium wing at the working or laboratory level is shown in Figure 7. As indicated in Figure 8, there is a full attic above and a full basement below this level, both of which are used only as service areas. But, at the working level shown in Figure 8, the larger rooms on either side of the central utility corridor are the laboratory modules, and the smaller rooms adjacent to the outer walls are used either as offices or as space for auxiliary "cold" laboratories.

Except for the plutonium storage vaults shown on either side of the wing, plutonium is taken only into the corridors and the central laboratory modules; these are the potentially contaminated, or "hot," areas. Under no circumstance is it permissible to take plutonium into any of the offices or the auxiliary laboratories; these outer modules are strictly uncontaminated, or "cold," areas. Thus, at the level of the laboratories and in a direction perpendicular to the length of the wing, there

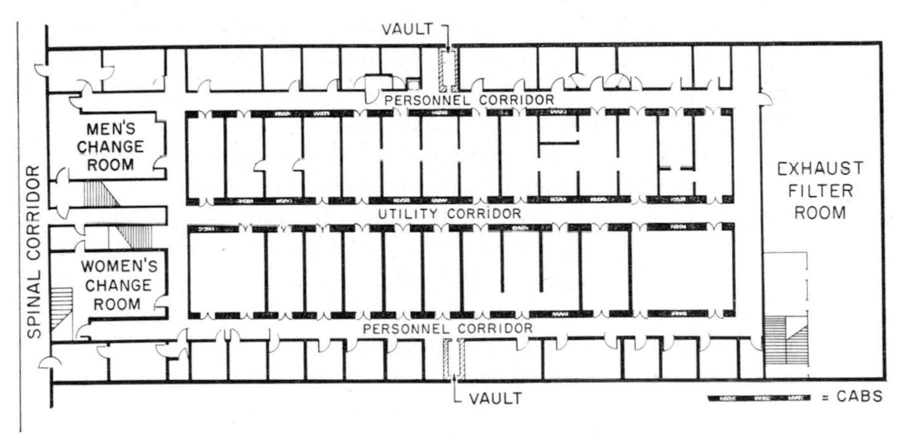

Fig. 7.—Plan view of a plutonium wing at the laboratory level

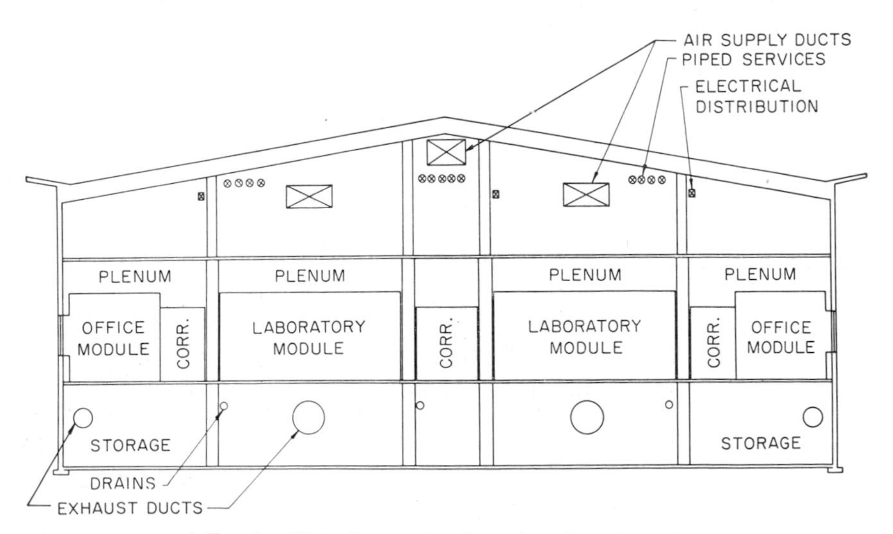

Fig. 8.—Elevation section through a plutonium wing

exist two gradients of contamination potential, each beginning at zero in the outer modules and increasing toward a maximum at the central utility corridor.

In keeping with these gradients of contamination potential, the following practices are observed: (1) as stated above, plutonium may never be introduced into any of the outer modules; (2) plutonium may be transported through the personnel or utility corridors, but only in sealed containers; (3) plutonium is removed from containers and worked with only in the central laboratory modules; (4) any contaminated material or equipment to be removed from the laboratories is transported out of them through the central utility corridor and in a direction toward the exhaust filter room.

Thus at the laboratory level there is also a gradient of contamination potential in the direction of the length of the wing; it begins at zero in the spinal corridor and increases toward a maximum in the exhaust filter room. In keeping with this gradient, the intake of supply air and the introduction of other uncontaminated services (water, gas, electricity, etc.) occurs at the spinal-corridor end of the wing, and the removal of exhaust air, drainage from sinks, or any potentially contaminated waste is effected at the opposite end of the wing.

Finally, in the direction of the third, or vertical, dimension of each plutonium wing, there is still another gradient of contamination potential. As indicated above, the sense of this gradient has been made compatible with gravity and the high density that characterizes plutonium and most of its compounds. The gradient therefore begins at zero throughout all the attic level (except in the exhaust filter room at this level) and increases in the downward direction toward a maximum in the basement below the laboratories. In keeping with this gradient, intake air and the "clean" services are distributed within the attic area to the various laboratories and offices and are introduced into these areas from above; exhaust air and other wastes are removed from the laboratories and offices by extraction into the basement, and their movement at this level is toward the exhaust filter end of the wing. Air extracted from the central laboratory modules is filtered before being discharged into the outdoor atmosphere, but this is not done with air removed from the "cold" offices and auxiliary laboratories.

Liquid wastes from the laboratories are drained into one of two large retention tanks (used alternately) located near the exhaust end of the basement area. Samples of these liquid wastes are analyzed for plutonium content before the ultimate disposition of each accumulated batch is decided. (Throughout a period of more than five years that the CMR Building has been in use, no concentration of plutonium high enough to justify recovery operations has ever been found in the liquid wastes; hence they have always been discharged into a special sewer system employed only for wastes potentially contaminated with plutonium.)

In conformity with the gradients of contamination potential established for the vertical and longitudinal dimensions of each plutonium wing, louvres for the intake of outdoor (supply) air are located at the attic level adjacent to the spinal corridor, i.e., at the locations designated *A* in Figure 6 and shown in Figure 5 on the sides of each of the blocklike portions of the building that encompass the spinal corridor. At the attic level these blocklike units contain equipment pertinent to providing the clean services for each wing, and at the laboratory level they contain the clothing-change rooms shown in Figure 7. A major portion of the clean service equipment consists of fans and air-conditioning units, as shown in Figure 9. In these units the supply air is cleaned by means of roughing filters followed by water-spray washing, and both the temperature and the humidity of the supply air are ultimately adjusted to be optimum for human comfort. The purpose of thoroughly cleaning the supply air is much the same as that discussed above with regard to the air-conditioning of the wartime D Building, plus the desirability of minimizing all possible contaminating influences in the analytical chemistry laboratories.

Figure 8 is a vertical section through a plutonium wing showing the three floor levels: (1) basement, (2) laboratory and office level, and (3) attic. Also shown in cross-section are the pipes, electrical conduits, and rectangular ductwork for the

FIG. 9.—Fan and air-conditioning unit for supply air

clean services supplied through the attic and the circular sections of drain pipes and exhaust air ducts in the basement. In addition, there are shown spaces approximately 4 feet in height that exist below the floor of the attic and above the ceilings of the corridors and the laboratory and office modules. These spaces are plenum chambers into which the air carried by the supply ducts is discharged and from which it passes on into the corridors, laboratories, and offices through $\frac{3}{32}$-inch holes in the acoustical metal pans that comprise the ceilings above these areas.

By automatic regulation of the rates of flow within both the supply and the exhaust air ducts, static pressures in the laboratories, offices, and corridors are so adjusted that there will be some movement of air from corridors and offices into the "hot" laboratories but none in the reverse direction, i.e., from "hot" to "cold" areas. Air pressures are further regulated so that the static pressure is higher in the attic than outdoor atmospheric; at the laboratory level it is lower than outdoor atmospheric; and at the basement level it is still lower. The hot laboratories have no windows, and the windows in the office modules consist of transparent glass blocks and cannot be opened (see, in Fig. 5, *center*, the single row of windows at the laboratory level).

Air is removed from the offices and auxiliary laboratories through grilles located close to the floor. This air is transported to the exhaust filter rooms through the smaller exhaust ducts shown in the basement in Figure 8 and is ultimately discharged into the outdoor atmosphere without being filtered. Air is removed from the hot laboratories either through hoods, glove boxes, or other inclosures in which plutonium is handled or through floor grilles in those laboratories containing no inclosures required to have air flowing through them at a prescribed rate. This potentially contaminated exhaust air is transported to the exhaust filter rooms through the larger exhaust ducts shown in Figure 8.

At the exhaust end of each plutonium wing there are two filter rooms, one at the laboratory floor level and one at the attic level. These filter rooms are located in the blocklike units of the building at the ends of the plutonium wings farthest away from the spinal corridor. In Figure 5 the two levels of the filter rooms are indicated by the two rows of windows at the right, and the metal panels across the end wall of every plutonium wing (see Figs. 5 and 6) cover large openings through which components of the air-filtering units may be moved into or out of the building in case replacements or repairs are needed.

A single filtering unit at each of the two levels filters the air carried by one of the two large exhaust ducts. Although these filtering units were designed and built to provide water-spray washing as well as dry filters for cleaning the air, it has been found that adequate removal of contamination can be accomplished by the use of dry filters alone. The dry filters are of two types: (1) roughing filters composed of glass wool and (2) specially designed mats made of inorganic fiber and having a high efficiency for the removal of particulate matter of small size. The filtering materials are replaced approximately every three months, the exact frequency depending on the rate at which they become blocked with dust, and are hauled away to a contaminated dump and buried. Both as it enters and as it leaves a filtering unit, the air is continuously sampled for assay of its radioactivity. The results obtained represent total alpha activity integrated over a period of several hours. A photograph of one end of a filtering unit is shown in Figure 10; a fan at the opposite end (Fig. 11) blows the filtered air directly into an exhaust stack.

Like the unfiltered air from the offices, the filtered air from the hot laboratories is ultimately discharged into the outdoor atmosphere through stacks (visible at the right in Fig. 5) located above the roof of the filter rooms. These stacks are made convergent at the top, in order to eject the air upward into the atmosphere with increased thrust and dispersion. Throughout the entire length of the exhaust ducts the flow is maintained at a high velocity, in order to avoid deposition of particulate matter at any point. In addition, provision is made by means of spray jets distributed along the ducts (see Fig. 12) to flush their inside surfaces with water at regular intervals (twice per week). The wash water is collected in sumps below the filter rooms and pumped to the retention tanks for liquid wastes.

How effectively clean the interiors of the exhaust air ducts are kept is demonstrated by experience with making extensive changes in the exhaust ductwork of

FIG. 10.—One end of the filtering unit for exhaust air

FIG. 11.—Opposite end of the filtering unit and the fan for exhaust air

Wing 2 after it had been in use for more than five years. In the course of making these alterations, a number of large openings were cut in one of the main ducts and in some of the connecting ducts. These remained open to the basement atmosphere for periods of several days with no precautions against spread of contamination being taken except careful and extensive monitoring to detect the smallest trace of any alpha activity that might have escaped from them. None whatever was found.

The rate of air flow through each of the five plutonium wings is approximately 80,000 cubic feet per minute. The total for the entire CMR Building is thus almost 500,000 cubic feet per minute, since Wing 1 and the Administration Wing are also

FIG. 12.—View of the basement in a plutonium wing, showing one of the two main ducts for exhausting air from the "hot" laboratories, connecting ducts from opposite ends of the laboratory modules, and piping to spray-jets used to flush the interior of the main duct.

air-conditioned. In the individual laboratory modules this rate of flow provides about 36 air changes per hour and insures that, at the open faces of hoods and other inclosures, an air-flow velocity of not less than 100 linear feet per minute is continuously available. The over-all layout of ventilating equipment in half of a plutonium wing is shown pictorially in Figure 13. Figure 12 is a photograph taken in the basement of one of the plutonium wings and shows the main exhaust duct on one side of the wing, with branch connections to the hot laboratories. These cylindrical ducts and the connecting "horses" are constructed of $\frac{3}{16}$-inch type 316 ELC stainless steel, arc-welded at all joints except where flange fittings are shown. All the welds were radiographed to insure the complete absence of any pinhole porosity through which contamination might escape.

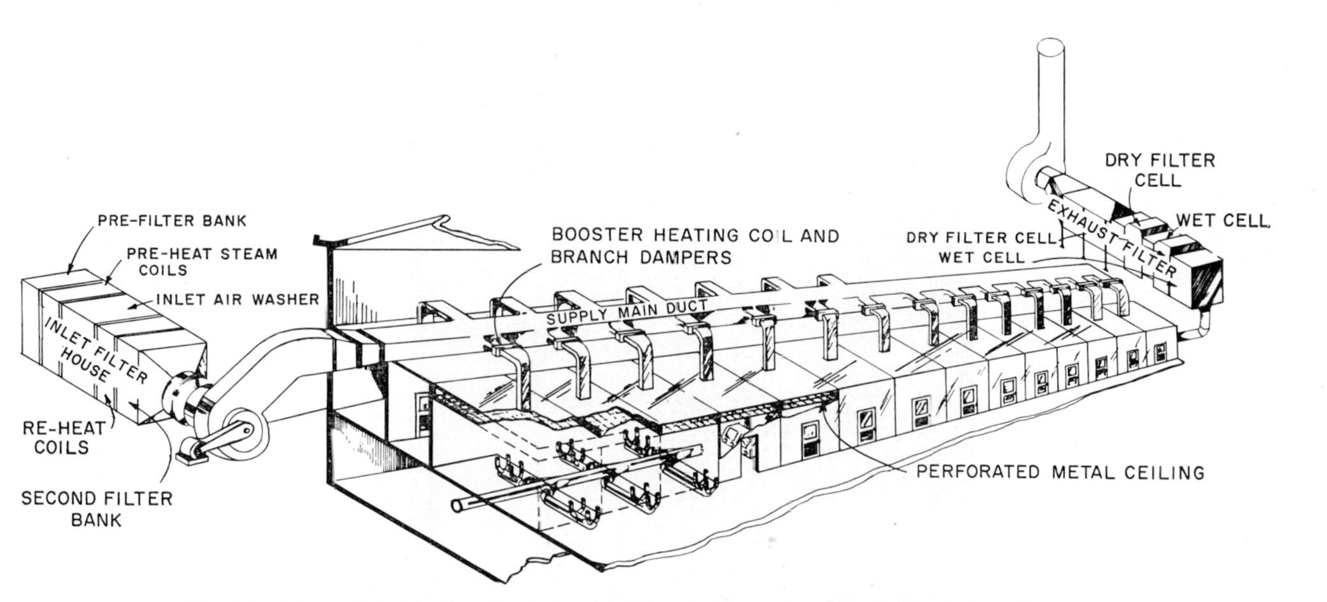

Fig. 13.—Pictorial section, showing the layout of ventilating equipment in half of a plutonium wing

The functioning of all ventilation equipment throughout the entire building is monitored at a central control room in the Administration Wing. The control panel, containing lights to signal the operational status of each piece of equipment, is shown in Figure 14. The cost (in 1950) of the entire CMR Building was between $11 and $12 million, and each plutonium wing cost about $2 million. Approximately half of this amount represents the cost of the ventilating system in each plutonium wing. Technical details of the ventilating and air-conditioning equipment in the CMR Building have been described by Glassmire and Wahlen (1).

Provision for the movement of personnel within the CMR Building and for the wearing and removal of protective clothing has been made to accord with the

FIG. 14.—Control panel for monitoring the ventilating equipment in all wings of the CMR Building.

gradients of contamination potential discussed above. Thus the building is normally entered through the Administration Wing, with passage to the plutonium wings being through the spinal corridor at the laboratory level; moreover, protective clothing is never worn in Wing 1, the Administration Wing, the spinal corridor at either the laboratory or attic level, or any part of the attic except the filter rooms at that level.

The plutonium wings are normally entered through the clothing-change rooms adjacent to the spinal corridor, and here protective clothing is put on before the laboratories, offices, or corridors of the wing are entered. Protective clothing is also worn in the filter rooms, throughout the basement in the plutonium wings, and in the spinal corridor at the basement level. Thus it is possible to go from one plutonium wing to another at the basement level without changing clothes or passing through the change rooms. Access to stockrooms and to shops for the repair of potentially contaminated equipment is also at the basement level.

On returning to the change room in order to leave a plutonium wing, the used protective clothing is removed and dropped through chutes into hampers in the basement below. It is collected at the basement level for transport to the laundry. On its return from the laundry, the clean protective clothing is returned to the change rooms through "cold" areas at the laboratory level.

The glove boxes and other inclosures used in working with plutonium at the Los Alamos Scientific Laboratory and at other establishments in North America and Europe have been described by many authors (2–7). In particular, Jette (2), Cof-

FIG. 15.—A typical glove-box train in Wing 2 of the CMR Building

finberry and Waldron (3), Schonfeld *et al.* (7), Maraman (7), and Metz (7) have described the inclosures employed at Los Alamos. Figure 15 illustrates a typical glove-box train employed in one of the hot laboratories of Wing 2, and in chapter xxix other inclosures used by CMF-5 are shown.

Group CMF-5 moved into the CMR Building in the spring of 1953. What has been its experience in working with plutonium throughout the years since then? In Wing 2 have the gradients of contamination potential discussed above become gradients of actual contamination? The answer to the latter question is emphatically negative. As exemplified by the experience related above regarding the internal cleanliness of the exhaust ventilation ducts, no permanent accumulation of plutonium contamination in significant (overtolerance) amount appears to exist at any point in Wing 2, and the condition of the other wings seems to be the same.

Although protective clothing continues to be worn, its wearing is now regarded as primarily a precaution against the possibility of the wearer's own clothing becoming contaminated in case of the rare occurrence of an accident involving plutonium. It is very seldom indeed that used protective clothing is returned to the laundry showing anything other than the complete absence of radioactivity that it had when it was first issued.

As a result of thorough and systematic monitoring, it is to be expected that in the laboratories, offices, and corridors overtolerance radioactivity would not be permitted to remain or accumulate on surfaces outside inclosures. It is with respect to housekeeping inside inclosures that practices at Los Alamos may, however, differ somewhat from procedures at other installations where plutonium is handled. Although it is not intended at Los Alamos that all glove boxes should be gas-tight, as is the aim at a number of other laboratories, in the design of the Los Alamos inclosures emphasis is placed on small size, simplicity, and easy accessibility to all points within them, in order that any accumulation of plutonous material may be most effectively kept either at zero or at a level as small as possible. The realization of optimum efficiency in handling plutonium with the least spreading or accumulation of alpha activity may at times take the form of working with the metal or its alloys in open hoods, or even completely uninclosed in the open laboratory, but such practices are based on knowledge gained from experience regarding the behavior of plutonium under well-understood conditions.

Los Alamos has never had a disastrous accident or incident involving the spreading of plutonium contamination, and it seems unlikely that such an incident will ever occur as long as present housekeeping practices are maintained. In cases of more serious incidents at other laboratories, there has always been involved a larger accumulation of plutonium-containing material than would be possible at Los Alamos under operating procedures in force since the war years. It is our feeling that gas-tight inclosures do not necessarily insure adequate protection against the plutonium health hazard if they are so designed as to discourage internal cleanliness. We believe that minimum spreading of plutonium to places where it is not wanted is most effectively realized through keeping accumulation to a minimum everywhere.

Admittedly, during the time that CMF-5 has occupied Wing 2, there have been occasional spills involving small quantities of plutonium, infrequent overtolerance air counts have occurred, and, on three occasions, small specimens of plutonium alloy spontaneously ignited and burned. Such events have required immediate and careful attention to thorough cleanup operations, but body intake of contamination by exposed personnel has not been detected, and apparently significant permanent deposition of plutonium occurred only in the filtering media of the exhaust air filters (in which the filtering media were later replaced).

Regarding overtolerance air counts, it should be noted that the definition of the tolerance limit for plutonium in air (as estimated in 1951: 10 disintegrations per minute per cubic meter of air at a pressure of 760 mm Hg) is that amount which if inhaled by a person continuously for 40 hours a week will, after 20 years, result in his body's containing 0.75 μg of plutonium, the maximum body burden considered to be safe. Typical experience with overtolerance air counts in Wing 2 has been reported by Cramer and Schonfeld (7) for three laboratory rooms in which metallography and the preparation of alloys are done. The total count obtained in each test is based on continuous air sampling for a period of 8 hours. For a total of 2,115 such daily tests made in the three rooms during 1953 through 1957, the average count was less than one-eighth of tolerance, and in only three tests was the count above tolerance. The highest of these three overtolerance counts was slightly less than twice the tolerance limit.

In addition, during a four-year period (1954–57) 25 "special" air tests were made under conditions considered likely to lead to overtolerance counts, but for 24 of these tests the average count was again only about one-eighth of tolerance, with only one count being three times tolerance. Comparison of experience of this type with the much poorer results of tests conducted in the old D Building has strongly reinforced the convictions of CMF-5 personnel with regard to the overriding importance of day-to-day housekeeping in the control of plutonium contamination.

From the foregoing discussion it is apparent that the CMR Building is a unique structure, designed to solve in a unique manner the problems of working with a unique substance. One of the building's most singular features is its ratio of service space to working space. The total volume of each plutonium wing is somewhat more than a million cubic feet, but, within this volume, only 90,000 cubic feet, or 8.5 per cent, is working space into which plutonium can be taken. Even with the offices and auxiliary laboratories counted as parts of the total useful volume, the ratio of service space (attic, basement, filter rooms, change rooms, and corridors) to working space is about 7 to 1.

Although suggestions relating to details of how the CMR Building should be built came from many sources, the excellence of the over-all result must be credited to Jette, who, in the supervision of its design and construction, brought to bear all the accumulated Los Alamos experience in working with plutonium. Thus this remarkable structure that is sure to endure for a very long time can in a sense be regarded as a monument to the administrative genius of Eric Randolph Jette, who, by successfully combating the problem of its health hazard, proved to the world that any reasonable operation can be performed with plutonium efficiently, safely, and—within the meaning of the word in nuclear technology—economically.

REFERENCES

1. Glassmire, S. H., and Wahlen, J. P. *Heating, Piping and Air Conditioning*, pp. 117–21, 1956.
2. Jette, E. R. "Plutonium Metal," in *Nuclear Metallurgy*, pp. 29–37. ("Institute of Metals Division Special Report Series," No. 1.) New York: American Institute of Mining and Metallurgical Engineers, 1955.
3. Coffinberry, A. S., and Waldron, M. B. "The Physical Metallurgy of Plutonium," chap. 4 of *Progress in Nuclear Energy*, Ser. V, Vol. **1**, pp. 354–410. London: Pergamon Press, Ltd., 1956.
4. Lord, W. B. H., and Waldron, M. B. *Journal of the Institute of Metals*, **86**:385–92, 1957–58.
5. Walton, G. N. (ed.). *Glove Boxes and Shielded Cells*. New York: Academic Press, Inc., 1958.

6. SCHONFELD, F. W., TATE, R. E., and MARAMAN, W. J. "Procedure at the Los Alamos Scientific Laboratory," in *Glove Boxes and Shielded Cells*, pp. 150–72.
7. UNITED NATIONS. *Proceedings of the Second United Nations International Conference on the Peaceful Uses of Atomic Energy* (Geneva, 1958), Vol. **17**: O. J. WICK and I. D. THOMAS, "Design and Operation of the Hanford Plutonium Metallurgy Installations," pp. 531–36; E. M. CRAMER and F. W. SCHONFELD, "Techniques for the Metallography of Plutonium," pp. 668–75; W. J. MARAMAN, "Recent Developments in Facilities for Handling Intense Alpha Sources," pp. 676–80; C. F. METZ, "Analytical Chemical Laboratories for the Handling of Plutonium," pp. 681–90.

CHAPTER VI

PLUTONIUM METALLURGY AT THE ARGONNE NATIONAL LABORATORY

Frank G. Foote

INTRODUCTION

Not only is plutonium ($_{94}Pu^{239}$) an extraordinarily interesting metal in its own right, but it is also one of the three major fissionable isotopes, the other two being U^{233} and U^{235}. Further, although it occurs in nature only in minute amounts, it is potentially a relatively abundant material, since it can be made by a series of nuclear reactions from U^{238}, the major isotope of natural uranium. The situation is summarized in Table 1. Thus U^{235}, the only naturally occurring fissionable isotope,

TABLE 1

FERTILE AND FISSIONABLE ISOTOPES

	Natural Element		Fissionable Isotope
Uranium.....	99.3 per cent U^{238}	→	Pu^{239}
	0.7 per cent U^{235}	=	U^{235}
Thorium.....	100 per cent Th^{232}	→	U^{233}

is a quite minor component of natural uranium and must be separated from the major component by physical, rather than by chemical, methods. Pu^{239} and U^{233}, made by nuclear reactions from the relatively abundant isotopes U^{238} and Th^{232}, respectively, can be chemically separated from the natural isotopes in which they are produced. These artificial, chemically separable, fissionable isotopes can then be used in the same manner as the physically separated natural isotope, U^{235}.

FAST-BREEDER REACTOR

By far the most attractive use of plutonium is as fuel in fast (unmoderated) reactors. Not only is plutonium fissionable by fast neutrons with the liberation of considerable energy, but an unusually large number (about three) of new neutrons are produced per fission event. One of these neutrons is required to continue the chain reaction; some leak out of the reactor or are lost to competing reactions; but, under proper conditions, from 1.5 to 1.7 can be made available to react with U^{238} to produce more plutonium (1). This process not only replaces the original plutonium atom but yields a nice bonus besides. Such a reactor-power system is thus

Frank G. Foote is at the Argonne National Laboratory, Argonne, Illinois.

both self-sustaining and self-expanding. The Argonne National Laboratory has been working in this field of fast reactors since 1945. EBR-I (Experimental Breeder Reactor No. 1) (2), fueled with U^{235} rather than Pu^{239}, was built at our Idaho site and went into operation on August 24, 1951, and first produced electric power on December 20, 1951. The Mk-I fuel loading (3) was wrought and beta-heat-treated unalloyed enriched uranium; the Mk-II core (4) was centrifugally cast U-2 w/o Zr alloy and was damaged by partial melting (5) during an experimental excursion stability test on November 29, 1955. The Mk-III loading, recently completed, consists of U-2 w/o Zr alloy rod clad with Zircaloy-II by coextrusion at Nuclear Metals, Inc. Three longitudinal ribs of zirconium are spot-welded to each fuel element, and the elements are bundled to form a rigid assembly. This core is being used to study the stability and safety characteristics of the reactor.

The EBR-II (Experimental Breeder Reactor No. 2) (6) is a prototype power reactor rated at 20 eMW. This reactor is in the final design stage, and construction should start next year at the Idaho site. This reactor system embodies a number of unique design features, including metallurgical reprocessing of the fuel and recovery of bred plutonium from the blanket. The Mk-I core will consist of injection-cast enriched U-5 w/o Fs alloy. The Mk-II core will be injection-cast U-20 per cent Pu–10 per cent Fs alloy. The Fs (fissium) is the mixture of certain fission products not removed or only partially removed by the metallurgical reprocessing operation. Many of these residual fission products are desirable alloying elements anyway, and if they had been removed by the reprocessing, some at least would have to be re-added to stabilize the fuel against distortion during subsequent irradiation. Additional information on these fuels is given by L. R. Kelman and R. J. Dunworth in chapter xxx. It should be pointed out that, while the EBR-II, Mk-II core is relatively rich in plutonium, it is still a long way from being pure plutonium, and this is characteristic of all power-reactor fuels. Only rarely is pure fissionable material used as fuel and then only for special purposes or tests. In power reactors operating at practical heat-generation rates, the fissionable material is diluted with fertile materials, such as U^{238} or Th^{232}, or with inert materials, such as aluminum, zirconium, or stainless steel.

THERMAL REACTORS

Since plutonium is a fissionable material, there is no reason why it cannot be burned in a thermal reactor in much the same manner as U^{235}. Since the fissionable material concentration in thermal-reactor fuels is ordinarily even less than that required for fast-reactor fuels, we are here dealing with materials whose properties even more closely resemble those of the diluent. It seems to be generally true that if a satisfactory fuel containing a few per cent of U^{235} can be made, then the same fuel containing the equivalent amount of Pu^{239} can be made by the same technology and will behave in much the same fashion during use. Some years ago, A. B. Shuck, with the co-operation of the Los Alamos Laboratory, prepared a group of irradiation specimens containing plutonium, the so-called Napkin Rings (7), some of which have been under irradiation in MTR for over three years and have been burned to essentially 100 per cent depletion of the plutonium. The core alloy consisted of Al-5 w/o Pu or Al-10 w/o Pu clad with aluminum by hot rolling. The technology used was essentially identical with that used for making MTR fuel plates using U^{235} as fuel. Similarly, B. Blumenthal and R. J. Dunworth have recently prepared Zr-5.6 w/o Pu alloys by arc melting and have cold rolled such alloys over 95 per cent without difficulty. Plutonium is soluble in alpha uranium to about 10 per cent, and uranium alloys containing a few per cent plutonium are but slightly different from uranium itself. These few examples illustrate the statement made earlier that plutonium can be substituted for U^{235} in a variety of materials useful as reactor fuels.

HAZARD CONTROL

Since plutonium has many advantages and can be used in many types of fuel, one may well ask why plutonium is not more widely used as a reactor fuel. The answer lies in the rather extraordinary toxicity of plutonium and the rather un-

TABLE 2

ENERGY, SPECIFIC ACTIVITY, AND MAXIMUM PERMISSIBLE LIMITS OF RADIOACTIVE ISOTOPES IN AIR

Isotope	Energy of α-Particle (Mev)	Specific Activity (gm/curie)	Maximum Permissible Limits in Air		
			NCRP* (μcuries/ml)	Calculated	
				(gm/m³)	(DPM/m³)
U-natural. . .	. . . .	. . . .	1.7×10^{-11}	5.1×10^{-5}	38
U^{238}.	4.18	3.0×10^{6}	1.7×10^{-11}	5.1×10^{-5}	38
U^{235}.	4.56	4.7×10^{5}	1.6×10^{-11}	7.4×10^{-6}	36
U^{233}.	4.82	1.1×10^{2}	1.6×10^{-11}	1.7×10^{-9}	36
Pu^{239}.	5.14	16.1	2.0×10^{-12}	3.2×10^{-11}	4.4
Po^{210}.	5.3	2.2×10^{-4}	7.0×10^{-11}	1.6×10^{-14}	156
Ra^{226}.	4.8	1.0	8.0×10^{-12}	8.0×10^{-12}	17.8

* National Committee on Radiation Protection.

usual precautions which must be taken to handle safely even small amounts of plutonium and plutonium-containing materials. Plutonium is a strong alpha-emitter with a half-life of 24,300 years; it is a bone-seeker, and the emitted alpha particles are very damaging to tissue in the immediate vicinity. This radiation is unusually damaging because the range in most materials is short and the energy of the alpha particle is transferred to the material through which it moves in a very short distance. Rubber gloves are sufficient to absorb these alpha particles, and plutonium can be safely handled with the gloved hand. The most likely path into the body is through the lungs, and the maximum permissible limit of plutonium in the air breathed by those working with plutonium is extremely low. Table 2 shows the currently accepted air concentrations for a number of hazardous materials.

At Argonne we have adopted the philosophy of complete confinement; i.e., an impervious barrier is established between the man and his working material, and at no time is the air the man breathes in contact with plutonium-containing materials. Gloves are sealed to the barrier, and most manipulations are carried out with the gloved hand (8). Considerable emphasis is placed on automation to relieve the tedium and, in some operations, the danger of gloved manipulation. Fires and explosions within the glove boxes could be extremely hazardous by damaging the barrier and scattering contamination into the personnel area. Consequently, many of the boxes are set up with inert atmospheres. It seems quite clear, then, that while plutonium-containing materials of interest as reactor fuels are, metallurgically speaking, not much different from similar materials containing uranium, the facilities in which plutonium can be safely handled are much more expensive to build and maintain and more cumbersome to operate.

FACILITIES

A considerable part of any plutonium metallurgy program must be the design and construction of facilities suitable for the safe handling of plutonium-contain-

FIG. 1.—View of the plutonium metallurgy research laboratory

ing materials (9). A brief description of existing and planned plutonium-handling facilities in the Metallurgy Division at ANL follows:

F-WING, BUILDING 200

Building 200 is a small and very crowded physical metallurgy laboratory installed some years ago in three modules in the F-Wing of the Chemistry Building. It consists of a series of interconnected stainless-steel glove boxes, most of which operate with a highly purified recirculated helium atmosphere. Containment of plutonium is complete, and there has been no detectable contamination of the personnel area. The laboratory contains equipment for (*a*) alloy preparation—melting and casting by resistance and induction heating and by arc melting; (*b*) sample preparation—by rolling, pressing, extrusion, and machining; (*c*) heat treatment; (*d*) metallography—cutting, mounting, grinding, polishing, etching, and photography; (*e*) thermal analysis and dilatometry; and (*f*) thermal conductivity. A general view of the laboratory is shown in Figure 1.

BUILDING 350, FUEL-FABRICATION FACILITY

Building 350 is a versatile plutonium-fabrication facility (10) still under construction. It is designed to enable us to develop the technology of making plutonium-containing fuel elements; to manufacture fuel elements, particularly fast-reactor fuels, at a reasonable rate; and to make the great variety of test objects

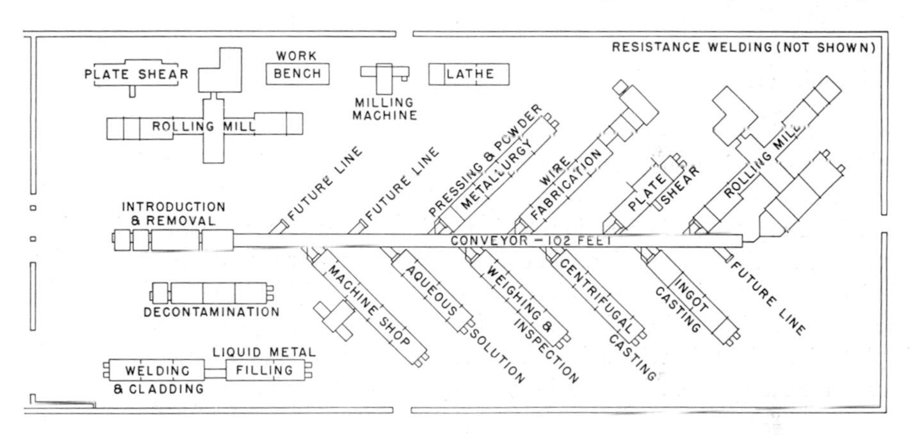

FIG. 2.—Fuel fabrication facility showing glove-box arrangement

required by physicists, chemists, chemical and reactor engineers; and for irradiation-damage studies. The facility consists of a series of glove boxes interconnected by means of a central conveyor system. Figure 2 shows the general plan of the working area. The equipment is housed in one large room, 70 × 160 feet, and again is a tight glove-box system with helium atmosphere in those boxes where fire is a hazard. The system consists of the following:

PF-3: Central conveyor system which interconnects the individual glove boxes
PF-4: An entry and exit port at one end of the conveyor
PF-5: A machine shop, containing a 50-ton hydraulic press, a toolmaker's lathe, a vertical boring mill, a horizontal milling machine, and a sheet-metal shear
PF-6: Aqueous solution and electrolysis
PF-7: Weighing and inspection
PF-8: Centrifugal casting
PF-9: Ingot and injection casting
PF-10: Blank
PF-11: Rolling mill (16-inch, 2-high/4-high)
PF-12: Plate shear
PF-13: A wire and small specimen fabrication facility, containing a hydraulic draw bench, swaging machines, and a 2-high/4-high rolling mill

PF-14: A powder metallurgy and ceramics facility, containing a 250-ton press, a ball mill, and a sintering furnace
PF-15: Blank
PF-16: Blank

The conveyor system and the glove boxes attached to it become alpha-contaminated during use, and during normal operation no attempt is made to decontaminate completely any part of the system. However, there are some operations, such as jacketing and cladding, in which the outside of the work piece should be free of contamination. Machines for these operations are set up in alpha-risk glove boxes, are normally kept free of contamination, and are isolated from the alpha-contaminated conveyor-interconnected glove-box system.

BUILDING 212, FUELS TECHNOLOGY CENTER

Building 212 is currently under design. Its primary purpose is to provide improved and considerably expanded facilities for study of the physical metallurgy of plutonium and plutonium-containing materials.

EBR-II REPROCESSING FACILITY

The reprocessing facility is currently under design and is to be built at Idaho as part of the EBR-II Reactor Complex. This is to be an alpha-gamma-active hot-laboratory type of operation. This facility, for the reprocessing of irradiated core alloy for the fast breeder, will contain equipment for the stripping of the fuel-pin jacket, the partial decontamination of the irradiated fuel by oxidative drossing, the reconstitution and casting of new fuel pins, and the rejacketing. The facility is plutonium-tight, heavily shielded to attenuate gamma radiation, and filled with inert-gas atmosphere. All operations are by remote control.

REFERENCES

1. ZINN, W. H. "A Review of Fast Power Reactors," *Proceedings of the International Conference on the Peaceful Uses of Atomic Energy*, **3**:198–204. New York: United Nations, 1956.
2. LICHTENBERGER, H. V., THALGOTT, F. W., KATO, W. Y., and NOVICK, M. "Operating Experience and Experimental Results Obtained from a NaK-cooled Fast Reactor," *Proceedings of the International Conference on the Peaceful Uses of Atomic Energy*, **3**:345–60. New York: United Nations, 1956.
3. SHUCK, A. B. *The Manufacture of Enriched Uranium Slugs for the EBR.* (USAEC Rept. ANL-4847, 1953.)
4. ———. *Development of Equipment and Methods of Centrifugally Casting Reactor Fuel Slugs.* (USAEC Rept. ANL-5123, 1953.)
5. KITTEL, J. H., NOVICK, M., and BUCHANAN, R. F. *The EBR-I Meltdown: Physical and Metallurgical Changes in the Core.* (USAEC Rept. ANL-5731, 1957.)
6. BARNES, A. H., KOCH, L. J., MONSON, H. O., and SMITH, F. A. "The Engineering Design of EBR-II: A Prototype Fast Neutron Reactor Power Plant," *Proceedings of the International Conference on the Peaceful Uses of Atomic Energy*, **3**:330–44. New York: United Nations, 1956.
7. SHUCK, A. B. *The Manufacture of Aluminum-clad Aluminum-Plutonium Alloy Irradiation Specimens.* (USAEC Rept. ANL-5043, 1952.)

8. Davis, D. A., Ayer, J. E., and Mayfield, R. M. *Gloves for Protective Enclosures.* (USAEC Rept. ANL-5743, 1957.)
9. Kelman, L. R., Wilkinson, W. D., Shuck, A. B., and Goertz, R. C. *The Safe Handling of Radioactive-pyrophoric Materials.* (USAEC Rept. ANL-5509, 1955.)
10. Shuck, A. B., and Mayfield, R. M. *Process Equipment and Protective Enclosures for the Fuel Fabrication Facility* (*D-350*). (USAEC Rept. ANL-5499, 1956.)

CHAPTER VII

STUDIES ON PLUTONIUM AT CHALK RIVER

O. J. C. Runnalls

INTRODUCTION

Only a modest effort has been expended at Chalk River on plutonium metallurgical research and development during the last ten years. The experience and technology gained during that period have, however, influenced other Canadian programs appreciably, such as those dealing with chemical reprocessing, reactor physics, and fuel-cycle studies. The past experience has, in addition, pointed the way to a probable future program in plutonium metallurgy.

The first studies on plutonium metallurgy at Chalk River were carried out by Carter, Langs, and Milsted (1) during the period 1947–51 as part of a joint British-Canadian program. The object of the work was to outline a process for the preparation of plutonium metal on a milligram scale, prior to establishing laboratories for larger-scale production of plutonium in the United Kingdom. Small metal beads weighing up to 200 mg were prepared, using a vapor-phase reduction technique similar to one described by Fried and Davidson (2) for the preparation of neptunium metal. A displacement method was developed to determine the density of metal beads weighing as little as 100 mg. Measured densities varied from 15.4 gm/cm^3 for a face-centered cubic phase (delta) to 18.5 gm/cm^3 for a "hard form" with a complex X-ray diffraction pattern that was probably a mixture of phases. Two melting-point determinations were made on metal beads weighing less than 100 mg. The observed melting points were 607° and 612° C. The poor agreement was ascribed to a difference in impurity content.

Early in 1951 a requirement arose for a plutonium-containing fuel element to enrich the NRX reactor and to supply, eventually, a source of plutonium rich in the higher isotopes for reactor-physics studies. Consequently, a program of plutonium-alloy development was begun at Chalk River.

PLUTONIUM-ALUMINUM ALLOYS

NRX Fuel Rods

Since aluminum has a relatively low thermal neutron-absorption cross-section and a well-established technology, it was chosen as a plutonium diluent for the NRX fuel rod. A proposed method for the fabrication of a plutonium-aluminum fuel rod was suggested by Carter *et al.* (3) in 1951. Plutonium metal was to be pre-

O. J. C. Runnalls is with Atomic Energy of Canada Limited, Chalk River, Ontario.

pared on the 100-gm scale by the bomb reduction of plutonium trifluoride with calcium chips. The resulting button was to be melted with aluminum and cast into an aluminum can. After casting in the alloy, an aluminum plug would be cast in on top, so that a completely inclosed alloy could be removed from the glove box for sheathing.

In late 1951, however, experiments on the 100-mg plutonium scale demonstrated the feasibility of reducing plutonium trifluoride with an excess of aluminum, thus forming the alloy directly. The glove-box line, which had been designed and partially built for the preparation of 100-gm plutonium-metal ingots, was modified to accommodate the aluminum reduction process. By early 1952 the plutonium-aluminum alloys for the first NRX enriched fuel rod had been prepared. A second, simpler, direct reduction process was then developed, namely, the aluminum reduction of plutonium dioxide in the presence of cryolite. The results of the first reductions were encouraging, i.e., plutonium yields were 99 per cent. Thus it was decided to construct a new extension, housing a laboratory of dimensions 25 × 50 feet, for the express purpose of fabricating plutonium fuel rods for the NRX reactor. The laboratory was ready for occupancy in mid-1953. The total cost of the extension with the necessary glove-box line, ventilation, and services was $55,000. The glove-box line, illustrated in Figure 1, "went active" in late 1953 and has been in contin-

FIG. 1.—Glove-box line for the fabrication of plutonium-aluminum alloys

uous use since that time. Alloys for thirty-five fuel rods for the NRX reactor have been prepared in the equipment thus far. In addition, special tubular and flat castings were made for physics studies. Altogether, about 10 kg of plutonium have been processed through the system by a staff of three, on the average.

Alloy Constitution

Since the main Chalk River plutonium-alloy requirement was for plutonium-aluminum alloys containing up to 20 w/o plutonium, the constitution of such alloys was of particular interest. Several alloys were examined by microradiography. A 10 w/o plutonium-aluminum alloy, for example, is shown in Figure 2. The

Fig. 2.—Microradiograph of a 10 w/o plutonium-aluminum alloy

white phase proved to be $PuAl_4$. Single crystals of $PuAl_4$ and $PuAl_3$ for X-ray diffraction studies were separated from 20 and 40 w/o plutonium alloys, respectively, by dissolving away the aluminum matrix in sodium hydroxide. Typical crystals are shown in Figure 3. Since the aluminum-containing solution proved to be virtually free of alpha contamination, it was concluded that the solid solubility of plutonium in aluminum was vanishingly small. Thermal analysis experiments indicated that a eutectic exists between aluminum and $PuAl_4$ at 13 ± 1 w/o Pu and $643° \pm 3°$ C and that $PuAl_4$ decomposes peritectically at $913° \pm 5°$ C. Thus it appeared that the aluminum-rich ends of the systems plutonium-aluminum and uranium-aluminum were somewhat similar. X-ray diffraction studies showed that $PuAl_4$ and $PuAl_2$ are isostructural with UAl_4 and UAl_2, respectively. Hexagonal $PuAl_3$ was found, however, to differ from cubic UAl_3. In addition, a phase with an approximate composition of PuAl was found that was not matched in the uranium-aluminum system.

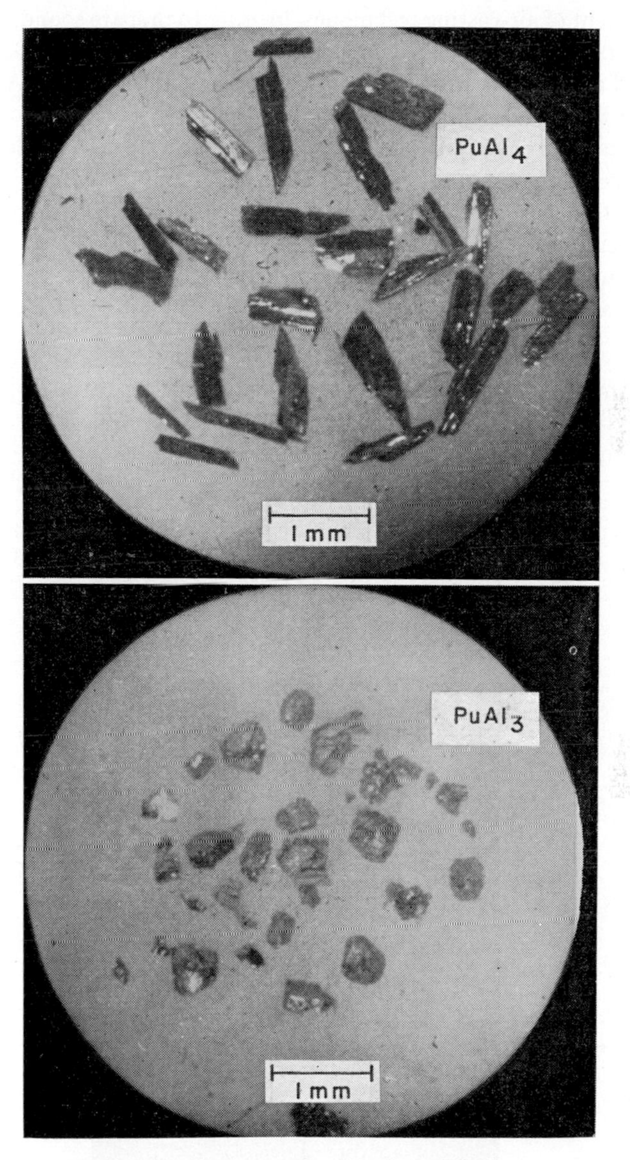

FIG. 3.—Single crystals of $PuAl_4$ and $PuAl_3$

FABRICATION STUDIES

The main effort on the fabrication of plutonium-aluminum alloys has been on the development of air-casting techniques, in order to prepare sound, homogeneous, solid cylinders for assembly into fuel rods. Initially, alloys were cast directly into water-cooled aluminum cans similar to the one shown in Figure 4. After an aluminum plug had been cast on top of the alloy and then machined out to the bottom of the threads, a number of cans were screwed together to form a full-length element. Since the outer diameter of the can was the same as that of the uranium metal used in NRX, 1.36 inches, a standard three-finned aluminum sheath could be drawn over the assembly by using existing equipment. Standard end-closure techniques and end fittings could also be used.

FIG. 4.—Aluminum casting can containing a trial aluminum pour

More recently, the method of assembly has been changed, in order to eliminate one of the two unbonded interfaces between the alloy and coolant that existed in the above design. The alloys now are cast to 1.4-inch diameter, machined to 1.36 inches, end-faced, and then stacked in the three-finned aluminum sheath. The assembly technique is described in more detail in chapter xxviii by Wauchope.

Some experience has been gained on the rolling properties of 14 w/o plutonium-aluminum alloy during the fabrication of 1,040 foils measuring 9.8 by 1.5 by 0.020 inches (4), which were required for reactor-physics experiments in the ZEEP reactor. The alloy was air-cast into rectangular blocks 3 by 1.6 by 0.250 inches in a twelve-cavity graphite mold. The castings were cold rolled to 0.210 inch on a two-high 1.5 by 2-inch mill and then hot rolled at 400°–450° C to 0.022 inch. The reduction to the finished size of 0.020 inch was done at room temperature. The foils were trimmed in a modified paper cutter and were inserted into a flattened aluminum tube which had been sealed at one end by hot pressure end-welding. The tube was rolled into contact with the foil prior to pressure-welding the other end. The total weight of alloy fabricated was 15 kg.

Irradiation Experiments

Some thirty-two plutonium-aluminum elements have been irradiated for varying periods in the NRX reactor during the last five years. It is only within the last two years, however, that a remote-handling cell has been available at Chalk River for detailed postirradiation examination. Thus little precise information has been obtained as yet on the dimensional stability of aluminum-sheathed alloys irradiated in cold water.

Plutonium-aluminum alloys are of potential value as fuel for pressurized water reactors. An irradiation program on alloys containing 5–20 w/o plutonium, canned in Zircaloy-2, has been devised by Jones and Robertson (5) for the Chalk River X-2 loop to test dimensional stability at alloy surface temperatures up to 500° C and surface heating rates up to 140 watts/cm^2. The test specimens were fabricated for a 4-month irradiation scheduled for early 1958. A later test on purposely defected elements will be carried out to determine the corrosion resistance of the fuel under power-reactor operating conditions.

STUDIES OF OTHER PLUTONIUM ALLOYS

Several other possible alloying processes have been investigated in some detail, notably (*a*) the beryllium reduction of plutonium trifluoride; (*b*) the silicon reduction of plutonium trifluoride; and (*c*) the reaction of plutonium trifluoride with gallium, indium, and thallium.

A few experiments have also been carried out on the reduction of plutonium trifluoride by magnesium, thorium, and uranium, as described in chapter xxvi. Small samples of plutonium/transition-metal alloys were prepared by heating a powdered mixture of the transition metal and plutonium trifluoride in lithium vapor at 900° C for X-ray diffraction studies.

An interest in the plutonium-silver system existed at Chalk River because of the pyrometallurgical processing studies being carried out by McKenzie on the extraction of plutonium from irradiated uranium by liquid silver. Plutonium-silver alloys were prepared by first reducing plutonium trifluoride with lithium vapor at 950° C in tantalum crucibles and then melting the resulting plutonium metal with silver at 1,050° C in vacuum. The silver-rich end of the system plutonium-silver appeared somewhat similar to lanthanum-silver but markedly different from uranium-silver. A eutectic reaction occurred between a compound $PuAg_3$ and silver at a composition of 21 ± 1 w/o plutonium and a temperature of 790° ± 10° C. The hexagonal $PuAg_3$ melted at 975° ± 50° C. Plutonium-silver alloys near the composition $PuAg_3$ proved to be chemically reactive and emitted sparks when scratched in air.

As a result of the work mentioned above on the preparation of plutonium-beryllium alloys, an interest arose in their use as portable neutron sources. The factors affecting the neutron yield from such alloys were therefore investigated in some detail. It was found that the neutron yield was markedly influenced by the particle size of the plutonium-containing intermetallic phase $PuBe_{13}$. For example, the neutron yield from an alloy with $PuBe_{13}$ particles approximately 2 μ in diameter was 20 per cent higher than that from a second alloy containing particles of 15-μ diameter. Polished sections of the two alloys are shown in Figure 5. Further work on the preparation of plutonium-beryllium neutron sources is currently under way at Chalk River. Michaud is investigating the reaction between plutonium dioxide and beryllium on the 1-gm plutonium scale. In addition, Wauchope has installed the equipment required for the preparation of neutron sources containing up to 100 gm of plutonium by the beryllium reduction of either plutonium dioxide or plutonium trifluoride.

PLANS FOR THE FUTURE

An outline of an irradiation program which would indicate the feasibility of using plutonium fuels in water-cooled thermal power reactors has been suggested recently by Robertson (8) as follows: (1) a continuation of irradiation and corrosion tests on plutonium-aluminum in Zircaloy-2; (2) a survey of irradiation and corrosion effects on alloys of plutonium with thorium, zirconium, silicon, and magnesium for plutonium concentrations up to about 0.5 gm/cm^3; (3) a study of the effects of additions of plutonium dioxide up to about 0.1 gm/cm^3 on the melting, thermal conductivity, corrosion, and irradiation behaviors of urania, thoria, alumina, magnesia, and beryllia; and (4) irradiation and corrosion tests on elements made by graphite impregnation techniques. We hope to do as much work on the above program as our limited resources will permit while maintaining an emphasis on the development of plutonium-aluminum fabrication techniques.

The plutonium re-irradiation program in NRX will be continued, in order to build up a large source of highly irradiated plutonium for reactor-physics studies. It is anticipated that 3 kg of fresh plutonium will be added to the reactor each year as plutonium-aluminum alloy. Plutonium-alloy elements of high surface area may be required for irradiation in the NRU reactor as well. Thus techniques will be developed for the fabrication of flat, thin, plutonium-aluminum plates sheathed in aluminum, aluminum-nickel alloys, or Zircaloy-2.

ALLOY A

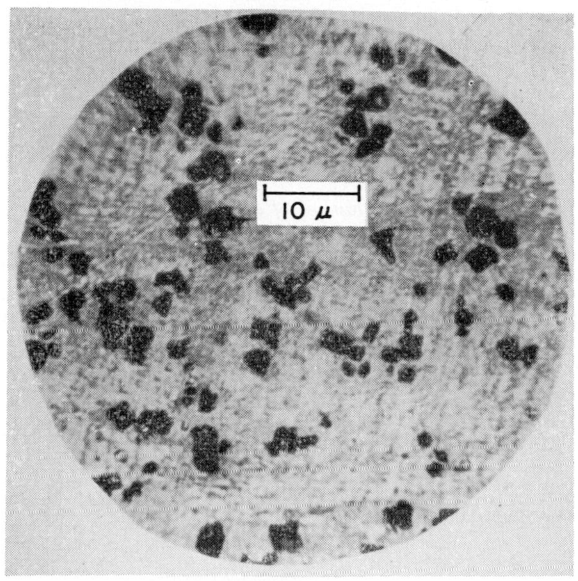

ALLOY B

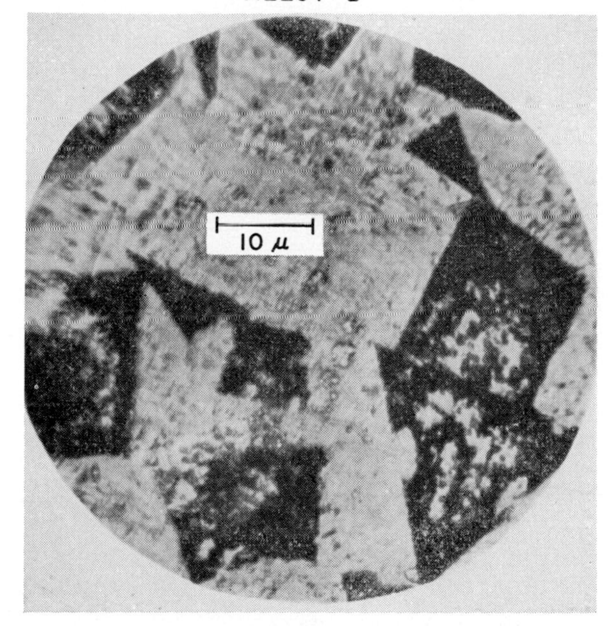

Fig. 5.—Polished sections of plutonium-beryllium alloys

The author gratefully acknowledges the collaboration of many colleagues at Chalk River in carrying out much of the work reported, in particular, Messrs. A. S. Bain, J. Baird, R. R. Boucher, T. I. Jones, G. G. Michaud, J. A. L. Robertson, and K. L. Wauchope.

REFERENCES

1. Carter, R. E., Langs, M. P., and Milsted, J. AECL Rept. CRC-490 (December, 1951).
2. Fried, S., and Davidson, N. R. *Journal of the American Chemical Society*, **70**:3539, 1948.
3. Carter, R. E., Bain, A. S., and Langs, M. P. AECL Rept. CRC-491 (January, 1952).
4. Huffman, H. R., and Wauchope, K. L. Unpublished data.
5. Jones, T. I., and Robertson, J. A. L. Chalk River Irradiation Proposal NRX-101 (June, 1957)
6. Robertson, J. A. L. AECL Rept. UK/C6/112 (September, 1957).

CHAPTER VIII

METALLURGICAL STUDIES ON PLUTONIUM IN GREAT BRITAIN

H. M. Finniston

The historical account of the isolation of plutonium metal in the United States and the determination of some of the physical and chemical properties of the metal and a number of its alloys are admirably described in preceding chapters. Collaboration between the United States and the United Kingdom in this field of metallurgy was suspended just after World War II, so that for some 13 years the United Kingdom had to make much the same mistakes and achieve—in the end—much the same technical successes, independently. It is a lesson that modern history continually repeats: highly industrialized nations will, given the same problem, independently of each other produce the same technical results—later, if not sooner. This chapter, therefore, to some extent repeats much of the discoveries which the early American workers found.

The first British effort in plutonium metallurgy was in association with the Canadians; Runnalls in chapter vii has described how Milsted of the then Department of Atomic Energy in the United Kingdom assisted in isolating the first beads of plutonium and determined the density of at least two of its phases. This chapter takes over from there and is concerned with subsequent developments in the United Kingdom. In describing these results, particular establishments of the complex of groups working on plutonium within the present United Kingdom Atomic Energy Authority organization are not mentioned by name. The establishments most concerned are the Atomic Energy Research Establishment at Harwell, the Industrial Group and especially its Windscale branch, and the Atomic Weapons Research Establishment at Aldermaston.

The program in the United Kingdom started in 1947. Its first task was to devise, construct, and operate experimental equipment for determination of the physical properties of plutonium metal, on the assumption that about $\frac{1}{2}$ gm of that material would be available in 1950/1951. The assumption that only relatively small quantities would be available meant that less conventional methods had to be used than if larger samples could be assumed. The $\frac{1}{2}$ gm was to be fabricated into wire. This could be expected to give some indication of the working properties of the material;

H. M. Finniston is at the Atomic Energy Research Establishment, Harwell, Berkshire, United Kingdom.

the experimental work was then to be based on lengths of wire as samples. There being no plutonium, two metals were used as stand-ins; these were uranium and cerium, the latter choice being based on the presumption that plutonium would perhaps be akin to the rare earths as well as its nearer and more readily available neighbor in the periodic table.

Concurrently, a small group of metallurgists and chemists was charged with the responsibility of preparing the metal on a larger scale, this development to constitute the basis of the technique for the preparation of plutonium metal from uranium irradiated in the Windscale piles. From experience with the extraction of uranium from its salts and following upon a series of process developments on the preparation of pure cerium, the calcium reduction of the tetrafluoride was chosen as the route for the manufacture of plutonium metal. The extraction of the metal on this larger scale was successful at the first attempt, and this metal and subsequent repeat extractions constituted the raw materials for the physical metallurgical experiments which followed.

The development of methods for the determination of the physical properties of plutonium on the small scale had gone much more slowly than anticipated, and, since the extraction of larger quantities of plutonium metal went unexpectedly rapidly, equipment based on more conventional methods applicable to larger samples was quickly made ready. The purity of the early plutonium metal samples was not as high as that which can now be achieved, and many of the anomalies and obscurities of the early experiments were undoubtedly related to impurities in these original samples. The first large piece of metal, about 250 gm, was obtained on December 15, 1951, and by February, 1952, the first exploratory physical metallurgy tests had been completed. Our knowledge then included a determination of the melting point, the existence of at least five phases between room temperature and the melting point, the large changes in density associated with the transformations from one phase to another, the considerable hysteresis between heating and cooling in these phase changes, the negative coefficient of expansion of the delta phase, and, most important, that our handling techniques and the box atmosphere imposed serious limitations on the speed and facility with which we could do experimental work on plutonium. In a succession of experiments, the dilatometric behavior of plutonium metal, its thermal conductivity, electrical resistivity, magnetic susceptibility, specific heat, elastic properties, and the crystallography of its simpler phases, were all established; these are discussed in later chapters. A comparison of United Kingdom results with those obtained by other workers in other countries will show that this early work was done with very considerable care and that the United Kingdom results will not depart significantly from the accepted values, which will be written into future textbooks.

The "shaping" characteristics of plutonium had also been the subject of some research; techniques of casting in vacuum (with its attendant problems of finding suitable crucible materials) and the know-how of extrusion and of machining were also determined.

At this stage (early 1952) there arose a real need in connection with the fast reactor, now almost completed at Dounreay, to make some nuclear measurements on plutonium assemblies. This necessitated the manufacture of a charge of plutonium-alloy fuel elements, sheathed in nickel, for the zero-energy fast-reactor experiment, "Zephyr." The metallurgists undertook the usual role of assisting the physicist to get on with his experiments by making this charge; their motives were not entirely altruistic, however, since this afforded an opportunity for modifying the existing handling techniques. Until then, the glove boxes used had been designed in the earliest days primarily for chemical experiments; these boxes backed onto a frog-suit corridor and therefore could be operated from only one face. Free-standing boxes which could be operated from all four faces (and the top, if necessary) were developed and thoroughly tested in the production line for manufacture of the Zephyr fuel elements. The lessons learned from this exercise have now become standard practice and have widely affected the philosophy of glove-box design in the United Kingdom; the free-standing box with improvements and standardization has been adopted increasingly for experiments involving alpha-active materials. The manufacture of the Zephyr charge was not all lost time scientifically; besides developing techniques of safe handling, it led to refinements in the techniques of casting, in sheathing and welding in confined spaces and special atmospheres, and to methods of decontaminating from plutonium. It proved, too, that plutonium-based fuel elements were not to be regarded as laboratory curiosities but that they could be made by semiskilled industrial labor under factory conditions. This work will stand in good stead in further developments involving the manufacture of fuel elements containing plutonium for power reactors of the future.

Meanwhile, continued work on the physical properties of plutonium had brought to light several new features. There were unexplained, but reproducible, kinks in the dilatometry and thermal-analysis curves between the delta and the epsilon phases. We had, in fact, found delta-prime, but the carefulness of our experimental work in dilatometry and thermal analysis was not matched by our courage; because of what we now know to be contamination from the silica capillaries, high-temperature X-ray studies revealed only delta and epsilon patterns, and the idea of a sixth phase was abandoned. More detailed work on the specific heat showed the atomic heat value to be higher than the Dulong and Petit value, which accorded with the behavior of uranium and thorium. These and the other unusual features of plutonium gave our theoreticians full scope for their imaginative powers and the experimentalists still more experiments to be done to prove them wrong.

Work was initiated early in 1953 on the investigation of a number of alloy systems mainly for their technological importance but with some attempt to gather information about the alloying characteristics of plutonium. Alloy additions included uranium, thorium, titanium, zirconium, hafnium, aluminum, bismuth, and iron. Since most of the work was in aid of the use of plutonium in civil reactors, the irradiation behavior of alloys from these binary and even ternary systems was investigated. The results of these studies are described in later chapters.

Although the four countries represented at this conference have presented reviews of their past and present activities no one has as yet commented on the future prospects, and I should like to end this chapter by crystal-gazing. The United Kingdom program on power development is based on the use of plutonium coming from existing and future graphite-moderated, natural-uranium, gas-cooled systems. Whether this plutonium will be incorporated in a fast or in a thermal system is open to question and experiment; but, so long as any doubt remains, the direction of research and development has to be devoted to answering the question for both types. The future program is based on three broad divisions. There is the improvement in our basic knowledge of the properties of plutonium; there is the search for reactor materials based on plutonium; and there are the problems associated with the improvement of the military potential. Dealing with the last first, it should be recognized that many features of direct interest to science and civil application originated in the military field. Plutonium itself is a good example of such a development, since it is questionable whether, without the military impetus to its production, we should be concerned with this metal today. One may therefore expect from those engaged with the military problems of plutonium a contribution to and understanding of its basic properties and behavior under special conditions.

On the fundamental aspects of plutonium metallurgy, there are various features for which we are looking for a more complete understanding. The fact that there are six phases between room temperature and the melting point requires explanation, as does the complexity of some of the crystal structures and the mode of transformations. The hysteresis on cooling involves a study of the kinetics of transformations and of their mechanisms. Many of the peculiarities that are so evident in this metal are due to the electronic characteristics of the plutonium atom, and a theoretical study of the electronic structure of plutonium is being undertaken, as well as experimental researches that will give better understanding of electronic behavior. This latter will include determination of low-temperature properties in general and low-temperature specific heat, magnetic susceptibility, and electrical resistivity in particular. The electronic behavior of plutonium also influences its alloying characteristics, and these will be examined first from the "classical" starting point of determining equilibrium diagrams and by measuring the properties of solid solutions. Subsequently a study of those points of similarity and difference between systems involving plutonium itself, as well as between corresponding systems of uranium, thorium, etc., will be entered into.

The dependence of many of the properties of plutonium on the degree of its purity demands the purest starting material. This is being sought by more refined techniques of reduction, whether thermal or electrolytic, and care in the preparation and handling of the raw materials for such reductions; application of zone melting techniques is also being developed.

For reactor fuels, increasing attention will be paid to finding those alloys which will resist irradiation and which will be compatible with the various coolants of future reactors using plutonium-bearing materials. This involves knowledge not only of metallic systems but of ceramics and cermets. Work is already proceeding on the plutonium-oxygen system, on the physical and chemical properties of PuO_2 and PuC, and on the properties of binary systems based on PuO_2 and Pu_2O_3 in UO_2, ThO_2, graphite, and ceramic and metal matrices. New techniques for the manufacture of such fuels on a production scale will have to be developed, and these are already being sought, not only through conventional techniques but by methods such as arc melting and powder metallurgy.

With all this activity, perhaps one should pause to ask the question: "Is it all really necessary?" I believe that there are many reasons for thinking that, far from being at an end before it has started, insufficient attention has been paid to plutonium in power programs. For many countries which have natural uranium but have not or cannot afford diffusion plants, plutonium is an obvious alternative to enrichment. However wasteful mankind may have been to date with its fossil fuels, it cannot seriously consider that burning merely 0.7 per cent of the fuel in natural uranium is sufficient or sensible; the 99.3 per cent of potential fuel—and this, in effect, means plutonium—still remains for us to use if we are sufficiently clever. I have no hesitation in predicting that we and many more will be attending an even larger conference on plutonium metallurgy two or three years hence.

CHAPTER IX

PLUTONIUM METALLURGY IN FRANCE

E. Grison

Work with metallic plutonium was initiated in France in January, 1956, when we had several grams of plutonium, extracted from the slugs of the EL_2 pile at Saclay. Since this reactor, having a thermal power of 2,000 kw, has so far been our only source of plutonium, we have worked only with experimental quantities sufficient for basic tests but not for studies of possible applications. It is this phase of preliminary research that will be described in this chapter. With the start of the plutonium-extraction plant at Marcoule, where the G1 reactor has been fully operating for more than a year, we shall shortly arrive at a more advanced stage, which will permit us to fabricate and test prototype fuel elements.

SPECIAL EQUIPMENT

The first problem posed is obviously that of the protective arrangements. A detailed description of what we have done will not be presented here because this subject was discussed at length at the Harwell symposium on glove boxes in February, 1957 (1, 2). It should be mentioned, however, that our equipment was designed before we knew the properties of plutonium, because up to then very little had been published on the subject. A visit to the Harwell laboratories provided us with our first information on the techniques of manipulation, but some errors of evaluation, which later practice allowed us to correct, could not be avoided initially. For instance, we overestimated the rate at which plutonium oxidizes and were agreeably surprised to find that the purity of the inert gas in glove boxes is of minor importance and that, treated by a passivating polish, plutonium can be kept a long time without any extraordinary precautions.

Nevertheless, since work in inert atmosphere is often necessary—usually for reasons of personal safety—we have adopted the free-standing glove-box system described by Pascard (2). In this system the boxes are made as leak-proof as possible (leakage is about 10^{-5} per hour under a pressure of 2 cm of water) and have a regulating device that maintains within them a negative pressure relative to the laboratory. The leak rate is small enough that the system can function several weeks without undue rise in pressure within the reservoir where the leakage accumulates. When the air content of the argon gas within the boxes reaches 1 per

E. Grison is with the Commissariat à l'Énergie Atomique, Centre d'Études Nucléaires de Fontenay-aux-Roses (Seine), France.

cent, this gas is circulated through a purification furnace. We use a rather large number of these inert-atmosphere systems, each of which feeds from one to four glove boxes. We have been satisfied with this equipment from every point of view—price, sturdiness, and ease of maintenance. There is nothing unusual about the methods of manipulation in the glove boxes or the construction of equipment within them.

METHOD OF METAL PREPARATION

The method of metal preparation used was published in 1956 by Anselin, Faugeras, and Grison (3) and consists of the reduction of PuF_3 by calcium in a CaF_2

FIG. 1.—Apparatus for reducing plutonium trifluoride to metal

crucible. The unique feature of this process—as compared with the methods used in other countries and about which we have as yet obtained very little information—is that we have no need to add a booster to furnish auxiliary heat to the reaction. The apparatus used for the reductions is shown in Figure 1.

Actually, high-frequency electric current is used to heat the metal contained in the crucible rather than to heat a metallic envelope outside the crucible. (The calcium is heated first, then the plutonium.) This explains why we obtain good separation of the metal and very good yield, even though we use trifluoride, which has a heat of reaction distinctly smaller than that of the tetrafluoride. So far, we have used this method only for small reductions—several grams of plutonium—but we have extrapolated the results to several hundreds of grams. Tests performed so far only with uranium fluoride justify the best hopes.

The same equipment as that shown in Figure 1 has also been used to prepare plutonium alloys either by reduction of PuF_3 with an excess of the alloying metal or by reduction of a mixture of fluorides. Use of the first method, extensively em-

TABLE 1

ANALYSIS OF PLUTONIUM METAL

Impurity	Parts per Million		Impurity	Parts per Million	
	In PuF_3	In Plutonium Metal		In PuF_3	In Plutonium Metal
Fe	0≪ 300 < 250*	~ 300 ≤ 300*	Ca	≤ 100	< 100
			Cr	< 300	< 100
			Mg	< 10	≤ 10
Ni	0≪1,000 < 250*	<1,000 ~ 200*	Mn	< 10	< 10
			Sn	< 10	< 10
U	~ 250*	~ 250*	Th	Undetected	...
Al	≤ 100	< 100	Zn	Undetected	...
Ba	< 100	< 100			

* By quantitative chemical analysis (wet methods). Other results are by spectrographic analysis.

FIG. 2.—Equipment for spectroscopic analysis. Electrodes are prepared in the glove box on the right; the sparking chamber is in the glove box at the left. The spectrograph is located behind the latter glove box.

ployed by Runnalls (4) (and described by him in chap. xxvi), has enabled us to prepare aluminum alloys containing from 5 to 50 per cent plutonium by weight. The second method has been used in the preparation of U-Pu alloys (5–20 per cent Pu) and Pu-Th alloys (5–15 per cent Th).

The purity of the plutonium metal obtained obviously depends on the purity of the fluoride. In Table 1 are listed typical impurities as determined spectrographically by the so-called "copper-spark" method (Fig. 2). The sensitivity of this method is not very high when used with the very small samples analyzed (approximately 0.1 mg of Pu). The metal is recast under vacuum in a crucible of tantalum or of fluorite, CaF_2.

EARLY RESULTS

We have not concentrated on the study of pure plutonium, the properties of which have been extensively described by others. Nevertheless, we have done some

FIG. 3.—Arc furnace for melting plutonium

work with the pure metal both to prove our methods and to verify the quality of our metal. Results obtained by dilatometry, X-ray diffraction, and micrography are reported in chapters xi, xiii, and xix.

We intend rather to put our main effort on the study of alloys, which is essential for the fabrication of fuel elements. We shall also soon start a study of ceramics-containing plutonium (mixed and sintered UO_2-PuO_2 or ThO_2-PuO_2).

Alloys are prepared by the conventional methods, either in a crucible or in a tungsten-arc furnace (Fig. 3) in argon; samples are cut either by hand or with a small laboratory lathe (Fig. 4).

Heat treating is done either in vacuum furnaces (Fig. 5) or in sealed containers.

Fig. 4.—Watchmaker's lathe for turning test specimens

Fig. 5.—Heat-treating furnace

FIG. 6.—Glove box for sealing containers under argon or vacuum

FIG. 7.—Heat-treating and quenching equipment

For the latter we have set up an installation (Fig. 6) wherein we can seal silica under vacuum or weld metallic containers (usually tantalum) under argon. Quenching is done in a glove box (Fig. 7) in the conventional manner.

We have mostly worked on aluminum-based alloys, since accounts published by Runnalls indicate a very close analogy to the corresponding alloys of uranium, thus justifying the hope that plutonium may be used as the nuclear fuel in the well-known dispersion-type fuel elements. Tests of such fuel elements in the form of massive slugs of aluminum alloy containing a few weight per cent of plutonium have already been made at Chalk River.

ALUMINUM-PLUTONIUM ALLOYS

The range of Pu-Al alloys studied is from 5 to 60 per cent plutonium by weight. Alloys containing from 5 to 25 per cent plutonium are prepared by direct dissolu-

FIG. 8.—Apparatus for preparing and casting Pu-Al Alloys. A resistance furnace is in the lower center with a crucible being withdrawn to the left. At the top right is a simple induction furnace.

tion of plutonium in aluminum melted in a tantalum crucible. We use a resistance furnace. The purity of the argon in our glove boxes is sufficient to avoid using a controlled-atmosphere furnace. We thus benefit because of the greater simplicity of the manipulations, especially for stirring the melt and for pouring it. The crucible, held by a horizontal rod movable along the horizontal axis of the furnace, can be rapidly withdrawn and tilted over a heated graphite ingot mold (Fig. 8, at the lower left corner). The castings produced are usually cylindrical bars, 4–10 mm in diameter and several centimeters in length. Micrographic study of different sections of these bars has shown that they are macroscopically homogeneous.

In this same glove box we have a simple arrangement for induction melting. To avoid extracting the crucible from the heating coil, pouring is done from the bottom. The crucible and ingot mold are made of graphite.

Alloys containing from 30 to 60 per cent plutonium are prepared by reduction of PuF_3 with an excess of aluminum in an induction furnace under argon. Crucibles are made of magnesia or graphite.

FIG. 9. Equipment for differential thermal analysis. The apparatus for non-radioactive specimens is at the right; the apparatus for plutonium, at the left.

RESULTS

1. Differential thermal analysis (see Figs. 9 and 10) has allowed us to fix the fusion temperature of the Al-$PuAl_4$ eutectic at 642° C. For alloys containing more than 50 w/o plutonium we have been able to fix 925° C as the temperature of the peritectic reaction: $PuAl_3 + \text{liquid} \rightarrow PuAl_4$.

2. In micrography the samples are covered with an "Araldite" resin and are mechanically polished by hand in the usual way. They are afterward polished electrolytically at 10–15 volts in the following solution: 20 ml of $HClO_4$ (density = 1.61 gm/cm^3) and 70 ml of $HC_2H_3O_2$ (glacial). The structure is revealed by electrolytic etching in a water solution of phosphoric acid at 10–15 volts or by chemical etching in a 0.5 per cent solution of hydrofluoric acid in water.

For low (hypoeutectic) plutonium contents we have studied the influence of annealing on the spheroidization of the Al-$PuAl_4$ eutectic structure and on the hardness of the alloys. Microstructural results for a 10 w/o plutonium alloy are shown in Figures 11, 12, and 13.

The cast 10 per cent alloy, shown in Figure 11, has a Vickers hardness of 37 kg/mm^2 under a 5-kg load. Annealing 8 hours at 600° C (Fig. 12) yields a Vickers hardness of 27 kg/mm^2. Annealing 38 hours at 600° C (Fig. 13) gives a Vickers hardness of 24 kg/mm^2. A comparative study of U-Al alloys showed that, with these alloys, much longer annealing times are required to produce similar changes.

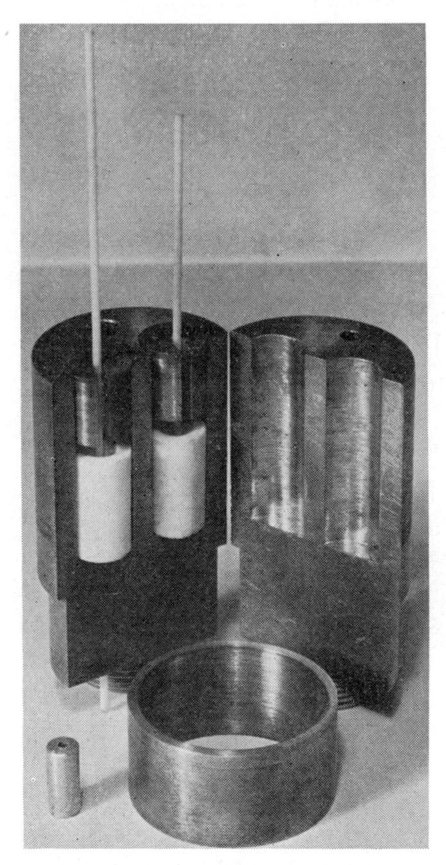

FIG. 10.—Assembly of samples and thermocouples for differential thermal analysis

The micrographs (Figs. 14–20) illustrate the as-cast microstructures of aluminum alloys having increasing plutonium content, from 10 to 50 w/o. The crystals of $PuAl_4$ (a phase whose existence is confirmed by X-ray diffraction) are birefringent, as shown by the photographs taken in polarized light (Figs. 21 and 22). Finally, it may be remarked that hardness increases with plutonium concentration.

These studies are being continued, to determine more precisely the form of the phase diagram in the aluminum-rich portion of the Pu-Al system, as well as to study the $PuAl_3 \rightarrow PuAl_4$ reaction in comparison with the $UAl_3 \rightarrow UAl_4$ reaction.

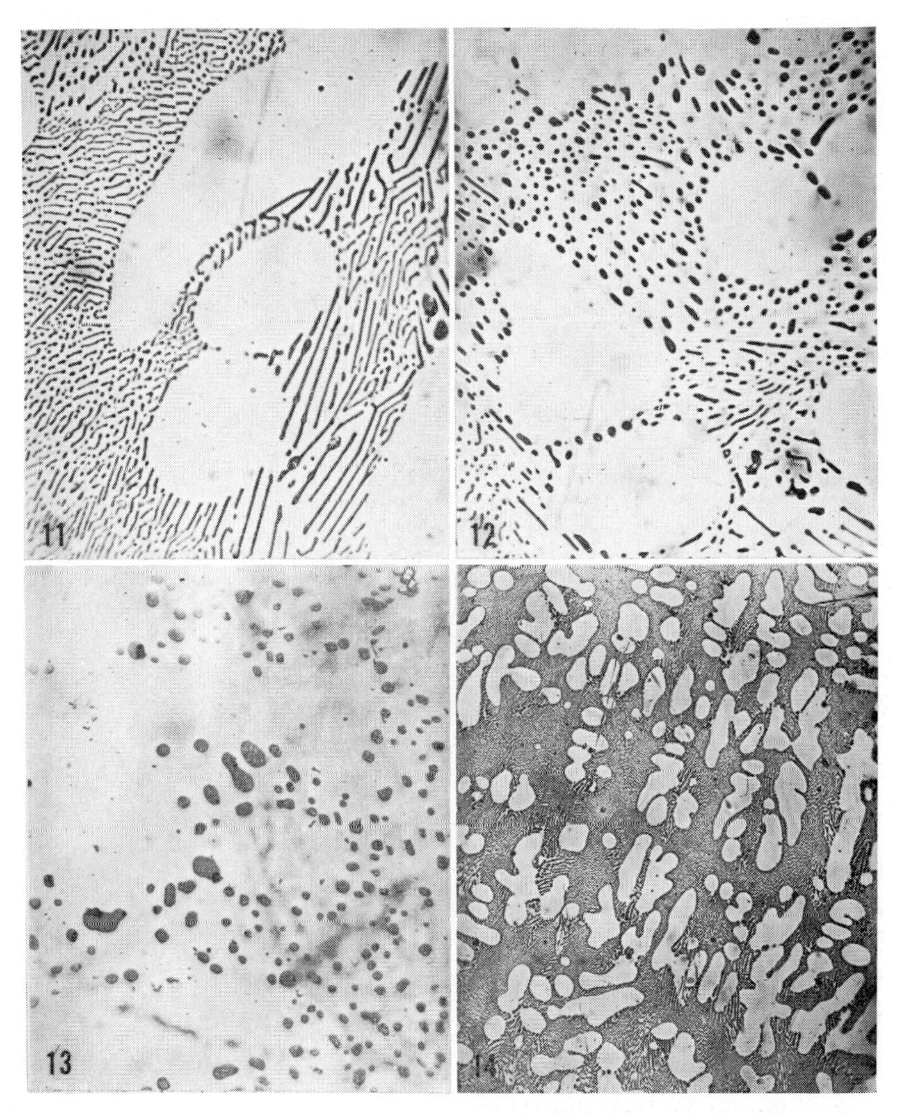

FIGS. 11–14.—Fig. 11: Pu-Al Alloy (10 w/o Pu), as cast; 1,020×. Fig. 12: Same alloy after annealing 8 hours at 600° C in argon; 1,020×. Fig. 13: Same alloy after annealing 38 hours at 600° C in argon; 1,020×. Fig. 14: Pu-Al alloy (10 w/o Pu), as cast; 120×.

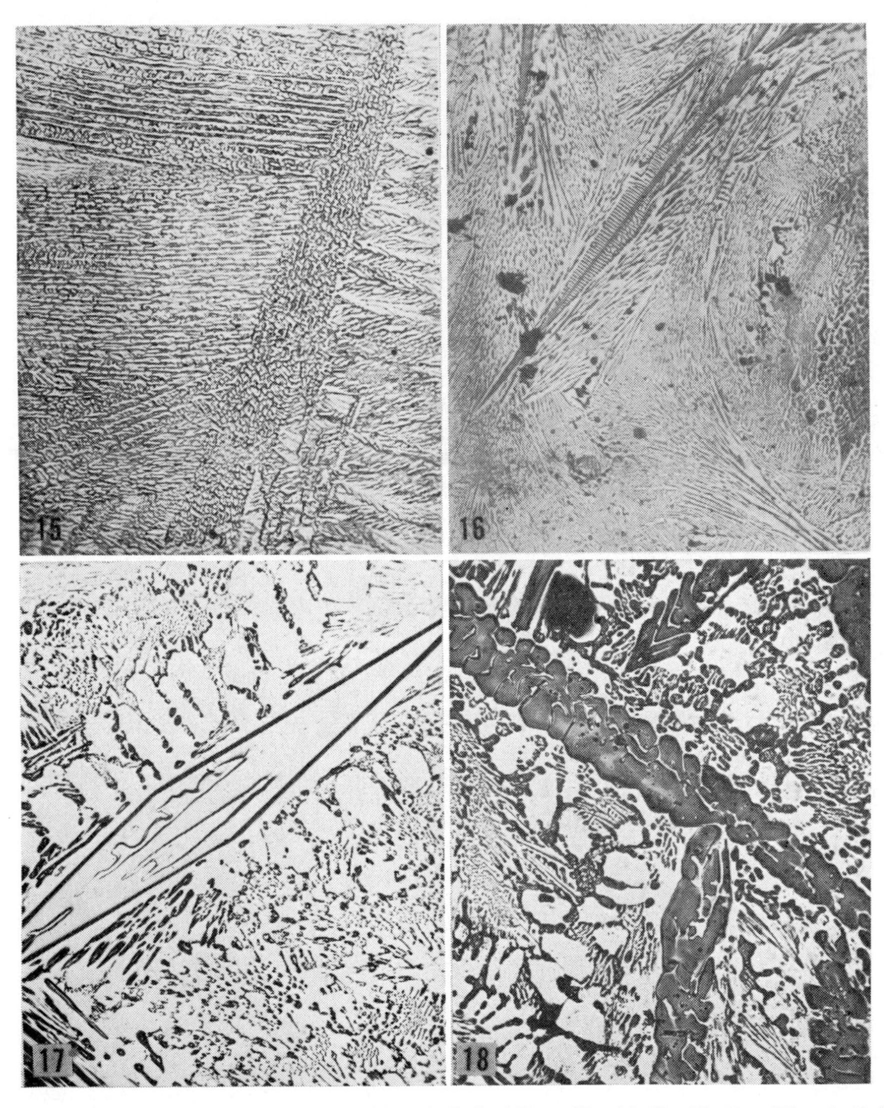

FIGS. 15–18.—Fig. 15: Pu-Al alloy (14 w/o Pu); 120×. Fig. 16: Pu-Al alloy (20 w/o Pu); 120×. Fig. 17: Pu-Al alloy (25 w/o Pu); 120×. Fig. 18: Pu-Al alloy (30 w/o Pu); 120×.

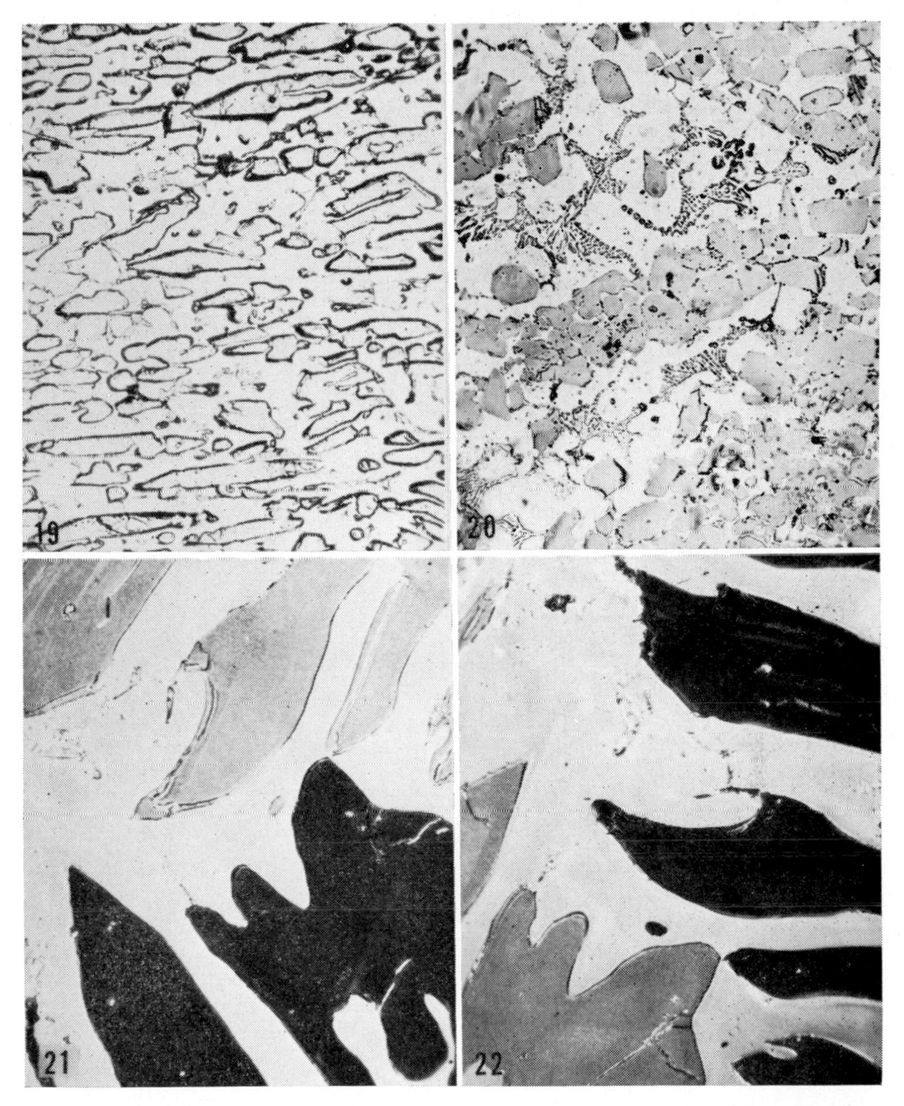

FIGS. 19–22.—Fig. 19: Pu-Al alloy (40 w/o Pu); 120×. Fig. 20: Pu-Al alloy (50 w/o Pu); 120×. Fig. 21: Pu-Al alloy (25 w/o Pu) under polarized light; 520×. Fig. 22: Pu-Al alloy (25 w/o Pu) under polarized light; 520×.

REFERENCES

1. Grison, E., and Pascard, R. "Construction of Glove Boxes for Plutonium Metallurgy in France," in *Glove Boxes and Shielded Cells*, ed. G. N. Walton, pp. 223–30. New York: Academic Press, Inc.; London: Butterworths Scientific Publications, 1958.
2. Pascard, R., and Fabre, R. "The Purification of Argon and Regulation of Pressure in Glove Boxes," in *Glove Boxes and Shielded Cells*, pp. 100–107.
3. Anselin, F., Faugeras, P., and Grison, E. *Comptes Rendus de l'Académie des Sciences*, **242**:1996, 1956.
4. Runnalls, O. J. C. *Progress in Nuclear Energy*, Ser. V, Vol. **2**, chap. 4-2, pp. 98–118. London: Pergamon Press, 1959.

PART II

PLUTONIUM METALLURGY

CHAPTER X

CRYSTAL-STRUCTURE STUDIES OF PLUTONIUM METAL

W. H. Zachariasen

THE CRYSTAL STRUCTURES

Complete crystal-structure information has been reported for five of the six phases of plutonium metal. For β-plutonium the unit cell dimensions are known, but not, as yet, the atomic positions. Although a thorough theoretical understanding of the structural features of plutonium and of other heavy-element metals is still lacking, it is possible to use the structure data as a basis for a general discussion leading to some results of interest.

The crystal-structure results for the various phases of plutonium metal are reported in scattered articles in the literature. As a convenience to the reader, the pertinent structural information will be summarized below. The linear coefficients of thermal expansion for all six phases of plutonium metal have been determined by X-ray methods. These results are given and discussed in a following section.

Further sections of this chapter are devoted to a discussion of the metallic radius of plutonium and of the electronic configuration in plutonium and the other heavy metals. For easy reference, data for the various transition temperatures on heating are listed in the accompanying table (1).

Transformation	Temperature (° C)	Transformation	Temperature (° C)	Transformation	Temperature (° C)
α–β........	122 ± 2	γ–δ.......	319 ± 5	δ'–ϵ.......	$476\ \pm5$
β–γ........	206 ± 3	δ–δ'.......	451 ± 4	ϵ–liq......	639.5 ± 2

α-PLUTONIUM (2)

The low-temperature form of plutonium metal is simple monoclinic with 16 atoms in the unit cell. The unit-cell dimensions at 25° C are as follows: $a = 6.182 \pm 0.001$ A; $b = 4.826 \pm 0.001$ A; $c = 10.956 \pm 0.001$ A; and $\beta = 101.74° \pm 0.01°$. The calculated density is 19.84 gm/cm³.

The space group is $P2_1/m$, with all atoms lying in the reflection planes with co-ordinates $\pm (x, \frac{1}{4}, z)$. The co-ordinate values for the eight kinds of plutonium atoms are given in the accompanying table.

W. H. Zachariasen is at the University of Chicago, Chicago, Illinois.

Atom	x	z	Atom	x	z
I..........	0.332	0.155	V.......	0.016	0.621
II.........	.774	.175	VI......	.465	.644
III........	.144	.341	VII.....	.337	.926
IV........	0.658	0.457	VIII....	0.892	0.897

The configuration about the various types of plutonium atoms is given in the accompanying table. The short bonds are sharply defined, all lying in the range

Atom	Short Bonds (Bond Length in A)			All Bonds (Bond Length in A)		
	No.	Range	Mean	No.	Range	Mean
I..........	5	2.52–2.69	2.62	12	2.52–3.51	3.09
II.........	4	2.62–2.69	2.68	14	2.62–3.62	3.21
III........	4	2.54–2.67	2.63	14	2.54–3.67	3.19
IV........	4	2.54–2.70	2.63	14	2.54–3.51	3.15
V.........	4	2.54–2.73	2.65	14	2.54–3.56	3.20
VI........	4	2.58–2.73	2.68	14	2.58–3.62	3.22
VII.......	4	2.52–2.70	2.65	14	2.52–3.56	3.21
VIII......	3	2.68–2.70	2.69	16	2.68–3.67	3.30

2.52–2.73 A, with all other bond lengths being 3.1 A or greater. The formation of short bonds, presumably of covalent nature, is not unique to α-plutonium. Similar short bonds have also been found in α- and β-uranium and in α- and β-neptunium.

β-Plutonium (3)

β-plutonium is also monoclinic, but in this form the translation lattice is body-centered, with 34 atoms in the unit cell.

At 190° C the unit-cell dimensions are as follows: $a = 9.284 \pm 0.003$ A; $b = 10.463 \pm 0.004$ A; $c = 7.859 \pm 0.003$ A; and $\beta = 92.13° \pm 0.03°$, corresponding to a calculated density of 17.70 gm/cm³. The probable space group is $I2/m$. The atomic positions are not yet known.

γ-Plutonium (4)

The γ-phase is orthorhombic, with a face-centered translation lattice. The unit cell contains 8 atoms and at 235° C has the following dimensions: $a = 3.159 \pm 0.001$ A; $b = 5.768 \pm 0.001$ A; $c = 10.162 \pm 0.002$ A. The calculated density is 17.14 gm/cm³.

The space group is Fddd, with the 8 equivalent plutonium atoms at: (0, 0, 0); $(\frac{1}{2}, \frac{1}{2}, 0)$; $(\frac{1}{2}, 0, \frac{1}{2})$; $(0, \frac{1}{2}, \frac{1}{2})$; $(\frac{1}{4}, \frac{1}{4}, \frac{1}{4})$; $(\frac{3}{4}, \frac{3}{4}, \frac{1}{4})$; $(\frac{3}{4}, \frac{1}{4}, \frac{3}{4})$; $(\frac{1}{4}, \frac{3}{4}, \frac{3}{4})$. Each plutonium atom has four neighbors at 3.026 A, two at 3.159 A, and four at 3.288 A.

δ-Plutonium (5)

In the δ-phase region the structure is face-centered cubic. At 320° C the unit-cube edge is $a = 4.6371 \pm 0.0004$ A, corresponding to a calculated density of

15.92 gm/cm³. The interatomic distance is 3.279 A, and the co-ordination number is 12.

δ′-Plutonium (5)

The δ′-phase has a tetragonally deformed cubic close-packed structure. It can be described as body-centered tetragonal with 2 atoms per unit cell. At 465° C the dimensions of the unit cell are $a = 3.327 \pm 0.003$ A, $c = 4.482 \pm 0.007$ A, giving a calculated density of 16.00 gm/cm³ (see chap. xxiv, however). Each plutonium atom has eight neighbors at 3.249 A and four at 3.327 A.

ε-Plutonium (5)

The structure of the ε-phase is body-centered cubic with $a = 3.6361 \pm 0.0004$ A at 490° C. The calculated density is 16.51 gm/cm³. Each plutonium atom has eight neighbors at 3.149 A.

THERMAL EXPANSION

X-ray methods have been used to determine the principal linear coefficients of thermal expansion for all six phases. Because of the complexity of the X-ray diffraction patterns, the values deduced for α- and β-plutonium are necessarily less precise than those for the other phases.

TABLE 1

Coefficients of Thermal Expansion

Phase	Range (° C)	×10⁶				φ	Reference
		α_1	α_2	α_3	$\bar{\alpha}$		
α	21–104	64 ±5	72±5	28±5	55 ±5	10°	6
β*	83–252	85 ±9	15±5	13±6	38 ±6	37°	3
γ	210–310	−20 ±1	39±1	84±1	34 ±1	......	4
δ	320–440	− 8.6±0.3	....	....	− 8.6±0.3	......	5
δ′	465–85	305 ±35	....	−659±67	−16 ±28	......	5†
ε	490–550	36.5±1.1	....	....	36.5±1.1	......	5

* The β-phase contained 2 per cent uranium.

† See chap. xxiv, however.

In the two monoclinic structures the orientation of the tensor ellipsoid of thermal expansion is not determined uniquely by symmetry. The second axis of the ellipsoid (corresponding to the principal coefficient α_2) must coincide with the crystallographic b-axis; but to describe the orientation of the ellipsoid completely, one has to specify the angle ϕ which the first tensor axis (associated with α_1) makes with the crystallographic a-axis in the plane of the obtuse angle β. This first tensor axis lies within the angle β and makes an angle β-ϕ with the crystallographic b-axis.

The experimental results for the linear coefficients of expansion are shown in Table 1. The striking feature is the anomalous behavior in the δ- and δ′-regions, where the mean linear coefficient is negative, instead of having a value of about $+35 \times 10^{-6}$ as one would expect from the results for γ- and ε-plutonium.

THE METALLIC RADIUS OF PLUTONIUM

In order to compare atomic sizes as found in different structures, it is necessary to make corrections for the effects of co-ordination number and thermal expansion. The standard co-ordination number will be taken as 12. Experimentally, it is known that the interatomic distances are 3.0 per cent larger for 12 co-ordination than for 8 co-ordination. The corrections to be applied to the observed distances to reduce them to 12 co-ordination will therefore be taken to be those given in the accompanying table.

Co-ordination No.	Correction (Per Cent)	Co-ordination No.	Correction (Per Cent)
16	−2.2	10	+1.4
14	−1.2	8	+3.0

Table 2 gives the results for the metallic radius of plutonium corrected to 12 co-ordination. For the high-temperature phases the metallic radius extrapolated to room temperature is also shown.

TABLE 2

THE METALLIC RADII OF PLUTONIUM

Phase	Temperature (° C)	Radius (A)	Radius Extrapolated to 25° C (A)	Phase	Temperature (° C)	Radius (A)	Radius Extrapolated to 25° C (A)
α	25	1.58	1.58	δ'	465	1.638	1.644
γ	235	1.601	1.589	ϵ	490	1.622	1.594
δ	320	1.640	1.644				

In the α-plutonium structure the interatomic distances in the first co-ordination sphere vary from 2.5 to 3.6 A. Accordingly, the metallic radius in this structure is not well defined, and no real significance can be attached to the difference between the room-temperature radii for α- and γ-plutonium.

As seen in Table 2, one finds two distinct radii for plutonium, one for the α-, γ-, and ϵ-phases and another appreciably larger radius for the δ- and δ'-phases. These two radii will be referred to as the "normal" and the "anomalous" radius, respectively. The difference between the two radii varies with the temperature, as shown in the accompanying table.

Radius	25° C	319° C	476° C
Normal	1.592 A	1.607 A	1.620 A
Anomalous	1.644 A	1.640 A	1.638 A

The normal radius behaves in the normal way in the sense that it increases with temperature, whereas the anomalous radius decreases with rising temperature in a most unusual fashion. A possible explanation for the existence of two radii and for their variation with temperature is discussed in the next section.

ELECTRON CONFIGURATION

The normal valences observed for plutonium in its compounds are 3, 4, 5, and 6. Hence there is no way of knowing a priori how many valence electrons (7s and 6d) and how many 5f electrons there are in the metal. It is possible, however, to make fairly reliable determinations of these numbers with the aid of the metallic radius, since the s and d electrons, on the one hand, and f electrons, on the other, affect the metallic radius quite differently. For this purpose it is useful to compare the results obtained for the metallic radius of plutonium with similar data for other elements in the same row of the periodic system.

Crystal-structure results have been reported for all the metals from actinium to americium. Values found for the metallic radii (for 12 co-ordination and room temperature) are shown in the accompanying table.

Metal	Radius (A)	Reference	Metal	Radius (A)	Reference
Ac.	1.88	7	Np.	1.55	11
Th.	1.79	8	Pu (normal). .	1.59	. . . .
Pa.	1.63	9	Am.	1.73 (or 1.82)	12, 13
U.	1.56	10			

The result for actinium is in doubt. It is highly unreasonable that the actinium radius should be only 0.01 A larger than that of lanthanum, and the work should therefore be repeated. The value of 1.82 A for the radius of americium reported by the workers at Berkeley (13) is also in doubt. The Los Alamos workers (12) find the much smaller radius of 1.73 A.

The normal valences, v, which the elements actinium to americium show in compounds and solutions are given in the accompanying table. Since thorium is a

Element	v = No. of 7s+6d Electrons	Element	v = No. of 7s+6d Electrons	Element	v = No. of 7s+6d Electrons
Ac.	3	U.	3, 4, 5, 6	Am.	3, 4, 5, 6
Th.	4	Np.	3, 4, 5, 6		
Pa.	4, 5	Pu.	3, 4, 5, 6		

typically tetravalent element, it must be stated without room for doubt that there are four valence electrons in thorium metal.

Table 3 shows the metallic radii (12 co-ordination, room temperature) for the first ten elements in the two long periods beginning with rubidium and with cesium, together with the well-established electron configurations in these metals. The entirely different manner in which s and d electrons, on the one hand, and the f electrons, on the other, affect the metallic radius is strikingly shown in the table.

The different behaviors are due to the fact that s and d electrons participate in bond formation while the f electrons do not. In Table 3 the symbol v denotes the numbers of s + d electrons, the symbol f the number of f electrons.

As electrons enter the s and d shells, with no electrons going into the f shell, the metallic radius varies with v in the same manner in the different periods except for a scale factor. Thus it becomes possible to predict the metallic radii of the elements

TABLE 3

METALLIC RADII IN OTHER PERIODS

v	f	Element	Radius (A)	v	f	Element	Radius (A)
1....	0	Rb	2.54	1.....	0	Cs	2.74
2....	0	Sr	2.15	2.....	0	Ba	2.23
3....	0	Y	1.80	3.....	0	La	1.87
4....	0	Zr	1.60	3.2...	0.8	Ce	1.83
5....	0	Nb	1.46	3.....	2	Pr	1.83
6....	0	Mo	1.39	3.....	3	Nd	1.82
7....	0	Tc	1.36	(3).....	(4)	Pm	....
8....	0	Ru	1.34	3.....	5	Sm	1.80
9....	0	Rh	1.34	2.....	7	Eu	2.05
10....	0	Pd	1.37	3.....	7	Gd	1.80

actinium to americium on the assumption that there are no 5f electrons present. These predicted values are obtained from the radii in the rubidium series, using a scale factor of 1.12, which gives the observed radius for thorium. The predicted and observed radii are given in the accompanying table.

ELEMENT	RADIUS (A)		ELEMENT	RADIUS (A)	
	Predicted	Observed		Predicted	Observed
Fr........	2.85	....	U........	1.56	1.56
Ra........	2.41	....	Np.......	1.52	1.55
Ac........	2.02	1.88	Pu........	1.50	1.59
Th........	(1.79)	1.79	Am.......	1.50	1.73 (1.82)
Pa........	1.64	1.63			

The conclusion is inescapable that there are no 5f electrons in any of the metals up to and including uranium. (The discrepancy for actinium will be passed over, since the experimental value for the radius of this metal is in doubt.) The lack of agreement between the predicted and observed radii for Np, Pu, and Am suggests the presence of 5f electrons in these metals, and the next task is to determine how many electrons are in the 5f shell.

An electron entering the f shell leaves the chemical bonding and the valence unchanged. There is, however, a small decrease in the size of the atom as compared with the preceding element, as a consequence of the increased nuclear charge. In the lanthanide series this 4f contraction is shown both in the metallic and in the ionic radii. The crystal-structure results for compounds of the heavy elements show the 5f contraction to be similar to the 4f contraction, although about 50 per cent larger in magnitude. It is thus possible to predict with considerable accuracy the metallic radii of the series actinium to americium on the assumption that there are 5f, as well as 6d and 7s, electrons. Conversely, the observed radius can be used to determine the electron distribution between the 5f and the 7s + 6d shells.

Consider the metallic radius of uranium as an illustration. The observed radius of 1.56 A agrees with that predicted for an empty 5f shell. However, suppose there were one 5f electron in uranium metal and thus five valence electrons, $v = 5$. Were this true, the metal radius would be that of protactinium decreased by the

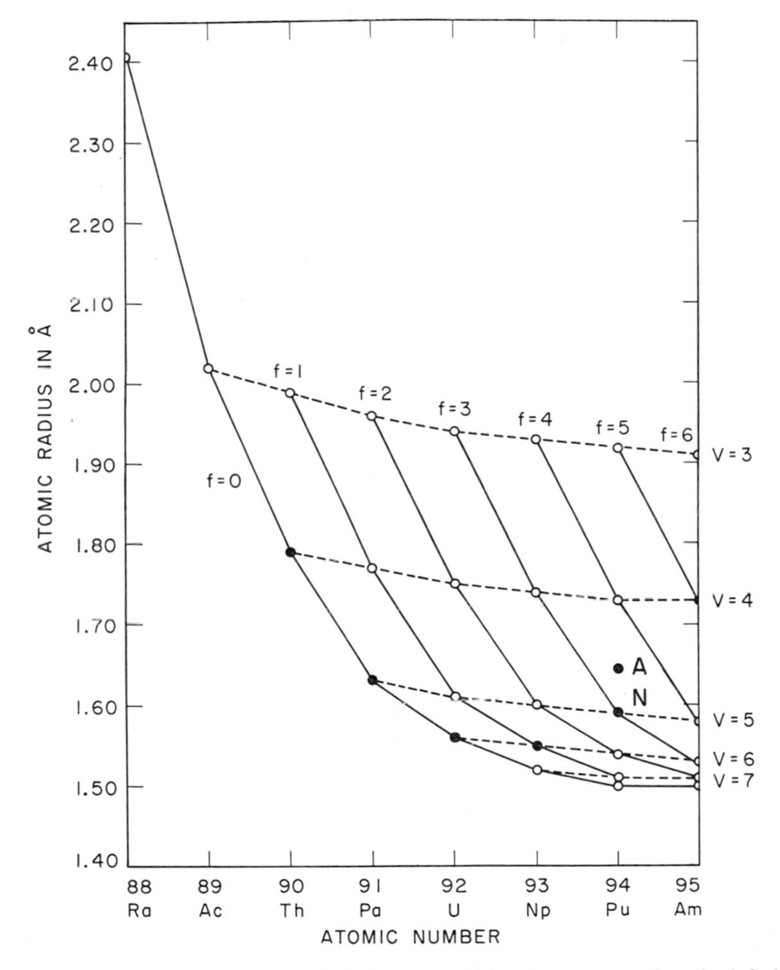

FIG. 1.—The predicted metallic radii of the transradium elements as a function of electron configuration. Here v is the number of valence (7s + 6d) electrons in an atom, and f is the number of 5f electrons. The solid circles represent the experimentally observed radii. The "normal" and "anomalous" metallic radii of plutonium are indicated by N and A, respectively. As exemplified by the anomalous radius of plutonium, a non-integral valence is derived by interpolation when the experimental value of the radius does not lie at an intersection in the network representing integral numbers of electrons.

contraction caused by the single 5f electron, which is about 0.02 A, giving an expected radius of 1.61 A if uranium metal were pentavalent. Similarly, were there two 5f electrons in uranium metal, requiring $v = 4$, the metallic radius would be that of thorium less the contraction due to two 5f electrons, or the metallic radius of uranium would be 1.75 A.

The predicted metallic radii arrived at in this manner are shown in Figure 1 and listed in Table 4, where the actually observed values are in italics. The electron

TABLE 4

PREDICTED METALLIC RADII (IN A) FOR DIFFERENT VALENCES

Element	v				
	3	4	5	6	7
Ac	2.02	......	......	......	......
Th	1.99	*1.79*	......	......	......
Pa	1.96	1.77	*1.63*	......	......
U	1.94	1.75	1.61	*1.56*	......
Np	1.93	1.74	1.60	*1.55*	*1.52*
Pu	1.92	1.73	*1.59*	1.54	1.51
Am	1.91	*1.73*	1.58	1.53	1.51

distribution deduced from the experimental data is, accordingly, as shown in the accompanying table (with v and f defined as above).

Element	v	f	Element	v	f
Th	4.0	0	Pu (normal)	5.0	3.0
Pa	5.0	0	Pu (anomalous)	4.6	3.4
U	6.0	0	Am (1.73 A)	4.0	5.0
Np	6.0	1.0	Am (1.82 A)	3.5	5.5

THE ANOMALOUS RADIUS OF DELTA-PLUTONIUM

Extrapolated to room temperature, the anomalous metallic radius of plutonium in the δ-phases is 1.644 A. It is 1.640 A at the lower limit and 1.638 A at the upper limit of the δ- and δ'-temperature ranges.

As shown above, the room-temperature value corresponds to $v = 4.6$, with 3.4 being the number of 5f electrons. Allowing for thermal expansion as applied to the predicted room-temperature radii of Table 4, the number of 5f electrons in δ-plutonium in the actual phase range can be deduced. At 319° C, one finds 3.24 and, at 476° C, 3.13, for the number of 5f electrons if their number is precisely 3.00 in the γ- and ϵ-phases of plutonium. There are two conclusions of significant interest as regards the δ-phases of plutonium. First, in the δ-phases plutonium has a smaller number of valence electrons and a larger number of 5f electrons than are found in the other phases. Second, in the δ-range, about 0.1 electron is promoted from the 5f shell to the valence shell as the temperature is increased.

In other words, the analysis based on the metallic radius leads to the conclusion that the negative thermal expansion in the δ-region is due to the gradual promotion of a small fraction of an electron from the 5f shell to the valence shell. Factors other than the number of 5f electrons, such as changes in magnetic interaction, may, of course, also affect the metallic radius. It is thus conceivable that such factors, rather than electron promotion, provide the true explanation for the anomalies in the δ-phases of plutonium.

REFERENCES

1. JETTE, E. R. *Journal of Chemical Physics*, **23**:365–68, 1955.
2. ZACHARIASEN, W. H., and ELLINGER, FINLEY. *Journal of Chemical Physics*, **27**:811–12, 1957.
3. ZACHARIASEN, W. H., and ELLINGER, F. H. *Acta Crystallographica*, **12**:175–76, 1959.
4. ———. *Ibid.*, **8**:431–33, 1955.
5. ELLINGER, F. H. *Transactions of the American Institute of Mining, Metallurgical and Petroleum Engineers*, **206**:1256–59, 1956.
6. ZACHARIASEN, W. H., and ELLINGER, F. H. Unpublished data.
7. FARR, J. G., GIORGI, A. L., MONEY, R. K., and BOWMAN, M. G. USAEC Rept. LA-1545 (1953).
8. HERMANN, C., LOHRMANN, O., and PHILIPP, H. *Strukturbericht*, **2**:182. Leipzig: Akademische Verlagsgesellschaft, 1928–32.
9. ZACHARIASEN, W. H. *Acta Crystallographica*, **5**:19–21, 1952.
10. JACOB, C. W., and WARREN, B. E. *Journal of the American Chemical Society*, **59**:2588, 1937.
11. ZACHARIASEN, W. H. *Acta Crystallographica*, **5**:660–64, 1952.
12. ELLINGER, F. H. Private communication.
13. GRAF, P., CUNNINGHAM, B. B., DAUBEN, C. H., WALLMANN, J. C., TEMPLETON, D. H., and RUBEN, H. *Journal of the American Chemical Society*, **78**:2340, 1956.

CHAPTER XI

THE EXAMINATION OF PLUTONIUM METAL BY X-RAY DIFFRACTION

F. Anselin

X-ray examinations of massive samples of plutonium and its alloys have been made by means of monochromatic X-rays and a Seeman-Bohlin camera having asymmetric focusing (see Fig. 1). The monochromator is a Johansson quartz plate

FIG. 1.—Photograph of the focusing X-ray camera and accessory equipment

curved elastically to a radius of 250 mm and placed at a 3° angle in relation to its $10\bar{1}1$ plane. The angle of incidence of the X-ray beam on the sample is 20°.

The camera of 360-mm circumference can be displaced horizontally with the aid

F. Anselin is with the Commissariat à l'Énergie Atomique, Centre d'Études Nucléaires de Fontenay-aux-Roses (Seine), France.

of a micrometer adjustment, thus, by careful regulation, providing an achromatic setting for a given region of the film. The sample can be made to move in either of the following two ways to compensate for a coarse grain size in the sample: (1) continuous rotation in a plane around a horizontal axis and (2) oscillation through 1°–7° around a vertical axis. The use of chromium K-α radiation enables one to obtain high resolution at small diffraction angles, where the first lines are numerous and intense. The whole apparatus is placed under an evacuated bell jar, in order to avoid absorption and diffusion of the X-rays by air. For operating convenience, we

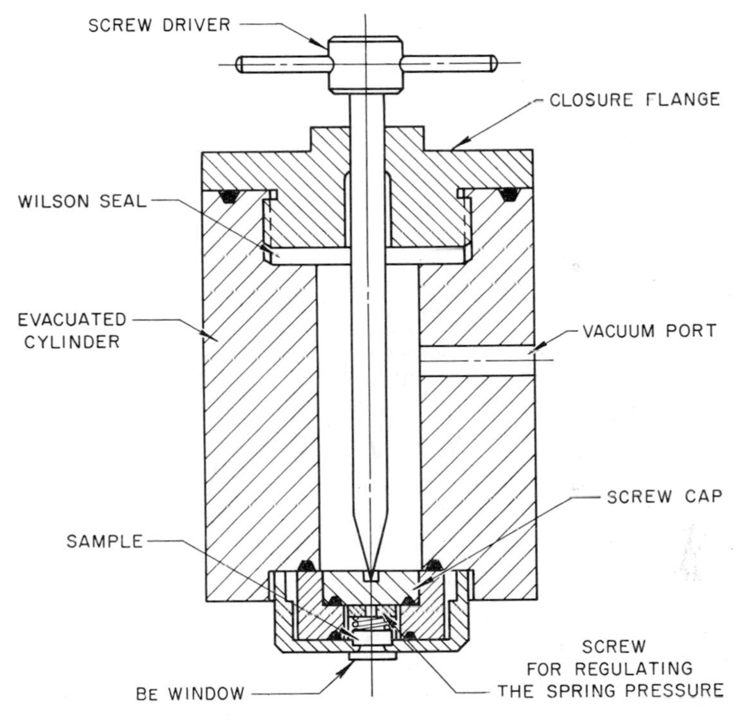

FIG. 2.—Drawing of the evacuated sample container, showing locations of the sample and the beryllium window.

have not inclosed the apparatus in a glove box but have placed over it a simple hood (removed for taking the photograph of Fig. 1).

The sample of electropolished plutonium is placed inside a container (Fig. 2) behind a beryllium window 0.3 mm thick. The container is hermetically closed under vacuum by the screw-down mechanism shown in Figure 2. All this preliminary manipulation is done in a glove box. The container is carefully decontaminated on the outside, and it is then taken out of the glove box and placed in the carriage on the diffraction chamber or camera.

The neutron flux resulting from the (*α*, *n*) reaction at the contact between the plutonium sample and the beryllium window is well below the maximum permissible level. The beryllium window is inconvenient, however, in that it superposes

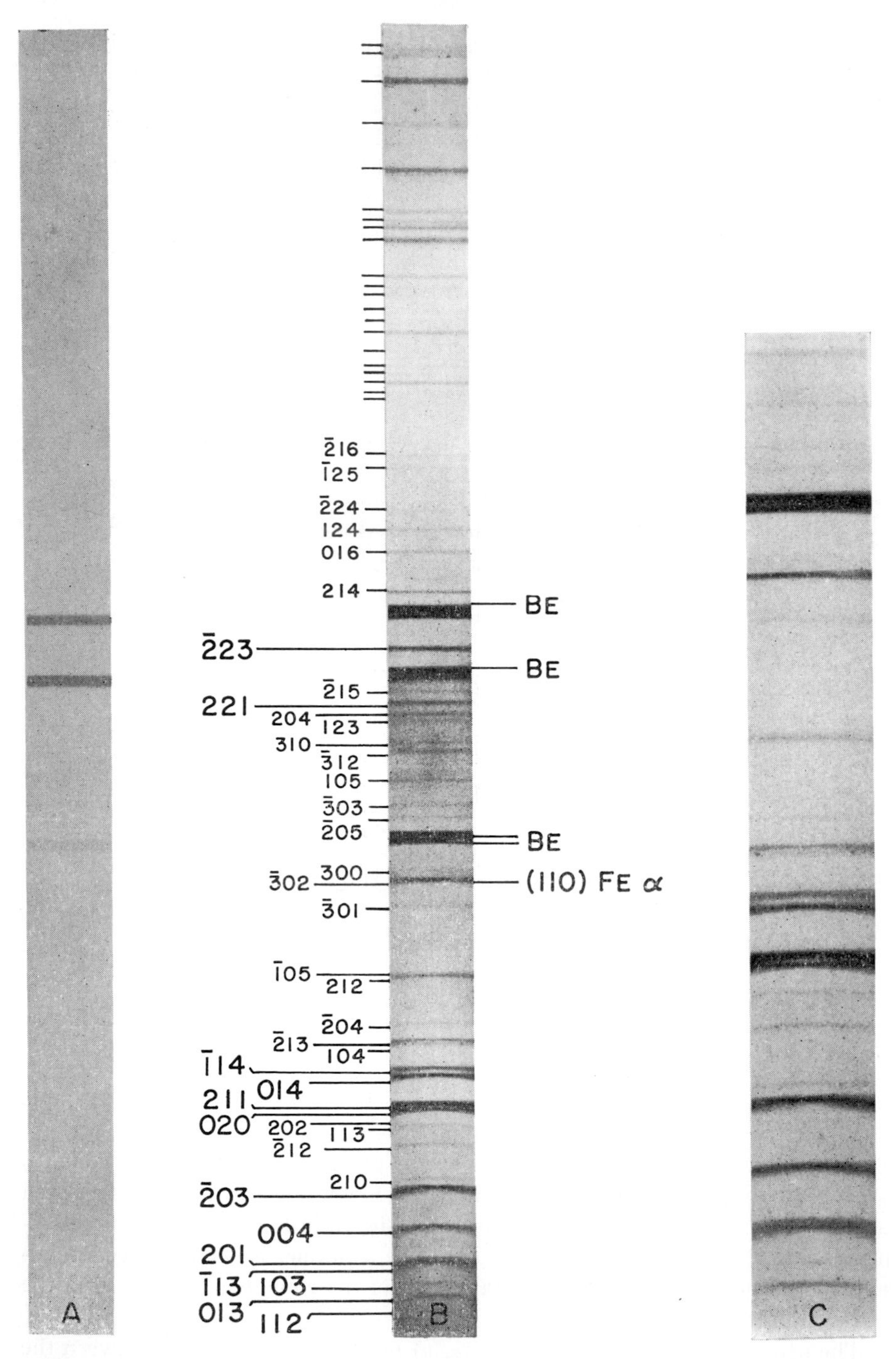

FIG. 3.—X-ray diffraction patterns of (*A*) the beryllium window with sample container empty; (*B*) alpha plutonium, the beryllium window, and an iron standard; (*C*) alpha plutonium (*low-angle lines*).

its own diffraction lines onto those of the sample, and these, being slightly out of focus, are rather wide and can mask some important lines of the material under study.

Figure 3 illustrates the pattern of alpha plutonium obtained with the equipment shown in Figure 1. Table 1 gives indices of the first lines and their interplanar spac-

TABLE 1

THE X-RAY DIFFRACTION PATTERN OF ALPHA PLUTONIUM

Miller Indices	Interplanar Spacings (A)	Observed Intensities*	Miller Indices	Interplanar Spacings (A)	Observed Intensities*
112....	2.930±0.015	W+	$\bar{3}01$....	2.064±0.004	W+
013....	2.887	M	$\bar{3}02$....	2.030	M
103....	2.850	W−	300....	2.017	VVW or trace
$\bar{1}13$....	2.787	S	Be line.	....	....
201....	2.774	M	015....	1.960	VW
004....	2.688±0.005	S	$\bar{2}05$....	1.949	W−
$\bar{2}03$....	2.588	S+	$\bar{3}03$....	1.930	W−
210....	2.565	W	105....	1.903	W
$\bar{2}12$....	2.487	W	$\bar{3}12$....	1.869	W+
113....	2.445	W	310....	1.861	W−
202....	2.433	VW	123....	1.836	W−
020....	2.412	VS	204....	1.830	W+
211....	2.403	VVS	221....	1.816	M
014....	2.343	S	$\bar{2}15$....	1.804	VW
$\bar{1}14$....	2.330	S−	Be line.	....	....
104....	2.282	W } resolved	$\bar{2}23$....	1.761	M+
$\bar{2}13$....	2.277	M− } resolved	Be line.	....	....
$\bar{2}04$....	2.248	VW	214, 222	1.710	M−
212....	2.166	W+ } resolved	016....	1.676	W+
$\bar{1}05$....	2.164	W+ } resolved	124....	1.657	W

* VS = very strong; S = strong; M = medium; W − weak; VW = very weak; VVW − very very weak.

ings, corresponding to the lattice dimensions of the monoclinic crystal structure described by Zachariasen and Ellinger (1). Calibration of the camera has been done by superposing on the alpha-plutonium pattern the diffraction lines of a high-purity standard, such as iron or aluminum.

REFERENCE

1. ZACHARIASEN, W. H., and ELLINGER, FINLEY. *Journal of Chemical Physics*, **27**:811–12, 1957.

CHAPTER XII

THE DILATOMETRY AND THERMAL ANALYSIS OF PLUTONIUM METAL

E. M. Cramer, L. L. Hawes, W. N. Miner, and F. W. Schonfeld

INTRODUCTION

A major contribution to the early technology of plutonium came through the use of thermal-analysis and dilatometric techniques. Studies of the first small specimens of the metal established the existence of five of the six known allotropes and yielded values for their transformation temperatures. Some of the early data and experiences of the investigators at Los Alamos were reported by Smith (1) in 1954. With the availability of larger specimens and purer metal in the immediate postwar years, the unique expansion characteristics of plutonium became better known, and the thermal points were determined more accurately. The experimental data suffered, however, from three persistent faults: sharp transformations were not evidenced in expansion curves, but, instead, the metal behaved like an alloy; the data were erratic and not reproducible; and the transformations on cooling were very sluggish, i.e., pronounced hystereses occurred between heating and cooling curves. The physical-property measurements of plutonium published in England in 1954 by Lord (2) and by Ball *et al.* (3) are representative of the values determined at Los Alamos in the period from 1945 to 1953 and appear to possess similar flaws.

In 1953, metal of a better purity was produced at Los Alamos in quantities suitable for physical testing, and a program to re-examine many of the known properties with the purer material was undertaken. Not only thermal and dilatometric characteristics but X-ray diffraction constants, resistivities, and other related properties were to be redetermined as well. In addition to using purer metal, the dilatometric method was modified to take advantage of the self-heating property of the metal. The radioactive decay of plutonium yields sufficient energy to heat massive specimens sensibly, and, since the metal is a relatively poor conductor of heat, it had been recognized for some time that significant thermal gradients could be present in dilatometer specimens during heating and would be unavoidable during cooling. By thermally shielding the specimen from its surroundings, a self-heating method was achieved in which the transfer of heat to or from the specimen was held to a minimum. Not only was the expansion curve improved as a result, show-

The authors are at the University of California, Los Alamos Scientific Laboratory, Los Alamos, New Mexico.

ing that the appearance of alloy-like behavior was caused by thermal gradients, but evidence was clearly discovered for a sixth allotrope occurring in a temperature range between those of the delta and epsilon phases. A comparison of expansion curves for metal of normal purity and for high-purity metal under conditions of self-heating has been published both by Jette (4, 5) in 1955 and by Coffinberry and Waldron (6) in 1956, together with many of the physical measurements that followed from the change in method of heating and the use of purer metal.

Although the sharpness of the transformations was greatly improved by the changes to self-heating and purer metal, the considerable variability of results obtained in successive dilatometer runs was altered very little. Moreover, since the self-heating method yields data only during the heating portion of a complete cycle of heating and cooling, no amelioration of the hysteresis effects was realized.

The hysteresis effects in plutonium are not appreciable for the phase transformations that occur at higher temperatures, but the temperature gaps for the alpha/beta, beta/gamma, and gamma/delta transitions may amount to several tens of degrees, depending on heating and cooling rates. It was, of course, recognized that the hysteresis commonly observed between the temperature of a phase transformation on heating and the temperature of the reverse transformation on cooling represents a departure from equilibrium during both heating and cooling (i.e., superheating and undercooling). Although of secondary interest compared with the objective of the self-heating experiments, it was thought worthwhile to undertake a preliminary investigation of isothermal transformation behaviors within the temperature ranges of the known hysteresis gaps. These isothermal experiments were conducted with the same dilatometric equipment as that employed for the self-heating runs, simply by holding the specimen at constant temperature for an extended time, even though it was recognized that the existence of a small temperature gradient within the specimen was not avoided.

The first measurements of thermal-expansion coefficients by the self-heating method were made at temperatures above room temperature, so that only a relatively small portion of the total stability range of alpha plutonium was investigated. Because, however, the coefficient of the alpha phase was known to be temperature-dependent, it was considered important to extend this range to temperatures below ambient. Hence a longer interval, from -180° to 100° C, was studied. In 1952 at Los Alamos, Elliott and Tate (7) had determined the linear coefficient of expansion of alpha plutonium over this same temperature interval. They reported the relation

$$\alpha = 48.39 \times 10^{-6} + 95.88 \times 10^{-9} t^\circ \text{ C}$$

for a massive specimen containing 99.86 weight per cent plutonium.

An attempt was made to achieve conditions of self-heating in the dilatometer after having cooled the specimen to -180° C with liquid nitrogen. The abnormally high initial heating rate observed was interpreted to indicate that the dilatometer would need drastic redesigning to prevent heat leaks to the specimen while it was at low temperatures. Instead of redesigning the dilatometer, however, it seemed simpler and safer to use a tubular specimen if one could be made with a wall so thin that a significant thermal gradient would not exist within it. The specimen that was finally used had a wall 0.017 inch thick, the minimum that could be machined from the brittle alpha plutonium without fracture and that could be heated or cooled at rates up to 3° C per minute without hysteresis. With this specimen, expansion data were obtained for both the alpha and the beta phases. But when the experiment was extended to include gamma, the metal became too soft to support the weight of the dilatometer push-rod, and obvious plastic deformation took place.

The method of self-heating was also adapted to the thermal analysis of plutonium, but less effectively than to dilatometry. Interposition of ceramic insulation between the thermocouples and the specimen reduced sensitivity and, consequently, decreased the responsiveness of the equipment. With somewhat less exacting control, however, thermal points were obtained at temperatures that were within the scatter of the dilatometric data, and the melting point of the metal was reproduced within the limits of accuracy of the determination.

EXPERIMENTAL DETAILS

A fused-silica tube and dial-indicator dilatometer, as shown in Figure 1, were used in the investigation. The instrument had been modified for vacuum and inert atmospheres by inserting the silica dilatometer tube through a water-cooled base plate provided with a vacuum seal and by inclosing the dial indicator in a bell jar. The specimens used in the self-heating experiment were cylinders approximately 0.500 inch in diameter and 0.500 inch long, machined from slightly oversize castings. The shape of the specimen was chosen to be equiaxed, in order to reduce the effect of longitudinal thermal gradients.

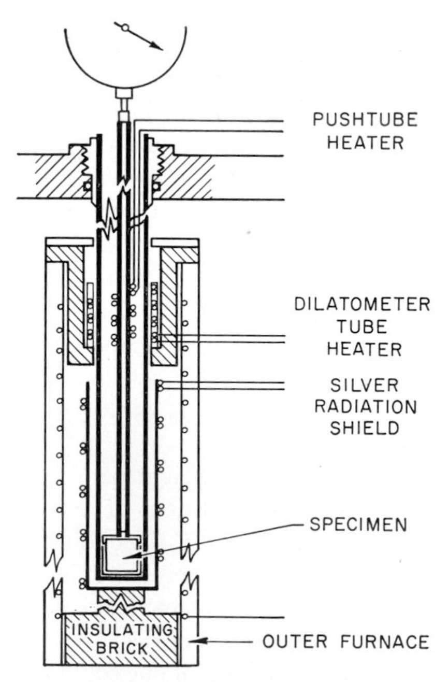

FIG. 1.—Diagram of the dial-indicator dilatometer with thermal shielding for plutonium

Heating elements surrounding the dilatometer tube were designed to minimize the transfer of heat to or from the specimen by its surroundings. The lower portion of this silica tube, where the specimen was located, was inclosed in a silver radiation shield around which was wound a resistance heater. Around higher sections of the dilatometer tube and of the push-tube, above the specimen location, were wound resistance heaters to prevent conduction of heat upward from the region of the specimen to the base plate. A resistance furnace surrounded the assembly, as shown in Figure 1. All resistance windings, except that of the outer furnace, were made bi-filar, in order to minimize inductive effects in the thermocouples.

During operation, control of the temperatures of each of the heating elements surrounding the specimen was obtained by manually changing the power input to the individual resistance heaters. The need for a change was indicated by observing

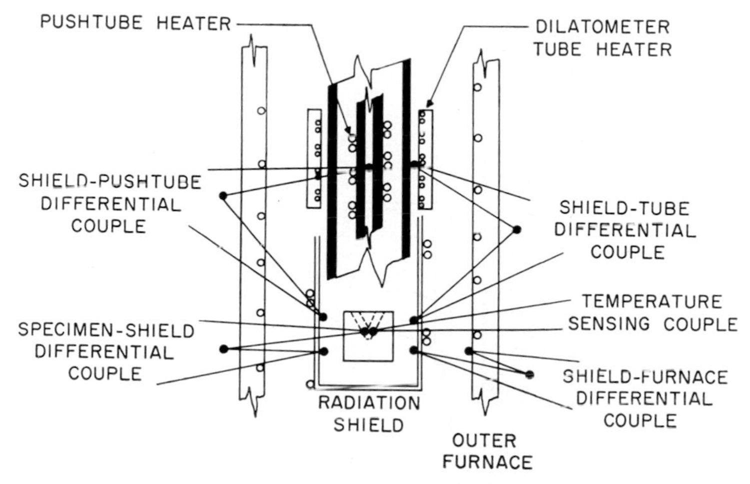

FIG. 2.—Thermocouple locations in the shielding and in the plutonium specimen

a difference in temperature between the specimen and a heating element in its surroundings.

Two chromel-alumel thermocouples were inserted into holes drilled to the center of the specimen. A portion of each thermocouple wire was wrapped around the push-tube, and the ends were brought through compression seals in the base plate. One of the thermocouples was used to measure the temperature of the specimen. One wire of the other couple terminated in a second junction that was in contact with the silver radiation shield, thus allowing the differential temperature between the specimen and shield to be measured directly. Because of lack of space within the dilatometer tube, the temperatures of this tube, the push-tube, and the outer furnace were compared with the temperature of the silver shield rather than with that of the specimen. Three additional differential couples were in contact with the shield and each of the other heating elements. A schematic diagram of thermocouple placement is shown in Figure 2. All couples were made from 28-gauge chromel and alumel wire and were calibrated against a standard platinum/platinum-rhodium couple, which had been certified by the National Bureau of Standards.

The dilatometer required the services of two operators. One recorded dilation versus temperature of the specimen, the other observed differences in temperature between elements in the system and changed the power input accordingly. It was found feasible to maintain the temperature of the radiation shield within $\pm 0.2°$ C of the specimen temperature and the temperatures of the heated areas of the dilatometer tube and push-tube within $\pm 0.6°$ C of the shield temperature when the temperature of the outer furnace was held from 0° to 1.2° C below that of the shield. A vacuum of the order of 10^{-4}–10^{-5} mm of mercury was maintained in the dilatometer system by continuous pumping.

The same basic dilatometer was used for both the self-heating and the low-temperature experiments. For the latter (see Fig. 3), all the thermal shielding was replaced with a simple resistance-heated copper jacket surrounding the dilatometer

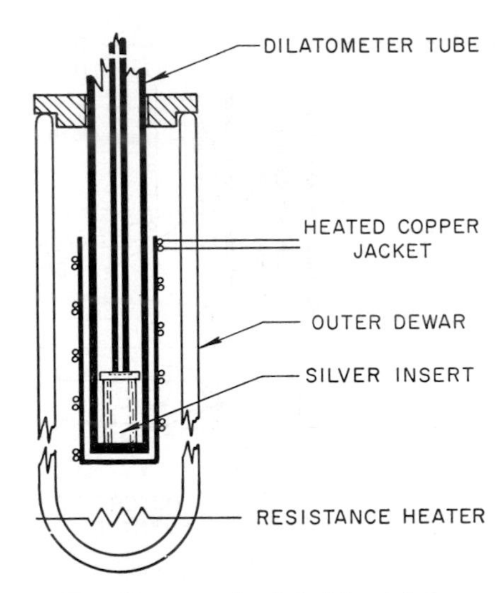

FIG. 3.—The dilatometer loaded with a tubular specimen

tube. This assembly, with a bare Nichrome heater under it, was placed in a Dewar flask. The specimen was heated or cooled at predetermined rates by manually varying the power input either to the resistance element on the copper jacket or to the Nichrome element in the bottom of the Dewar, which was kept immersed in liquid nitrogen during the cooling cycle. Expansion of the tubular specimen was transmitted to the push-tube of the dilatometer by means of a silver cylinder inserted into the specimen and supported on it by a projecting shoulder. The dimensions of the body of the silver insert were such that it would not tightly contact the specimen at $-190°$ C, and the shoulder was such that it would center the specimen in the dilatometer tube without tight contact at 100° C. A recess was machined into the top of the insert to receive the push-tube and to center it with respect to the specimen. The silver insert also served to compensate for the expansion of a length of silica equivalent to the length of the specimen. To accomplish this purpose, the push-tube recess was machined to a depth less than the thickness of the shoulder, thus adding a known length of silver to the combined lengths of the specimen and push-tube. It was calculated that any uncorrected expansion in the dilatometer system would be less than the sensitivity of the instrument, and, consequently, no corrections of this nature were made to the experimental data. The temperature of the specimen was not measured directly; instead, the temperature of the silver insert was measured with a copper-constantan thermocouple. A helium atmosphere was maintained at approximately 1 mm of mercury pressure in the system to facilitate the transfer of heat between the specimen, insert, and dilatometer tube.

A check for the absence of thermal gradients and the satisfactory operation of the dilatometer consisted of a series of runs at different rates of heating and cooling. Since the effect of thermal gradients appears as hysteresis where no transformations are involved, the criterion for satisfactory operation was a reversible curve within

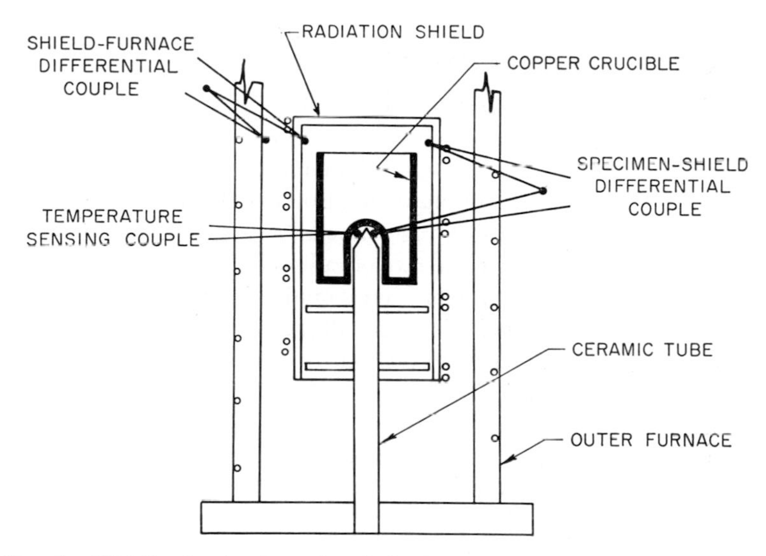

FIG. 4.—Shielding for the thermal analysis of plutonium specimens by self-heating

the accuracy of the instrument. Reversible curves were obtained at rates up to 3° C per minute, where a slight separation of the heating and cooling traces was observed. Consequently, a rate of temperature change of about 1.5° C per minute was adopted.

For thermal analysis the equipment was similar to that used for dilatometry but was simplified (see Fig. 4). Since no external mechanical connections were needed, all the thermal shielding was placed within the vacuum furnace housing, and the spot heaters were eliminated. The metal was contained in a copper crucible having an integral thermocouple well and protected inside and out with a wash of magnesia. The crucible was supported from a Lavite base plate on a pedestal consisting of a ceramic tube. The tube also served to position two thermocouples within the well in the crucible. Inclosing the crucible was a cylindrical resistance-heated silver shield similar to the one used in the dilatometer. The shield, closed at one end, rested on the crucible but was insulated from it with mica sheet and was closed at the bottom by several silver disks concentric with the crucible support. The shield was in turn surrounded by a ceramic tube wound with a resistance-heating element. Again, two specimen thermocouples were employed, but their actual contact was with the crucible. One was used to determine the temperature of the crucible, and the other formed part of a differential couple between the crucible and the shield. The temperature of the outer furnace was referred to the shield in turn by another differential couple.

In operation, the silver shield was maintained within 0.5° C of the temperature of the crucible and the outer furnace at such a temperature, lower than that of the shield, that power could be applied continuously to the shield element. Pressure in the furnace housing was maintained at about 10^{-4} mm of mercury by continuous pumping.

RESULTS OF THE SELF-HEATING EXPERIMENTS

Stout and Jones (8) have reported the rate of radioactive self-heating of plutonium to be $(1.923 \pm 0.019) \times 10^{-3}$ watt per gram. The determination of this value was made with plutonium of low irradiation level, so that it was considered to contain only negligible amounts of plutonium isotopes other than Pu^{239}. However, the plutonium used in the self-heating dilatometric experiments contained sufficient Pu^{240} and Pu^{241} to increase the rate of radioactive energy output to about a 10 per cent greater value. This rate of energy production resulted in a self-heating rate of approximately 0.75° C per minute for all the plutonium allotropes, although the exact rates varied somewhat with differences in the heat capacities of the different allotropes and with variations at different temperatures of the effectiveness of minimizing heat transfer to or from the specimen. Because of the latter consideration, the exact heating rates were not considered to be accurate measurements of the specific heats of the allotropes, although it may be noted that rates approximating 0.75° C per minute correspond to heat capacities of the order of 9–10 cal/° C/gram-atom.

The purity of the metal studied was such that the total impurities approximated 0.05 weight per cent.

The operation of the dilatometer was checked by determining the thermal expansion, from room temperature to 600° C, of a silver specimen 1 inch long. A comparison of these results with Bureau of Standards values for the expansion of silver indicated that an accuracy within 2.5 per cent in dilation measurement throughout the range from room temperature to 600° C could be expected. The chromel-alumel thermocouples used in the self-heating experiments were calibrated against a standard platinum/platinum-rhodium couple, and the error of specimen temperature measurement is considered to be within ± 1° C.

The observed scatter of dial-indicator readings and transformation temperatures was greater than the accuracy of their determination; hence the limits of error in the following data are the actual maximum variations in values obtained from seven independent measurements involving three different specimens. Seven determinations were made of each transformation temperature on heating, and each expansion coefficient, except that of the epsilon phase, was measured seven times. Only one determination of the expansion of epsilon was made.

The Alpha Phase

At the beginning of each dilatometer run in the self-heating experiments, the temperature of the specimen was always above room temperature, and an additional temperature increase occurred while temperature control was being established. Consequently, the expansion of alpha plutonium was obtained over only a narrow range of temperature. The alpha phase was found to be stable below 122 $(+1, -2)^\circ$ C, and for the range of 58–106° C the average coefficient of linear expansion was determined to be $\alpha_\alpha = 58\ (+3, -3) \times 10^{-6}/^\circ$ C. The density of alpha at 25° C is about 19.8 gm/cm^3.

The Beta Phase

On transforming from alpha to beta, the specimen dilated 288 $(+25, -36) \times 10^{-4}$ inch per inch of length at temperature. Beta was stable from 122 $(+1, -2)^\circ$ to 206 $(+3, -2)^\circ$ C; $\alpha_\beta = 35\ (+2, -3) \times 10^{-6}/^\circ$ C from 144° to 201° C. The calculated density at 122° C was 17.8 gm/cm^3.

The Gamma Phase

The isothermal expansion in transforming from beta to gamma was 78 $(+6, -11) \times 10^{-4}$ inch per inch of length at temperature. Gamma was stable from 206 $(+3, -2)^\circ$ to 319 $(+6, -3)^\circ$ C; $\alpha_\gamma = 33\ (+2, -2) \times 10^{-6}/^\circ$ C from 224° to 300° C. The calculated density was 17.2 gm/cm^3 at 206° C.

The Delta Phase

The isothermal expansion in transforming from gamma to delta was 205 $(+22, -11) \times 10^{-4}$ inch per inch of length at temperature. Delta was stable from 319 $(+6, -3)^\circ$ to 451 $(+4, -2)^\circ$ C; $\alpha_\delta = -16\ (+2, -2) \times 10^{-6}/^\circ$ C from 336° to 447° C. The calculated density was 16.0 gm/cm^3 at 319° C.

The Delta-Prime Phase

Evidence for the existence of the delta-prime phase had not been seen clearly before high-purity metal was examined by self-heating. However, anomalies in the expansion of delta in the vicinity of 450° C were known to exist. It appears that impurities in the metal used earlier, rather than thermal gradients, were responsible for the failure to detect delta-prime in metal of ordinary purity. Isothermal expansion in transforming from delta to delta-prime was found to be $-12\ (+7, -5) \times 10^{-4}$ inch per inch of length at temperature. Delta-prime was stable from 451 $(+4, -2)^\circ$ to 476 $(+3, -5)^\circ$ C; $\alpha_{\delta'} = -118\ (+53, -21) \times 10^{-6}/^\circ$ C from 459° to 474° C. The calculated density was 16.2 gm/cm^3 at 451° C. It is not known to what extent the specimen may have deformed plastically while it was contracting thermally within the delta-prime temperature range.

The Epsilon Phase

The isothermal expansion in transforming from delta-prime to epsilon was $-126\ (+10, -12) \times 10^{-4}$ inch per inch of length at temperature. Epsilon was stable from 476 (+3, −5)° to 640° C; $\alpha_\epsilon = 27 \times 10^{-6}$/° C from 490° to 527° C for one specimen only. In all other instances the curves showed unmistakable evidence of plastic deformation of the specimen, and the data were discarded. It is possible that extensive deformation also took place in this instance, and some indication of the amount may be found by comparing the dilatometric value of α_ϵ with the value determined by X-ray diffraction, $\alpha_\epsilon = (36.5 \pm 1.1) \times 10^{-6}$/° C. The calcuated density of epsilon at 476° C was 17.0 gm/cm³.

THE EXPANSION OF THE TUBULAR SPECIMEN

As explained above, in order to extend the range of measurement of the thermal expansion of the alpha phase, a tubular specimen was studied in the temperature interval from −186° to 101° C. In addition, by extending the heating of this specimen into the range of 133°–197° C, a more accurate value was obtained for the thermal expansion of the beta phase.

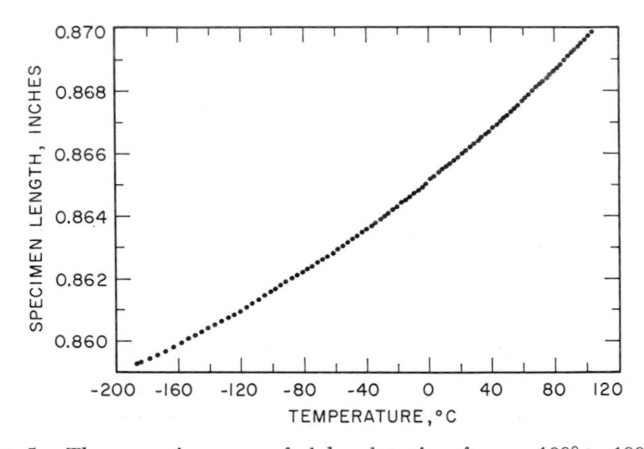

FIG. 5.—The expansion curve of alpha plutonium from −180° to 100° C

A copper-constantan thermocouple was used with the tubular specimen in the lower-temperature range. It was calibrated with fresh boiling nitrogen, solid CO_2, melting ice, and boiling water. The accuracy of temperature measurements below 0° C is believed to be within 2°, and above 0° C within 1°. A chromel-alumel thermocouple was used in the interval from 133° to 197° C. This thermocouple was calibrated against a standard noble-metal couple, and the measurements in this interval are believed to be correct within 1° C.

Expansion of the specimen was recorded as increments of ΔL in units of hundred-thousandths of an inch. (The dial of the Starrett model 655-T6 indicator that was used was graduated in units of ten-thousandths of an inch, the value in the fifth decimal place being estimated.) These data were then corrected for dial-indicator error as determined by calibrating the indicator with a Clarkator calibrating device. Average corrected values of ΔL from two cycles in the alpha phase were combined into tables of $t_\alpha + \Delta t$ versus $L_0 + \Delta L_\alpha$, and the values from four cycles in the beta phase were combined into tables of $t_\beta + \Delta t$ versus $L_0 + \Delta L_\alpha + \Delta L_{\alpha\to\beta} + \Delta L_\beta$.

The expression, $L = L_0(1 + at + bt^2 + \ldots)$, for the length of a solid body as a function of temperature (9) was fitted to the data by the method of least squares, and the constants were evaluated. Best fits to the data yielded the following expressions for linear thermal expansion. For alpha plutonium, $\alpha = 46.847 \times 10^{-6} + 0.055881 \times 10^{-6}\, t^\circ$ C; and, for beta plutonium, $\alpha = 33.86 \times 10^{-6}$. The maximum divergence of observations from the values derived from the above expressions was 7×10^{-5} for alpha and 3×10^{-5} for beta at any temperature within the applicable temperature intervals. Plots of the experimental points obtained in these two determinations of expansion behavior are shown in Figures 5 and 6.

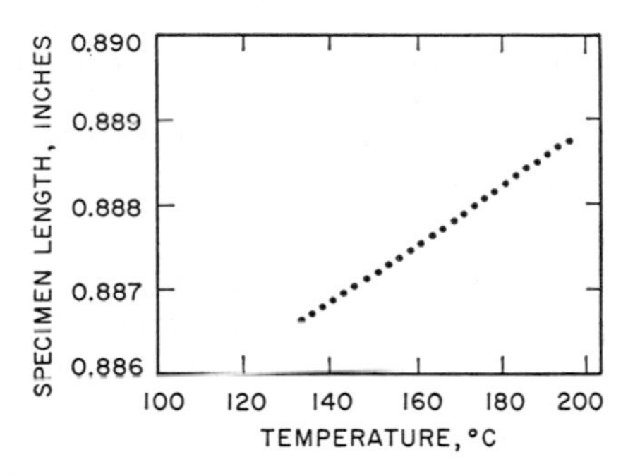

FIG. 6.—The expansion curve of beta plutonium from 133° to 197° C

THE ISOTHERMAL TRANSFORMATION EXPERIMENTS

The experiments performed in attempting to narrow the hysteresis gaps of the alpha/beta, beta/gamma, and gamma/delta transformations were only preliminary in character, and only limited success was represented by the results obtained. Holding times longer than 1 week did not yield acceptable data because of difficulties in maintaining constant temperature. For this reason, no results obtained for the alpha/beta transition were considered satisfactory.

In runs lasting a week, the beginning of the beta/gamma transformation was obtained at 199° ± 1° C, and the beginning of the reverse gamma/beta transition at 174° ± 1° C. Thus the temperature span of the hysteresis in this case was reduced to 25° C from a gap of about 70° C observed in early dilatometer runs at convenient heating and cooling rates. Holding times of 5 days narrowed the hysteresis of the gamma/delta transformation to only 3° C, the difference between 311° ± 1° C on heating and 308° ± 1° C on cooling. In all the successful isothermal runs, temperatures were held constant to within 1° C, but the thermocouples were not recalibrated following completion of the runs.

THERMAL-ANALYSIS RESULTS

Only one thermal-analysis run was made by the method of self-heating, except that two determinations of the melting temperature were made. The results of these two melting-point determinations gave an average value of 639.5° C, with experimental accuracy estimated to be of the order of ±2° C. The temperatures of the solid-state transformations observed by thermal analysis all fell within the ranges of temperatures found for the same transformations in the seven self-heating dilatometer runs. These temperatures of thermal arrests have been reported elsewhere (4, 5, 6).

REFERENCES

1. SMITH, C. S. *Metal Progress*, **65**:81–89, 1954.
2. LORD, W. B. H. *Nature*, **173**:534–35, 1954.
3. BALL, J. G., ROBERTSON, J. A. L., MARLON, P., LEE, J. A., and ADAMS, E. T. *Nature*, **173**:535, 1954.
4. JETTE, E. R. *Journal of Chemical Physics*, **23**:365-68, 1955.
5. ———. "Plutonium Metal," in *Nuclear Metallurgy*, pp. 29–37. ("Institute of Metals Division Special Report Series," No. 1.) New York: American Institute of Mining and Metallurgical Engineers, 1955.
6. COFFINBERRY, A. S., and WALDRON, M. B. "The Physical Metallurgy of Plutonium," chap. 4 of *Progress in Nuclear Energy*, Ser. V, Vol. **1**, pp. 354–410. London: Pergamon Press, Ltd., 1956.
7. ELLIOTT, R. O., and TATE, R. E. *A Determination of the Coefficient of Thermal Expansion of Alpha Plutonium*. (USAEC Rept. LA-1390, 1952.)
8. STOUT, J. W., and JONES, W. M. *Physical Review*, **71**:582–85, 1947.
9. HIDNERT, P., and SOUDER, W. *The Thermal Expansion of Solids*. (National Bureau of Standards Circ. No. 486, March 15, 1950.)

CHAPTER XIII

A DILATOMETRIC STUDY OF PLUTONIUM

R. Abramson

In this chapter are summarized the results of various dilatometric measurements made on plutonium in our laboratories previous to 1958. The equipment and procedures employed for fourteen successive tests of the dilatometry of plutonium are described very briefly, and results obtained for a single specimen are reported. The curves indicate clearly the existence of the alpha, beta, gamma, delta, and epsilon phases, as well as the delta-prime phase. Transformation temperatures and expansion coefficients are given in Tables 1 and 2 and are compared with values published by other authors.

TABLE 1

TRANSFORMATION TEMPERATURES (° C)

Test No.	$\alpha \to \beta$	$\beta \to \gamma$	$\gamma \to \delta$	$\delta \to \delta'$	$\delta' \to \epsilon$	$\epsilon \to \delta'$	$\delta' \to \delta$
1....	105	....	....	....	....	....	....
2....	115	202	....	....	....	....	....
3....	115	....	....	....	....	....	....
4....	109	195	300	458	....	477	456
5....	115	197	315	457	....	477	455
6....	110	210	322	465	....	....	....
7....	113	200	312	454	....	477	456
8....	107	200	315	460	477	477	455
9....	105	207	310	454	475	473	450
10....	....	....	306	....	....	....	....
11....	....	200	311	....	....	....	....
12....	....	210	320	463	....	483	463
13....	....	212	320	466	....	486	461
14....	107	....	....	....	....	....	....

EQUIPMENT

The dilatometric equipment is shown in Figures 1–3. Dilatation curves are obtained by photographic means with an improved Chevenard dilatometer (1). The thermal expansions of the sample and a reference standard are indicated and magnified by the reflection of a light beam from a mirror mounted on a tripod, such as that shown at the end of the dilatometric head in Figure 3. The tripod is attached

R. Abramson is with the Commissariat à l'Énergie Atomique, Centre d'Études Nucléaires de Fontenay-aux-Roses (Seine), France.

TABLE 2

MEAN COEFFICIENTS OF EXPANSION (×10⁶ PER ° C)

Test No.	On Heating				On Cooling	
	α	β	γ	δ	δ'	ϵ
4....	61	35	36	−7.5	....	30
5....	50	30	32.5	−8.2	−61	27
6*...	58	33.7	36.6	−8.2	....	{ 23.6 26.5
7*...	61	32	31.3	−6.6	−64	{ 23 22.4 21.4
8....	62	34.4	32	−10	−71	29.8
9....	63.2	29	35.8	−9.7	−58	29.8
10....	....	27.9	31.9	....	....	27.5
11....	....	....	33.4	....	....	....
12....	....	23.6	32.2	−10.8	−65	25.8
13....	....	27.6	30.9	−9.6	−57	26.5

* In tests Nos. 6 and 7, more than one value of the expansion coefficient of the ϵ-phase was obtained by cycling the specimen at high temperature (see Fig. 4).

FIG. 1.—The Chevenard dilatometer and accessory equipment. Part of this equipment is inclosed within and part is placed outside the glove box.

Fig. 2.—View through the glove-box window, showing the evacuated silica sample tube, with furnace removed.

Fig. 3.—Photograph of a dilatometric head, showing the tripod and mirror on the end toward the optical system.

to the dilatometric head in such a way that the expansion of the *pyros* standard displaces the luminous spot following a vertical y-axis (*ordinate*), while the expansion of the sample is translated into a displacement along the x-axis (*abscissa*). Pyros is a nickel-chromium-tungsten alloy having a nearly constant coefficient of thermal expansion. Since the expansion coefficient of this standard is accurately known, the temperature scale (*ordinate*) is calibrated by measurements along its standardization curve (1). An evacuated silica tube containing the sample (Fig. 2) is heated by an Adamel furnace mounted on rails. Chevenard-Joumier equipment is used for regulation of the heat and temperature programs.

Magnification of the dilatations of the sample and standard depends on the dimensions of the dilatometric head and tripod holding the mirror (Fig. 3). The magnification factors are designated K_1 and K_2 for the ordinate and abscissa, respectively. We have used the following two dilatometric heads: head 1; $K_1 = 308$; $K_2 = 159.6$; head 2: $K_1 = 312$; $K_2 = 302.6$. Head No. 1 was used for tests Nos. 4–9, in which we wished to record the complete expansion cycle of the metal, while head No. 2 was used for tests Nos. 10–14, in which we were interested in studying more particularly a detail of the total dilatometric curve within a limited range of temperature.

THE SAMPLE

All measurements were made on a single sample. The tests were numbered from 1 to 14, and Figures 4, 5, and 6 illustrate the curves that were obtained in tests Nos. 7, 9, and 13, respectively. (In these figures the traces obtained in the dilatometer

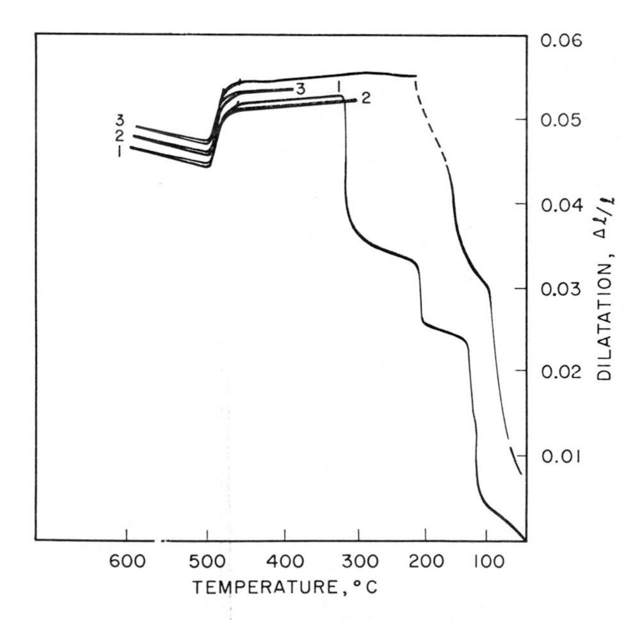

FIG. 4.—The dilatometric trace for run No. 7. The sample was cycled three times above 300° C, in the δ and ϵ temperature ranges. Heating rate: 55° C per hour. Dilatometric head No. 1: $K_1 = 308$; $K_2 = 159.6$. Length of sample: 14.2 mm. Vacuum: 8×10^{-6} mm Hg.

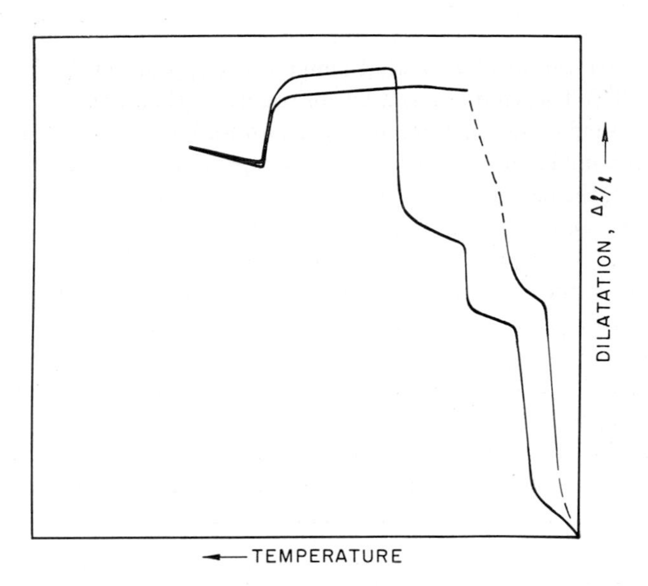

FIG. 5.—The heating and cooling traces for run No. 9. Heating rate: 60° C per hour. Cooling rate: 40° C per hour. Dilatometric head No. 1: $K_1 = 308$; $K_2 = 159.6$. Length of pyros standard: 40 mm. Length of sample: 14.2 mm. Vacuum: 1×10^{-5} mm Hg.

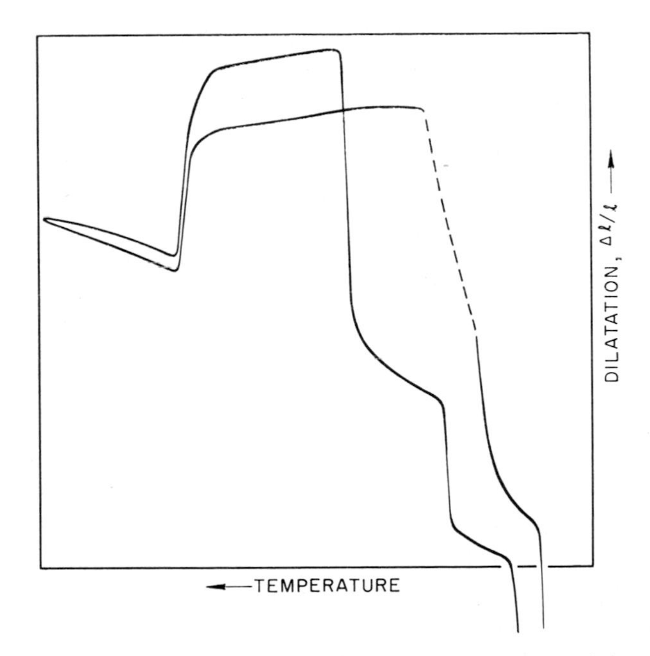

FIG. 6.—Heating and cooling traces for run No. 13. Note displacement of the origin and the larger scale along the dilatation axis caused by the larger value of K_2. Heating: 20°–350° C at 70° per hour; 350°–560° C at 30° per hour. Cooling: 560°–350° C at 30° per hour; 350°–20° C at 70° per hour. Dilatometric head No. 2: $K_1 = 312$; $K_2 = 302.6$. Length of pyros standard: 50 mm. Length of sample: 14.2 mm.

have been rotated through 90°, so that temperature appears as abscissa and dilatation as ordinate.) The sample was a bar of square section, melted by a very brief high-frequency induction heating, and cast into an ingot mold made of CaF_2. It was slightly pyramidal in shape because of the taper of the mold. After squaring the ends on a lathe, the bar measured 14.2 mm in length, and its weight was 4.234 gm. Before the series of tests the sample was electrolytically passivated. As a result of this treatment, its color was gray and lacked metallic luster.

After 14 tests, the bar had become gray-black in color and quite dull, but, in spite of its weight loss (0.026 gm), no oxide dust was found inside the evacuated silica sample tube. The sample was slightly elongated, measuring 14.5 mm, and slightly bent. Its specific gravity, close to 18.8, varied little throughout the complete series of tests.

Spectrographic and chemical analyses of the metal gave the results shown in the accompanying table. Although the relative sensitivity of the analyses is rather poor

	Ppm		Ppm		Ppm
Al.	<100	Ca.	<100	Fe.	~300
Ba.	<100	Cr.	<100	Ni.	<100

because of the very small weight of the assay samples, we estimate that the purity of the metal must have been near that indicated by Jette (2): 99.85–99.90 weight per cent plutonium.

OPERATING CONDITIONS

All tests were run in high vacuum, of the order of 10^{-5} mm Hg. Heating and cooling rates varied in successive runs, but they were always low, from 30° to 70° per hour for heating between 20° and 600° C, and from 50° to 90° per hour for cooling between 600° and 150° C. At temperatures below 150° the thermal inertia of the furnace itself insures a cooling rate of about 60° per hour without the necessity for regulation.

THE RESULTS AND INTERPRETATIONS OF MEASUREMENTS

Transformation Temperatures

All the allotropic transformations of plutonium appear clearly in Figures 4, 5, and 6. The transformation temperatures observed in the 14 tests are listed in Table 1. These temperatures were indicated clearly for all the transformations on heating, but only for the $\epsilon \rightarrow \delta'$ and $\delta' \rightarrow \delta$ transitions on cooling.

The $\alpha \rightarrow \beta$, $\beta \rightarrow \gamma$, and $\gamma \rightarrow \delta$ transitions on heating.—The $\alpha \rightarrow \beta$ and $\gamma \rightarrow \delta$ transitions begin less abruptly than do those of the $\beta \rightarrow \gamma$ and $\epsilon \rightarrow \delta'$ changes, but all show a character recognizable as usual for transformations of the first order.

The $\delta \rightarrow \delta'$ transition on heating.—This is a clearly visible change of slope on the heating branch of the curves. However, no marked discontinuity is observed that would correspond to an expansion typical of a change of the first order (as for $\alpha \rightarrow \beta$, $\beta \rightarrow \gamma$, and $\gamma \rightarrow \delta$). There seems to be only a clear change in value of the coefficient of expansion.

The $\delta' \rightarrow \delta$ transition on cooling.—Figure 5 shows this change to be identical in character with that of $\delta \rightarrow \delta'$. The change of slope is very clear.

The $\delta' \rightarrow \epsilon$ and $\epsilon \rightarrow \delta'$ transitions.—It is possible, by closely observing the thinning-out of the photographic trace, to find the transformation temperature, although it cannot be established as precisely as for the $\delta \rightleftharpoons \delta'$ transition. Values have been noted in tests Nos. 8 and 9 on heating and in Nos. 4, 5, 7, 8, 9, 12, and 13 on cooling (see Table 1). In contrast to the $\delta \rightleftharpoons \delta'$ changes, the $\delta' \rightleftharpoons \epsilon$ transformations are represented by a very pronounced vertical displacement in the dilatometric trace.

The $\delta \rightarrow \gamma$, $\gamma \rightarrow \beta$, and $\beta \rightarrow \alpha$ transitions on cooling.—These transitions on cooling have a very different aspect from the reverse transitions on heating and also manifest a strong temperature hysteresis, notably for the $\delta \rightarrow \gamma$ transformation, which begins around 230°–215° C (instead of 313° C, as on heating). The $\delta \rightarrow \gamma$ transformation is characterized by a very sharp discontinuity, which begins a series of small, slightly inclined steps separated by sharp contractions, suggesting a very high transformation speed as in a martensitic transformation. These discontinuous phenomena disappear around 170°–150° C, and the trace then passes slowly and continuously to the β-phase. The γ-phase thus seems to be completely obscured when cooling occurs at the rates employed in our tests. This discontinuous character is clearly shown in curves published by Jette (2), and we have also seen it indicated on curves recorded point by point at Harwell (3).

The $\beta \rightarrow \alpha$ transformation begins, also with a large hysteresis, around 80°–90° C and seems to continue very slowly to ambient temperature. In test No. 13 (Fig. 6) the light spot returned to its initial zero position only after cooling from 150° to 20° C throughout all of one night.

COEFFICIENTS OF EXPANSION

Measurements of the thermal-expansion coefficients of various phases are reported in Table 2. For any one phase the observed values are rather widely scattered. Nevertheless, the average values agree quite well with figures reported by Jette (2, 4), and by Lee (4) (Table 4).

Although the coefficients listed in Table 2 for the δ'- and ϵ-phases represent behavior on cooling, an average coefficient for the δ-phase on cooling was not calculated. With regard to the behavior of the δ-phase on cooling, it is important to note the existence in all the curves of a slight "knee" around 280°–290° C, where the abnormal negative expansion coefficient changes to a normal positive coefficient.

The over-all elongation on heating from room temperature to the maxima shown in Figures 4, 5, and 6 (the linear expansions plus the length changes accompanying the phase transformations) amounts to 5 or 6 per cent of the initial length. This is a little less than the value quoted by Jette (2, 4).

Cyclic Tests

Figure 4 shows cyclic tests of the δ- and ϵ-phases above 300° C. For each of these phases the metal appears to have a reproducible behavior, the expansion coefficients (the slopes of the traces) being about the same in each cycle. However, the length changes on transformation vary from one cycle to another, and the existence of remnant expansions or contractions may be noted. Thus successive curves are parallel but not superimposed. Transition temperatures remain stable within the limits of probable error indicated below.

THE ACCURACY OF THE MEASUREMENTS

Temperature

As indicated above, the temperature is determined to an accuracy of ±2° C by the elongation of the pyros standard (1 mm = 4° C on the curve of the standard). However, because of the difference between the specific heats of the various phases

TABLE 3

Comparison of Transformation Temperatures

Transformation	Jette, 1955 (2, 4)		Ball *et al.*, 1954 (4, 5)		Konobeevsky, 1955 (4, 6) Dilatometry and Thermal Analysis	Abramson, 1957 Dilatometry	Pascard, 1957 (7) Thermal Analysis
	Dilatometry	Thermal Analysis	Dilatometry	Thermal Analysis			
$\alpha\rightarrow\beta$	122±2	122	135	125	119	110±4	135
$\beta\rightarrow\gamma$	206±3	203	225	235	218	203±6	222
$\gamma\rightarrow\delta$	319±5	317	315	325	310	313±7	320
$\delta\rightarrow\delta'$	451±4	453	450	450	450	460±5	....
$\delta'\rightarrow\epsilon$	476±5	477	480	475	472	476	474
$\epsilon\rightarrow\delta'$	....	....	....	....	....	479±5	....
$\delta'\rightarrow\delta$	....	....	....	....	....	457±4	....
Melting	....	639.5±2	....	640	640	....	638

of plutonium and that of the pyros standard, it is not certain that the two are at exactly the same temperature at all times. Moreover, when operating in a vacuum (and especially at temperatures below 200°), equalization of temperature takes place very slowly. For this reason, we have estimated ±4° C as the probable error for each transformation temperature in a series of tests. Obviously, this affects each value in Table 1.

Using statistical methods on the rather small amount of data, we have calculated the standard deviation for each of the transformation temperatures by the formula

$$\sigma = \frac{\Sigma(\chi - \chi_m)^2}{n - 1},$$

where σ = the standard deviation, χ = the value of any measurement, n = the total number of measurements, and χ_m = the average value of all n measurements.

The average values of the transformation temperatures as found in our experiments (as well as the corresponding standard deviations) are indicated in the next to the last column of Table 3, wherein our figures are compared with the results of other investigators. (σ is not calculable for the $\delta' \rightarrow \epsilon$ transformation temperature, because we made only two determinations at this temperature.) The standard deviations found (4°–7° C) are greater than the 4° C given above as the anticipated probable error, thus indicating that causes of accidental error other than that mentioned above should be taken into consideration.

Expansion Coefficients

Mean coefficients of thermal expansion, $\bar{\alpha}$, for the temperature range θ_1 to θ_2 are calculated according to the relation

$$\bar{\alpha}\Big]_{\theta_1}^{\theta_2} = \frac{\Delta l\Big]_{\theta_1}^{\theta_2}}{(\theta_2 - \theta_1)\,K_2},$$

where θ_1–θ_2 is the temperature range in which a phase exists, $\Delta l\Big]_{\theta_1}^{\theta_2}$ = the change in length recorded for this phase as measured along the sample dilatation axis of the photographic trace, and K_2 = the magnification factor for this axis.

TABLE 4

Comparison of Mean Coefficients of Thermal Expansion ($\times 10^6$ per ° C)

Phase	Jette, 1955 (2, 4) by Dilatometry	Ellinger, 1956 (4, 8) by X-Ray Diffraction	Lee, 1956 (4) by Dilatometry	Abramson, 1957 by Dilatometry
	At 25°	At 25° C	Under 80° C	
α	49.6	50.8	52	59 ±5
			Over 80° C	
			58–75	
	At 164° C	At 164° C		
β	33.8	38	41	30.3±3.8
γ	....	34.7	31	33.3±2.1
δ	− 10 ±0.5	− 8.6±0.3	−7 to −13	− 8.8±1.4
δ'	−120	−16 ±28	....	−63 ±5*
ϵ	27.5±2	36.5±1.1	23.3–25.9	25.6±2.8*

* Values obtained on cooling.

The accuracy with which the coefficients can be determined is limited by the precision with which Δl can be measured (approximately 0.5 mm). This is why the accuracy is of the order of ±10 per cent for the α-, β-, γ-, and ϵ-phases, for which Δl is of the order of 10 mm. For the δ- and δ'-phases, for which the values of Δl are negative (contraction on heating), the accuracy is only about ±20 per cent. Statistical computations yield the average uncertainties indicated in the last column of Table 4. These standard deviations are seen to be well within the limits of accuracy to be expected for the expansion coefficients.

CONCLUSIONS

In Tables 3 and 4, the results of this first series of tests are compared with other published data. Within the limits of accuracy indicated above, we see that only the temperature of the $\alpha \longrightarrow \beta$ transformation (110° ± 4° C) has been found to be significantly lower than the corresponding values reported by Jette (2), the temperatures for the other phase changes being in agreement within the ranges of the standard deviations.

Qualitatively, our curves are all rather similar to one another. From one test to another, plutonium seems to retain the same expansion characteristics. From one cycle of heating and cooling to the next, given sufficient time, the specimen appears to return to its initial length. For the whole series of tests, however, a permanent elongation of 0.3 mm, or 2 per cent, was noted. Our tests have been too few in number to establish any correlation between this elongation and the number of cycles. Nor can our measurements give correct indications regarding the influence of impurities in the metal on the temperatures or sharpness of the transformations. Our curves are comparable with those published by Jette (2) for metal of 99.87 per cent purity, but they do not have the sharpness of those for metal of greater purity, such as his 99.97 weight per cent plutonium.

REFERENCES

1. Chevenard, P. *Journal de physique et le radium*, Ser. 6, **7**:240–49, 1926.
2. Jette, E. R. *Journal of Chemical Physics*, **23**:365–68, 1955.
3. Waldron, M. B. Atomic Energy Research Establishment, Harwell, private communication.
4. Coffinberry, A. S., and Waldron, M. B. "The Physical Metallurgy of Plutonium," chap. 4 of *Progress in Nuclear Energy*, Ser. V, Vol. **1**, pp. 354–410. London: Pergamon Press, Ltd., 1956.
5. Ball, J. G., Robertson, J. A. L., Mardon, P., Lee, J. A., and Adams, E. T. *Nature*, **173**:535, 1954.
6. Konobeevsky, S. T. "Phase Diagrams of Some Plutonium Systems," in *Session of the Division of Chemical Sciences (of the U.S.S.R. Academy of Sciences), Conference on the Peaceful Uses of Atomic Energy, Moscow, 1955*, pp. 362–75; also pp. 207–14 in English translation by Consultants Bureau (for sale by Superintendent of Documents, Washington, D.C., price $1.00).
7. Pascard, R. Centre d'Études Nucléaires de Fontenay-aux-Roses, private communication.
8. Ellinger, F. H. *Transactions of the American Institute of Mining, Metallurgical and Petroleum Engineers*, **206**:1256–59, 1956.

CHAPTER XIV

SOME PHYSICAL PROPERTIES OF PLUTONIUM METAL STUDIED AT HARWELL

J. A. Lee and P. G. Mardon

INTRODUCTION

Plutonium for metallurgical studies first became available in Great Britain in 1951. The initial samples were pellets weighing about 100 mg and, like the first plutonium produced in the United States and Canada, showed no phase transformation up to the melting point. Very shortly thereafter, large-scale reductions were carried out at the Atomic Energy Research Establishment (1), and the metal so produced was first examined by Lord (2) of the Atomic Weapons Research Establishment, using a dilatometer he had installed in a glove box at A.E.R.E. He found that five allotropic modifications of the metal existed between room temperature and its melting point. This result was soon confirmed by thermal-analysis, electrical-resistance, and dilatometric measurements carried out by the A.E.R.E. staff under J. G. Ball.

During 1952, plutonium of greater purity than that already examined became available for experimental work. The thermal-analysis and dilatometric behavior in the region of 450° C varied considerably from one billet to another, some samples showing signs of a further possible phase change at this temperature while others did not. X-ray examination in this field showed only lines attributable to the delta and epsilon phases of the metal as previously determined on the less pure samples, and the discrepancy was left unsolved until Ellinger (3) at Los Alamos published his discovery of a sixth phase, delta-prime.

Several of the early experiments at A.E.R.E. were designed to employ only a thin wire (less than 1 gm) of the metal because of its scarcity; only the thermal-analysis and dilatometric specimens were larger than this. It soon became apparent, however, that, because of the unusual electrical-conductivity behavior of plutonium, determinations of specific heat and thermal conductivity by electrical heating of such a wire would be extremely difficult; more recent determinations of these properties have used much larger specimens.

Up to the present time, no serious investigation has been made of the behavior of plutonium below room temperature, but A.E.R.E., acting in conjunction with

J. A. Lee and P. G. Mardon are with the Atomic Energy Research Establishment, Harwell, Berkshire, United Kingdom.

the Clarendon Laboratory at Oxford, hopes to have apparatus for such examination ready shortly.

The present authors wish to point out that they are acting as rapporteurs for the work carried out at A.E.R.E. and have tried to include the names of the principal workers against each experiment. The early work was carried out under J. G. Ball, while the more recent experiments have been directed by M. B. Waldron.

THE HANDLING OF PLUTONIUM

Owing to the high toxicity associated with the α-particle activity of plutonium, special handling techniques are required to minimize inhalation and ingestion. The practice for handling is essentially the same in all laboratories, the material being kept in either fume hoods or glove boxes. The latter are usually maintained at about -1 inch static water-gauge pressure relative to atmosphere by means of a fitted extract system; all handling is done through shoulder-length rubber or neoprene gloves sealed onto ports suitably placed in the sides of the boxes.

The practice varies slightly from one laboratory to another in Britain, but they all accept the idea of total containment, at all times, in contrast to some aspects of Los Alamos practice. Thus it is a cardinal principle of the health physicists of the United Kingdom Atomic Energy Authority that the maximum permitted air-tolerance levels recommended by the International Commission on Radiological Protection should never be exceeded, even during maintenance. As a result of this condition, it becomes impossible to open up an active glove box in the laboratory area

FIG. 1.—Frog-suited operator

without extensive decontamination of the interior of the box, and an important feature of all United Kingdom plutonium metallurgy laboratories is therefore an active maintenance area to which access is possible only in frog suits (Fig. 1).

At Harwell, the glove boxes are usually of the single-skinned free-standing type; one face can be clamped against an adaptor plate on the outside of the double wall of the frogman area, to form a sealed surface. The double wall acts as an air lock, allowing access to this face, which can thus be removed, leaving the box as an extension of the frog-suit area (Fig. 2). If work is necessary on the structure of the

FIG. 2.—A free-standing glove box connected to the frog-suit area

box itself, this system cannot be used, of course, and the box is then taken right into the frogman area through a series of air-lock doors.

During normal use, complete box integrity is achieved by the use of a heat-sealed polyvinyl chloride bag technique for transference, together with double-grooved ports to facilitate the changing of gloves and transfer bags. Attempts to emulate Los Alamos practice by bringing polished plutonium samples into a fume hood for microscopic examination have not proved successful; a rapid spread of activity over the environs of the specimen takes place. It is believed that this is associated with the much higher relative humidity of the air in the British laboratories, since a nonadherent oxide coat forms on the metal under moist conditions.

FIG. 3.—View of the metallurgical area, radiochemical building, Harwell

To leave the floor area as clear as possible, all services are brought to the boxes from overhead (Fig. 3). Several different forms of box are in use, the choice being governed by consideration of leak-tightness and vacuum requirements. Thus, for large vacuum systems, it is customary to use a two-compartment box (Fig. 4); the upper compartment is the working area and is normally filled with argon, while the lower compartment contains the vacuum system and is filled with nitrogen. The use of the latter gas is designed to reduce any fire hazard associated with electrical failures in the pumping system. For small vacuum systems, it is possible to include the pumps in the working compartment and also to pass the constituent parts in and out through the transfer ports as required, thus allowing easy maintenance of the system.

The question of leak-tightness arises in conjunction with the problem of argon purity. High-purity argon (approximately 5 ppm oxygen) is available from a recirculating plant working by liquefaction. It is essential that boxes connected to the plant should have a low leak rate—first, to maintain a high degree of purity in the box and, second, because there is an upper limit to the amount of oxygen in the returning argon which the plant can handle. The maximum allowable leak rate has been fixed at 0.5 per cent of box volume per hour at -1 inch static water gauge. This can be achieved quite easily with the type of box in which the perspex (lucite) faces are bolted onto the frame, but it is difficult with the cheaper mass-produced boxes in which the perspex face is held in place by merely a rubber sealing strip. For many metallurgical operations, high-purity argon is not necessary, and the latter type of box can then be used either with a supply of bottle argon or by sending plant argon into the box and then straight to extract (exhaust ducts) instead of recirculating it.

FIG. 4.—A typical free-standing glove box

A recent development has been the incorporation of a simple automatic regulating value on the argon inlet and extract lines (4). This is designed to maintain the box pressure at -1 inch static water gauge under all operating conditions and will rapidly correct any sudden change in box pressure, making it impossible to pressurize the box, which previously was a hazard with boxes run on continuously recirculating argon. It also rapidly corrects pressure changes due to glove movement, which is important when using small boxes.

EXPERIMENTAL METHODS AND RESULTS

THERMAL ANALYSIS (E. T. ADAMS, P. G. MARDON, J. M. NORTH, AND J. H. PEARCE)

Method.—The early experiments in this field by Adams were, of necessity, carried out on comparatively small specimens (1–10 gm), and, for this reason, a differential thermocouple method was chosen as being of high sensitivity. The method employed automatic recording and is still in use at the present time with only minor changes.

TABLE 1

AVERAGE PHASE-TRANSFORMATION TEMPERATURES FOR ZONE-MELTED PLUTONIUM
(Heating Rate 1° C/Minute)

Transformation	Heating (° C)	Cooling (° C)	Transformation	Heating (° C)	Cooling (° C)
$\alpha \rightleftharpoons \beta$	126 ± 2	Not determined	$\delta \rightleftharpoons \delta'$	456 ± 5	Not certain
$\beta \rightleftharpoons \gamma$	209 ± 3	*Ca.* 150	$\delta' \rightleftharpoons \epsilon$	484 ± 2	487 ± 2
$\gamma \rightleftharpoons \delta$	316 ± 2	*Ca.* 230	$\epsilon \rightleftharpoons$ liquid	641 ± 2	637 ± 2

The specimen, in the form of a short right cylinder, is drilled out to allow a sheath containing the specimen thermocouple to be placed inside it. The specimen is held in a thin-walled crucible which is mounted side by side with a similar crucible holding a block of tantalum, which acts as a standard. A second thermocouple is placed inside the standard block and connected to the first, but in opposition. The assembly is mounted in a small furnace, the temperature of which can be raised at an approximately linear rate; the resultant EMF produced by the differential couple is amplified by a factor of 10–100 times and fed to a recorder simultaneously with a record of the specimen temperature. To prevent the standard couple from interfering with the specimen couple, it is necessary to open the circuit of the latter when the former is plotting.

Results.—The early results obtained by Adams have been reported elsewhere (5). Since that work was done, progressively purer samples of plutonium have become available, and in recent months metal purified by zone melting has been prepared at A.E.R.E. (6). The total impurity content of this material is in the region of 250 ppm, equivalent to better than 99.8 atomic per cent plutonium, so that its purity is directly comparable to that of United States high-purity metal. The thermal analysis results listed in Table 1 were obtained with this material.

In addition, the effect of heating rate on the changes has been examined. As can be seen from Table 2, to within experimental accuracy, the $\delta \rightarrow \delta'$ and $\delta' \rightarrow \epsilon$ changes are unaffected, but the lower transitions are all sensitive to the heating rate.

TABLE 2

EFFECT OF HEATING RATE ON TRANSFORMATION TEMPERATURES

Transformation	Heating Rate (4° C/Minute)	Heating Rate (1° C/Minute)	Transformation	Heating Rate (4° C/Minute)	Heating Rate (1° C/Minute)
$\alpha \rightarrow \beta$......	142±5	126±2	$\delta \rightarrow \delta'$......	451±5	456±5
$\beta \rightarrow \gamma$......	238±5	209±3	$\delta' \rightarrow \epsilon$......	481±3	484±2
$\gamma \rightarrow \delta$......	325±5	316±2			

THERMAL EXPANSION (J. A. LEE)

Method.—The thermal-expansion characteristics of plutonium metal were first determined at Harwell by means of a simple silica tube and push-rod vacuum dilatometer in which the specimen dilatation was measured by means of a calibrated dial gauge sensitive to 10^{-5} inch.

Results.—Figure 5 shows the type of curve obtained with one of the first samples available, which contained about 1 per cent total impurities. It shows the presence of five allotropic modifications between room temperature and the melting point. The transformation temperatures were in good agreement with those already obtained by thermal analysis. With the exception of the $\delta \rightarrow \epsilon$ transformation, the

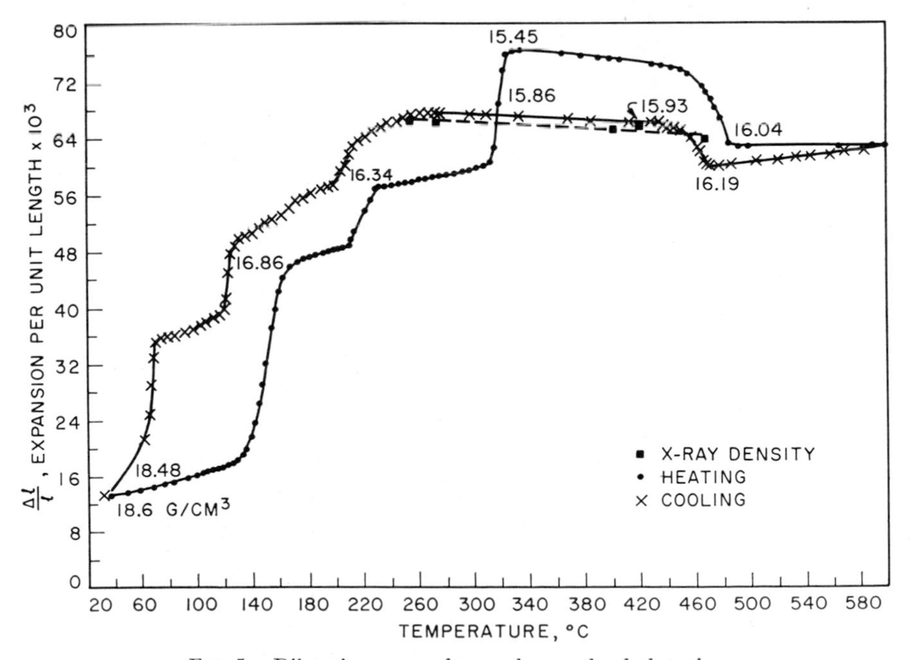

FIG. 5.—Dilatation curve of an early sample of plutonium

transitions on heating took place over temperature ranges that were relatively narrow. The $\delta \rightarrow \epsilon$ transformation of this sample appeared sluggish, however, the wide temperature spread of the latter transition giving the curve a form more characteristic of the two-phase region of an alloy than of an allotropic transformation in a one-component system. Almost immediately after this sample was examined, purer samples became available which gave traces indicating the presence of a sixth allotropic modification existing over a small range of temperature between δ- and ϵ-plutonium (Fig. 6).

Tables 3 and 4 give the mean transformation temperatures, thermal-expansion coefficients, and phase densities derived from a number of thermal cycles of the earlier samples (7).

For comparison, a dilatometer curve for a recent high-purity plutonium sample, which had an as-cast density of 19.63 gm/cm³, is shown in Figure 7, with the relevant data given in Table 5. The heating rate was approximately 1°/minute.

In the experimental arrangement used for this sample the specimen loading was

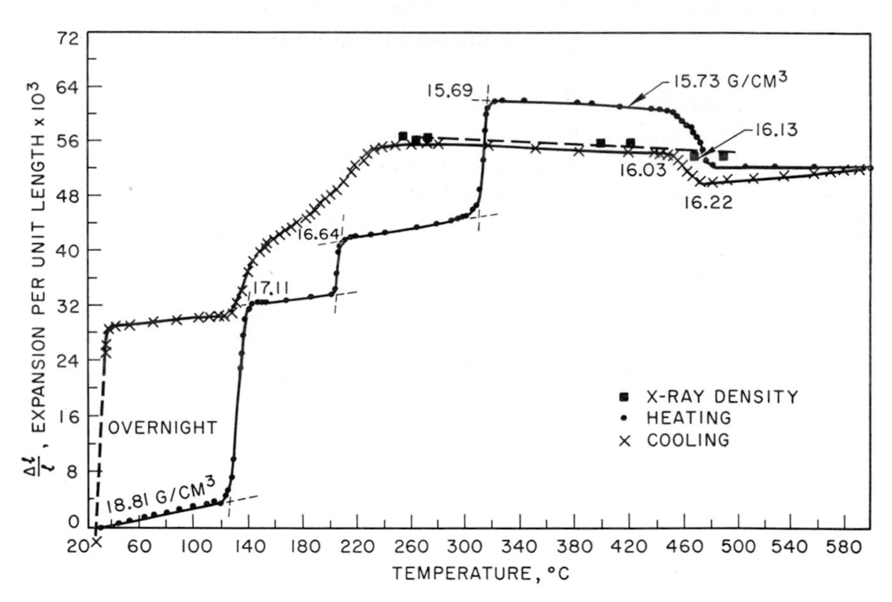

Fig. 6.—Dilatation curve of pure plutonium (sample purer than that of Fig. 5)

TABLE 3

DILATOMETRIC TRANSFORMATION TEMPERATURES

Transition	Temperature On Heating (° C)	Temperature On Cooling (° C)	Transition	Temperature On Heating (° C)	Temperature On Cooling (° C)
$\alpha \rightleftharpoons \beta$.......	128	70	$\gamma \rightleftharpoons \delta$.......	310	280
$\beta \rightleftharpoons \gamma$.......	206	140	$\delta \rightleftharpoons \epsilon$.......	472	468

greater than in the arrangement for the previous figures, and the magnitudes of the $\delta \rightarrow \delta'$ and $\delta' \rightarrow \epsilon$ changes have been exaggerated by creep at the transformation temperature. It will be seen that the transformation temperatures and expansion coefficients are in good agreement with those for earlier samples. A more detailed examination of the dilatometer traces reveals many interesting features.

1. The lower-temperature transformations show considerable temperature hys-

TABLE 4

THERMAL-EXPANSION COEFFICIENTS OF THE PLUTONIUM PHASES

Phase	Linear Coefficient of Thermal Expansion on Heating (×10⁶/° C)	Density (gm/cm³)	Phase	Linear Coefficient of Thermal Expansion on Heating (×10⁶/° C)	Density (gm/cm³)
α.........	+50 → +65	19.25	δ.........	−27	15.7
β.........	+43	17.3	ε.........	+20	16.3
γ.........	+39	16.7			

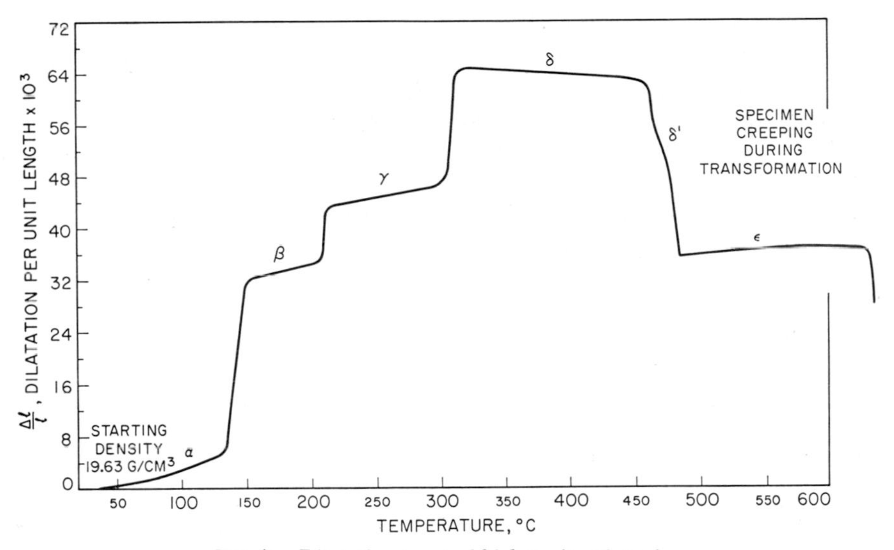

FIG. 7.—Dilatation curve of high-purity plutonium

TABLE 5

DILATOMETRIC BEHAVIOR OF HIGH-PURITY PLUTONIUM

Phase	Linear Coefficient of Thermal Expansion (×10⁶/° C)	Transformation Temperature on Heating (° C)	Phase	Linear Coefficient of Thermal Expansion (×10⁶/° C)	Transformation Temperature on Heating (° C)
α......	67 (80°–120° C)	133	δ'......	−596* (470° C)	476
β......	41 (160°–200° C)	208	ε......	15 (490°–550° C)	....
γ......	35 (220°–280° C)	306	Melting point.	....	634
δ.......	− 8.6 (340°–440° C)	460			

* Value probably too high because of creep deformation accompanying the negative expansion.

teresis. At a heating rate of 1° C/min, the $\alpha \longrightarrow \beta$ transformation takes place some 40°–50° C higher on heating than on cooling. A similar degree of hysteresis is present in the $\beta \longrightarrow \gamma$ transformation. The $\gamma \longrightarrow \delta$ transformation again shows hysteresis to a marked degree. In addition, the transformation on cooling takes place in marked steps, typical of the martensitic type of transformation. The $\delta \longrightarrow \delta' \longrightarrow \epsilon$ transitions of pure samples show little temperature hysteresis.

2. The changes in density at the transformations are very large. The $\alpha \longrightarrow \beta$ transformation is accompanied by a volume change of 10 per cent, and the total expansion from α to δ represents a volume change of 20 per cent, comparable with the volume change at the gray-tin/white-tin transformation. Thus the α-phase, which Zachariasen (8) has shown to be a complex monoclinic structure, has a density 20 per cent higher than the δ-phase, which has the "close-packed" face-centered cubic structure. The δ-phase itself transforms at higher temperatures to a more dense body-centered cubic structure.

3. Another feature of the δ-phase is the negative expansion coefficient—unique in an isotropic metallic structure. This is discussed later.

ELECTRICAL RESISTIVITY (P. G. MARDON AND J. A. L. ROBERTSON)

Method.—The electrical-resistance/temperature behavior was determined on specimens from several of the early plutonium billets. The specimens were in the form of wire up to 6 inches long and either 0.040 or 0.020 inch in diameter, prepared

TABLE 6

ELECTRICAL RESISTIVITY AND TRANSFORMATION TEMPERATURES OF PLUTONIUM

Phase	Resistivity (Microhm-cm)	Transformation Temperature (° C)		Phase	Resistivity (Microhm-cm)	Transformation Temperature (° C)	
		Heating	Cooling			Heating	Cooling
α	150	....	....	δ	108	325	225
β	117	135	90	ϵ	123	480	480
γ	115	220	160				

by the extension of small billets in the δ-phase at about 350° C (9). Separate current and potential connections were made by spring-loading the wire onto grooves in four copper blocks; the grooves prevented the wire from being deformed into a thin ribbon at the contact points when heated into the very soft δ-phase.

The resistance was determined either by simple comparison of the voltage drop across the wire with the voltage drop across a 1-ohm standard resistance or, more commonly, by maintaining a steady current through the wire, backing off most of the potential drop across the wire, amplifying the residual EMF, and feeding it onto a recorder, which also carried the temperature trace. This method allowed the specimen to be continuously heated or cooled and gave a complete cycle from room temperature to 500° C and back in a single day.

The runs were carried out in a vacuum furnace at a pressure of 10^{-5} mm of mercury, but even this pressure was insufficiently low to prevent some oxidation of the wire at high temperatures.

Results.—The average resistivities for the plutonium samples are given in Table 6, as are also the phase transformation temperatures to the nearest 5° C. The resistivity values apply to the low-temperature end of the respective temperature ranges and have been corrected for volume changes on the basis of isotropic expansions from the α-phase.

In general, heating runs were stopped at about 400°–450° C because of difficulty encountered with increase in oxidation rate above these temperatures, so that no information is available for the δ-prime phase. The value for ϵ was obtained from samples too impure to show δ-prime.

Table 7 gives the temperature coefficients of resistance of each phase. The wide variations in magnitude occurred not only from specimen to specimen but also from run to run on the same specimen. The values assume a linear dependence of

TABLE 7

TEMPERATURE COEFFICIENTS OF RESISTANCE ($\times 10^5/° C$)

Phase	Temperature Coefficient	Phase	Temperature Coefficient
α.	−22 → −40	γ.	− 4 → − 8
β.	− 5 → −11	δ.	+ 8 → +18

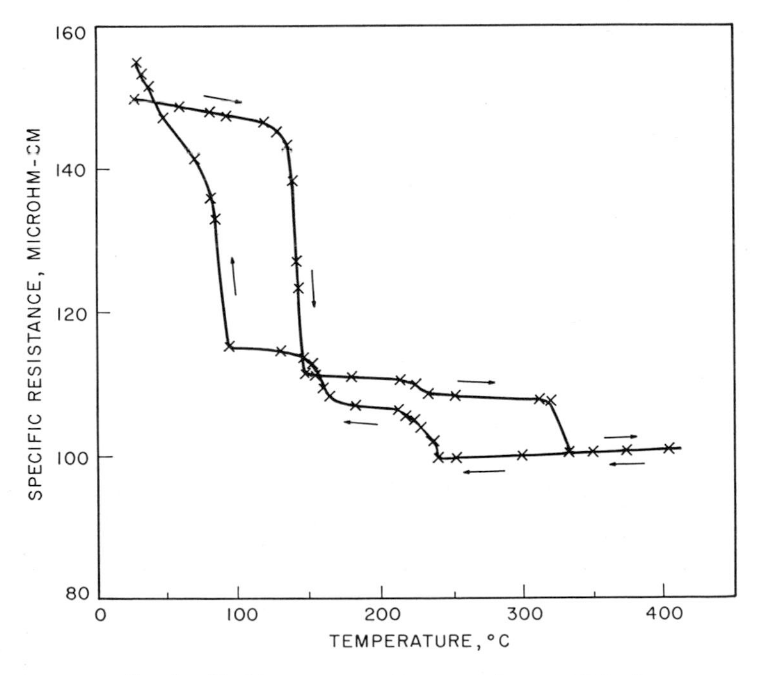

FIG. 8.—Resistance-temperature curve for pure plutonium

resistance on temperature in each phase, but a careful examination shows that this is not true in the case of α-plutonium, for which the temperature coefficient appears to increase in magnitude with temperature from about $-25 \times 10^{-5}/° C$ at room temperature to $-40 \times 10^{-5}/° C$ at around 100° C. For the other phases, the values appear to be much closer to linearity than for the α-phase. A typical run for one of the purer plutonium samples is illustrated in Figure 8.

X-Ray Examination (P. Greenfield, J. A. Lee, P. G. Mardon, and J. A. L. Robertson)

Method.—X-ray photographs of all the plutonium phases have been taken with a standard Unicam high-temperature camera to which minor modifications have been made. These include narrowing the window gap with brass inserts from $\frac{1}{2}$ to $\frac{1}{4}$ inch, to reduce the probability of the window shattering, and replacing (1) the original control thermocouple with two chromel-alumel couples to control the top and bottom furnaces separately; (2) the original specimen-measuring couple with a ring thermocouple connected to the camera body by very thin leads; and (3) the original magnetic drive for the specimen rotation with a direct drive. The ring thermocouple is regularly calibrated against silver, so that accurate specimen temperatures may be obtained. In addition, for room-temperature photographs, a standard Philips 11.4-cm camera and a specially built Guinier-type focusing camera (10) have been used.

Results.—In spite of the evidence of the thermal-analysis and dilatometric work for the existence of a possible sixth phase in plutonium, distinct X-ray patterns for only five phases were obtained from the early plutonium samples. Only the δ- and ε-phases gave simple patterns, being face-centered cubic and body-centered cubic, respectively. The unit-cell edges were found to be $a_\delta = 4.631 \pm 0.001$ A (420° C) and $a_\epsilon = 3.64 \pm 0.01$ A (530° C).

The δ- and ε-phases were found to coexist over the temperature range where a sixth phase was thought to exist. No detailed examination of the expansion coefficients was undertaken, although sufficient photographs of the δ-phase were taken to confirm the negative coefficient of thermal expansion shown by the dilatometry and to give an approximate value of $-10 \times 10^{5}/° C$ for this function.

The patterns of the α-, β-, and γ-phases all proved complex, and the present authors were unable to solve them. When Zachariasen and Ellinger published their solutions, first for the γ-phase (11) and later for the α-phase (8), our patterns were checked against these solutions. For the α-phase, using values obtained from a photograph taken in the focusing camera, agreement was very good. But for the γ-phase, the result was less satisfactory, certain lines on our pattern being unaccounted for. The quality of our photographs was not high, however, and it is possible that the extra lines were due to the presence of phases other than γ-plutonium.

Recently some high-purity plutonium has been produced by refinement of the normal reduction technique (12), and this has been investigated in the δ-prime field as reported by Jette (13). The results confirm the structure published by Ellinger (3) for this phase, but there are two unexplained lines on the photograph in addition to the coexisting lines of δ-, δ-prime, and ε, plus traces of plutonium oxides. As in Ellinger's work, no line of the δ-prime pattern was visible with indices higher than 222.

SPECIFIC HEAT (P. G. MARDON AND J. A. L. ROBERTSON)

Method.—Up to the present time, the only work on this property at A.E.R.E. has been an approximate determination of the specific heat of α-plutonium, although an elaborate method based on the self-heating property of the plutonium is in hand.

The approximate determination was carried out by quenching a small cylinder of plutonium from about 80° C into a calorimeter containing carbon tetrachloride at room temperature. The rise in temperature was given by a Beckmann thermometer, and corrections had to be made for evaporation of the liquid, the self-heating of the plutonium, and heat transferred to the bath by aluminium foil inclosing the specimen. In preliminary experiments with lead serving as a stand-in for plutonium, this foil inclosure had been shown to be necessary for a body having high emissivity, as otherwise an excessive amount of heat was lost during transfer of the specimen into the calorimeter.

Results.—The value obtained from a series of runs for the average specific heat over the range 20°–80° C was $C_p = 0.036 \pm 0.001_5$ cal/gm/° C (mean deviation). Allowing for possible error in determining the water equivalent of the calorimeter, this gives a value of $0.036 \pm 0.002_5$ cal/gm/°C. Laquer (chap. xvi) has published figures for the compressibility of plutonium, so that it is possible to correct C_p to C_v by the well-known thermodynamic relationship and hence obtain the gram atomic heat, 7.6 ± 0.6 cal.

THERMAL CONDUCTIVITY (J. P. EVANS AND P. G. MARDON)

Method.—An early attempt by Robertson (14) to determine thermal conductivity was based on Kannaluik's method for a wire specimen. This requires a knowledge of the temperature coefficient of resistance of the wire and the slope of the resistance-power curve for the wire under steady-state conditions. The initial value so obtained has to be multiplied by various correction factors, of which by far the largest in the case of plutonium is the correction for emissive loss of heat from the surface of the wire. Because this correction increases both with decrease in conductivity and with decrease in the diameter of the specimen, it was so large for the thin wire of very low-conductivity metal being used that the equation for conductivity was not solvable. However, the value obtained without attempting to apply this correction gives an upper limit to the thermal conductivity, and for α-plutonium this was found to be 0.01_3 cal/cm²/sec/unit temperature gradient.

It was realized that such an abnormally low conductivity for a metal would present numerous experimental difficulties, whatever method was chosen to measure it, especially in view of the difficulty of keeping an oxide-free surface on the plutonium. A guard-tube method was finally chosen, employing specimens 3 inches long and ¼ inch in diameter. During initial dummy runs on uranium it was made an absolute method, but, because of the difficulty of insuring that all the heat flowed through the bar, it was converted to a comparative method, using stainless steel as the standard material. This proved very satisfactory for uranium, but, on changing to a plutonium specimen, the method broke down because of the difficulty of getting the heat into and out of the plutonium and the fact that the self-heating proved to be a larger factor than anticipated. The present form of the apparatus uses a plutonium rod alone, the self-heating acting as the source of power. Under these conditions, the solution of the heat-transfer equation is a parabolic temperature gradient in the specimen. Knowing the distance of the measuring thermocouples from the top of the bar, it is possible to solve this equation by using only two thermocouples; but, as a check in practice, three couples are used, to avoid any need for an end correction. The plutonium sample has been increased in size to a bar 5 inches long and 1 inch in diameter, with the bottom end projecting through a copper heat sink into a recirculating coolant. This arrangement has proved neces-

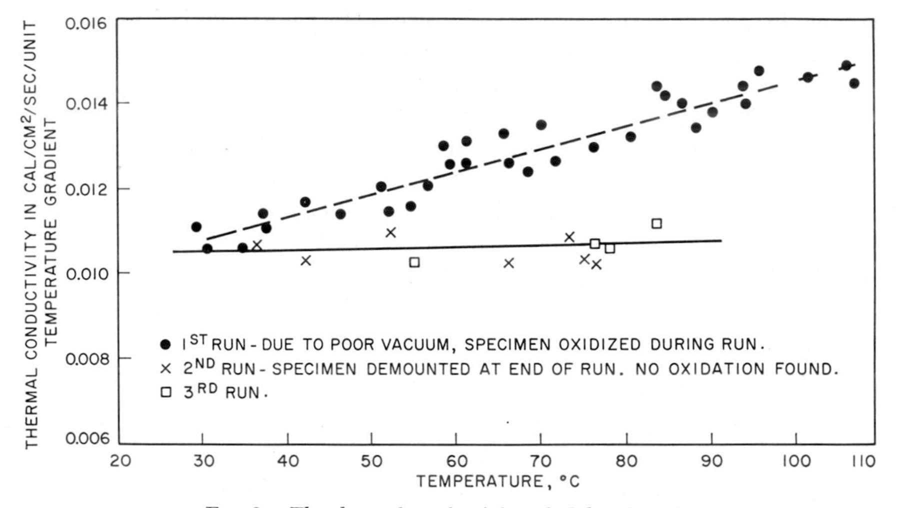

FIG. 9.—The thermal conductivity of alpha plutonium

sary to insure satisfactory removal of the heat from the plutonium, previous attempts to solder the plutonium to the copper with low-melting-point materials such as Wood's metal having been unsuccessful.

Only the α-phase has so far been investigated, and the reproducibility of the results is not so good as had been hoped for. This is believed, however, to be due in part to oxidation of the surface of the bar during the first runs, when the vacuum in the system was poor. An initial attempt to determine the conductivity of the β-phase failed because of severe bending of the specimen at the $\alpha \rightarrow \beta$ transformation because large temperature gradients were present in the bar at that time.

Results.—The most recent work has given a value of 0.010_5 cal/cm^2/sec/unit temperature gradient at 40° C, with almost no dependence on temperature (Fig. 9). An earlier run on the same specimen had given considerably higher values at temperature (Fig. 9), but the specimen had heavily oxidized, and this is believed to have caused unreliable thermocouple readings. The latest values proved to be reproducible after demounting the specimen to check for oxidation, reloading it, and repeating the experiment. This result suggests that these values are probably correct to within the experimental scatter of ± 5 per cent, although more work will be needed to confirm this. An attempt will be made shortly to obtain a check value at room temperature from a Lee disk apparatus, but considerable doubt is felt as to whether sufficiently good contact can be established between the plutonium and the copper plates.

KINETICS OF THE $\alpha \rightleftharpoons \beta$ TRANSFORMATION (J. GARSTONE)

Method.—Very little work has as yet been done at A.E.R.E. on the nature and rate of the various plutonium transformations, but preliminary work has been undertaken on the kinetics of the $\alpha \rightleftharpoons \beta$ transformation. The method employs a rod specimen mounted in a simple dilatometer. The dilatometer is mounted within a

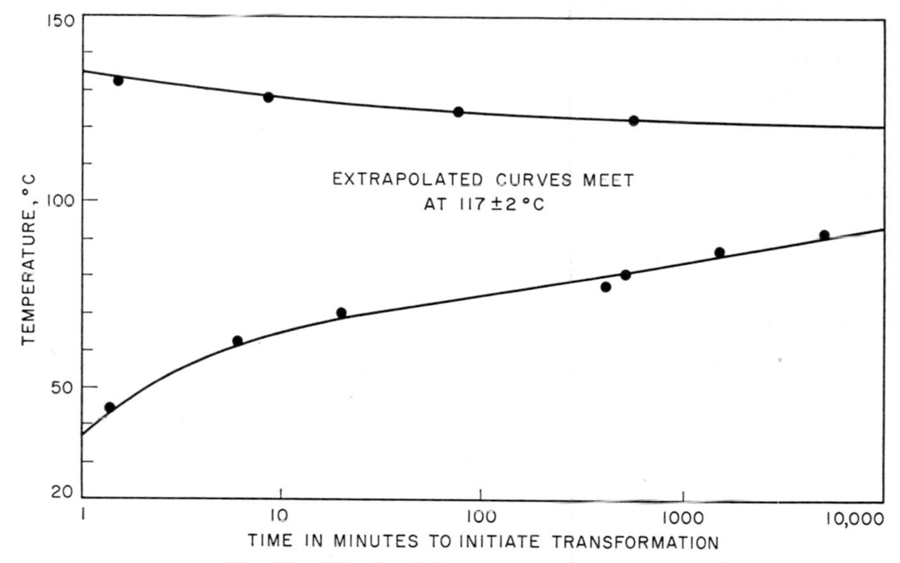

FIG. 10.—Kinetics of the $\alpha \rightleftharpoons \beta$ transformation

removable furnace. Moving the furnace upward, away from the dilatometer and specimen, simultaneously brings a thermostatically controlled oil bath into position around the specimen. Thus the specimen can, in effect, be rapidly transferred from furnace to oil bath. The furnace was normally run at 150° C for examining the $\beta \rightarrow \alpha$ change or, for investigating the $\alpha \rightarrow \beta$ change, simply switched off, so that the specimen was at room temperature initially. A rotation of the dial gauge equivalent to 10 per cent of the theoretical magnitude of the $\alpha \rightarrow \beta$ transition was arbitrarily chosen as indicating the start of the transformation.

Results.—The time interval between transfer to the bath and initiation of the transformation was determined for a series of bath temperatures both above and below the expected $\alpha \rightleftharpoons \beta$ transition temperature. The results are plotted on a logarithmic scale in Figure 10. They extrapolate to an ideal equilibrium transformation temperature of 117° $\pm$ 2° C, in good agreement with Jette's values (13) for high-purity plutonium.

THE MAGNETIC MOMENT OF THE Pu^{239} NUCLEUS (J. BUTTERWORTH)

It has been shown by electronic paramagnetic resonance studies (15) that the nucleus of Pu^{239} has a spin of $\frac{1}{2}$ and a nuclear magnetic moment in the range of 0.2–0.6 nuclear magneton. To determine the magnetic moment more accurately, a search was made for the nuclear magnetic resonance of Pu^{239}, with a nuclear magnetic induction spectrometer.

Because all the known salts of plutonium are paramagnetic, theory indicates that the resonance from such salts should be undetectable because of gross broadening. The possibility remains, however, of detecting the resonance in plutonium metal and, accordingly, a sample of finely divided plutonium metal weighing about 9 gm was used. The sample had a total impurity content of less than 1 atomic per cent and, in particular, contained less than 0.03 atomic per cent of any ferromagnetic metal.

The magnetic field applied to the sample had a strength of 4,200 gauss, and the applied radio-frequency field had an amplitude of 0.5–2.0 gauss, depending on frequency. In order to achieve the best signal/noise ratio, the band width of the spectrometer was reduced to $\frac{1}{64}$ cps, and the dispersion mode of the resonance was sought. The sensitivity of the equipment was such that the dispersion signal from H^2 in a 0.4 vol. per cent solution in H_2O was observed, with a band width of $\frac{1}{16}$ cps, to be twice the noise level.

The range examined so far is 0.27–0.58 nuclear magneton, and no resonance has been detected with the present sensitivity.

YOUNG'S MODULUS (J. A. LEE AND N. H. HANCOCK)

The adiabatic Young's modulus of pure plutonium was measured at room temperature by determining the resonance frequency of a bar in transverse oscillation in the "free-free" mode. For α-plutonium,

$$E = 8.60 \times 10^{11} \text{ dynes/cm}^2$$
$$= 12.5 \times 10^6 \text{ lb/in}^2 .$$

DISCUSSION

There are many unusual features among the physical properties of plutonium, but to the present authors the most striking ones appear to be the following: (1) the large number of allotropic transformations in the comparatively small temperature range between room temperature and the melting point; (2) the very considerable hysteresis associated with the transitions, even in high-purity material; (3) the negative coefficient of thermal expansion of the isotropic δ-phase of the metal; (4) the low density of δ-plutonium compared with the densities of the other phases; (5) the very low, or negative, temperature coefficients of resistance of all the phases; and (6) the unusually high atomic heat of the α-phase.

It is clear that there are six allotropic forms of the metal, but the present work has confirmed Ellinger's (3) observation that the δ-prime phase is very sensitive to impurities, being readily replaced by a mixture of δ and ϵ. As yet, no one has reported an X-ray pattern of δ-prime free from the lines of the δ- and ϵ-phases, presumably because of pickup of traces of impurity from the silica capillaries.

The considerable hysteresis accompanying the lower-temperature transitions appears to be as large in the recently prepared zone-melted plutonium as in the earlier less pure samples, and it had already been observed by earlier British workers (5) that the hysteresis decreased much more slowly than expected as the heating rate was lowered.

The δ-phase of the metal is particularly interesting, in that it has a negative coefficient of thermal expansion; this behavior is unique for a pure isotropic metal and has aroused considerable speculation as to the probable mechanism. The present authors have suggested that this is due to continuous redistribution of the outermost electrons between the 6d and 5f subshells. A decrease in the number of f electrons could lead to a decrease in the spatial extent of the 5f subshell (16) and also to a decrease in atomic volume due to increased bonding between atoms. Associated with this could be an increase in the number of electrons in the conduction band, which would imply a tendency for the electrical conductivity of the δ-phase to increase with temperature, thus counteracting the normal decrease in conductivity due to increased scatter of electrons by the larger thermal vibrations of the lattice. Some confirmation of this suggestion is provided by the fact that the temperature coefficient of resistance of the δ-phase is approximately zero at the low-temperature end of the δ-field.

More recently, Varley (17) has shown that a negative coefficient of volume expansion is theoretically possible for larger atoms if certain conditions relating to their electronic configurations are satisfied. The essential condition is that the density of states of the electrons rise very sharply with increase in energy in the neighborhood of the Fermi surface. This can occur when there are a large number of overlapping, partially filled electron bands, and such a system would have a high electronic specific heat. Varley also predicted that alloying additions that lower or raise the electron concentration in the δ-phase could change the expansion coefficient from negative to positive without a change in phase. Both dilatometry and X-ray parameter measurements have shown this to take place in certain alloy systems (18).

As mentioned in the section on dilatometry, the X-ray density of the close-packed δ-phase is 20 per cent lower than that of the monoclinic α-phase. This unusual behavior is further complicated by the fact that the apparent δ-phase density obtained by linear thermal-expansion measurements during the heating portion of dilatometer runs was well below the X-ray density in the case of early samples of plutonium but was in good agreement with the X-ray value on cooling. The recent high-purity plutonium, in which the starting α-density is much closer to theoretical, gives apparent densities on heating for the γ- and δ-phases that are in excess of theoretical, presumably due to creep, or plastic distortion, at the phase transitions. Creep is very obviously taking plase during the $\delta \longrightarrow \delta$-prime $\longrightarrow \epsilon$ transitions of this material, the apparent ϵ-density being greatly in excess of theoretical.

The low electrical conductivities of the plutonium phases are not so anomalous as might at first appear, since—as Mott and Jones (19) have pointed out—one should compare not the conductivity, σ, but, instead, the function $\sigma/M\theta^2$, where M is the atomic weight and θ is the Debye temperature. Under these conditions, elements of any one group or class of the periodic table give similar values. Among the actinides, values are available only for uranium (1.17×10^{-2}) and thorium (0.68×10^{-2}); these are not too dissimilar from the value of 0.17×10^{-2} for α-plutonium. A more unusual feature of the electrical-resistance properties of plutonium is the fact that the temperature coefficients of resistance of the α-, β-, and γ-phases are negative and small in magnitude, and even for the δ-phase, the temperature coefficient is either zero or of a small positive magnitude. The slope of the resistance-temperature curve for the α-phase suggests that it is an intrinsic semiconductor. The value of the thermal conductivity of α-plutonium is extremely low for a metallic element, even when compared with uranium; it is, nevertheless, of the right order of magnitude, assuming that the metal obeys the Wiedemann-Franz law.

The atomic heat of plutonium is considerably higher than would be expected from the simple Dulong and Petit law, and this is also true for uranium and thorium at similar temperatures. Recent measurements on the specific heat of neptunium (20, 21) have shown that this metal also has a very high atomic heat, suggesting that this property is common to the actinides. There are some signs of a similar, but less well-marked, trend in the case of the rare earths. Since Varley's theory of thermal expansion has indicated that the negative thermal expansion coefficient of δ-plutonium may be associated with a high electronic specific heat, it is possible that a large electronic contribution is responsible for the high atomic heat, but, in general, electronic specific heats are small in comparison with the lattice specific heats at the relatively low temperatures involved in the present measurements, if one judges from the behavior of the transition metals.

The present work points to the need for a great deal more information if any coherent explanation is to be obtained of the unusual properties shown by plutonium. The work in hand at A.E.R.E. is concentrated on some of the obvious gaps in our present knowledge—the behavior of plutonium at low temperatures, detailed examination of the specific-heat/temperature curve, and changes in the sign and magnitude of the temperature coefficients of expansion and resistance of the δ-phase on alloying.

The authors wish to thank Dr. M. B. Waldron for his helpful criticism of the work.

REFERENCES

1. HEDGER, H. J., HURST, R., and MYERS, R. H. Unpublished work.
2. LORD, W. B. H. *Nature,* **173**:534, 1954.
3. ELLINGER, F. H. *Transactions of the American Institute of Mining, Metallurgical and Petroleum Engineers,* **206**: 1256, 1956.
4. NORTH, J. M. *Nuclear Engineering,* **3**:293, 1958.
5. BALL, J. G., LEE, J. A., MARDON, P. G., and ROBERTSON, J. A. L. To be published in *Revue de Métallurgie.*
6. NORTH, J. M. Unpublished work.
7. LEE, J. A. Unpublished work.

8. ZACHARIASEN, W. H., and ELLINGER, FINLEY. *Journal of Chemical Physics*, **27**:811, 1957.
9. BIDDLE, W. F. Unpublished work.
10. WILLIAMSON, G. K., POOLE, D. M., and MARPLES, J. A. C. *Journal of the Institute of Metals*, **85**:431, 1956–57.
11. ZACHARIASEN, W. H., and ELLINGER, F. H. *Acta Crystallographica*, **8**:431, 1955.
12. MOWAT, J. A. S. Unpublished work.
13. JETTE, E. R. *Journal of Chemical Physics*, **23**:365, 1955.
14. ROBERTSON, J. A. L. Unpublished work.
15. BLEANEY, B., LLEWELLYN, P. M., PRYCE, M. H. L., and HALL, G. R. *Philosophical Magazine*, **45**:773, 991, 1954.
16. ZACHARIASEN, W. H. "The Crystal Chemistry of the 5f Elements," in *The Actinide Elements*, ed. G. T. SEABORG and J. J. KATZ, chap. 18, pp. 769–96. ("National Nuclear Energy Series," Div. IV, Vol. **14A.**) New York: McGraw-Hill Book Co., Inc., 1954.
17. VARLEY, J. H. O. *Proceedings of the Royal Society of London*, Ser. A, **237**:413, 1956.
18. LEE, J. A., and MARDON, P. G. Unpublished work.
19. MOTT, N. F., and JONES, H. *Theory of the Properties of Metals and Alloys*, p. 246. Oxford: Oxford University Press, 1936.
20. EVANS, J. P., MARDON, P. G., and WALDRON, M. B. Unpublished work.
21. ELDRED, V. W., and CURTIS, G. C. *Nature*, **179**:910, 1957.

CHAPTER XV

RESULTS OF MEASUREMENTS OF PHYSICAL PROPERTIES OF PLUTONIUM METAL

Thomas A. Sandenaw

INTRODUCTION

It has been a hope that, if sufficient measurements of the physical properties of plutonium were completed, greater insight into the anomalous behavior of this so-called metal would result. Measurements made with specimens of plutonium metal regarded as being of high purity have not yielded data that are completely reproducible from specimen to specimen in phases which occur at higher temperatures than the alpha phase. At low temperatures, the specific-heat values for plutonium have not proved to be reproducible in certain temperature regions. Although measurements of some of the physical properties of plutonium have been made throughout a considerable temperature range, the over-all picture of plutonium has not been clarified to our satisfaction.

Members of our group (CMF-13) at the Los Alamos Scientific Laboratory have investigated the electrical resistivity, thermal conductivity, magnetic susceptibility, and specific heat of plutonium and have also checked for superconductivity.

EXPERIMENTAL WORK

Negative results were obtained in a check for superconductivity in plutonium made by C. E. Olsen and B. T. Matthias (1). No superconductivity was found on cooling a specimen of plutonium to approximately 1.3° K. The material remained strongly paramagnetic at this temperature.

The property of plutonium studied over the widest temperature range was electrical resistivity (2). The temperature range covered was 750° C. The measurements were made only for the warming portion of a complete temperature cycle because of known hysteresis effects when plutonium is cooled through the different phases. The resistivity found at 273° K was 146.5 μohm-cm. Measurements of electrical resistance made above room temperature were attempted under self-heating conditions (2).

Averaged values of the electrical resistivity, R, of plutonium are plotted as $100R/R_{273}$ versus temperature in Figure 1. (R_{273} is the resistivity at 273° K.) For the portion of the resistivity curve occurring above room temperature, the almost

Thomas A. Sandenaw is at the University of California, Los Alamos Scientific Laboratory, Los Alamos, New Mexico.

temperature-independent nature of the resistivity in all phases is apparent. Very little change in resistivity occurs in the $\beta \longrightarrow \gamma$ and $\delta \longrightarrow \delta'$ phase transitions. Temperature arrests at these latter phase transitions were negligible, indicating small heats of transition. The increase in resistance in the vicinity of the $\delta' \longrightarrow \epsilon$ transition should be noted. The slope of the resistance-temperature curve for the delta-prime phase approaches that found at 50°–60° K.

The values of thermal conductivity (2) were determined over a much shorter temperature range than was investigated for electrical resistivity. Only a few values of thermal conductivity were determined for the beta phase. The lower limit of measurements was set by the poor thermal conductivity observed as 100° K was approached on cooling. A minimum at about 100° K appeared to be a possibility. The thermal conductivity at the lowest point on the curve approached the value 0.0035 cal/cm/sec/° K. Values at the $\alpha \longrightarrow \beta$ transition were about ten times greater.

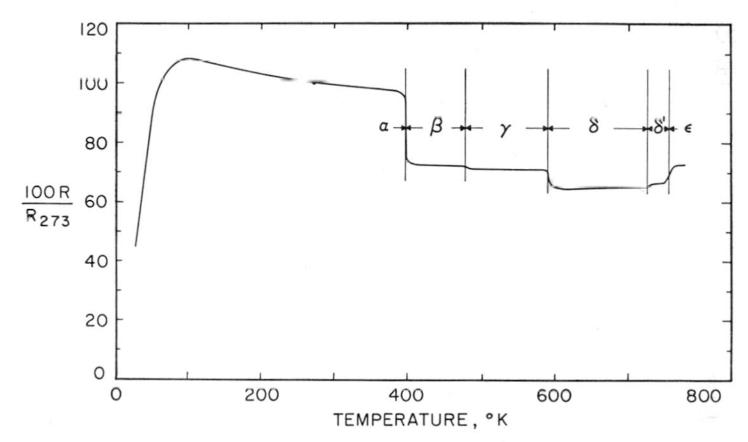

FIG. 1.—Electrical resistance, plotted as $100\ R/R_{273}$ versus absolute temperature

Averaged values, from determinations of thermal conductivity on three specimens of plutonium, are shown in Figure 2.

The magnetic susceptibility of solid plutonium, in phases occurring above room temperature, was studied by A. A. Comstock (3) in 1952, using the Gouy method. This method was chosen by Comstock because, as pointed out by Bates (4), if the specimen quantity is adequate, the Gouy method permits absolute measurements of a high degree of accuracy with ordinary laboratory equipment.

The values determined by Comstock for two specimens are shown in Figure 3. His specimens were selected for minimum iron, nickel, and cobalt content and had purities of 99.9 per cent or better. Since the susceptibility was found to be independent of field strength, it was concluded that the iron and nickel (135 ppm average for iron plus nickel) were not present in a ferromagnetic state.

One of Comstock's specimens was hot pressed to give maximum density, and the other was measured in the as-cast condition. The χ_g values shown in Figure 3 have been corrected for thermal expansion. Although there is an appreciable difference in the thermal-expansion behaviors of hot-pressed and cast plutonium, the scatter of values shown for the alpha phase is quite similar for both specimens. A solid curve has been drawn to show the trend of his data through the various phases. He allowed from 1 to 2 hours for completion of each phase transition before continuing measurements in the phase range at the next higher temperature. Completion of a run, up to the maximum temperature reached, required approximately 21 hours. Comstock's cylindrical specimens were quite large, having a length of 5.5

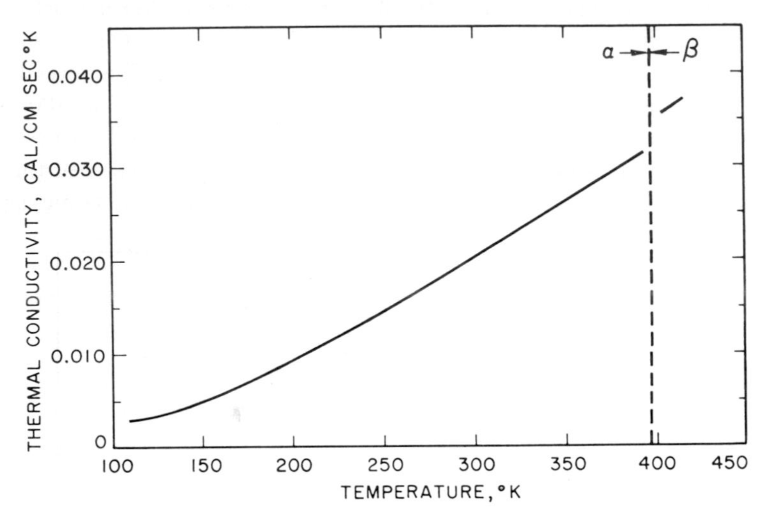

FIG. 2.—Thermal conductivity as a function of absolute temperature

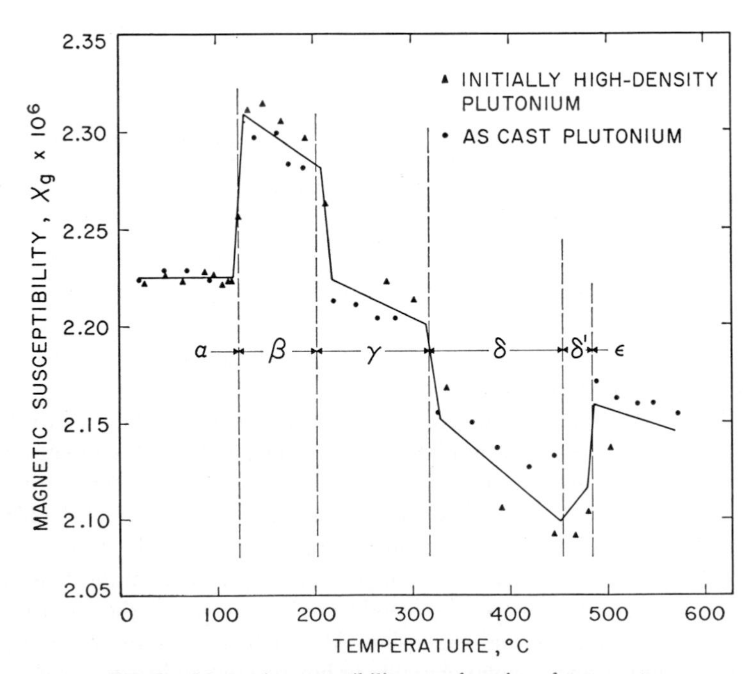

FIG. 3.—Magnetic susceptibility as a function of temperature

inches, a diameter of slightly less than 0.250 inch, and a weight of approximately 70 gm.

Magnetic susceptibility values for plutonium have been reported by J. K. Dawson (5) and S. T. Konobeevsky (6). Comstock's values are lower than those reported by these investigators. The temperature dependences of susceptibility in the beta phase and other higher-temperature phases are quite similar, however; i.e., the slopes found by the different workers are essentially the same.

The magnetic susceptibility curve of plutonium, taken as a whole, has a striking similarity to the same curve for manganese metal. The gram-atomic susceptibility values for the two metals at room temperature are almost identical: manganese, 533×10^{-6}; plutonium, 532×10^{-6}.

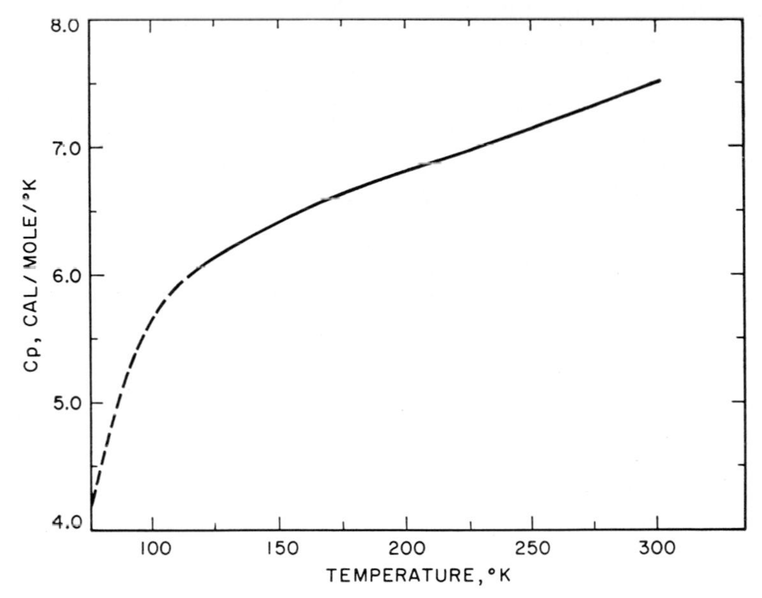

FIG. 4.—Heat capacity (C_p) as a function of absolute temperature

Low-temperature specific-heat measurements of plutonium have been found to give anomalous results. In the curve of heat capacity as a function of temperature, spikes and/or humps have been encountered in the region below 160° K. The specific-heat curve for a high-purity plutonium specimen is shown in Figure 4. This curve represents values for a rapidly cooled (quenched) specimen and was obtained by means of a least-squares analysis of values obtained in three runs. The dotted portion of the curve, below 110° K, is intended to imply that, in this region, results were found not to have good reproducibility.

Gibney and Sandenaw (2) found the latent heat of the $\alpha \rightarrow \beta$ transition to be 938 ± 10 cal/mole, with the transition apparently being of the first order. From the heat-capacity curve shown in Figure 4, calculations show that C_v becomes equal to $3R$ at 127° K, i.e., that 127° K is the Debye characteristic temperature of alpha plutonium. Calculations by Laquer based on sound-velocity measurements at room temperature (see chap. xvi) yield, however, a value of 178° K as the Debye characteristic temperature of the alpha phase.

DISCUSSION OF RESULTS

The hump in the electrical-resistivity curve of plutonium found at 100° K is very suggestive of a spin-disorder effect on resistivity, such as is found in the case of manganese and certain of the rare earths at low temperatures. The decrease in thermal conductivity as the temperature is lowered to 100° K may also have its origin in a spin-disorder phenomenon.

As a consequence of the noted anomalous specific-heat behavior in the vicinity of 100° K and the possibility of spin-disorder effects pointed out above, one might suspect the presence of an antiferromagnetic Néel or Curie temperature. Remanence and hysteresis appear in antiferromagnetism, and it has been suggested that traces of hysteresis might even survive above an antiferromagnetic Curie point. The low-temperature specific-heat behavior suggests hysteresis.

The negative thermal expansion, found in the delta and delta-prime phases of plutonium, has led Chandrasekhar (7) to speculate on the possibility of an ordering of atomic spins in these phases. Antiferromagnetism is inferred.

Only the delta and delta-prime phases of plutonium and the alpha phase below 100° K have positive temperature coefficients of electrical resistivity. As noted above, the slope of the resistance-temperature curve in the vicinity of the $\delta' \rightarrow \epsilon$ transition approaches that found at 50°–60° K. At higher temperatures, only the delta-prime phase shows an increase in magnetic susceptibility with increasing temperature. The electrical-resistivity and magnetic-susceptibility behaviors of these phases may thus be confirmatory evidence for Chandrasekhar's speculation. Moreover, the tetragonal crystal structure of the delta-prime phase ($c/a = 0.95$ for the f.c.t. unit cell; see chaps. x and xxiv) is a type that is frequently associated with antiferromagnetism.

REFERENCES

1. Olsen, C. E., and Matthias, B. T. Unpublished data.
2. Sandenaw, T. A., and Gibney, R. B. *Journal of Physical and Chemical Solids*, 6:81–88, 1958.
3. Comstock, A. A. (deceased). Unpublished data.
4. Bates, L. F. *Modern Magnetism*, p. 95. 2d ed. Cambridge: Cambridge University Press, 1948.
5. Dawson, J. K. *Journal of the Chemical Society*, Part III, pp. 3393–96, 1954.
6. Konobeevsky, S. T., "Phase Diagrams of Some Plutonium Systems," in *Session of the Division of Chemical Sciences (of the U.S.S.R. Academy of Sciences), Conference on the Peaceful Uses of Atomic Energy, Moscow, 1955*, pp. 362–75; also pp. 207–14 in English translation by the Consultants Bureau (for sale by Superintendent of Documents, Washington, D.C., price $1.00).
7. Chandrasekhar, B. S., *Acta Metallurgica*, 6:212–14, 1958.

CHAPTER XVI

SOUND-VELOCITY MEASUREMENTS ON ALPHA-PHASE PLUTONIUM

Henry L. Laquer

INTRODUCTION

Although the numerical results on sound velocities and elastic constants of plutonium have previously been reported (1), the following discussion will attempt to present in some detail an account of the extensive studies carried out at the Los Alamos Scientific Laboratory between 1948 and 1954. A major portion of the effort was devoted to the development of mechanical and electronic techniques required for precision sound-velocity determinations. Since other groups at LASL were simultaneously improving the purity and metallurgical quality of available plutonium specimens, our results were obtained on samples of variable perfection. In most cases the precision of individual measurements of a few parts in 10,000 exceeds the metallurgical reproducibility of the metal samples, and sometimes the observed variations can serve as a semiquantitative indication of the imperfections present in a specimen.

MATHEMATICAL RELATIONSHIPS

According to the theory of elasticity (2), there should be only two independent constants describing the elastic behavior of an isotropic material. A polycrystalline metal of sufficiently small grain size and randomly oriented grains should be essentially isotropic. However, there are a number of quite different physical quantities related to elasticity that can be observed experimentally. They are as follows:

E = Young's modulus = stress/strain (in elongation or compression),
μ = Shear modulus = stress/strain (in torsion or shear),
σ = Poisson's ratio = lateral contraction/longitudinal elongation,
λ = Lamé's constant,
v_0 = Velocity of compressional wave or disturbance in an infinitesimally thin rod,
v_L = Velocity of compressional wave in a laterally unbounded "infinite" medium,
v_T = Velocity of shear, transverse, or torsional wave in any medium,
χ_S = Adiabatic compressibility,
χ_T = Isothermal compressibility.

Henry L. Laquer is at the University of California, Los Alamos Scientific Laboratory, Los Alamos, New Mexico.

If, in addition, the density, ρ, the linear thermal expansion coefficient, α, and the specific heat at constant pressure, c_P, are known, the following relations can be established between the various quantities (2):

$$v_0 = \sqrt{\frac{E}{\rho}}, \tag{1}$$

$$v_L = \sqrt{\frac{(\lambda + 2\mu)}{\rho}} = v_0 \sqrt{\frac{(1-\sigma)}{(1+\sigma)(1-2\sigma)}}, \tag{2}$$

$$v_T = \sqrt{\frac{\mu}{\rho}}, \tag{3}$$

$$\sigma = \frac{E}{2\mu} - 1 = \frac{1}{2}\left(\frac{v_0}{v_T}\right)^2 - 1, \tag{4}$$

$$\chi_S = \frac{3(1-2\sigma)}{E} = \frac{9}{E} - \frac{3}{\mu} = \frac{3}{3\lambda + 2\mu}, \tag{5}$$

$$\chi_T = \chi_S \frac{9\alpha^2 T}{\rho c_P}. \tag{6}$$

Any experimental determination of more than two elastic quantities then provides a check on the internal consistency, on the absence of systematic errors, and on the homogeneity and isotropy of the sample.

METHODS OF MEASURING ELASTIC CONSTANTS

Of the commonly used methods for measuring elastic constants, static stress-strain tests are the best known. In the hands of careful experimenters (3) they can yield quite accurate values for Young's modulus and the torsion modulus. However, they often involve fairly large deformations and can at times be destructive. In the present work, dynamic methods were employed to measure the sound velocities in the metal at infinitesimal deformations. From these velocities the moduli were then calculated according to the relations given above. Two entirely different dynamic methods were employed, so as to reduce the likelihood of systematic errors.

The first method was the free-specimen resonance technique as developed by Bancroft and Jacobs (4) and previously used by us for measurements on uranium (5). Since it entails the setting-up of standing waves in usually cylindrical, rod-shaped specimens and the measurement of a series of harmonically related resonance frequencies, it has great inherent accuracy, particularly if frequencies are measured by modern counter techniques.

The second or pulse technique, which is a by-product of ultrasonic radar simulator development (6), involves the measurement of the transit time of a high-frequency ultrasonic pulse through a sample. It is the more direct and quicker method, if it works at all.

In the former method, sound velocities, v, are derived from the resonance frequencies, ν, through the well-known relationship

$$v = \nu\lambda, \tag{7}$$

where the wavelength, λ, is proportional to the length of the specimen, L, and inversely proportional to the harmonic number of the resonance, n; thus

$$\lambda = \frac{2L}{n}. \tag{8}$$

With any specimen of finite, rather than infinitesimal, diameter there is a dispersion or decrease in longitudinal sound velocity. This decrease is larger at the higher frequencies, for larger rod diameters, d, and for greater values of Poisson's ratio. The exact magnitude of this effect has been calculated and tabulated by Bancroft (7). A plot of this dispersion of the reduced longitudinal resonance frequencies, ν_n/n (or of the sound velocities calculated from various harmonics), against harmonic number or wavelength allows an independent estimate of Poisson's ratio, σ_{Disp}, often to ± 0.01. This then provides still another check on the self-consistency of the results. The scatter of the points from the smooth curve for reduced longitudinal harmonics and from constancy for reduced torsional harmonics serves as an indication of sample inhomogeneity or coarse-grainedness (see, e.g., Figs. 8 and 9).

Figure 1 is a schematic diagram of the apparatus used with the resonance tech-

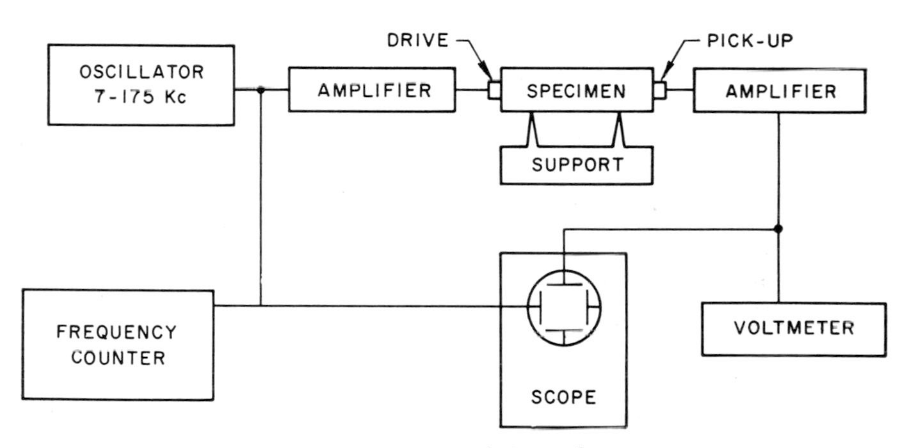

FIG. 1.—Resonance technique, schematic

nique. The specimen, which can have a length between 2 and 6 inches, is supported by four adjustable pins. The position of the pins is not critical, and it affects the resonance frequency of a $\frac{1}{4}$-inch diameter specimen by less than 1 part in 10,000. Two of the pins are made from thermocouple materials (chromel and alumel), which facilitates the measurement of the surface temperature of the specimen. The most critical parts of the equipment are the drive and pickup mechanisms. Originally, Rochelle salt crystals were attached to the specimens to produce both longitudinal and torsional vibrations (and unavoidably a large number of bending modes and satellite resonances). The electrostatic drive and pickup (4) eliminate all spurious resonances and allow observation of pure longitudinal modes, but considerable care is required in positioning the electrodes. Too close a spacing causes electrical breakdown in the gap and extraneous mechanical damping, thereby reducing the frequency of resonance by an amount which (with some difficulty) can be made as large as 2 parts in 1,000. Too wide a spacing reduces the signal intensity below usable levels. To produce torsional resonances without loading the specimen, the eddy-current method of Randall, Rose, and Zener (8) was used. With this method, the spacing is not at all critical, but, since large coils are needed to pick up a usable signal, the apparatus functions as a low-pass filter, and it is impossible to see resonances much above 40 or 50 kc. Figure 2 shows the relative signal strength observable with the various methods as a function of frequency.

In the earlier part of the work, nickel-coated plutonium specimens were measured in the open laboratory; but, since there is always some uncertainty as to the coating thickness and corresponding correction, the drive, support, and pickup components of the equipment were eventually installed in a glove box, and meas-

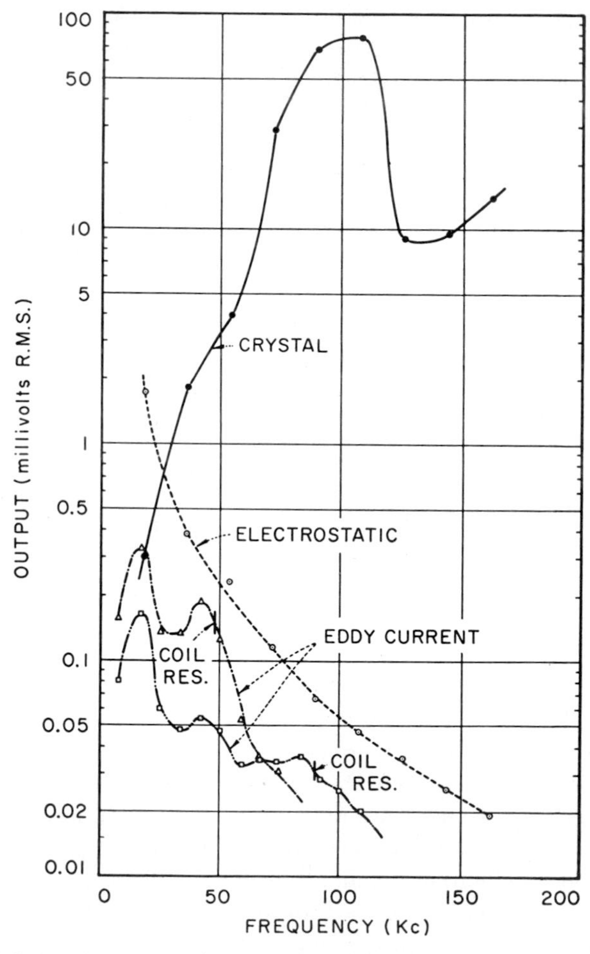

FIG. 2.—Relative signal strengths observable with different methods. —●— Rochelle salt crystal drive and pickup. • • • ○ • • • Electrostatic drive and pickup. -•-△-•- Eddy-current drive, series coil pickup. -• • -□-• • - Eddy-current drive, parallel coil pickup.

urements were made on bare specimens. The necessary coaxial leads were brought into the glove box through "pressurized" UG-30/U connectors, and the thermocouple leads were taken through Kovar seals. Having to use a glove box naturally made the setting-up and adjustment of the equipment more tedious and time-consuming.

Figure 3 presents a block diagram of the pulse-technique apparatus. The specimens should have a diameter of at least $\frac{3}{4}$ inch. Their length can have almost any value (depending on the magnitude of the attenuation coefficient), but we found 1–5 inches convenient. Our electronic techniques as developed in co-operation with C. W. Johnstone differ from published descriptions (6) of this method, primarily in that a step-wave is applied to transducer A, which is then allowed to ring at its natural frequency, and that the detailed structure of the signal rather than its

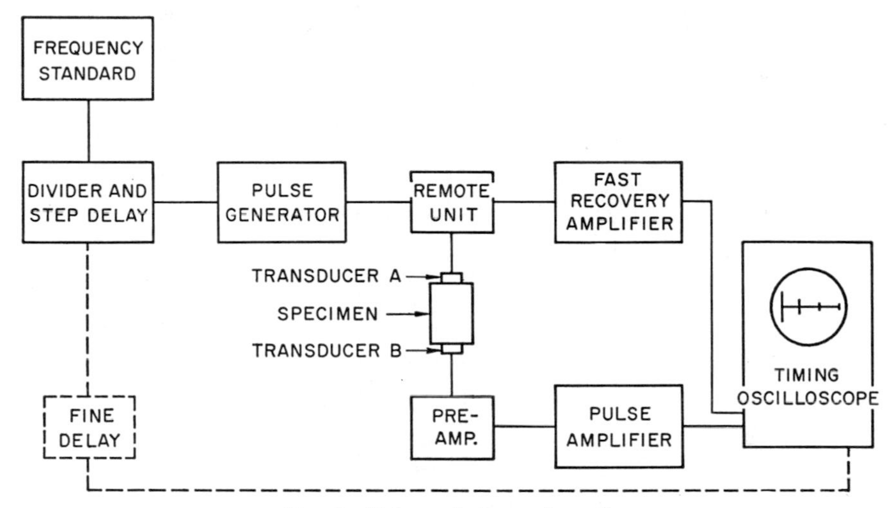

FIG. 3.—Pulse technique, schematic

envelope is displayed on the timing oscilloscope. This greatly increases precision, particularly in the measurement of the pressure coefficients of the sound velocity. Figures 4, 5, and 6 are schematic circuit diagrams, respectively, for the pulse generator, the remote unit, and the fast-recovery amplifier. The latter permits observations as close as 20 μsec after the relatively large (100-volt) initial pulse. Both reflection (transducer A only) and transmission measurements (transducers A and B) can be made, and frequency is changed simply by changing transducers. X-cut quartz crystals were used as transducers for longitudinal waves and Y- and AC-cut crystals as transducers for shear waves. Frequencies of 1, 2, and 5 Mc were employed, and the crystals were coupled to the specimens with mineral oil for compressional waves, with "Vistac" polyisobutylene for shear waves at atmospheric pressures, and with a thick aqueous sugar syrup for shear measurements at high pressures, since kerosene used as the hydraulic fluid dissolves Vistac. Measurements were made in the open laboratory, in the high-pressure apparatus with coated specimens, and in a glove box with uncoated specimens.

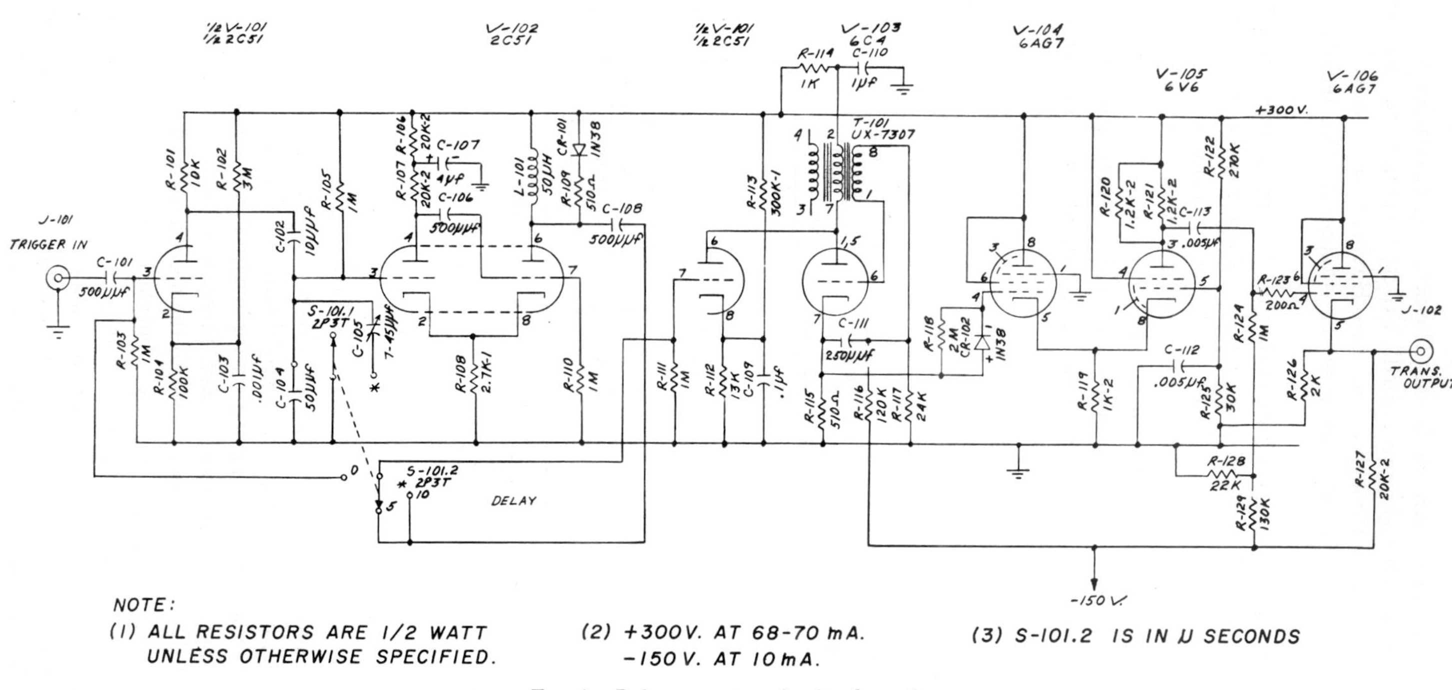

FIG. 4.—Pulse generator, circuit schematic

With the pulse technique the sound velocities are obtained from the time interval Δt required for the pulse to traverse the specimen of length L; thus

$$v = \frac{L}{\Delta t}. \tag{9}$$

The actual experimental results consist merely of the arrival time t_0, t_1, t_2, t_3, . . . , t_n for a series of pulses or echoes with integrally varying numbers of transits. When the same crystal serves as transmitter and receiver, only times with even subscripts are seen, corresponding to 2, 4, 6, etc., transits. When transmitter and receiver crystals are on opposite ends of the specimen, the times with odd subscripts are observed, corresponding to 1, 3, 5, etc., transits. One can also see t_0, the time of the

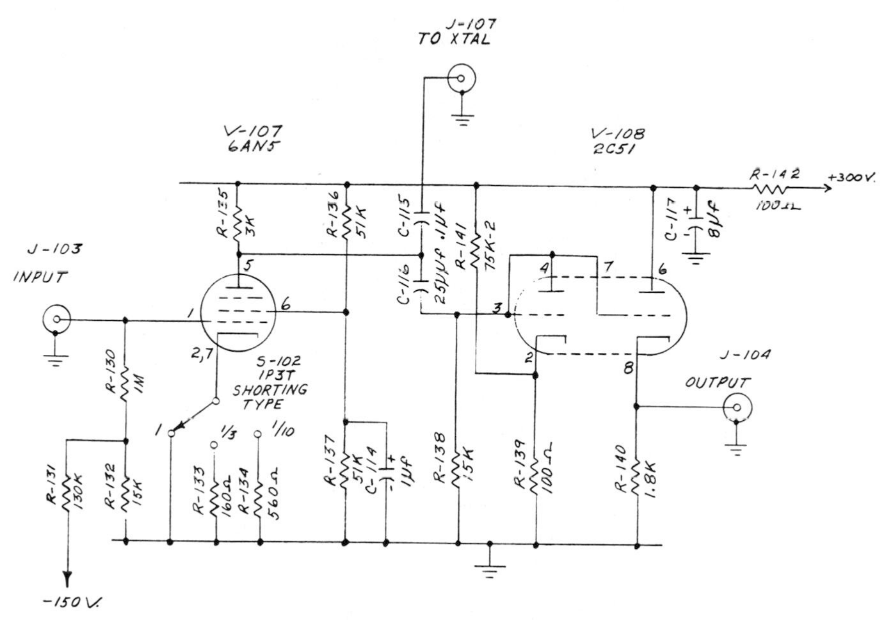

FIG. 5.—Remote unit, circuit schematic

initial pulse, which comes through by electrostatic pickup. Since the observed times are afflicted with probable experimental errors of about ± 0.05 μsec, some method is needed for determining an average Δt. In principle, this could be done by plotting the various t's against the number of transits and measuring the average slope of this line. However, we have found it more convenient first to subtract a series of *exactly* linearly increasing numbers, a_0, a_1, a_2, . . . , a_n, from the corresponding t_n's and then to plot the differences, $\delta_0 = t_0 - a_0$, $\delta_1 = t_1 - a_1$, . . . , $\delta_n = t_n - a_n$, against the number of transits. If the increment, Δa, of the a series is chosen close enough to the correct value of Δt, the δ values will not vary greatly. Hence the δ plot can be quite compact, and the best average Δt is obtained as the sum of Δa and the slope of the δ plot. This method also allows an easy visual averaging and weighting of the arrival times of various echoes and quickly shows up gross errors.

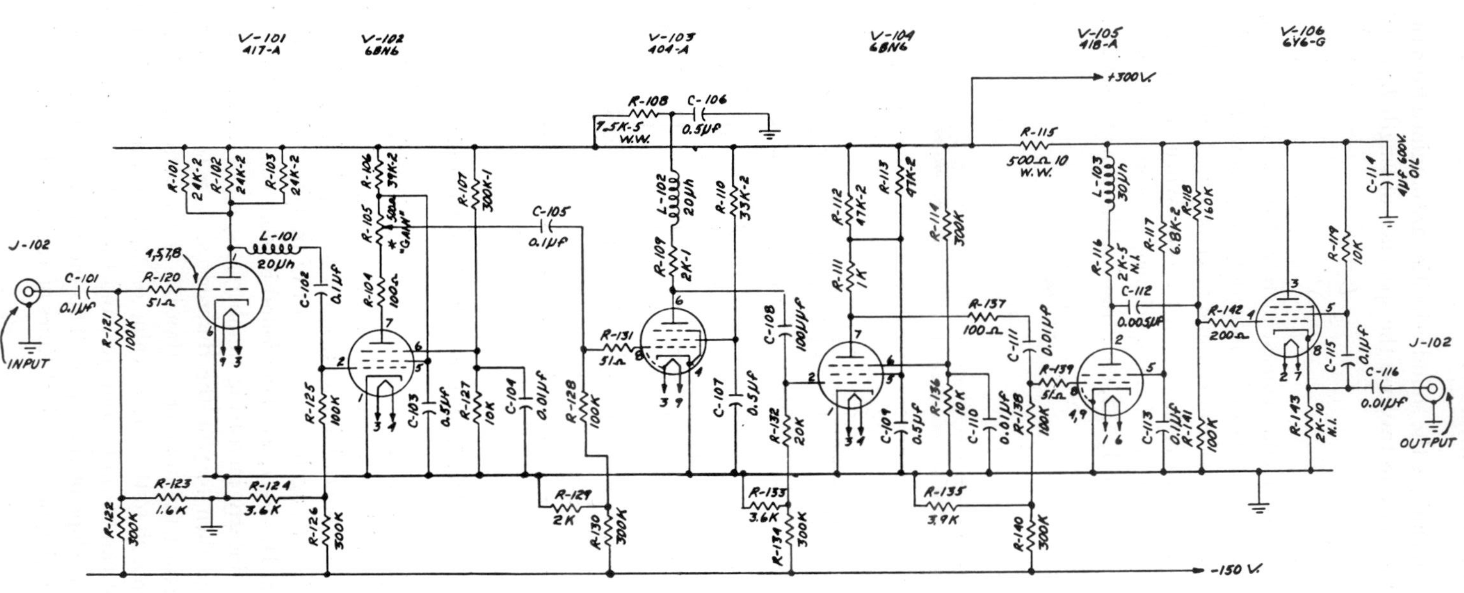

Fig. 6.—Fast-recovery amplifier, circuit schematic

Our pulse-technique results are usually based on the arrival times of 5 to 7 echoes under both transmission and reflection conditions, a total of 10 to 15 points. In all cases the interval t_1–t_0 was found to be larger than the average Δt. This extra delay in the first transit varies with pulse frequency and couplant. It can be explained as time spent by the pulse in traveling through part of each crystal and through the couplant film. Since subsequent reflections occur at the sample-film interface, the extra transit time, although added to all subsequent arrival times, cancels in the difference between the higher-order echoes. It is apparent that this method thus eliminates the need for the so-called end-corrections sometimes applied by other investigators (6).

DESCRIPTION OF SPECIMENS

Successful measurements were made on ten separate specimens. Four of these specimens were also remachined and measured again. Table 1 summarizes the fabrication procedure for all specimens. Table 2 lists their dimensions, their weights, and their densities as determined from the dimensions, ρ_{DIM}, and from immersion-weighing in bromobenzene, ρ_{IM}.

The problem of precision density determination on plutonium has been discussed in detail by the author (9). It appears that the two major corrections which should be made in precision work—namely, for air-buoyancy effects and for convection-current effects—are of about the same magnitude (4 parts in 10,000) but of opposite sign, so that they happen to cancel to within 1 part in 10,000. It should also be noted that, because of its large coefficient of expansion (10), plutonium decreases in density by 0.030 gm/cm^3/10° C and that a 10° temperature variation is not unlikely with some of the larger specimens. Finally, it should be pointed out that the errors reported for the dimensions in Table 2 include their variations from machining limitations, as well as the uncertainty of the measurement; thus in some cases the errors calculated for ρ_{DIM} are too pessimistic, and the agreement between ρ_{DIM} and ρ_{IM} is usually much better than one would expect on the basis of the quoted experimental uncertainties.

Table 3 summarizes the results of chemical and spectrochemical analyses on various samples taken from the specimens or, in some cases, from the melting stock or button used in fabricating a specimen. Extreme variations are evident in the carbon content of some of the samples. Specimen 135 shows the greatest variations. Sample 135X, taken from one end of the rod 135-S, gave generally high carbon values, whereas sample 135Y, taken from the other end of the rod, gave very low values. In fact, the end from which the high-carbon X-sample was taken appeared very hard and much more brittle than ordinary alpha plutonium. From the results given in Table 3, one can draw the conclusion that all early (up to 135) samples contained less than 0.1 w/o impurities, of which about 0.03 w/o was carbon. This, however, does not say anything about gaseous impurities such as oxygen and nitrogen, which might be of considerable importance. The recent very high-purity samples (1928 and above) probably contained less than 0.03 w/o of impurities, which would correspond to 0.2–0.3 a/o.

The two specimens which were radiographed, 135-S and E-302-S, showed some very fine radial cracks but no large-scale irregularities. This was somewhat surpris-

TABLE 1

SPECIMEN FABRICATION AND HISTORY

E-296	Vacuum cast from 910° C; extruded at 340° C and a pressure of 28,000 psi; machined; nickel-coated at 62 ± 3° C
E-296-S	Same specimen with coating stripped off
E-296-A	Stripped specimen machined down
E-296-B	Same specimen machined further to reduce diameter; measured heat output: $2.05_0 \pm 0.01 \times 10^{-3}$ watt/gm
E-302	Cast, extruded, machined in the same manner as E-296; density 19.39 gm/cm³, length 5.25 inches; diameter 0.492 inch; pressed at 335° C with end-on pressure of 75,000 psi (cooled to 55° C with pressure maintained) to: length 5.06 inches, diameter 0.500 inch; machined; imperfect coating attempts below 75° C
E-302-S	Same specimen with coating stripped off
E-302-A	Stripped specimen machined down; measured heat output: $2.03_7 \pm 0.01 \times 10^{-3}$ watt/gm
Z-13	Large machined cylindrical casting heated in double-acting die to 160° C, kept there for 30 minutes before applying pressure of 52,000 psi for 10 minutes; cooled under load to 38° C, $\rho = 19.61$ gm/cm³; refrigerated storage for 11 months, then coated at a temperature below 70° C
107	Irregular cylindrical casting with porous top surface ($\rho = 18.81$ gm/cm³); pressed at 350° C and 45,000 psi end-on, cooled under load of 25,000 psi; $\rho = 19.62 \pm 0.02$ gm/cm³; nickel-coated at below 70° C
135	Long, thin cylindrical casting (0.375 in. O.D.) with much surface porosity; machined to remove porosity, $\rho = 18.78$ gm/cm³; imperfect coating attempts
135-S	Same specimen with coating stripped off
135-A	Stripped specimen machined to reduce diameter, $\rho = 19.04 \pm 0.05$ gm/cm³
1928	High-purity stock, vacuum (0.04–1 μ) remelted at 850° C and cast into a CaO crucible, 0.25 I.D. by 4 inches long; furnace-cooled in 1 hour; machined, $\rho = 19.64$ gm/cm³
2020	High-purity button, vacuum (0.05–5 μ) cast in CaO crucible; maximum skull temperature 1,025° C; pressed at 325° C and 50,000 psi, cooled under load to 50° C in about 4 hours; machined lightly
2020-A	Same specimen remachined
2021	High-purity button, vacuum (0.03-0.5 μ) cast as No. 2020; maximum casting temperature 1,000° C; pressing and machining procedure identical with No. 2020
2021-A	Same specimen remachined
2202	High-purity button, vacuum (0.08–10 μ) cast into a CaO crucible, 0.281 inch I.D. by 4 inches long; maximum casting temperature 950° C; machined
3001	High-purity button plus high-purity stock, vacuum (0.08-0.4 μ) cast into CaO crucible of 0.375 inch I.D.; maximum casting temperature 1,250° ± 50° C; sudden gas evolution (100 μ) during pouring; sample kept molten for ½ hour, furnace cooled to room temperature in 4 hours; top surface filed square
3001-A	Same specimen remachined, analysis and micrography samples taken from both ends
3001-B	Same specimen further remachined

ing because the original casting, 135, had shown a considerable number of surface holes.

EXPERIMENTAL RESULTS

Sound Velocities and Elastic Constants

Table 4 summarizes the results of all the resonance-technique measurements with L-0 and T-0 being the extrapolated zero-order reduced frequencies which would apply to specimens of infinitesimal diameter. The corresponding values of the thin-rod and shear velocities calculated from the equations of the section on mathematical relationships are also given in this table.

Typical experimental results are shown in Figures 7, 8, and 9. Figure 7 presents reduced longitudinal and torsional resonance frequencies (ν_n/n) as a function of the harmonic number, n, for the extruded specimen E-296-A. The dispersion of the longitudinal frequencies is clearly apparent. There is also some systematic trend or variation in the reduced torsional resonance frequencies. The magnitude of this

TABLE 2

Bare Specimen Dimensions and Densities

Sample	Length (Inches)	Diameter (Inches)	Weight *in Vacuo* (Gm)	ρ_{DIM}	ρ_{IM}
E-296 coated....	5.532 ± 0.002	0.560 ± 0.002	431.498	$19.3_3 \pm 0.1_4$	19.43 ± 0.05
E-296-S........	$5.512 \pm .002$	$0.550_5 \pm .001$	418.40	$19.46 \pm .07$	$19.47_7 \pm .04$
E-296-A........	$5.038 \pm .001$	$0.496_8 \pm .000_2$	311.25	$19.45 \pm .02$	$19.47_6 \pm .04$
E-296-B........	$5.001_2 \pm .001$	$0.248_0 \pm .000_3$	76.975	$19.44 \pm .05$	$19.44_6 \pm .05$ $19.42_6 \pm .02$
E-302..........	....	....	....	....	$19.50 \pm .05$
E-302-S........	$4.997 \pm .001$	$0.477 \pm .001$	284.92	$19.47 \pm .08$	$19.53_3 \pm .05$
E 302 A........	$4.962_8 \pm .001$	$0.250_2 \pm .000_3$	78.091	$19.53 \pm .05$	$19.47_4 \pm .05$ $19.52_4 \pm .02$
Z-13 coated.....	$4.853 \pm .005$	$1.172 \pm .003$	1681.58	$19.60 \pm .05$	$19.61 \pm .04$
107 coated......	$1.023_5 \pm .001$	$0.745 \pm .001$	142.26	$19.46 \pm .06$	$19.60 \pm .05$ $19.64 \pm .05$
135-S..........	$6.63 \pm .01$	$0.370 \pm .001$	218.84_8	$18.7_3 \pm .1_1$	$18.78_3 \pm .05$
135-A..........	$5.011_6 \pm .001$	$0.368 \pm .002$	166.21	$19.0_3 \pm .2_1$	$19.03_7 \pm .05$
1928...........	$3.511 \pm .001$	$0.216_2 \pm .001_5$	41.544	$19.6_7 \pm .3$	$19.64_2 \pm .05$
2020...........	$0.9720 \pm .0002$	$0.7600 \pm .0002$	142.412	$19.70_9 \pm .01$	$19.71_2 \pm .02$ $19.73_9 \pm .04$
2020-A.........	$0.9331 \pm .0001$	$0.7487 \pm .0003$	132.5758	$19.69_4 \pm .01$	$19.72_6 \pm .04$ $19.737 \pm .005$
2021-A.........	$0.9385 \pm .0003$	$0.7475 \pm .0007$	132.6321	$19.65_2 \pm .02$	$19.69_2 \pm .04$ $19.717 \pm .005$
2202...........	$3.501 \pm .001$	$0.2502 \pm .0008$	55.55	$19.69 \pm .06$	$19.63 \pm .05$
3001...........	$4.49 \pm .01$	$0.375 \pm .005$	161.62	$19.8_9 \pm .27$	$19.52 \pm .05$
3001-A.........	$4.036_7 \pm .001$	$0.3478 \pm .0004$	123.33	$19.62 \pm .02$	19.67 ± 0.05
3001-B.........	4.030 ± 0.01	0.250 ± 0.0005	63.44	19.57 ± 0.06	....

TABLE 3

CHEMICAL AND SPECTROCHEMICAL ANALYSES (ppm)

SAMPLE	CHEMICAL					SPECTROCHEMICAL										
	C	O	Fe	F	S	Na	Mg	Al	Si	Ca	Cr	Mn	Ni	Cu	La	Pb
E-296X	255–60	...	...	...	...	10	8	20	17	5	100	200	200	10	125	5
E-296U	215	60	35	4	< 5	< 5	< 3	25	20	< 3	< 5	20	35	8	50	10
E-302X	250–370	...	...	...	...	8	20	8	100	8	...	...	...	...	80	...
E-302U	230	60–70	115	<1	< 5	< 5	< 3	30	25	< 3	< 5	30	50	10	200	10
Z-13	215–25	...	170	...	...	<10	5	15	50	< 5	10	...	100	10	40	6
107	300–630	...	...	...	...	6	10	6	35	16	...	...	...	...	60	...
135	100–820	...	...	...	...	4	4	10	120	2	...	...	...	...	<10	...
135X	60–575	...	...	...	...	< 5	4	< 2	17	5	10	150	50	5	90	6
135Y	65–95	...	...	...	...	10	< 5	20	15	< 5	100	...	100	5	<20	6
135U	70–85	30–40	170	3	20	< 5	< 3	40	30	5	< 5	35	20	10	<10	30
135V	65–70	40–130	6460	1.5	...	< 5	< 3	50	25	10	100	100	75	3	10	10
Stock	20	...	10– 40	...	65	< 5	< 3	< 5	10	< 3	25	15	3	15	<10	3
1928-X	45–55	60–190	130–215	...	...	<10	< 5	< 5	3	< 5	5	...	3	5	<10	< 1
Button	30–60	...	80	...	<10	< 5	< 3	< 5	20	20	25	30	2	8	<10	1
2020-X	25–25	720–1040	30	4	7	< 5	< 3	< 5	8	350*	20	30	10	10	<10	3
2020-Y	45–50	400–60	150	...	...	...	...	5	5	30	< 5	35	< 1	10	...	...
Button	15–30	...	70– 80	...	<10	< 5	< 3	< 5	12	15	< 5	30	< 1	6	<10	< 1
2021-X	55–55	270–410	60	6	< 5	< 5	< 3	5	5	50	< 5	25	1	3	<10	< 1
2021-Y	75–85	520–620	160	...	...	...	...	10	5	3000*	< 5	30	< 1	5	...	...
Button	20	...	50	...	<10	< 5	< 3	< 5	...	40	...	...	...	...	10	...
2202-X	320–60	...	70	...	...	<10	< 5	< 5	3	150–800*	...	...	< 1	1	<10	3
2202-Y	80–90	...	35	...	...	<10	< 5	< 5	< 1	5	...	...	< 1	3	<10	3
3001-U	25–40	130	60	<1	<10	<10	< 5	20	5	< 5	5	20	60	3	<10	3
3001-V	205–90	440	55	<1	<10	<10	< 5	20	5	< 5	5	20	100	3	<10	7
3001-W	40	50–100	50	...	...	<10	<10	< 5	< 1	< 5	< 5	25	15	2	<10	< 1
3001-X	<10	90–200	60	...	...	<10	<10	< 5	< 1	< 5	< 5	25	15	2	<10	< 1

* Erratic high Ca values probably caused by occlusion of CaO crucible fragments in the sample.

TABLE 4

Resonance-Technique Results

Sample	L-0 (Kc/Sec)	v_0 (Cm/Sec × 10^{-5})	Temperature (° C)	T-0 (Kc/Sec)	v_T (Cm/Sec × 10^{-5})	Temperature (° C)
E-296	$7.96_3 \pm 0.01$	$2.237_8 \pm 0.002_9$	33 ± 3	$5.19^{+0.10}_{-.01}$	$1.458_5{}^{+0.028}_{-.002_9}$	33 ± 3
E-296-S	$7.993 \pm .005$	$2.238_1 \pm .001_6$	30.7	...	...	...
E-296-A	$8.740 \pm .005$	$2.236_8 \pm .001_3$	27.6	$5.73_5 \pm .01$	$1.467_8 \pm .002_6$	25.3
E-296-B	$8.775 \pm .008$	$2.229_4 \pm .002_1$	26.6 ± 0.3	...	...	...
	$8.785 \pm .008$	$2.231_9 \pm .002_1$	25.1 ± 0.2	...	...	...
E-302-S	$8.73_5 \pm .01_5$	$2.217_4 \pm .003_8$	28.1	$5.772 \pm .002$	$1.465_2 \pm .000_6$	26.7
	$8.74_0 \pm .01$	$2.218_6 \pm .002_6$	28.8	$5.755 \pm .002$	$1.460_9 \pm .000_6$	30.9
E-302-A	$8.81 \pm .01_3$	$2.221_1 \pm .003_3$	26.0 ± 0.3	...	...	...
Z-13	$8.97_9 \pm .01$	$2.213_6 \pm .003_4$	50 ± 5	$5.93_0{}^{+.03}_{-.005}$	$1.461_9{}^{+.007_6}_{-.001_9}$	51 ± 5
135-S	...	...	...	$4.08^{+.10}_{-.01}$	$1.374_2{}^{+.034}_{-.004_0}$	26.9
135-A	$8.53 \pm .02$	$2.171_6 \pm .005_1$	26.8	$5.53_8{}^{+.03}_{-.01}$	$1.409_9{}^{+.007_6}_{-.002_5}$	27.3
1928	$12.75 \pm .03$	$2.274_1 \pm .005_4$	26 ± 1	$8.27 \pm .08$	$1.475_0 \pm .014$	26 ± 1
2202	$12.77_5 \pm .02$	$2.272_0 \pm .003_6$	26 ± 1	$8.39 \pm .04$	$1.492_2 \pm .007_2$	26 ± 1
3001	...	...	...	$6.43^{+.05}_{-.02}$	$1.466_6{}^{+.012}_{-.006}$	28 ± 2
3001-A	$10.95 \pm .02$	$2.245_4 \pm .004$	27.2	$7.16^{+.03}_{-.01}$	$1.468_2{}^{+.006}_{-.002}$	29.2
3001-B	$10.96_5 \pm 0.01$	$2.244_8 \pm 0.005$	28.5	$7.20^{+.03}_{-0.01}$	$1.474_0{}^{+.007}_{-0.004}$	26.5

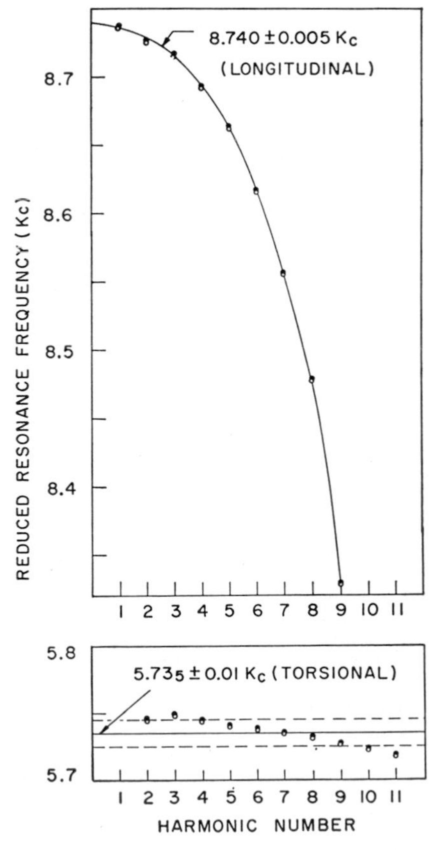

Fig. 7.—Variation in reduced resonance frequencies with harmonic number, specimen E-296-A. ● Run 1; ○ Run 2.

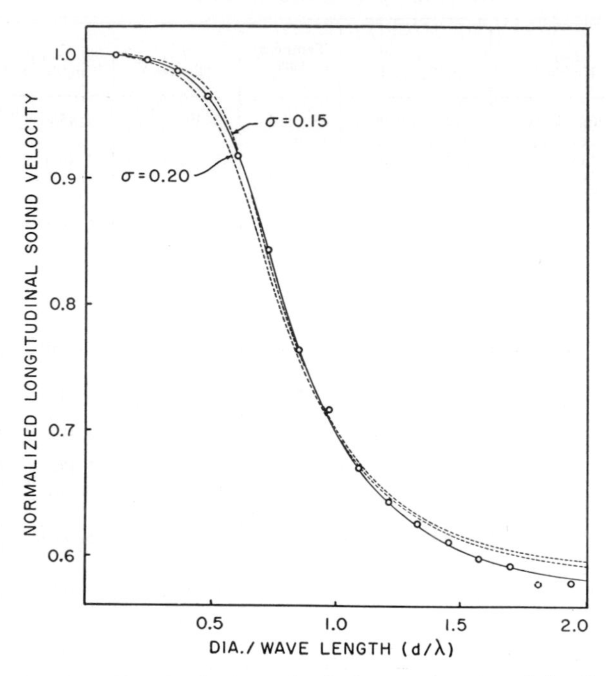

FIG. 8.—Dispersion of longitudinal sound velocity as a function of the diameter/wavelength parameter, specimen Z-13.

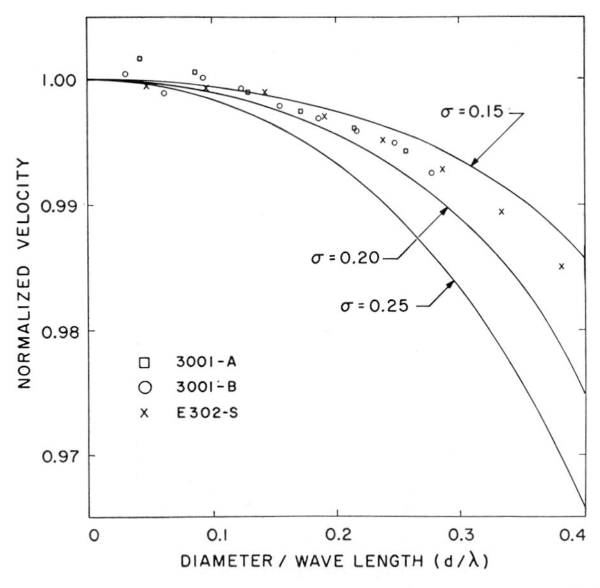

FIG. 9.—Dispersion of longitudinal sound velocity. × Specimen E-302-S. □ Specimen 3001-A. ○ Specimen 3001-B.

variation is less than 0.2 per cent for this specimen but reaches a value of as much as 2 per cent in the worst specimens. The range of extrapolated zero-order frequencies (*L*-0 and *T*-0) given in Table 4 includes both the scatter of the results and the range of this apparently systematic variation for the torsional resonances, in which case the average value does not lie at the center of a symmetrical Gaussian error distribution. For this reason the plus and minus probable errors in *T*-0 (Table 4) are not equal, differing by factors of 2–10.

Figure 8 for the large pressed and coated specimen Z-13 is similar, except that, in order to facilitate comparison with the theoretical dispersion curves of Bancroft (7), the reduced frequencies have been normalized to the *L*-0 value of Table 4 and the harmonic numbers have been replaced by the ratio of the specimen diameter to the wavelength. The normalized reduced frequencies are, of course, identical in

TABLE 5

PULSE-TECHNIQUE RESULTS

Sample	Crystal			$2\Delta t$ (μsec)	v (cm/sec $\times 10^{-5}$)	Temperature (° C)
	Freq.	Cut	Diam.			
Z-13	$2\frac{1}{4}$ Mc	X	$\frac{1}{2}$ sq.	108.24±0.06	$2.277_6 \pm 0.002_7$	40±5
	5 Mc	X	$\frac{1}{2}$ sq.	108.06± .06	$2.281_4 \pm .002_7$	40±5
	5 Mc	X	$\frac{3}{4}$	107.60± .05	$2.291_2 \pm .002_6$	38±2
107	$2\frac{1}{4}$ Mc	AC	$\frac{3}{4}$	35.4 ± .4	$1.46_9 \pm .01_7$	30±5
	5 Mc	X	$\frac{1}{2}$	22.71± .05*	$2.289_5 \pm .005_5$	30±5
	5 Mc	Y	$\frac{1}{2}$	35.19± .04*	$1.477_5 \pm .002_2$	30±5
2020	5 Mc	X	$\frac{1}{2}$	21.44± .04	$2.303_1 \pm .004_3$	28±1
	5 Mc	Y	$\frac{1}{2}$	33.26± .03	$1.484_6 \pm .001_4$	28±1
2020 A	5 Mc	X	$\frac{1}{2}$	20.56± .02	$2.305_5 \pm .002_3$	26±1
	5 Mc	Y	$\frac{1}{2}$	31.93± .03	$1.484_5 \pm .001_4$	26±1
2021 A	5 Mc	X	$\frac{1}{2}$	20.68± .04	$2.305_4 \pm .004_6$	26±1
	5 Mc	Y	$\frac{1}{2}$	32.24±0.04	$1.478_8 \pm 0.001_9$	26±1

* High-pressure measurements extrapolated to zero pressure.

value with the normalized sound velocities (eqs. [7] and [8]). The agreement between measured and calculated velocity dispersion, assuming a Poisson ratio of 0.17, is excellent for the first 10 per cent of velocity drop and remains quite good for most of the range. The 2.5 per cent discrepancy at the extreme right (for a 40 per cent velocity drop) might possibly be caused by the coating or by slight density gradients.

Figure 9 is another similar plot for several much thinner specimens and consequently covers a much more limited dispersion of only 2 per cent and, of course, also a correspondingly more limited range of the diameter/wavelength parameter. The agreement with the theoretical dispersion curves is again apparent, as is the scatter of individual resonance frequencies, particularly for the lower-order harmonics. The magnitude of the scatter is, however, no more than ±0.3 per cent.

Results of the pulse-technique measurements are to be found in Table 5, where

TABLE 6

SOUND VELOCITIES AND ELASTIC CONSTANTS AT 30° C

Sample	v_0 (Cm/Sec × 10^{-5})	v_L (Cm/Sec × 10^{-5})	v_T (Cm/Sec × 10^{-5})	ρ (Gm/Cm³)	E (Dynes/Cm² × 10^{-11})	μ (Dynes/Cm² × 10^{-11})	χS (Cm²/Dyne × 10^{12})	σ_{Disp}	σ
E-296	$2.241_6 \pm 0.004_8$	$(2.33_2{}^{+0.01}_{-.04})$	$1.461_3{}^{+0.028}_{-.004_1}$	19.45 ± 0.05	9.77 ± 0.05	$4.15_3{}^{+0.16}_{-.02_5}$	$1.99{}^{+0.28}_{-.06}$	0.20 ± 0.02	$0.177{}^{+0.005}_{-.046}$
E-296-S	$2.239_0 \pm .001_6$	...	...	$19.45 \pm .05$	$9.75_0 \pm .02_9$	...	...	$.17 \pm .02$	...
E-296-A	$2.233_8 \pm .001_4$	$(2.311 \pm .005)$	$1.463_3 \pm .002_6$	$19.45 \pm .05$	$9.70_5 \pm .02_8$	$4.16_5 \pm .01_8$	$2.07_0 \pm .02_9$	$.23 \pm .02$	$.165 \pm .005$
E-296-B	$2.225_2 \pm .002_1$	...	...	$19.43 \pm .03$	$9.62_1 \pm .02_5$	...	...	$.17 \pm .03$	...
	$2.225_9 \pm .002_1$	...	...	$19.43 \pm .03$	$9.62_7 \pm .02_5$	...	...	...	...
E-302-S	$2.215_0 \pm .003_8$	$(2.274 \pm .008)$	$1.4620 \pm .0007$	$19.50 \pm .05$	$9.57 \pm .04$	$4.16_8 \pm .01_1$	$2.21_0 \pm .03_4$	$.17 \pm .01$	$.148 \pm .004$
	$2.217_1 \pm .002_6$	$(2.278 \pm .005)$	$1.4618 \pm .0006$	$19.50 \pm .05$	$9.58_5 \pm .03_3$	$4.16_7 \pm .01_1$	$2.19_0 \pm .02_3$	$.17 \pm .01$	$.150 \pm .003$
E-302-A	$2.216_1 \pm .003_3$	...	...	$19.53 \pm .03$	$9.59_1 \pm .03_3$	...	...	$.20 \pm .05$	...
Z-13	$2.238_4 \pm .007_3$	$(2.29_2 \pm .01_8)$	$1.481_9{}^{+.009_2}_{-.005_4}$	$19.64 \pm .05$	$9.84 \pm .07$	$4.31_2{}^{+.06}_{-.03_4}$	$2.19{}^{+.11}_{-.08}$	$.17 \pm .01$	$.141{}^{+.005}_{-.012}$
	...	$2.288_5 \pm .006$	...	$19.64 \pm .05$	...	...	...	...	...
	...	$2.292_3 \pm .006$	...	$19.64 \pm .05$	...	...	...	...	...
		$2.300_0 \pm .003_5$							
107	...	...	$1.46_9 \pm .01_8$	$19.62 \pm .05$	...	$4.2_3 \pm .1_0$	...	...	...
	$2.234 \pm .005$	$2.289_5 \pm .007_8$	$1.477_5 \pm .005_3$	$19.62 \pm .05$	$9.79 \pm .05$	$4.28_3 \pm .03_3$	$2.18_6 \pm .03_7$	...	$.143_2 \pm .001_5$
135-S	...	...	$1.371_4{}^{+.034}_{-.004_0}$	$18.75 \pm .05$	...	$3.52_6{}^{+.17}_{-.02_2}$	...	...	...
135-A	$2.167_7 \pm .005_1$	$(2.26_6{}^{+.01_4}_{-.02_0})$	$1.407_4{}^{+.007_6}_{-.002_6}$	$19.01 \pm .05$	$8.93 \pm .05$	$3.76_6{}^{+.04}_{-.01_7}$	$2.11{}^{+.10}_{-.06}$	$.20 \pm .02$	$.186{}^{+.007}_{-.014}$
1928	$2.265_6 \pm .005_8$	$(2.37_2 \pm .03_1)$	$1.46_9 \pm .01_4$	$19.63 \pm .05$	$10.08 \pm .06$	$4.24 \pm .08$	$1.8_5 \pm .1_4$	?	$.18_9 \pm .02_4$
2020	$(2.242 \pm .003)$	$2.299_0 \pm .004_8$	$1.481_7 \pm .002_1$	$19.72 \pm .01$	$9.91_2 \pm .02_4$	$4.32_9 \pm .01_0$	$2.15_0 \pm .02_0$	...	$.144_8 \pm .001_3$
2020-A	$(2.239 \pm .002)$	$2.297_4 \pm .003_1$	$1.478_6 \pm .002_1$	$19.722 \pm .005$	$9.88_6 \pm .01_7$	$4.31_2 \pm .01_0$	$2.14_6 \pm .01_2$	...	$.146_4 \pm .0008$
2021-A	$(2.235 \pm .003)$	$2.297_3 \pm .004_9$	$1.472_9 \pm .002_1$	$19.702 \pm .007$	$9.83_9 \pm .02_6$	$4.27_4 \pm .01_2$	$2.12_8 \pm .02_0$	...	$.151_0 \pm .001_4$
2202	$2.263_5 \pm .004_1$	$(2.33_6 \pm .01_4)$	$1.486_3 \pm .007_4$	$19.63 \pm .05$	$10.06 \pm .04$	$4.34 \pm .04$	$2.03 \pm .07$	?	$.16_0 \pm .01_2$
3001	...	...	$1.463_7{}^{+.01_2}_{-.006}$	$19.55 \pm .05$	...	$4.19{}^{+.07}_{-.04}$		...	...
3001-A	$2.239_6 \pm .004_2$	$(2.31_7{}^{+.01}_{-.02})$	$1.466_9{}^{+.006}_{-.002}$	$19.64 \pm .05$	$9.85 \pm .04$	$4.22_6{}^{+.04}_{-.01_6}$	$2.04{}^{+.13}_{-.06}$	$.18 \pm .02$	$.166{}^{+.005}_{-.011}$
3001-B	$2.241_7 \pm 0.006_0$	$(2.31_8 \pm 0.02)$	$1.468_9{}^{+.007}_{-0.004}$	19.60 ± 0.07	9.85 ± 0.06	$4.22_9{}^{+.04}_{-0.03}$	2.04 ± 0.13	0.18 ± 0.01	$0.165{}^{+.004}_{-0.010}$

we have listed the specimen, the crystal used, the time for two transits (one echo), the corresponding sound velocity, and the temperature of the specimen. The 5-Mc data for sample 107 are extrapolations to zero pressure from the later pressure-coefficient study (p. 178).

Before one can compare the data collected in Tables 4 and 5, one has to bring them to the same temperature. This has been done by using the temperature coefficients given in Table 8. The resultant 30° C values are collected in Table 6. This procedure may be somewhat questionable, especially for the large-diameter specimen Z-13, since it is assumed that its surface temperature is equal to its average temperature. However, a more detailed analysis of the thermal-gradient problem would not appear justifiable at this point, in view of the large uncertainties in the temperature coefficients themselves. Velocities in parentheses were not measured directly but were calculated from the two other velocities on the same line, assuming isotropy. The probable errors for these calculated velocities are, of course, larger than those for the direct measurements.

Table 6 also includes, for most specimens, the best value of the density, ρ, the elastic moduli, E and μ, and the adiabatic compressibility, χ_S, as well as values for Poisson's ratio, σ_{Disp}, obtained from the velocity-dispersion curves, and σ, calculated directly from the velocities by equation (4). It will be noted that, in general, the values of Poisson's ratio obtained by the two methods overlap within their reported uncertainties, the only conspicuous exception being the numbers listed for E-296-A. The most probable value for σ_{Disp} is, however, usually about 0.02 unit larger than the corresponding value for σ.

The velocity measurements and calculations for the various specimens obtained by the various techniques are then compared in Figures 10, 11, and 12, where the velocities are plotted as a function of the density of each sample. Solid symbols represent direct observations, whereas hollow symbols correspond to the numbers in parentheses in Table 6. It is immediately apparent that there is a clear-cut increase in all velocities with density and that the "non-density" spread or scatter of the data covers a band of about 2 per cent for the thin-rod velocity, 1 per cent for the infinite-medium velocity, and 1.5 per cent for the shear velocity. It also appears that only the cast high-purity samples 1928 and 2202 deviate far enough from the mean to be suspect of directionality. This is not surprising, however, in view of the small diameter of the molds used in casting these specimens (Table 1). Unfortunately, for these two samples the scatter of the reduced frequencies exceeded the expected small dispersion, so that it was impossible to determine σ_{Disp}.

There is no evidence of lack of isotropy in any of the shear velocities, in spite of the previously discussed trend in the reduced torsional-resonance frequencies. Thus it is extremely gratifying to note that, in spite of the obvious structural defects and anisotropies evident in the raw data, the density variation alone is responsible for the major variations in the sound velocities and elastic moduli.

The systematic variation of velocity with density also permits extrapolation to the ideal X-ray density of 19.79 gm/cm^3 at 30° C, obtainable from the work of Zachariasen and Ellinger (11). This extrapolation can be made in an empirical manner without saying whether the low densities seen with most specimens are caused by retained β-phase, microcracks, or even larger voids. The solid lines in Figures 10, 11, and 12 have been drawn with a slope such as to make each velocity directly and linearly proportional to the density, and their intercepts with the ideal density lines yield an internally consistent set of values. Table 7 summarizes

TABLE 7

SUMMARY OF RESULTS AND EXTRAPOLATION TO X-RAY DENSITY AT 30° C

Quantity	Symbol	Range of Observed Values	Extrapolation to X-Ray Density	Units
Density	ρ	(18.8) 19.0 –19.72	19.79_0	gm/cm$_3$
Rod velocity	v_0	2.17– 2.26	$2.25_5 \pm 0.01_5$	$\times 10^5$ cm/sec
Inf. med. velocity	v_L	2.27– 2.30	$2.31_3 \pm .01$	$\times 10^5$ cm/sec
Shear velocity	v_T	(1.37) 1.40– 1.49	$1.49_0 \pm .007$	$\times 10^5$ cm/sec
Poisson's ratio	σ	0.14– 0.20	$0.14_5 \pm .01$	....
Young's modulus	E	8.9 –10.1	$10.0_2 \pm .1_3$	$\times 10^{11}$ dynes/cm^2
		12.9 –14.6	$14.5_4 \pm .2$	$\times 10^6$ psi
Shear modulus	μ	(3.5) 3.8 – 4.3	$4.37_6 \pm .04_1$	$\times 10^{11}$ dynes/cm^2
		(5.1) 5.5 – 6.2	$6.34_6 \pm .06$	$\times 10^6$ psi
Adiabatic compressibility	χ_S	1.8 – 2.2	$2.1_2 \pm .1_4$	$\times 10^{-12}$ cm^2/dyne
		1.8 – 2.2	$2.1_5 \pm .1_4$	$\times 10^{-6}$ (atm)$^{-1}$
Isothermal compressibility	χ_T	2.1 – 2.5	$2.4_1 \pm 0.2$	$\times 10^{-12}$ cm^2/dyne

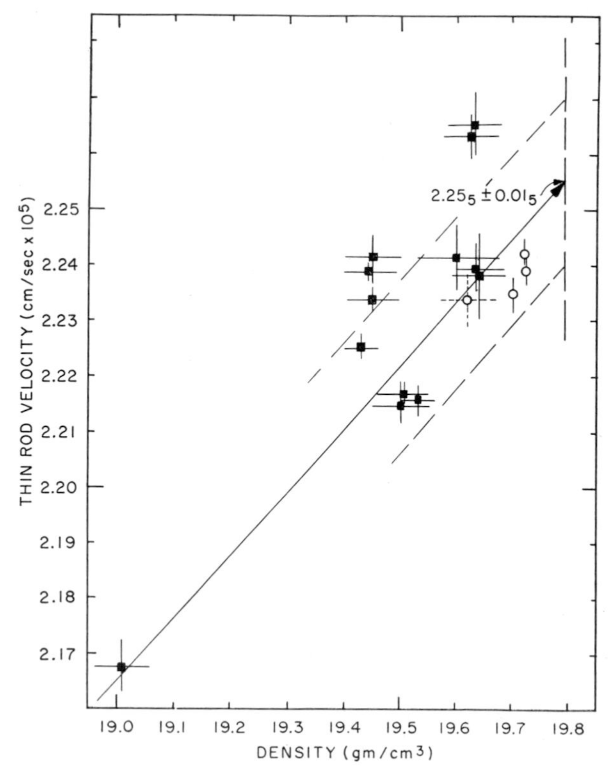

FIG. 10.—Variation of longitudinal thin-rod velocity with density. ■ By resonance technique, measured. ○ By pulse technique, derived.

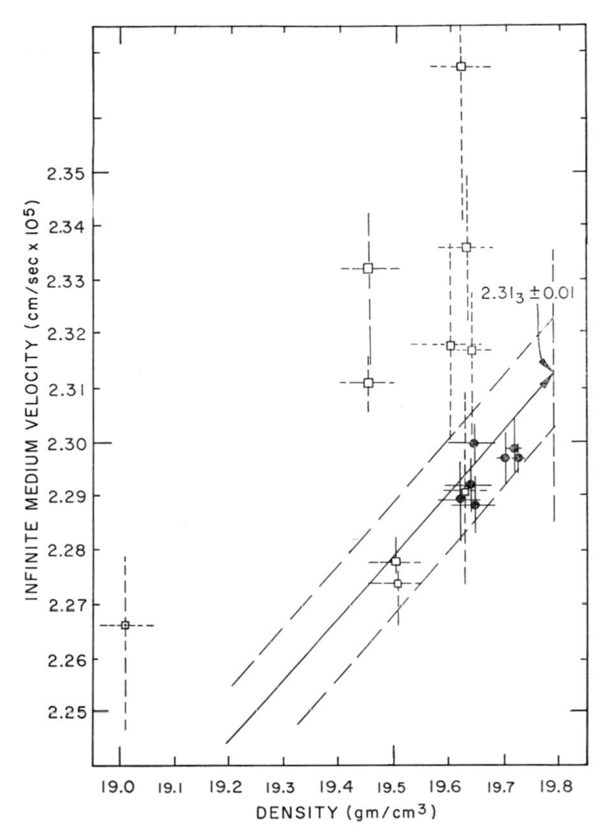

FIG. 11.—Variation of infinite medium compressional velocity with density. ● By pulse technique, measured. □ By resonance technique, derived.

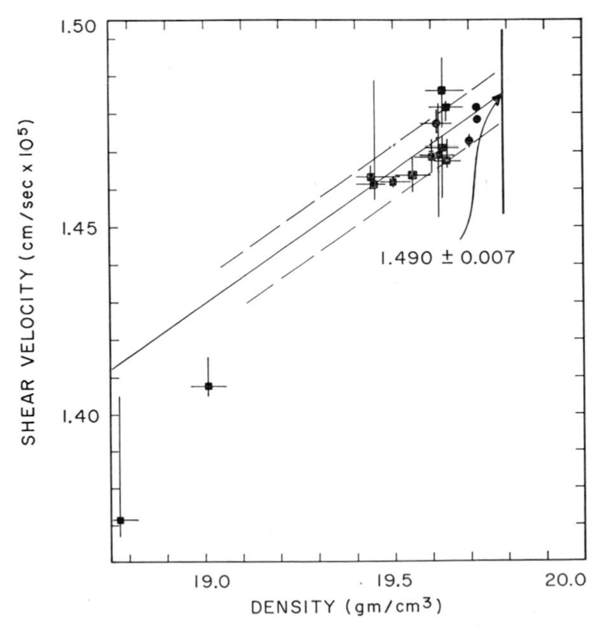

FIG. 12.—Variation of shear velocity with density. ■ By resonance technique, measured. ● By pulse technique, measured.

both the ranges of our measured velocities and elastic constants and the extrapolated values for a perfect sample of α-phase plutonium. The isothermal compressibility was obtained from equation (6), using Sandenaw's heat-capacity value (chap. xv) of 7.5 cal/mole (1.31×10^6 ergs/gm for c_p) and Elliott and Tate's (10) thermal expansion. The difference between the isothermal and adiabatic compressibilities is 2.9×10^{-13} cm²/dyne, which makes the ratio of the heat capacities $\gamma = C_P/C_V = \chi_T/\chi_S = 1.14$.

Temperature Coefficients

In the subsequent discussion, temperature coefficients are reported as the fractional change per degree centigrade of any one quantity, x—i.e., $(1/x)(dx/dt)$—which has the dimensions of reciprocal degrees.

TABLE 8

Measurement of Temperature Coefficients

Sample	Resonance	Temperature Interval (° C)	Frequency ν (Kc/Sec)	Slope $d\nu/dT$ (Cps/° C)	$(1/\nu)(d\nu/dT)$ [(° C)$^{-1}\times 10^5$]	$(1/v)(dv/dT)$ [(° C)$^{-1}\times 10^5$]
Z-13........	*L*-1	37–50	9.1	− 5.5±0.3	− 61± 4	− 56± 4
E-296-B....	*L*-1	25–27	8.8	− 5.7±1.3	− 65±15	− 60±15
3001-B.....	*L*-2	19–25	22.0	−22.0±1.2	−100± 6	− 95± 6
	L-3	9–25	33.2	−32.2±1.5	− 97± 5	− 92± 5
E-302-S.....	*T*-2 to *T*-9	26.7–30.9	11.5–52	....	− 70± 5	− 65± 5
3001-A.....	*T*-2	21.7–27.2	14.4	−13.1±1.5	− 91±10	− 86±10
	T-2	10–25	14.5	−15.7±0.7	−108± 5	−103± 5
	T-3	− 5–15	22.0	−23.2±1.2	−105± 6	−100± 6
	T-3	10–28	21.6	−20.8±0.6	− 96± 3	− 91± 3
3001-B.....	*T*-2	5–27	14.5	−15.8±0.4	−110± 3	−105± 3
	T-3	25–27	21.6	−19.2±2.4	− 89±13	− 84±13

No complete study of the temperature coefficients of the sound velocities in plutonium over an extended range has as yet been undertaken, and the numbers presented in this section are strictly preliminary ones. They were obtained on the assumption that the surface temperature of a specimen at any time is equal to the average sample temperature plus a constant, since one cannot readily measure the average specimen temperature. The validity of this assumption is particularly questionable for specimens that are rapidly changing temperature, as was the case during these measurements.

The measurements are summarized in Table 8. The coated sample Z-13 was cooled under tap water, and its subsequent rise in surface temperature was followed simultaneously with the decrease in its first longitudinal-resonance frequency. A plot of frequency against temperature (8 points between 37° and 50° C) gave a straight line of slope -5.5 ± 0.3 cps/° C. Since the uncorrected value of *L*-1 was 9.06 ± 0.04 kc, this corresponds to a temperature coefficient for *L*-1, and presumably also for *L*-0, of $(-61 \pm 4) \times 10^{-5}/°$ C at 40° C. In order to obtain the temperature coefficient of the velocity from this value, one has to allow for the appreciable thermal expansion of plutonium. The necessary data for a sample of density 19.58 ± 0.02 gm/cm³ are found in a report of Elliott and Tate (10), who state that $\alpha_{40°\,C} = 52.23 \times 10^{-6}/°$ C. Thus $(1/v_0)(dv_0/dt) = (-56 \pm 4) \times 10^{-5}/$ ° C. Measurements on E-296-B happened to cover a temperature interval of about 2° C because of variations in room temperature, and the coefficients agree with the results obtained for Z-13. On the other hand, considerably higher coefficients were observed for the second and third longitudinal resonances of the thin, uncoated, high-purity specimen 3001-B. These measurements were made with the longitudi-

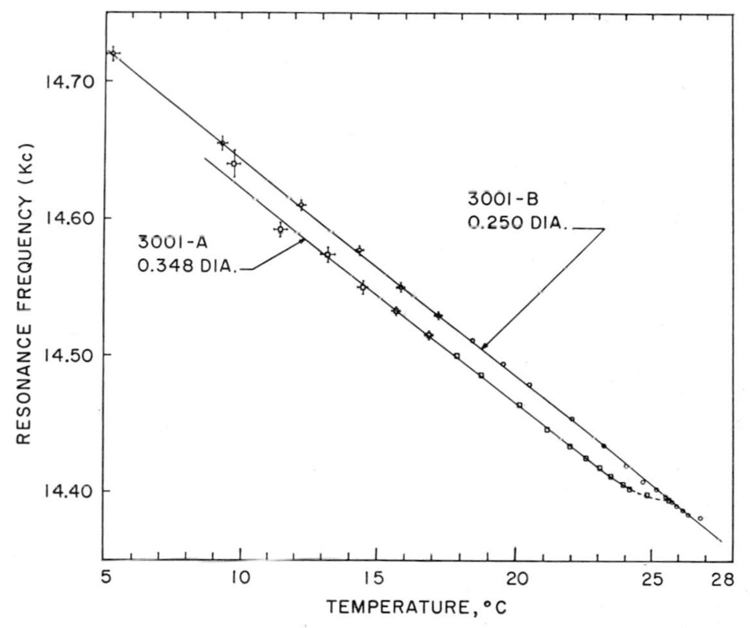

Fig. 13.—Measurement of temperature coefficients, using second torsional harmonic. □ Specimen 3001-A, 0.348 inch in diameter. ○ Specimen 3001-B, 0.250 inch in diameter.

nal eddy-current drive and pickup, and the specimen was cooled five times with dry ice and then allowed to warm up. Since the specimen had a diameter of only ¼ inch, it would seem less likely that there could be large temperature gradients within the sample.

A little more extensive data are available for the coefficient of the shear velocity. Torsional measurements on E-302-S covered a 4.2° temperature interval and led to a value for the temperature coefficient of *T*-0 of $(-70 \pm 5) \times 10^{-5}/°$ C at 29° C, which then leads to $(1/v_T)(dv_T/dt) = (-65 \pm 5) \times 10^{-5}/°$ C. The dry-ice cooling and subsequent warmup method was also applied to samples 3001-A and 3001-B. A set of the warmup curves for these samples is shown in Figure 13. Within the experimental uncertainty, there is no difference between the coefficients obtained with the two samples having different diameters and hence different ratios of heat generation to surface area. However, there is again a large increase above the measurements made on the specimen of lower purity (E-302-S).

Since there are such large variations in the measurements, Table 9 lists approximate temperature coefficients both as reported earlier (1) on the basis of the lower-purity samples (Z-13, E-302-S) and as obtained on the high-purity and high-density samples 3001-A and 3001-B. The latter are probably the more reliable results, giving $(-93 \pm 7) \times 10^{-5}$ as the temperature coefficient of the longitudinal thin-rod velocity and $(-99 \pm 5) \times 10^{-5}$ as the coefficient of the shear velocity. The derived coefficients in Table 9 are, of course, even more uncertain, and, in particular, the values for σ and χ_S may represent more of an exercise in logarithmic differentiation than in physical reality, with the value of the coefficient of σ being uncertain even as to its sign. All we can say here is that Poisson's ratio is about constant with temperature, i.e., it changes no more than +0.2 or −0.05 per cent per degree C. The signs of all the other temperature coefficients are the same as

TABLE 9

APPROXIMATE TEMPERATURE COEFFICIENTS AT 30° C

(Reciprocal ° C × 10^5)

Quantity	Low Purity	High Purity	Source
L	+ 5.127	+ 5.127	Measured (10)
ρ	− 15.381	− 15.381	Derived from above
v_0	− 56 ± 4	− 93 ± 7	Measured
v_T	− 65 ± 5	− 99 ± 5	Measured
v_L	− 48 ± 6	− 88 ± 11	Derived
σ	+138 ± 70	+ 90 ± 130	Derived
E	−127 ± 8	−201 ± 14	Derived
μ	−145 ± 10	−213 ± 10	Derived
χ_S	+ 68 ± 30	+166 ± 70	Derived

those of the corresponding coefficients of any normal metal, but the magnitudes of the coefficients are two to five times as large as for common metals. However, this should not be too surprising, in view of the uncommonly large thermal-expansion coefficient of plutonium (10).

PRESSURE COEFFICIENTS

During the first half of 1945 Professor P. W. Bridgman studied inter alia the compressibility of several plutonium samples. The work has only recently been published (12). He was furnished a number of samples by the Los Alamos Laboratory. His first measurement on a sample of relatively low density (ρ = 19.44 gm/cm^3) seemed to exhibit a reversible phase change at about 1,000 atm. However, subsequent measurements on a sample of higher initial density (ρ = 19.66 gm/cm^3) did not show such a phase change. Since it seemed highly desirable to check on the details of this possible allotropic transformation, equipment was constructed with the help of C. E. Faneuff to measure sound velocities by the pulse technique at pressures up to 30,000 psi. Measurements were made on the coated sample 107.

The precision of the measurements was increased greatly by following on the oscilloscope a specific peak or valley of a second, third, or fourth echo as a function of pressure. In a total of eight runs, no definite break or discontinuity in echo arrival times, and hence in sound velocity, was observed, which leads us to conclude that no phase change occurs below 2,000 atm for material of density 19.6 gm/cm^3 or higher. It is perhaps possible that the low-density material sent to Professor Bridgman was compacted in his apparatus at about 1,000 atm, and, since his equipment was built to go to extreme pressures, it may have been difficult to estimate the reversibility of a small (approximately 0.5 per cent) density change. Our pressure study also furnished values for the pressure coefficients of the sound velocities. The raw data expressed as change in echo arrival time per echo per 10,000 psi are given in the bar graphs of Figure 14. Weighted average values based primarily on the last two runs give 0.154 ± 0.006 μsec/echo/10,000 psi for the compressional wave and $0.16_5 \pm 0.01_5$ μsec/echo/10,000 psi for the shear wave. From these numbers one obtains the pressure coefficients reported in Table 10, making due allowance for the

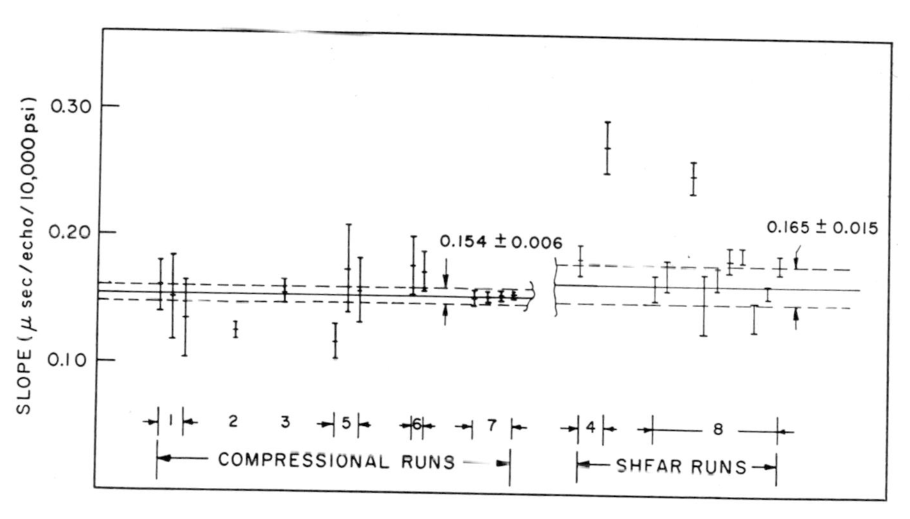

Fig. 14.—Results of study of pressure coefficients of sound velocity

TABLE 10*

Pressure Coefficients, 0–2,000 Bars, at 30° C

(Reciprocal Bars × 10^6)

L	$-0.72_9 \pm 0.01_4$	Derived
ρ	$+2.18_8 \pm 0.04_2$	Derived
v_0	$+7.7_7 \pm 0.5_5$	Derived
v_T	$+6.1_2 \pm 0.6_5$	Measured
v_L	$+9.2_1 \pm 0.4_2$	Measured
σ	$+26._6 \pm 6._6$	Derived
E	$+17._7 \pm 1._1$	Derived
μ	$+14._4 \pm 1._3$	Derived
χ_S	$-28._4 \pm 2._9$	Derived

* 1 bar = 10^6 dynes/cm^2 = 0.98692 atm = 14.504 psi.

change in specimen dimensions produced by the pressure, i.e., the compressibility. In Figure 15 we have compared our own measurements of the instantaneous compressibility of sample 107 over a limited range of pressures with compressibility values calculated from Bridgman's data, neglecting any phase change. The agreement is probably within the sum of the experimental uncertainties involved in the two rather different methods of investigation, particularly since our compressibilities are obtained from equation (5) as the difference between two numbers of about equal magnitude and are hence of considerably lower accuracy than the elastic moduli.

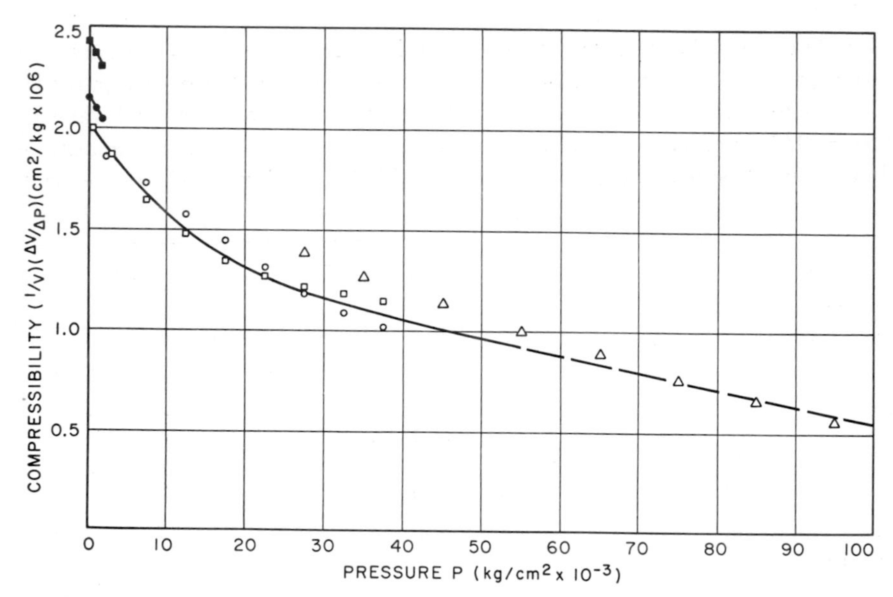

FIG. 15.—Variation of compressibility with pressure. ● Adiabatic compressibility, specimen 107. ■ Isothermal compressibility, specimen 107. □ Bridgman, $\rho = 19.44$ gm/cm³. ○ Bridgman, $\rho = 19.66$ gm/cm³. △ Bridgman (3 piezometers), $\rho = 19.57$ gm/cm³.

DEBYE THETA

A knowledge of the sound velocities permits the calculation of the Debye characteristic temperature, θ, which is useful in describing the shape of the low-temperature specific-heat curve. In Debye's model it also determines the cutoff frequency, ν, of permissible thermal modes of vibration, so that (13)

$$\theta = \frac{h\nu}{kT} \tag{10}$$

and

$$\theta = \frac{h}{k}\left(\frac{9}{4\pi}\right)^{1/3} N^{1/3}\left(\frac{\rho}{M}\right)^{1/3}\left(\frac{2}{v_T^3}+\frac{1}{v_L^3}\right)^{-1/3}. \tag{11}$$

Putting in the accepted values for the various physical constants, equation (11) becomes

$$\theta = 3.605 \times 10^{-3}\left(\frac{\rho}{M}\right)^{1/3} v_L v_T (2\,v_L^3 + v_T^3)^{-1/3}. \tag{12}$$

Using the room-temperature velocity values for α-phase plutonium having a density equal to the X-ray value (Table 7), we calculate

$$\theta = 178._2 \pm 1^\circ \text{ K} .$$

In principle, one should use velocities at the absolute zero for the calculation of Debye theta. This would increase the value of θ. However, in general, this does not lead to a better agreement with values obtained from heat-capacity measurements (see chap. xv).

CONCLUSIONS

Precision sound-velocity measurements were made on an assortment of plutonium samples of varying metallurgical history. A number of the samples exhibited some internal flaws and/or directional properties. All had densities less than 99.7 per cent of the theoretical value. If one assumes that low density is entirely due to retained β-phase and not to voids and microcracks, our best samples would have contained about 3.8 per cent of β-phase, and our sample of density 19.0 gm/cm^3, 44 per cent of β-phase. However, there appears to be a clear trend in the variation of velocity with density, so that the extrapolation to X-ray density can be undertaken with considerable confidence. From a practical point of view it should be remarked that, as long as "industrial-grade" plutonium samples are of widely varying and frequently low density, it is more appropriate to quote a range of values for their sound velocities and elastic constants rather than a single set of numbers.

Thus Young's modulus was found to range from 8.9 to 10.1 $\times$ 10^{11} dynes/cm^2 (12.9–14.6 $\times$ 10^6 psi) and the torsion modulus from 3.8 to 4.3 $\times$ 10^{11} dynes/cm^2 (5.5–6.2 $\times$ 10^6 psi). However, aside from this variation with density, the elastic properties of plutonium appear to be well behaved and much less erratic than were those found for uranium (5).

W. E. McGee helped to set up the resonance-technique equipment and C. E. Faneuff set up the high-pressure apparatus. Professor D. Bancroft made many helpful suggestions during the earlier part of the work, and C. W. Johnstone designed the electronic circuitry. Dr. A. S. Coffinberry's group prepared most of the specimens; in particular, F. W. Schonfeld worked out the hot-pressing procedures and V. O. Struebing did all the vacuum casting. K. W. R. Johnson painstakingly prepared the high-purity plutonium, and a number of machinists equally painstakingly machined samples to 0.0001-inch tolerances inside a glove box. All the analytical results were provided by Dr. C. F. Metz's group. Dr. A. S. Coffinberry editorially scrutinized the manuscript, and, finally, Dr. E. F. Hammel and Dr. E. R. Jette encouraged the work in every possible way.

REFERENCES

1. Coffinberry, A. S., and Waldron, M. B. "The Physical Metallurgy of Plutonium," chap. 4 of *Progress in Nuclear Energy*, Ser. V, Vol. **1**, p. 383. London: Pergamon Press, Ltd., 1956.

2. Love, A. E. H. *A Treatise on the Mathematical Theory of Elasticity*, pp. 103 ff. and 297 ff. 4th ed. Cambridge: Cambridge University Press, 1927.
3. Grüneisen, E. *Annalen des Physik*, [4], **22**:801–51, 1907.
4. Bancroft, D., and Jacobs, R. B. *Review of Scientific Instruments*, **9**:279–81, 1938.
5. (*a*) Laquer, H. L., McGee, W. E., and Kilpatrick, M. F. *Transactions of the American Society for Metals*, **42**:771–84, 1950.
   (*b*) Laquer, H. L. *Nuclear Science and Engineering*, **5**:197–99, 1959.
6. (*a*) Huntington, H. B. *Physical Review*, **72**:321–31, 1947.
   (*b*) Arenberg, D. L. *Journal of the Acoustical Society of America*, **20**:1–26, 1948.
   (*c*) Lazarus, D. *Physical Review*, **76**:545–53, 1949.
   (*d*) Hughes, D. S., Pondrom, W. L., and Mims, R. L. *Physical Review*, **75**:1552–56, 1949.
   (*e*) Pochapsky, T. E. *Physical Review*, **84**:553–58, 1951.
7. Bancroft, D. *Physical Review*, **59**:588–93, 1941.
8. Randall, R. H., Rose, F., and Zener, C. *Physical Review*, **56**:343–48, 1949.
9. Laquer, H. L. USAEC Rept. LA-1656 (April, 1954).
10. Elliott, R. O., and Tate, R. E. USAEC Rept. LA-1390 (March, 1952).
11. Zachariasen, W. H., and Ellinger, F. *Journal of Chemical Physics*, **27**:811–12, 1957.
12. Bridgman, P. W. *Journal of Applied Physics*, **30**:214–17, 1959.
13. Fowler, R. H. *Statistical Mechanics*, pp. 119 ff. 2d ed. Cambridge: Cambridge University Press, 1936.

CHAPTER XVII

SOME PHYSICAL AND PHYSICOCHEMICAL PROPERTIES OF PLUTONIUM METAL

A. E. Kay

The following is a description of some of the physical research which has been carried out in recent years on plutonium metal by a number of workers at the Atomic Weapons Research Establishment, Aldermaston, England.

THE SPECIFIC HEAT OF α-PLUTONIUM

This study was made by D. J. Dean and A. E. Kay.

INTRODUCTION

It is difficult to make accurate thermal measurements on plutonium metal by conventional methods, because the metal generates heat at the rate of $1.923 \pm 0.019 \times 10^{-3}$ watt/gm (1) by absorbing its own alpha-particle radiation. An estimate of the thermal conductivity, using published data for the electrical conductivity and the Wiedemann and Franz relation, indicates that it is probably very low, being about 0.013 cal/cm/sec/° C. Thus it should be appreciated that significant temperature gradients, which will affect thermal measurements, might be present in the metal.

Before we were aware of the use of self-heating for dilatometry at Los Alamos (2), we had decided that specific-heat measurements of reasonable accuracy might be made on plutonium at high temperatures (near its melting point), as well as at low temperatures, by using an adiabatic method with the "self-heat" as the constant heat source. If H calories is the heat generated per second per gram of plutonium and C_p is the specific heat at constant pressure, then the rate of change of temperature of the plutonium under adiabatic conditions, $d\theta/dt$, is given by

$$H = C_p \frac{d\theta}{dt}. \qquad (1)$$

It is thus possible to deduce C_p from measurements of $d\theta/dt$ at various temperatures. The accuracy could never be better than ± 1 per cent because this is the accuracy claimed for the value of the self-heating rate (1). The adiabatic method has been used frequently for specific-heat measurements of non-radioactive speci-

A. E. Kay is at the Atomic Weapons Research Establishment, Aldermaston, Berkshire, United Kingdom.

mens at very low temperatures, and Sykes (3) has used it at higher temperatures, up to about 500° C. All these measurements have necessitated the use of an external heating source, which gives the method a wide measure of control. With plutonium, however, this degree of control is impossible, since the heating rate is controlled completely by the specimen and any interference merely upsets equilibrium conditions. The use of self-heating insures uniform heating, which is an important consideration when making dynamic measurements on a material of relatively low thermal conductivity like plutonium.

Apparatus

A specimen weighing about 10 gm and having a heat output of approximately 5×10^{-3} cal/sec was held in a thin-walled tantalum crucible inside a vacuum

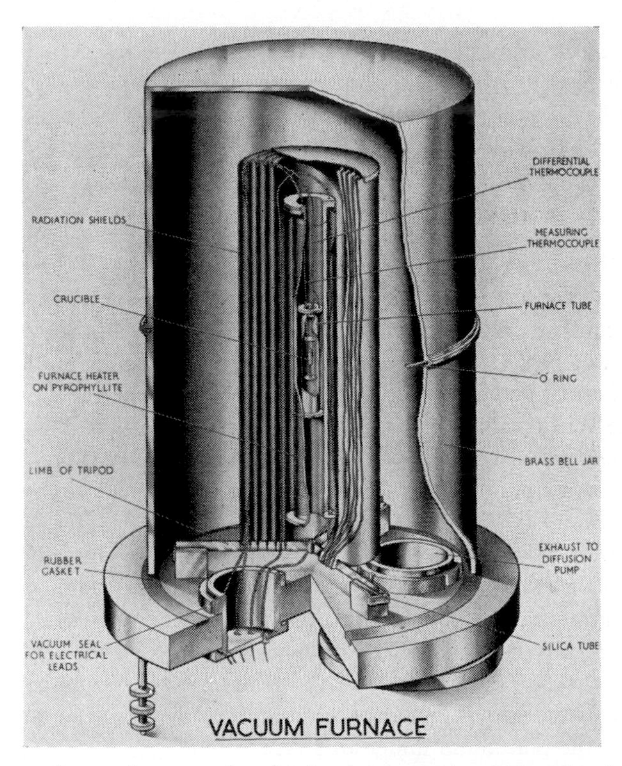

Fig. 1.—Vacuum furnace for measuring the thermal properties of plutonium by self-heating

furnace, as shown in Figure 1. The method of support insured that the specimen was effectively insulated from the furnace wall, so that heat losses by conduction were very small. One junction of a differential thermocouple and the hot junction of the measuring thermocouple were pressed into a thin tantalum tube which was welded onto the side of the crucible. The other junction of the differential thermocouple was fixed into a cavity in the furnace wall, which, in this instance, was made of copper. The thermocouples were iron-constantan, and only the measuring thermocouple was calibrated accurately. The inner surface of the copper furnace tube was highly polished to minimize radiative heat transfer between the crucible and the furnace. For measurements at high temperatures it is preferable to have the measuring thermocouple fixed to the furnace wall, since this will reduce the heat losses and the heat capacity of the specimen and crucible.

The furnace was designed specifically for glove-box work and can be dismantled and reassembled with little effort. The whole assembly is supported on a tripod, each limb of which fits into the groove formed by two short lengths of silica tubing

Fig. 2.—The vacuum furnace of Fig. 1 and the accessory vacuum pumping equipment

placed side by side (Fig. 1). The radiation shields are slotted and fit on the tripod, which thus locates them in a concentric system. A brass bell jar, made in an upper and a lower section fits over the whole assembly. It is possible to operate the furnace at a pressure somewhat lower than 1×10^{-4} mm Hg with the pumping equipment shown in Figure 2. The vacuum furnace had a uniform temperature zone about 2 inches long, the temperature gradient here being estimated to be about 0.5° C per inch for temperatures between 20° and 130° C.

Temperature control was achieved with the electronic unit diagramed in Figure 3. This unit consists of two phase-controlled thyratrons coupled in a full-wave circuit with respect to the heater winding of the furnace. One thyratron provides a current that is controlled by the absolute temperature of the furnace wall, while the other increases the heater current just sufficiently to achieve adiabatic conditions, as indicated by the differential thermocouple. In practice, the furnace temperature was made to oscillate slowly about the temperature of the specimen with an amplitude of about $\pm 0.5°$ C. The use of a galvanometer and photoelectric cell to control the current supplied to a furnace by a thyratron, for temperature regulation, is in itself not new (4), but the use of two to give a coarse and fine control is novel.

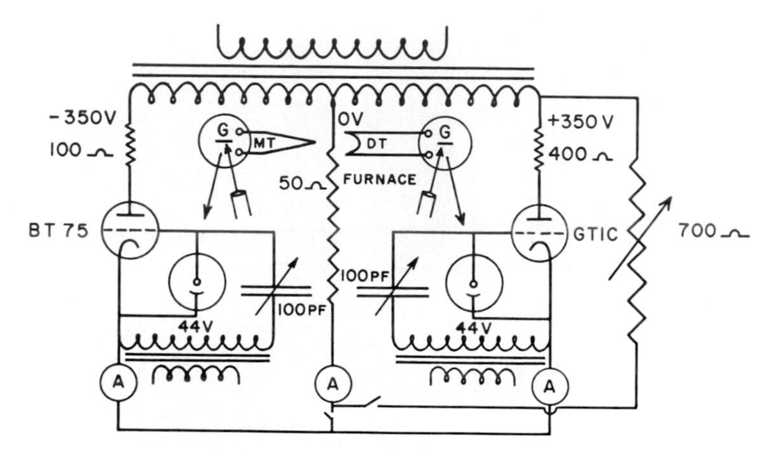

FIG. 3.—Diagrammatic representation of the temperature controller. MT = measuring thermocouple; DT = differential thermocouple; G = galvanometer suspension.

EXPERIMENTAL METHOD

The layout of the measuring equipment is shown in block form in Figure 4. Normally, the specimen, when suspended in the vacuum furnace at room tempera-

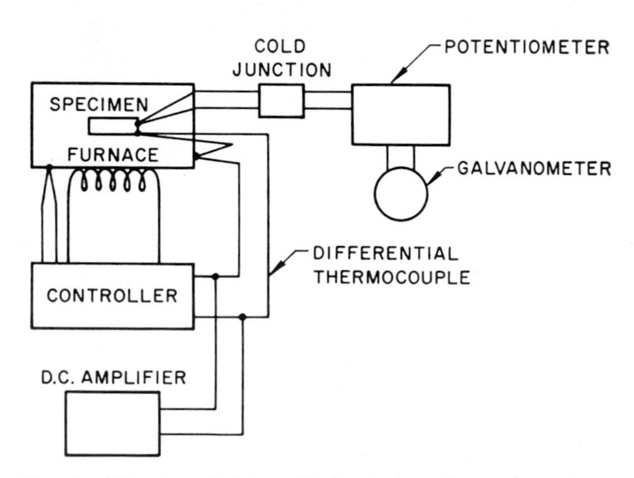

FIG. 4.—Circuitry of the specific-heat-measuring equipment

ture, was hotter than the furnace by about 16°–20° C, depending on the conduction loss through the thermocouples. It was therefore necessary to increase the furnace temperature until the controller could take over the regulation. A 700-ohm variable resistance connected in parallel with the thyratrons was used for this purpose. Measurements were taken as soon as the controller had established the required degree of control. The d.c. amplifier indicated the temperature difference between the specimen and the furnace, and measurements occupied an interval extending over about 0.1° C on either side of zero, to compensate for heat losses. During such intervals the change in the deflection of the galvanometer connected to the potentiometer was timed, using an accurate stop watch. The corresponding rate of change of the differential thermocouple output, recorded by the d.c. amplifier, was also noted. From these measurements it was possible to calculate the specific heat of plutonium at the temperature indicated by the potentiometer.

To obtain measurements at room temperature the specimen was quenched in liquid nitrogen and then speedily placed in the furnace, which was then evacuated. The quenching provided a sufficiently long time for obtaining a good vacuum before the specimen heated up to room temperature, when measurements were begun.

The following procedure was adopted for measuring the latent heat at the $\alpha \rightarrow \beta$ phase transition. At a temperature of about 100° C, absolute times were recorded against temperatures at intervals of about 2 minutes, using a second stop watch, and these observations were continued into the β-phase. From a record of results of this type the calculation of the latent heat is straightforward.

The experimental procedures described above are essentially those that were employed by Dempsey and Kay (5) in measuring the specific heat of α-plutonium and the latent heat of the $\alpha \rightarrow \beta$ transformation. By improving temperature control and the thermal insulation of the specimen, better values of these thermal properties have been obtained, as well as heat capacities and transition latent heats at all higher temperatures up to the melting point.

Results

If M is the mass of the plutonium specimen and h is the heat capacity of the crucible, then

$$\left(MC_p + h\right)\frac{d\theta}{dt} = MH\,, \qquad (2)$$

which can be used to calculate values of C_p. The results are shown in Table 1 and have been reported elsewhere in graphical form (6). The accuracy of both the specific-heat and the latent-heat measurements is about 5 per cent.

Discussion

There is good agreement between the room-temperature values for C_p and the result found by Sandenaw and Gibney at Los Alamos (7), namely, 0.033 cal/gm/° C at 25° C. Moreover, our value of ΔH for the $\alpha \rightarrow \beta$ transition agrees well with their value of 925 cal/gm-atom.

According to classical theory, C_v, the specific heat at constant volume, for plu-

TABLE 1

THE SPECIFIC HEAT AND HEATS OF TRANSFORMATION OF PLUTONIUM

Temperature (° C)	Specific Heat (Cal/gm × 10^3)
− 62.0	33.4
− 58.5	33.4
− 50	34.2
− 44	35.2
− 32	37.5
− 25	38.9
− 21	38.7
− 13	38.8
− 1.5	35.7
− 1.0	35.5
5.0	32.4
13	35.5
15	33.5
20	33.8
29	34.1
38	34.5
44	34.9
52	36.6
62	36.6
65	35.0
70	38.7
79	38.6
81	41.3
85	41.5
90	41.3
96	42.5
104	44.9
111	44.9
113	47.8
118	52.8
121	56.0
122	60.9
124	76.7
126	101.7
126	129.2
127 = superheat; the true transformation temperature = 121° C; the latent heat of the $\alpha \rightarrow \beta$ transition = 958 ± 10 cal/gm-atom	
127	161.9
130	99.0
132	72.2
133	57.9
135	50.9
140	43.1
145	40.5
152	39.9
158	41.1
167	41.9
171	42.0
176	42.5
184	41.6
190	41.9
192	43.3
197	43.4
200	42.3
204	42.8
208	43.4
210	49.7
211	90.3
$\beta \rightarrow \gamma$ transition; latent heat = 140 ± 15 cal/gm-atom	
212	66.5
212	59.9
213	42.6
218	40.8
226	42.3
230	44.0
236	43.0
242	44.8
243	44.4
250	45.0
254	46.3
260	45.2
267	46.9
272	45.9
279	46.1
284	46.8
288	45.3
291	47.8
297	47.9
303	49.9
305	50.5
307	60.7
312	89.6
315	128.1
$\gamma \rightarrow \delta$ transition; latent heat = 156 ± 5 cal/gm-atom	
318	87.3
320	82.0
322	73.9
324	62.3
326	56.8
327	52.8
332	48.5
334	47.1
343	44.6
346	45.5
350	45.5
356	48.1
361	46.5
368	46.9
374	46.4
376	46.9
379	47.1
383	46.4
389	48.3
397	45.7
407	45.6
412	47.8
420	48.6
426	48.6
433	45.0
442	47.0
449	49.7
450	50.2
452	51.0
454	53.9
456	56.6
457	64.1
459	86.8
$\delta \rightarrow \delta'$ transition; latent heat = 17 ± 10 cal/gm-atom	
461	62.9
464	58.6
468	65.8
470	75.0
473	79.4
473	100.6
477	141.6
480	155.2
481	157.8
483	268.1
$\delta' — \epsilon$ transition; latent heat = 470 ± 10 cal/gm-atom	
486	55.6
488	43.8
491	43.4
498	43.1
505	44.0
513	42.2
518	39.4
527	41.7
533	35.4
538	36.9
546	37.6
$\epsilon \rightarrow$liquid transition at 639° C; latent heat = 940 ± 150 cal/gm-atom; total heat of plutonium in solid state = 11.7 ± 0.5 Kcal/gm-atom	

tonium should be about 0.025 cal/gm/° C. From thermodynamic considerations we have

$$C_p - C_v = \frac{1}{J}\frac{9\alpha^2 VT}{K}, \tag{3}$$

where α = the coefficient of linear expansion, K = compressibility, J = mechanical equivalent of heat, and T = temperature (° K).

Using the measured values, $K = 2.5 \times 10^{-6}$/bar (7), $V = 0.051$ cm³/gm (7), and $\alpha = 47 \times 10^{-6}$/° C (7), we find that $C_p = 0.028$ cal/gm/° C at room temperature. The measured value thus exceeds the theoretical one by about 14 per cent. In addition, the experimental results indicate that the specific heat increases relatively rapidly with temperature, at temperatures below about 70° C, suggesting that the specific heat is significantly larger than that expected from consideration of lattice vibrations alone, since estimates of the Debye temperature using compressibility data—Einstein relation (8)—and the melting point—Lindemann relation (8)—give values between 92° and 130° K. It is possible that this excess arises from an electronic contribution.

It is interesting to observe that the Grüneisen constants given by the relationship $\gamma = 3\alpha V/C_v K$, where α is the coefficient of linear expansion, V is the atomic volume, C_v the specific heat, and K the compressibility, for thorium, uranium, and plutonium are given in the accompanying table, while the rare-earth metals (9)

Th	α-U	α-Pu	β-, γ-, ε-Pu	δ-Pu
1.65	3.6	2.64	1.57	−0.52

have values ranging from 0.3 to 0.9. It has been assumed that the compressibility is about 2.2×10^{-6} per dyne/cm² (7) for the γ-, δ-, and ε-phases of plutonium. These constants suggest some correspondence in metallic behavior between thorium, uranium, and the α-, β-, γ-, and ε-phases of plutonium and between some of the rare earths and δ-plutonium.

ELASTIC PROPERTIES

Measurements of Young's modulus have been made at room temperature on cast plutonium metal. Tensile measurements made on a Hounsfield tensometer by W. B. H. Lord gave a value for Young's modulus of 8.2×10^{11} dynes/cm².

Compression measurements made by A. E. Kay and B. W. Lowthian, using stresses up to 70 kg/cm², gave a value of 9.9×10^{11} dynes/cm². Both longitudinal and lateral strains were measured, using resistance strain gauges, and comparison of the strains gave a value of 0.13 for Poisson's ratio. The strains varied linearly with stress, and the magnitudes were very reproducible. The specimen had been examined previously for internal flaws by radiography. These measurements agree reasonably well with Los Alamos and Harwell results (7).

OXIDATION OF PLUTONIUM IN AIR

Introduction

An initial survey of the behavior of plutonium when exposed to atmospheric conditions at different temperatures has been carried out by E. Dempsey and R. F. Powell, prior to undertaking a more detailed study of the corrosion characteristics of the metal.

Ordinary Temperatures

Observations were made at temperatures somewhat higher than room temperature, so that reasonable temperature control could be achieved. A disk of metal with a surface area of about 3 cm^2 was hung in a small aluminum can inside a desiccator. The air in the desiccator was maintained at the desired temperature by means of an electrical heater and thermostat, and the humidity was controlled at either dry or wet conditions. Weighings were carried out by removing the specimen in its can to a glove box containing a conventional balance. The specimen was cleaned on emery cloth prior to exposure in the desiccator.

The results are shown in Figure 5. Comparison of curves *2* and *4* indicates the

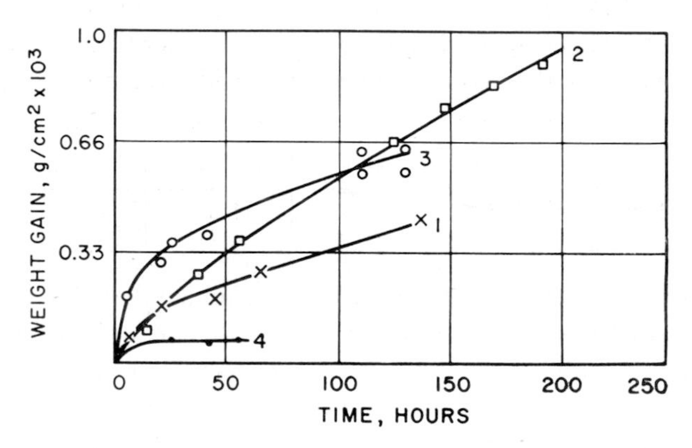

Fig. 5.—The oxidation of plutonium in air. The conditions are as follows:

Curve	Temperature of Test (° C)	Relative Humidity (Per Cent)	Curve	Temperature of Test (° C)	Relative Humidity (Per Cent)
1.	40	100	3.	40	100
2.	50	100	4.	50	0

importance of moisture in the oxidation of plutonium. After exposure to dry air, the specimen appeared to be dark red in color because of interference effects, and its general appearance was consistent with the formation of a very thin protective layer of oxide. A rough estimate of the thickness of the layer suggested that it did not exceed 600 A. At the end of test 2 the specimen was coated with a thick layer of powdery primrose-yellow oxide. The almost linear nature of the curve seems to indicate that, from the beginning of the test, the layer of oxide remained porous. The oxide was identified as PuO_2 by X-ray diffraction. The variation between curves

1, *2*, and *3* is attributed to different degrees of tarnishing of the specimens, because they were stored in the polished state for different lengths of time in an argon atmosphere, before being exposed to the test conditions. The oxide formed in tests 1 and 3 consisted of an outer black layer with some powdery yellow oxide beneath, in contrast to the completely yellow oxide produced in test 2.

HIGH TEMPERATURES

A balance has been developed which enables weight changes to be observed continuously as oxidation proceeds. The instrument, which will be described in more detail elsewhere (Dempsey, to be published), is shown in Figure 6. A crossed pair of beryllium-copper springs acts as a suspension, and the deflection is observed on a

FIG. 6.—Balance for the study of oxidation at high temperatures

scale viewed through a telescope via a plane mirror attached to the beam of the balance. The instrument has a sensitivity of about 1 mm/mg, and it is possible to estimate to 0.1 mm. Before using the balance for plutonium, it was tested on copper, and good agreement with the published results of other workers was obtained.

The plutonium specimens were disks weighing about 1.5 gm each and having a surface area of approximately 2 cm^2. Before being exposed in the furnace, the specimens were polished with fine emery cloth. Oxidation temperatures were measured by means of a thermocouple located in a hole in a dummy specimen that was suspended at the appropriate level in the furnace. The relative humidity of the air in the glove box was observed and remained between 50 and 55 per cent during all the tests.

The results are given in Figure 7. Between 200° and 300° C the oxidation proceeded linearly, indicating the formation of a porous oxide coating. The change in the rate of oxidation after about 120 minutes at 205° C might be associated with the specimen transforming from the β- to the γ-phase during the test, because of the unfortunate choice of temperature. At 303° C the results suggest that the specimen oxidized initially according to a linear law, followed by a region where the behavior seemed to be tending toward the parabolic form. In the δ-phase the oxidation proceeded slowly compared with that observed at 303° C, and the oxidation relationship was parabolic, having a rate constant of 1.2×10^{-5} gm²/cm⁴/hour. This result was so striking that the test was repeated with a different specimen, the surface of which had been prepared by machining instead of polishing. The results reproduced almost exactly and are indicated in Figure 7 by crosses. It was thus inferred that

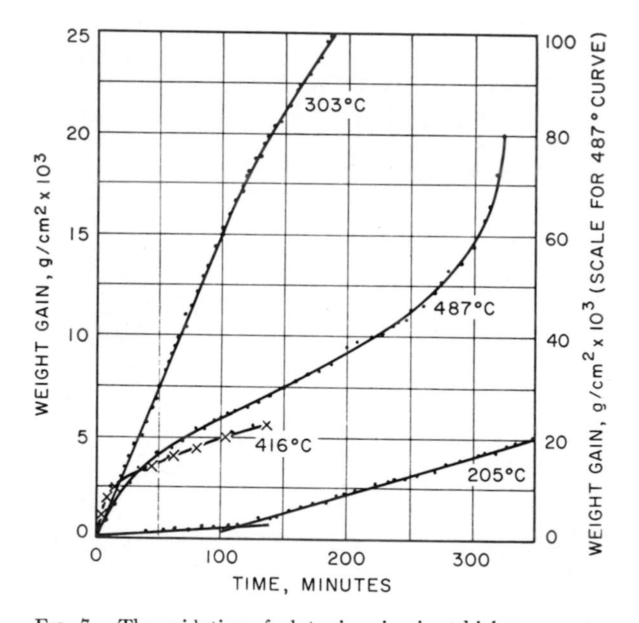

FIG. 7.—The oxidation of plutonium in air at high temperatures

the oxide forms a protective coating. At 487° C the specimen oxidized according to the parabolic law initially, then the rate slowly increased and was followed after 4 hours by a period during which the metal was oxidizing rapidly. It is suggested that after the oxidation had proceeded to a certain point the oxide layer became porous. Heat produced by the exothermic reaction of oxidation, which would be rapid because of porosity, was now adequate to raise the temperature of the metal, consequently increasing the rate of oxidation. In this way the specimen would soon reach a temperature at which spontaneous combustion could occur.

The oxide formed in all tests except that at 487° C was black, slatelike, and coherent in appearance. The residue after the test at 487° C consisted of a brown powdery oxide with large pieces of porous material dispersed in it.

Conclusions

At ordinary temperatures, moisture plays a significant part in determining the rate of oxidation of plutonium in air. At higher temperatures the oxidation law is probably linear for the β- and γ-phases, but parabolic for the δ- and ϵ-phases. Oxidation in the δ-phase is anomalous, because the rate is much slower than that observed in the γ-phase. At temperatures above 490° C small pieces of metal might ignite spontaneously.

The author wishes to thank Mr. W. B. H. Lord for advice and encouragement during the course of the work.

REFERENCES

1. Stout, J. W., and Jones, W. M. *Physical Review*, **71**:582–85, 1947.
2. Jette, E. R. *Journal of Chemical Physics*, **23**:365–68, 1955.
3. Sykes, C. *Proceedings of the Royal Society, A*, **148**:422, 1935.
4. Zabel, R. M., and Hancox, R. R. *Review of Scientific Instruments*, **5**:28–29, 1934.
5. Dempsey, E., and Kay, A. E. *Journal of the Institute of Metals*, **86**:379–84, 1957–58.
6. Dean, D. J., Kay, A. E., and Loasby, R. G. *Journal of the Institute of Metals*, **86**:464, 1957–58.
7. Coffinberry, A. S., and Waldron, M. B. "The Physical Metallurgy of Plutonium," chap. 4 of *Progress in Nuclear Energy*, Ser. V, Vol. **1**, pp. 354–410. London: Pergamon Press, Ltd., 1956.
8. Mott, N. F., and Jones, H. *Theory of the Properties of Metals and Alloys*, p. 13. London: Oxford University Press, 1936.
9. Barson, F., Legvold, S., and Spedding, F. H. *Physical Review*, **105**:418–24, 1957.

CHAPTER XVIII

THE CORROSION OF PLUTONIUM

J. T. Waber and E. S. Wright

INTRODUCTION

With regard to many of its properties, plutonium is similar to uranium; this is particularly true of its resistance to corrosion. It oxidizes slowly in dry air but relatively rapidly in a moist atmosphere. Oxygen tends to reduce the rate of attack by water vapor. There is some evidence for the differential aeration phenomenon. Strong nitric acid passivates the surface, although plutonium dioxide is readily soluble in this acid. In these ways the two metals are similar, and no significant differences have been noted. This is not to say, however, that differences in details and mechanism do not occur.

Plutonium is a rather reactive metal of the actinide group. It became apparent when only small amounts of the metal were available that it was appreciably attacked in ordinary laboratory atmospheres. When freshly cleaned, plutonium is bright and has the appearance of nickel. In ordinary air it darkens rapidly, forming interference colors, and, if exposed long enough to aggressive conditions, it acquires a powdery surface. Eventually, an olive-green powder, identified by X-ray diffraction as PuO_2, is formed. Apart from the inventory losses which would occur during handling, the extremely toxic nature of this material renders it imperative that oxidation be minimized and the corrosion products be contained. Fortunately, the massive metal is relatively inert to dry air. The attack is accelerated by water vapor, and atmospheric oxidation proceeds rapidly at even moderate relative humidities. The oxidation reaction usually proceeds to the formation of dioxide, but in some cases there is evidence that detectable amounts of lower oxides such as PuO or Pu_2O_3 are formed.

Most of the information relating to the corrosion of plutonium has been obtained in long-term tests under conditions which approach atmospheric. In a reactor, plutonium would undoubtedly have to be clad, in order to prevent the spread of fission products, even though the problem of rapid corrosion by hot water could perhaps be solved. Thus it is relatively unimportant to discuss other than the basic problems attending the storage of machined fuel elements.

Experience gained at Los Alamos strongly indicates that the best all-around at-

J. T. Waber is at the University of California, Los Alamos Scientific Laboratory, Los Alamos, New Mexico. E. S. Wright is at Lockheed Aircraft Corp., Palo Alto, California.

mosphere for storage is freely circulating air. Attempts to inclose the metal or reduce the access of oxygen seemingly promote corrosion attack. Numerous attempts have been made to store turnings and metal scrap in inert helium gas, but such attempts have been abandoned in favor of storage in dry air after strongly accelerated corrosion rates, as well as pyrophoric products, had been observed. These results may be attributed to moisture in the containers and in the helium. The relatively unpredictable pyrophoric properties of metallic uranium have been discussed by Smith (1).

REACTIVITY OF PLUTONIUM

The metal reacts with the halogens at moderate temperatures to form the corresponding trihalide. Ammonia gas and nitrogen react only in the vicinity of 1,000° C, forming PuN. A carbide may be formed by reaction between plutonium vapor and carbon monoxide.

Qualitative information on the rate and extent of chemical attack on plutonium metal by several common reagents is summarized in Table 1. The reaction of water

TABLE 1

TYPICAL REACTIVITIES OF PLUTONIUM METAL

Reagent Solution	Remarks
Acetic acid	No visible attack by glacial acetic acid even at elevated temperatures; very slow dissolution in dilute acid
Trichloroacetic acid	Vigorous attack in concentrated acid; rate of solution decreases upon dilution
Perchloric acid	Fairly rapid gas evolution in 72 per cent acid
Phosphoric acid	Fairly rapid attack in 85 per cent acid
Sodium hydroxide	No attack by 2.5 M base

with plutonium metal is slow in comparison with its rate of dissolution in acids like HI or $HClO_4$. Although attack is known to become more aggressive in boiling water, no study of the actual rate has been made.

The attack by moderately strong solutions of nitric acid is very slight, and, since the dioxide readily dissolves in strong nitric acid to form plutonyl nitrate, the metal can be cleaned of its oxide film. After a few minutes' immersion in warm 5 N nitric acid, the oxide is dissolved or detached, and the filmed metal regains its metallic luster.

Plutonium does not dissolve readily in sulfuric acid solutions, although the Pu(III) and Pu(IV) sulfates are easily soluble. Cunningham (2) suggests that these sluggish dissolution reactions may be due to the formation of an insoluble protective film. However, PuO_2 and the hydroxides are soluble in both strong nitric and sulfuric acid solutions. It is more reasonable to assume that passivation of the metallic surface occurs in these two acids.

Much insight into the corrosion behavior can be gained from examining the oxidation and reduction reactions of Pu(III) and Pu(IV) ions in solution. Plutonium is unique among metals, in that appreciable concentrations of ions in four oxidation states can coexist in equilibrium; this fact arises from the similarity of the potentials appropriate to the several oxidation-reduction couples.

Although the Pu(III) ion is not significantly hydrolyzed in solutions, the +4 ion begins to form $Pu(OH)^{+3}$ at pH values above 1.4, and precipitation of a polymeric hydroxide occurs between pH 2.5 and pH 3.0. In moderately acidic solutions, Pu(III) is stable against oxidation by air. However, the moist hydroxide is rapidly oxidized to the +4 state. This is understandable because the oxidation potential for the Pu^{+3} - Pu^{+4} couple is −0.96 volts in acidic solutions, whereas in basic media the potential of the $Pu(OH)_3$ - $Pu(OH)_4$ couple is +0.95 volts (3). Positive potentials are assigned to reactions that will occur spontaneously from left to right to yield the oxidized form. In addition, Pu(IV) on standing will undergo disproportionation to Pu(III) and Pu(VI). For these reasons, attack by neutral aqueous media may be expected to yield the Pu(IV) hydroxide. This compound will contain, as well, small amounts of the +3 state. This hydroxide can exist as a very stable colloid, similar to the hydroxides of quadrivalent cerium and uranium. The X-ray diffraction pattern of the $U(OH)_4$ colloid is diffuse but similar to that of UO_2. Analogously, the plutonium hydroxide, air-dried at 70° C, is said to contain "crystallites of PuO_2" having linear dimensions of about 30 A (4).

Plutonium hydride differs from UH_3 in many ways. It tends to be coherent and to retain the shape of the metal from which it forms. Although quite pyrophoric, plutonium hydride has been observed at times to be relatively inert with respect to moist air. The oxidation reactions are apparently slow at room temperature. This fact probably accounts for the presence of both the hydride and the oxide or "hydroxide" in the corrosion products. The properties and preparation of the plutonium hydrides have been reported by Mulford and Sturdy (5) and by Brown, Ockenden, and Welch (6).

CORROSION BEHAVIOR

The corrosion behavior of plutonium and its alloys has not been studied in detail. In many ways, however, it is so similar to the corrosion behavior of uranium that valid qualitative deductions can be made by analogy. For example, alloying elements that stabilize the delta phase of plutonium confer substantial corrosion resistance in accord with the fact that the stable (or metastable) gamma-phase alloys of uranium have good corrosion resistance. In many cases, analogous compounds, such as the hydride and dioxide, are the common corrosion products. Further important similarities in behavior of both fissionable elements have been mentioned above.

The bulk of the corrosion data obtained to date relates to long times and low ambient temperatures. In dry air at room temperature, both uranium and plutonium oxidize slowly at rates of only a fraction of one-thousandth of an inch per year. This rate is strongly dependent on the humidity of the air.

The oxide on plutonium may be tightly adherent or quite loose and powdery. In both cases, only X-ray diffraction patterns of the dioxide are obtained. Plutonium does not, however, exhibit the formation of thick oxide films, which then fracture into large flakes, as does uranium, but rather tiny oxide particles are formed. The loose oxide formed under aggressive corrosion conditions may occupy twenty to thirty times the volume of the original metal.

One of the major problems to be anticipated in using canned plutonium fuel elements is pinhole corrosion, such as has been studied extensively with uranium fuel elements. Small blisters form near the defects in the can or its closure. Plutonium is known to form hydride during atmospheric corrosion. Apparently, a larger amount of hydride is formed when the access of oxygen is restricted by the nearly perfect protective coatings or cans. Corrosion products form near the defect and create sufficient internal pressure, because of their relatively lower density, to bulge the metal, as well as any coating or cladding. These products are frequently pyrophoric. Experimentally, such blisters have been shown by the late Eugene Staritzky to contain PuH_{2+x} and some trivalent plutonium (possibly as the hydroxide). The presence of oxygen in the corroding gas minimizes the amount of these undesirable products. The proportion of hydride in a mixture of PuH_{2+x} and PuO_2 is difficult to determine, because both pure PuH_{2+x} and PuO_2 have the fluorite crystal structure with quite similar lattice dimensions. Moreover, both compounds have extensive composition, or solubility, ranges, which cause the lattice dimensions to be variable.

Numerous examples of blistering have been observed with nickel-coated plutonium pieces after several months of storage under slightly humid conditions. After several months of storage, similar blistering has been noticed at Argonne National Laboratory on uranium-plutonium-alloy fuel slugs canned in stainless steel.

Detailed results regarding the effect of oxygen on unalloyed plutonium have not been obtained, but they are thought to be in accord with the data relating to the corrosion of uranium. The attack by oxygen-saturated water is about 120 times less aggressive than attack by air-saturated water (7).

The important effect of small amounts of moisture in air or oxygen has been established more firmly for the corrosion of uranium than for plutonium. It can be said, however, that if a desiccant is only partially effective in removing the moisture, the corrosion rate will be essentially as high as though no attempt were made to dry the gas. Above 120° F, many commercial desiccants, such as silica gel (and alumina), will act not as desiccants but as potential sources of water vapor. Evaluation tests indicate that anhydrous magnesium perchlorate (a commercially available desiccant) is definitely superior to anhydrous calcium sulfate (Drierite), which reduces the moisture content of air to roughly 1 per cent relative humidity.

In 1945, Morris Kolodney (8) briefly investigated the relative influences of the temperature and humidity of air on the corrosion rate of unalloyed plutonium. His results are summarized in Table 2. He noted that a test at 25° C was "largely academic, since massive pieces of plutonium possess sufficient internal energy to raise the temperature well above ambient." More recent data are summarized in Figures 1, 2, and 3 and show that oxide formation in dry air is much less rapid than in moist air.

The conversion factors relating metal penetration to weight gain are as follows: 1 mg/cm^2 of oxide formation corresponds to the consumption or penetration of 3.77 μ of alpha plutonium; and 1 mil penetration of alpha plutonium corresponds to 6.74 mg of oxygen gained per square centimeter of surface.

TABLE 2

EFFECT OF TEMPERATURE AND HUMIDITY ON OXIDATION OF ALPHA PLUTONIUM, AFTER KOLODNEY (8)

Temperature (° C)	Relative Humidity (Per Cent)	Time (Hours)	Weight Gain (Mg/Cm^2)
25	40	900	0.01–0.07
	100	900	6.5
50	7	900	4.3
	7	200	0.6
65	5	200	1.0
	0	200	0.015

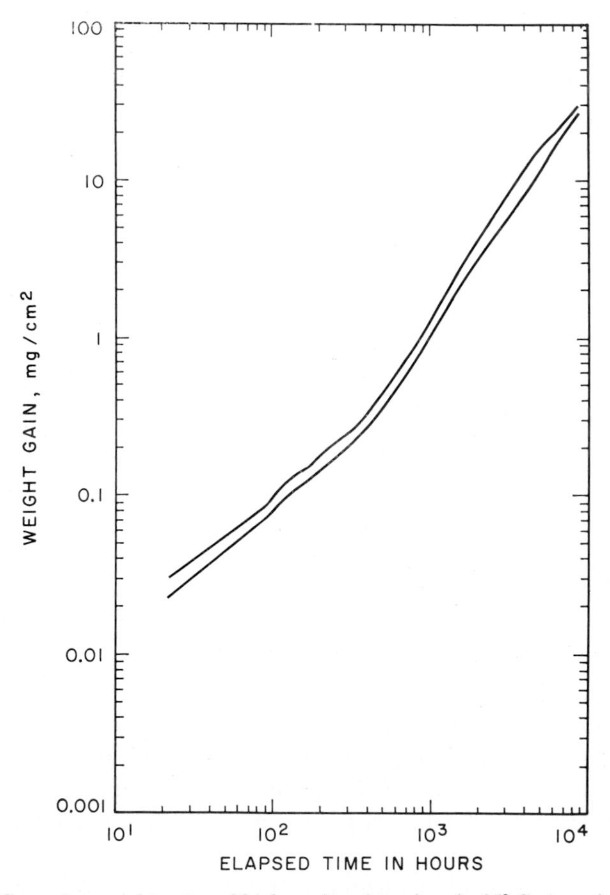

FIG. 1.—Corrosion weight gains of high-purity plutonium in 35° C air maintained at approximately 20 per cent relative humidity by a saturated solution of $LiNO_3$.

The attack by moisture follows an approximately linear growth law, in contrast to the nearly parabolic law obtained for oxidation by dried air. This difference in time dependence results, within a year, in almost one hundred times as much oxide, or hydroxide, being formed on the specimens exposed to moist air. Comparative tests on pure plutonium were not performed with moist helium; however, judging from our experience with uranium and its alloys, substantially more attack would be anticipated, since oxygen acts in a manner not fully explained to polarize anodically the metal and thereby reduce the corrosion rate.

Inasmuch as the curves for attack by moisture are approximately linear with time, the weight increase at long times is essentially proportional to the rate constants; that is, the effect of any small constant term in the linear law becomes negligible. The anticipated weight increases at 10,000 hours are plotted in Figure 4 against the reciprocal of the absolute temperature. The temperature dependence estimated from this slope is equivalent to an activation energy of 17 kcal. The

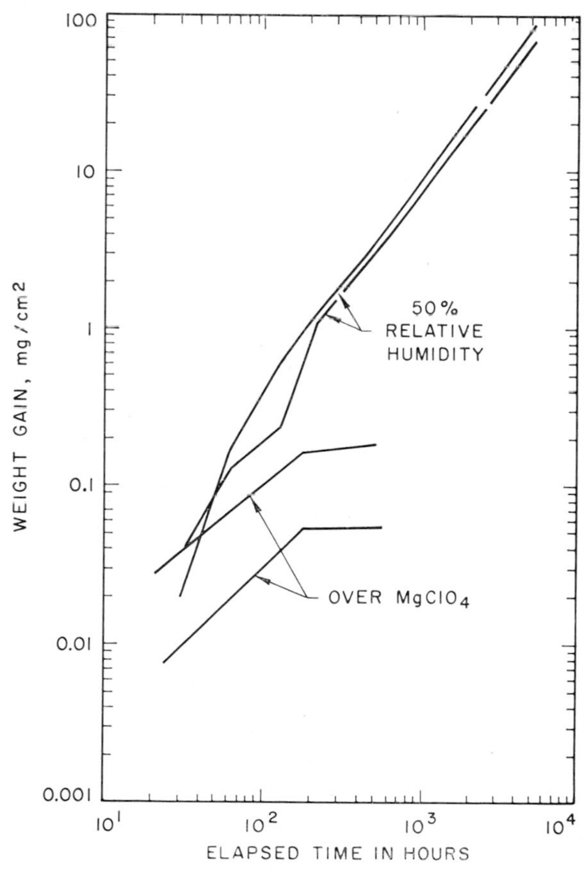

FIG. 2.—Comparison of the corrosion weight gains in dry and moist air at 55° C. The approximately 50 per cent relative humidity was maintained with a saturated solution of NaBr. The dried air was continuously in contact with $MgClO_4$.

temperature effect has not been estimated from the data for dried air, but apparently it is small.

HIGH-TEMPERATURE OXIDATION

Plutonium is not very reactive toward the nitrogen in air. For example, plutonium has been melted in a nitrogen atmosphere and held for several hours at 1,000° C, with the formation, on cooling, of only a few nitride dendrites near the surface. Thus the high-temperature properties in air can be reduced to its behavior toward pure oxygen at equivalent temperatures.

The oxidation of plutonium in pure oxygen in the range 315°–400° C has been briefly reported by the authors (9) elsewhere. Specimens made from 20-mil ribbon were oxidized to completion in about 20 minutes at 343° C and in about 6 minutes at 400° C, as shown in Figure 5. The rate data are represented quite well by the following equation:

$$W = k_n t^n , \tag{1}$$

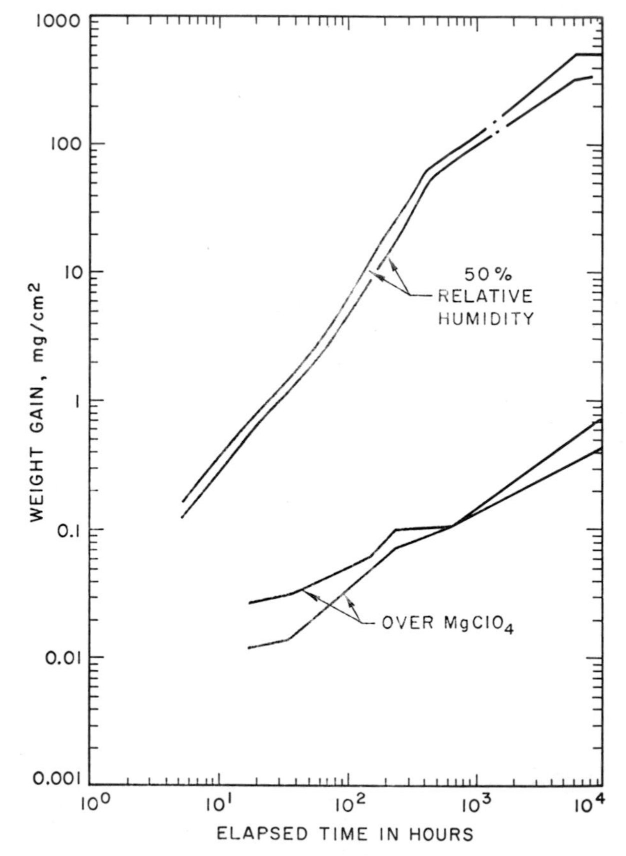

FIG. 3.—Comparison of the corrosion weight gains in dry and moist air at 75° C. The approximately 50 per cent relative humidity was maintained with a saturated solution of NaBr. The dried air was continuously in contact with $MgClO_4$.

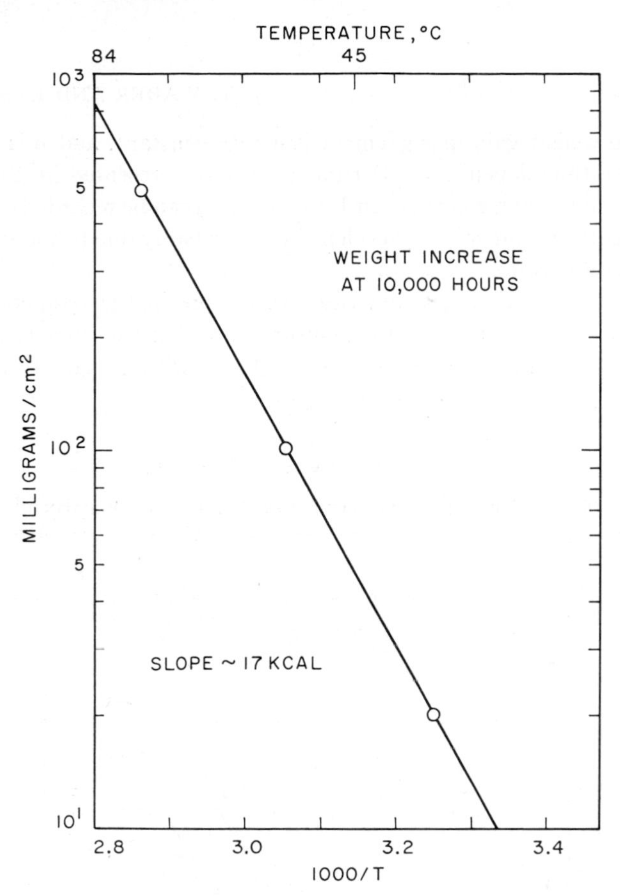

FIG. 4.—Approximate determination of the temperature dependence of the reaction rate constant for moist air. The weight gains of Figs. 1, 2, and 3 have been extrapolated to 10,000 hours.

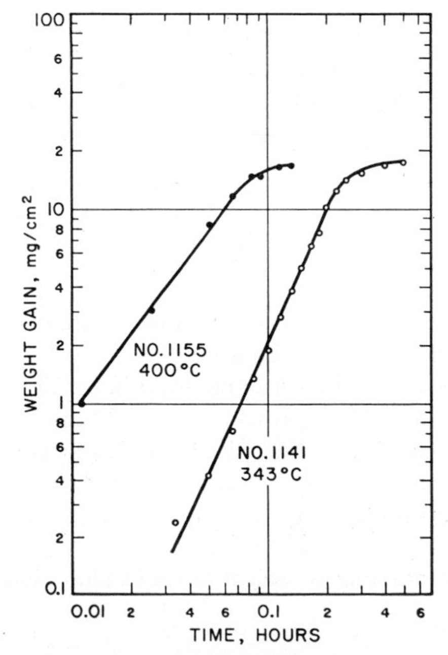

FIG. 5.—High-temperature oxidation of high-purity plutonium in purified oxygen. The slopes in excess of unity characterize the reaction as accelerating with time.

where W is the weight gain in mg/cm², k is a rate constant, and n is a constant representing the time dependence. Pertinent data are presented in Table 3. The exponent of the time being greater than 1, the reaction can be regarded as accelerating in this temperature range. The acceleration is probably due to the spallation of the oxide during the run.

In contrast, the reaction rate decreases with time in the temperature range 315°–325° C and probably at lower temperatures also. For the three runs shown in Figure 6, the rate data can be represented by the logarithmic growth law:

$$W = k_L \log_{10}(bt + c), \qquad (2)$$

TABLE 3

KINETIC DATA FOR THE OXIDATION OF UNALLOYED PLUTONIUM

Test Temp. (° C)	Rate Constant (k_n)	Exponent of Time (n)	Test Temp. (° C)	Rate Constant (k_n)	Exponent of Time (n)
345......	320	2.2	400......	480	1.4

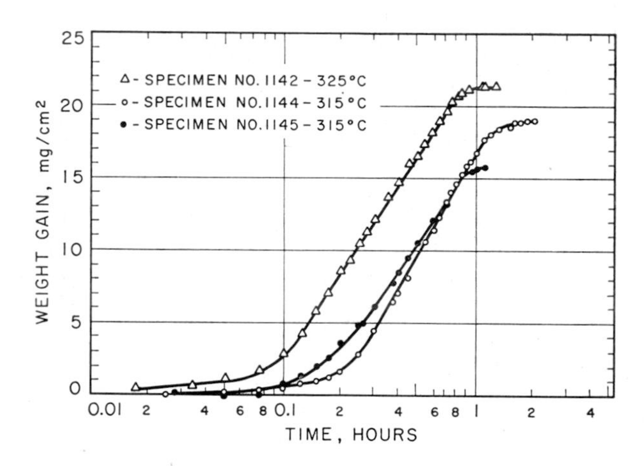

FIG. 6.—Semilogarithmic plots of the gas-consumption data for pure plutonium and oxygen. The gas and metal were preheated in a molten salt bath before they were brought into contact. The sigmoid shape of the curves results from an increase in surface temperature at the beginning of the reaction, followed by depletion of metal at its end.

where k_L and b are the appropriate kinetic constants. Values of these are presented in Table 4. It was found that c could be neglected.

The reactions went to completion on 20-mil sheet in a little over an hour at these temperatures. Inasmuch as the oxygen was preheated to the reaction temperature prior to contacting the metal, the slow initial reaction rates seen in Figure 6 are probably not due to temperature equilibration of the gas. The absence of similar delays in the higher-temperature tests shown in Figure 5 offers confirmation of this interpretation.

An aluminum alloy containing 10 weight per cent plutonium, which consists of the intermetallic compound $PuAl_4$ dispersed in a matrix of aluminum, was held 1 week at 400° C in pure oxygen. The pickup of oxygen was barely detectable by the manometric method that had been employed in the oxidation of pure plutonium. An upper limit on the amount of oxide formed is 8 μ. An X-ray diffraction study of the oxidized surface showed only a weak pattern of Al_2O_3 and the patterns of the underlying alloy phases, but no PuO_2. Comparison with the rate of oxidation of pure aluminum suggests that plutonium may slightly reduce the protection afforded aluminum by its oxide.

TABLE 4

KINETIC CONSTANTS FOR LOGARITHMIC GROWTH LAW APPLIED TO OXIDATION OF PLUTONIUM

Specimen No.	Test Temp. (° C)	k_L (Mg/Cm²)	b (Hr⁻¹)
1144....	315	24.6	4.81
1145....	315	20.2	6.49
1142....	325	20.6	12.96

SUMMARY

The corrosion behavior of plutonium has been investigated in moist and dry air for temperatures up to 75° C. The slow, steady oxidation by dried air is significantly accelerated by water vapor. The parabolic or diffusion-controlled reaction in dry air is changed to a linear or interface-controlled reaction by moisture.

At high temperatures, the behavior in pure oxygen is complex. Complications may arise from the increase in specimen surface temperature above that of the bath. This is one plausible explanation for the "delays" or slow initial rates observed for the runs in the range 315°–325° C. The "delays" may be the time intervals needed to heat the specimen surface. The higher-temperature runs of Figure 5 do not exhibit this phenomenon, presumably because the rate at which heat is released by the reaction exceeds the rate of its loss to the surroundings. The accelerating rates, characterized by the time dependences greater than unity shown in both Figure 5 and Figure 6, could also be attributed to the increase in surface temperature.

It is a pleasure to acknowledge the assistance of many of our colleagues in the Plutonium Metallurgy Group, especially D. D. Whyte and Virginia G. Shadden. Particularly important was the work of Gladys E. Sturdy in developing the technique for handling corrosion specimens. A number of X-ray diffraction determinations of oxide products were made by Finley H. Ellinger, Marian Gibbs, and Petasho Vigil.

REFERENCES

1. SMITH, R. B. *Fire Properties of Metallic Uranium.* (USAEC Rept. TID-8011, April, 1956.)
2. CUNNINGHAM, B. B. "Preparation and Properties of the Compounds of Plutonium,"

in *The Actinide Elements*, ed. G. T. SEABORG and J. J. KATZ, chap. 10, pp. 371–434. ("National Nuclear Energy Series," Div. IV, Vol. **14A.**) New York: McGraw-Hill Book Co., Inc., 1954.

3. LATIMER, W. *Electrochemical Constants*, p. 165. (National Bureau of Standards Circ. No. 524.) 1953.
4. DAVIDSON, N. B. USAEC Rept. CK-2089 (September, 1944).
5. MULFORD, R. N. R., and STURDY, G. E. *Journal of the American Chemical Society*, **77**: 3449–52, 1955; also **78**:3897–3901, 1956.
6. BROWN, F., OCKENDEN, H. M., and WELCH, G. A. *Journal of the American Chemical Society*, **77**:3932–36, 1955.
7. MOLLISON, W. A., ENGLISH, G. C., and NELSON, F. USAEC Rept. CT-3055 (June, 1945).
8. KOLODNEY, M. *Protection of Plutonium against Atmospheric Oxidation*. (USAEC Rept. LA-314, June, 1945.)

CHAPTER XIX

THE MICROGRAPHY OF PLUTONIUM

H. Monti

At the Centre d'Études Nucléaires we have devised techniques for the preparation and examination of metallographic specimens which allow us to work with very small samples of plutonium, some of them weighing no more than 250 mg. A

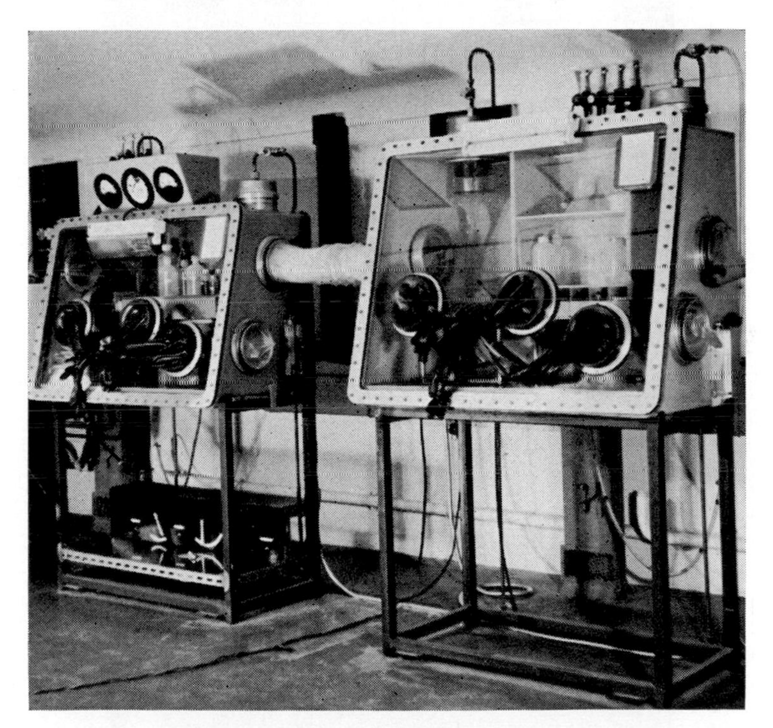

FIG. 1.—Polishing equipment for metallography

sample is mounted, polished first mechanically and then electrolytically, and finally examined under polarized light. Our micrographic equipment, contained in glove boxes, is illustrated in Figures 1–3.

H. Monti is with the Compagnie Saint-Gobain, Centre d'Études Nucléaires de Fontenay-aux-Roses (Seine), France.

Fig. 2.—The microscope shown at the left end in Fig. 1

Fig. 3.—The photomicrographic equipment

MECHANICAL POLISHING

The sample is mounted in "Araldite," a cold-setting resin. Polishing is done by hand on conventional apparatus by means of abrasive papers lubricated with kerosene. Abrasive papers made to be used under water behave very poorly under kerosene, so that it is difficult to obtain a good mechanical polish with very fine paper.

ELECTROLYTIC POLISHING

None of the polishing baths used for uranium has given satisfactory results with plutonium. Baths containing sulfuric acid—for example, 20 ml H_2SO_4 (density = 1.83 gm/ml), 10 ml H_3PO_4 (density = 1.71 gm/ml), and 10 ml glycerine—produce a glossy appearance, but later washing is accompanied by very rapid corrosion, which covers the sample with a grayish layer. The same thing happens when baths based on perchloric acid are used. Polishing in a 30 per cent solution of magnesium perchlorate in ethanol gives better results, but we ordinarily prefer (for safety reasons) to avoid using volatile and inflammable baths.

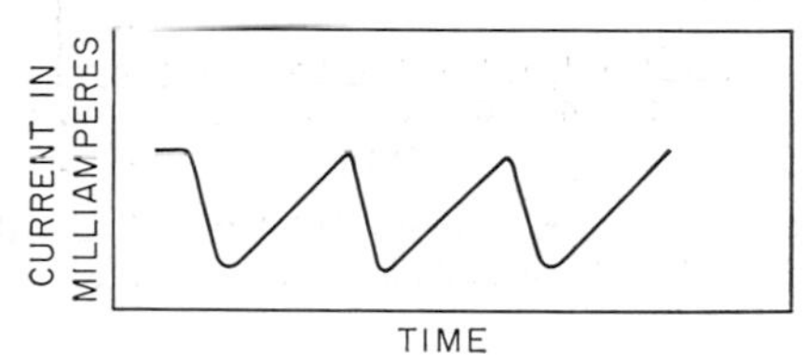

FIG. 4.—Oscillating curve obtained by electrolyzing plutonium in a phosphate bath diluted with ethanol.

To remove a strongly oxidized layer from the surface of plutonium, we use a bath that we had developed for polishing uranium. This bath is prepared by diluting 100 ml of the following solution with 50 ml of ethanol: 100 ml H_3PO_4 (density = 1.71 gm/ml), 40 gm $H_4P_2O_7$, 2 gm CrO_3, and 50 ml H_2O. Surfaces obtained in such a bath are brilliant but are unsuitable for micrographic examination. In Figure 4 is shown a curve representing the behavior of plutonium when electrolyzed in this bath or in mixtures of phosphoric acid and ethanol or methanol. It portrays a succession of polishing and oxidation steps.

Polishing in a bath containing 10 ml of HNO_3 (density = 1.33 gm/ml) and 90 ml of diethylene glycol at 7 volts is rapid (several minutes without agitation). It produces a very brilliant surface, free from pits, in which inclusions are retained. Unfortunately, as was true of all other polishing methods that we had developed previously, the surfaces obtained by this method are optically isotropic and thus do not reveal structure when examined under polarized light.

The only bath we have found that prepares surfaces appropriate for micrographic examination under polarized light is composed of 50 ml H_3PO_4 (density = 1.71 gm/ml) plus 50 ml diethylene glycol. Figure 5 shows the polishing curve obtained when this bath was used with a sample having 0.125 cm^2 of surface. In this bath the current density is very weak—0.056 amp/cm^2—and the polishing time is very long (30 minutes). After too long a time, however, inclusions at the surface are etched out, and the surface becomes pitted. All our efforts to increase the conductivity of the bath by adding water, ethanol, or methanol have been unsuccessful; they induce the oscillating behavior shown in Figure 4, and the polished surfaces are unsuitable for examination.

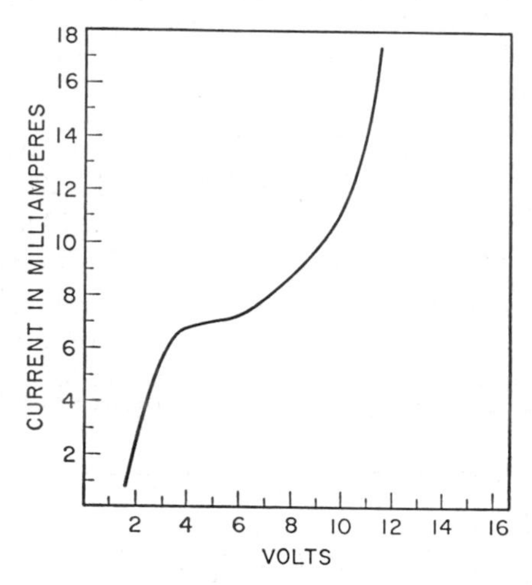

FIG. 5.—Polishing curve for a 0.125-cm^2 sample of plutonium in a bath containing equal parts of H_3PO_4 and diethylene glycol. Total time, 30 minutes.

MICROGRAPHIC EXAMINATION

After successful electrolytic polishing, as just indicated, the structure is clearly revealed by examination under polarized light. Within 1 or 2 days after the sample is polished, it becomes covered with a film of bluish oxide, which considerably increases the contrast.

All the plutonium samples that we have examined, as reduced or after annealing, are characterized by a very fine and fragmented grain structure (see Figs. 6 and 7). This confused structure is not easy to study and interpret. By heating the sample for 32 hours in the temperature range of the beta phase and cooling it slowly in the furnace, we have obtained a considerably larger grain size, as shown in Figures 8 and 9.

MICROGRAPHIC ETCHING

In view of the chemical similarity between plutonium and uranium, we tried the etchants developed for the latter. Especially in a bath of phosphoric acid and diethylene glycol, plutonium is, like uranium, anodically oxidized at the electrolytic decomposition potential. The layer of oxide obtained is uniform on both metals, but whereas on uranium a secondary chemical oxidation develops layers of epitaxial oxide that reveal the structure (1), we have never been able to obtain the same phenomenon with plutonium. The layers of oxide are uniform and, as long as they are not too thick, have only the effect of improving the contrast under polarized

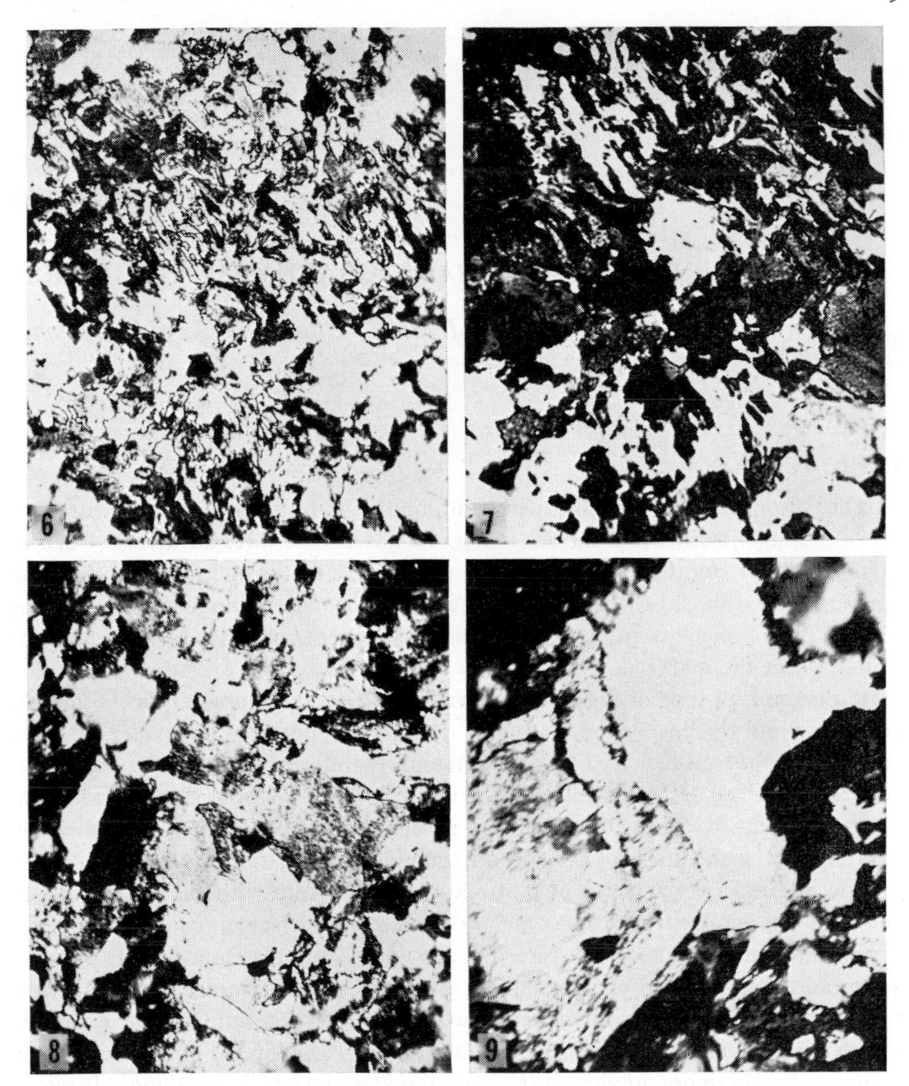

FIGS. 6–9.—Fig. 6: The microstructure of forged alpha plutonium under polarized light. 520×. Fig. 7: The same as Fig. 6. 1,020×. Fig. 8: The microstructure of alpha plutonium after annealing in the beta range (32 hours at 190° C), polarized light. 520×. Fig. 9: The same as Fig. 8. 1,020

light—the same effect, in fact, as occurs with natural oxidation. Likewise, contrary to the behavior of uranium, neither chemical nor natural oxidation of plutonium after electrolytic polishing in perchloric acid baths develops epitaxial layers.

REFERENCE

1. MONTI, H., and BLOCK, J. "Attaque micrographie de l'uranium par oxidation chemique," *Métaux*, No. 375 (November, 1956), pp. 444–50.

CHAPTER XX

RADIOGRAPHY AND AUTORADIOGRAPHY OF PLUTONIUM

Dana E. Elliott and Gerold H. Tenney

INTRODUCTION

The various methods of fabricating plutonium and its alloys are of a nature that precludes any justification for assuming the absence of internal defects, such as voids, flaws, or cavities. In the ultimate uses of this metal such imperfections can constitute a major cause of failure. Therefore, means must be applied for inspecting plutonium components prior to their final assignment, in order to guarantee the soundness of the material.

Radiography is one of the obvious means of inspecting material for its internal condition, and this inspection method is described in detail in this paper. Because of the high physical density of plutonium and, therefore, its high absorption characteristics toward external radiation, various energies of X- and gamma radiations have been studied. These energies were in the range from 300 kv to 22 Mev, depending on the thickness of the specimen under investigation.

The internal radioactivity of plutonium can interfere with the application of conventional non-destructive test methods, and its influence on such tests has to be investigated, not only to devise remedial procedures, but also to utilize its presence as an additional source of information. The use of radiographic films as a permanent recording device introduces the problem of determining the extent to which radiation emanating from the metal influences the quality of the radiographic image. Furthermore, in view of the contamination hazard in handling this metal, investigations have been performed to learn how well the radiation of the metal can be utilized to determine the quality of protective coatings or containers.

DESCRIPTION OF THE SPECIMENS AND THEIR CHARACTERISTICS

Three specimens were selected for this investigation. Two were plutonium step-wedges, illustrated in Figure 1. One wedge consisted of eight steps ranging from $\frac{1}{32}$ to $\frac{1}{4}$ inch, with $\frac{1}{32}$-inch increments. The other wedge consisted of five steps ranging from 0.5 to 1.5 inches, with 0.25-inch increments. Both wedges had a protective nickel coating about 0.006 inch thick. The chemical purity of the plutonium was

The authors are at the University of California, Los Alamos Scientific Laboratory, Los Alamos, New Mexico.

about 98.7 per cent by weight. Impurities included 0.97 per cent aluminum and small amounts of iron, nickel, silicon, and magnesium, to mention only a few of the elements present (1). The physical density of these two specimens was 15.41 gm/cm³.

The third specimen was a rectangular ingot of plutonium-aluminum alloy about 2.5 × 6 × 0.75 inches. This specimen did not have a protective metal coating but was placed in a plastic container (see Fig. 2). This container was sealed, and its outer surface was free of contamination. The alloy specimen contained about 13 per cent plutonium and about 87 per cent aluminum by weight. Its density was about 3.05 gm/cm³.

Fig. 1.—Plutonium step-wedges with 0.006-inch protective nickel coating. One step-wedge ranges from 0.5 to 1.5 inches, with 0.25-inch increments. A smaller step-wedge ranges from $\frac{1}{32}$ to $\frac{1}{4}$ inch, with $\frac{1}{32}$-inch increments. To the left is a lead compensator for the larger step-wedge, used when making radiographs for exposure techniques.

The first two specimens were manufactured for the specific purpose of radiographic exposure experiments and for study of the influence on radiographic films of the radiation emanating from the metal, whereas the alloy in the plastic container was used mainly for the latter type of investigation (autoradiography).

It is therefore appropriate to discuss briefly the radiation from plutonium. Table 1 (2) lists the most important gamma and X-radiations from this metal. Since the absolute abundance of the various energies is primarily a function of the duration of the neutron bombardment to which the parent uranium was exposed in a plutonium-producing reactor, it changes from specimen to specimen. For this reason, the relative abundance is listed. The alpha radiations are purposely omitted because it is assumed that any plutonium metal to be inspected non-destructively has

TABLE 1

GAMMA AND X-RADIATIONS FROM PLUTONIUM

$$\text{Pu-239} \xrightarrow[\text{alpha}]{24{,}400\ \text{y}} \text{U-235}$$

Energy (Kev)	Relative Abundance
39	0.29
53	1.00
100	0.79
124	0.36
384	0.21
677	0.01
20 (X-ray)	1.9×10^{3}

Relative abundance:

$$1=1.6\times10^{5}\times\frac{\%\,\text{Pu-239}}{100}\ \text{gammas/gm/sec}$$

$$\text{Pu-240} \xrightarrow[\text{alpha}]{6{,}600\ \text{y}} \text{U-236}$$

$$\text{Pu-241} \xrightarrow[\text{alpha}]{1.4\times10^{6}\ \text{y}} \text{U-237}$$

Energy (Kev)	Relative Abundance
100	570
145	< 50

Relative abundance:

$$1=1.6\times10^{5}\times\frac{\%\,\text{Pu-241}}{\%\,\text{Pu-239}}\ \text{gammas/gm/sec}$$

$$\text{Pu-241} \xrightarrow[\text{beta}]{14\ \text{y}} \text{Am-241}$$

$$\text{Am-241} \xrightarrow[\text{alpha}]{470\ \text{y}} \text{Np-237}$$

Energy (Kev)	Relative Abundance
60	1.00
26	0.07
21	0.16
17	0.56
13.5	0.32
20 (X-ray)	2.90

Relative abundance:

$$1=2.6\times10^{9}\times\frac{\%\,\text{Pu-241}}{\%\,\text{Pu-239}}\times\text{age in years}$$

been previously covered with a protective layer and successfully checked against any alpha leakage. On the other hand, the presence of gammas ranging from 39 to 677 kev can well influence the quality of a radiographic image. Some elaborate studies have been performed to obtain data on this subject.

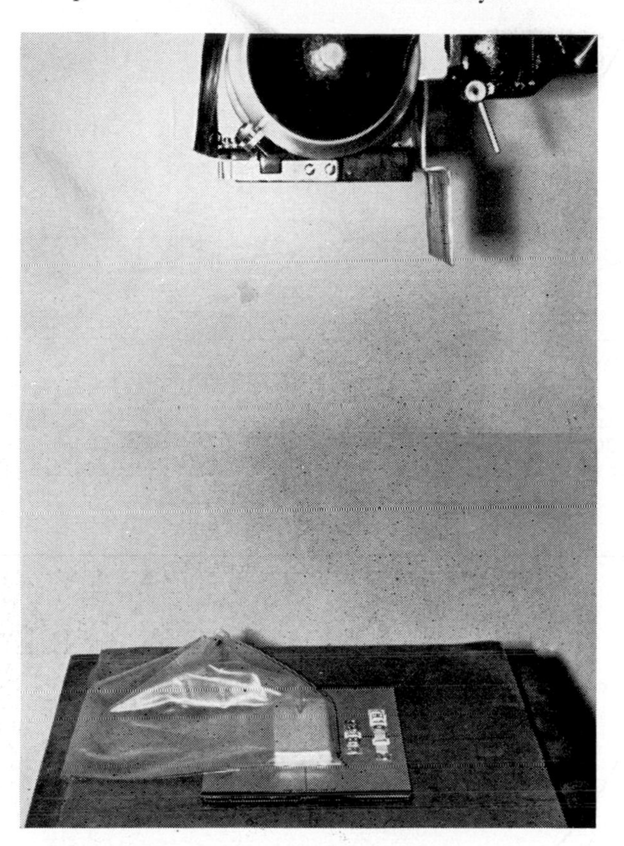

FIG. 2.—Photo of Pu-Al alloy in plastic container to be radiographed with 150-kvp X-ray unit

AUTORADIOGRAPHY WITH PLUTONIUM

Because of the relatively high physical density of the metal, the use of fine-grain and, therefore, slow radiographic film would not seem to be economical when inspecting specimens beyond a certain thickness; a relatively fast, but at the same time medium-grain, film has therefore been chosen for all autoradiographic experiments. It is the newest film that Eastman Kodak Company has added to its list: the Industrial Type AA film.

Since in many cases the plutonium part is of such geometry that the distance from the various points of its surface to the film varies, when being autoradiographically inspected, investigations were performed to study the influence of various object-to-film distances when exposing the film to the radiation of plutonium. For an arbitrarily chosen exposure time of 30 minutes, data were obtained as plotted in Figure 3. It is of interest to note that, when autoradiographing the specimen in the

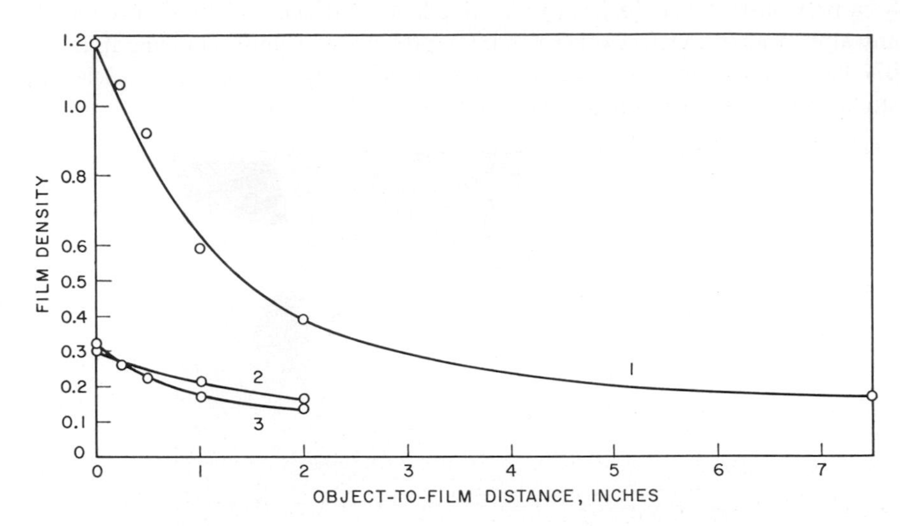

FIG. 3.—Autoradiography with plutonium. Film density versus object-to-film distance. Kodak Type AA film; 30-minute exposure time. *1*, Pu-Al alloy in plastic container; *2*, Pu-Al alloy in plastic container screened by 0.006 inch of nickel; *3*, plutonium with 0.006-inch protective nickel coating.

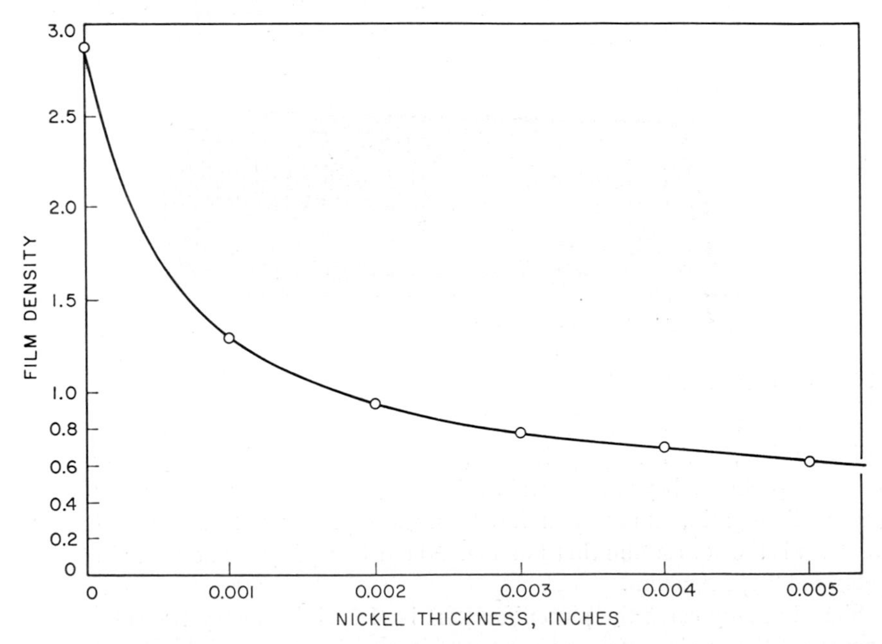

FIG. 4.—Autoradiography of Pu-Al alloy in plastic container. Film density versus nickel thickness. Kodak Type AA film; 30-minute exposure time.

plastic container, the absorption of radiation by 3 inches of air decreased the original film blackening by about 75 per cent. On the other hand, the influence of increased distance is almost negligible when 0.006 inch of nickel is placed between the alloy and the film or when the metal has a 0.006-inch nickel coating. These data, of course, indicate that soft gammas contribute appreciably to the film fogging and are almost completely absorbed by 6 mils of nickel.

The next experiment dealt with the investigation of the influence of increased nickel thickness between the ingot in the plastic bag and the radiographic film. Again an arbitrary exposure time of 30 minutes was selected. As presented in Figure 4, the data obtained indicate that the first mil of nickel decreases the film density by about 55 per cent, whereas the second mil decreases it by about an additional 22 per cent, and the third mil only by about another 10 per cent.

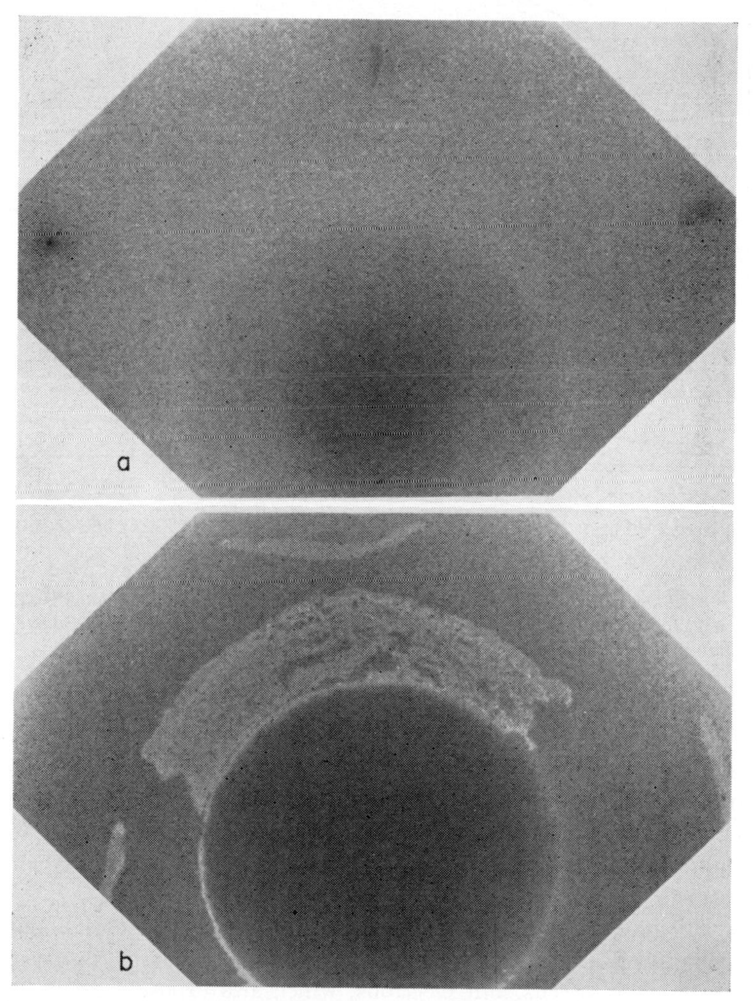

FIG. 5.—Autoradiograph of plutonium with non-homogeneous protective nickel coating: *a*, areas with thinner coatings than surroundings; *b*, areas with thicker coatings than surroundings.

This relation between nickel thickness and film density can also be utilized for a relatively accurate measurement of the thickness of a protective nickel coating. Although 1 mil of nickel definitely absorbs all the alphas and therefore prevents any contamination and eliminates any health hazard, nevertheless thicker protective nickel coatings might be desirable because of their capability of withstanding mechanical forces. In such cases it might be desirable to measure the uniformity and thickness of such coatings. Autoradiography can be applied for this purpose. Figure 5 is a typical example of this type of application.

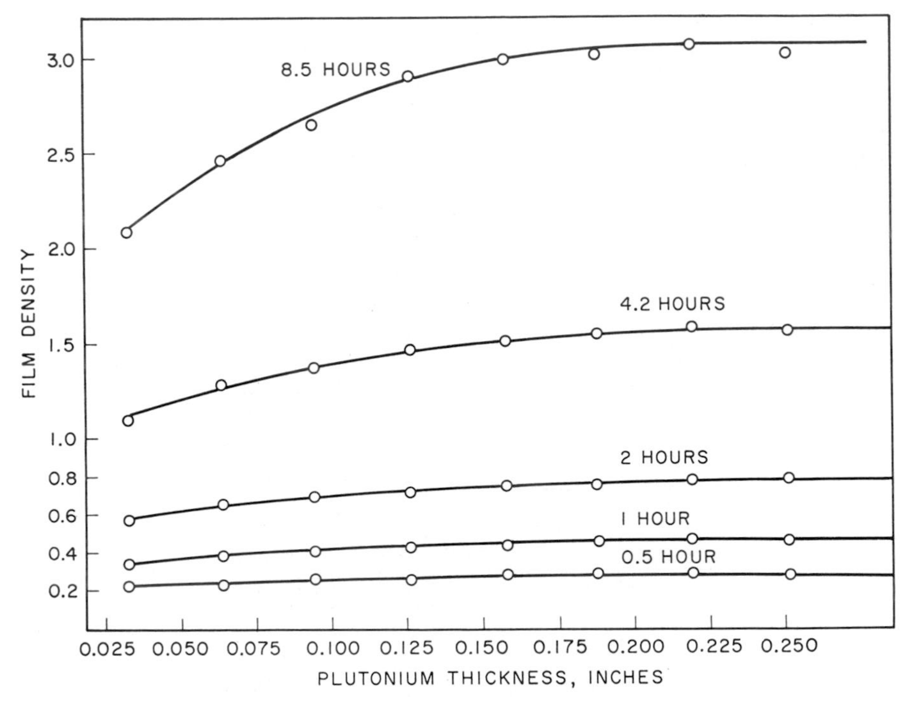

FIG. 6.—Autoradiography with plutonium and 0.006-inch protective nickel coating. Film density versus plutonium thickness for 0.5-, 1-, 2-, 4.2-, and 8.5-hour exposure times. Kodak Type AA film and 0.005-inch lead front screen.

Of further interest was the determination of the "infinite plutonium thickness" necessary to absorb the radiation of the metal originating beyond this thickness. The smaller step-wedge was used for these experiments. Radiographic films in a cardboard holder having 0.005-inch lead front screens were exposed for different times. These data are presented in Figure 6. The various exposure curves flatten out beyond a plutonium thickness of about 0.2 inch. It can therefore be deduced that this is the "infinite thickness" for practical purposes.

When radiographing material with radiation sources of energies of about 120 kv or more, films are placed between lead screens which serve a dual purpose. In this energy range these screens emit electrons, which also expose the film emulsion, thereby decreasing the needed exposure time. For this reason they are called "intensifying screens." At all times and independent of the radiation energy, they also act as filters toward any scattered radiation, whether it comes from the object under investigation (object-scatter) or from the room and other objects placed in the surrounding areas (room-scatter). Thus it is of interest to investigate the influence of various lead front screens as filters for the radiation emanating from plutonium. Experiments were performed using 0-, 20-, 80-, and 120-mil lead front screens with 5-mil lead back screens in all cases. A step-wedge and the ingot were used in this experimental series. Figure 7 contains the data obtained. The film den-

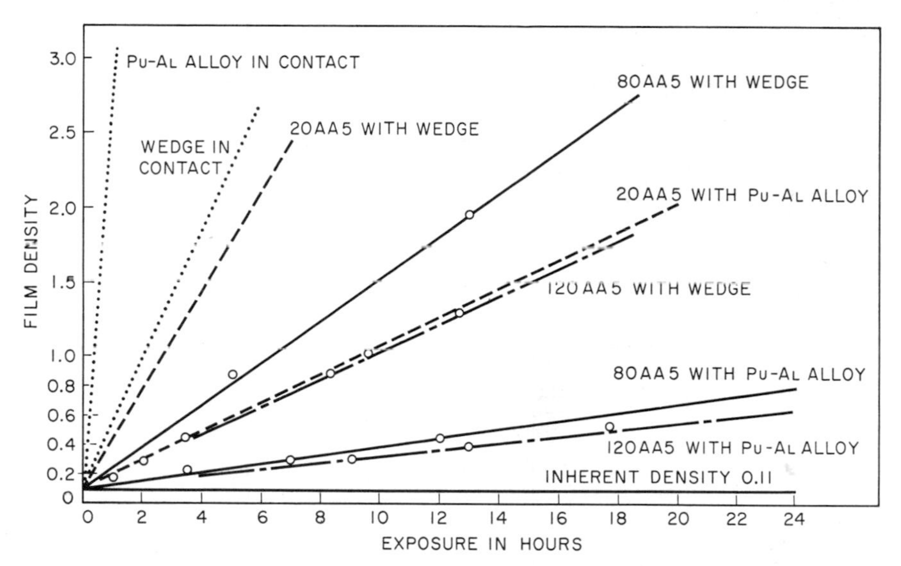

FIG. 7.—Autoradiography of plutonium and Pu-Al alloy. Influence of various thicknesses of lead front screens on film density versus exposure. Kodak Type AA film.

sity for all these cases is plotted against exposure time. This information can be of value when radiographing plutonium or its alloys. If the required exposure time is known, the respective thickness of the lead front screen can be selected to minimize the influence of the plutonium radiation on the radiograph.

RADIOGRAPHY OF PLUTONIUM

It is well known that radiographic exposure techniques vary with the plutonium content of various alloys. It is therefore impractical to investigate all probable combinations. On the other hand, it is felt that radiographic exposure data for almost pure plutonium for various energies of radiation might be of some value in establishing the range within which certain specific problems can be solved.

In order to develop these techniques, the two previously described step-wedges were used as specimens. Various sources of radiation were investigated, including a 300-kv constant-potential X-ray unit, a 1,000-pkv resonance-transformer type of X-ray unit, a 22-Mev betatron, and the following isotopes: cesium-137 and cobalt-60. The respective radiographic exposure conditions are listed in Table 2, which also includes the projected source areas, the radiation intensity, the source-to-film distances, and the respective film-screen combinations. In all cases, vacuum cassettes were used, thereby insuring intimate screen-to-film contact and consequently the utmost sharpness of the radiographic image (3).

Figure 8 gives the exposure techniques for plutonium from $\frac{1}{32}$ to $\frac{1}{2}$ inch in thickness. Because of this lower thickness range, the slower but finer-grain Kodak Industrial Type M film was used. All films were developed in General Electric liquid developer for 6 minutes at 68° F. The curves of Figure 8 are plotted for a film density of 1.5.

TABLE 2

RADIOGRAPHIC EXPOSURE CONDITIONS FOR VARIOUS SOURCES OF RADIATION

Source of Radiation	Projected Source Area (Mm)	Intensity	Source-to-Film Distance (Inches)	Film-Screen Combination
300-kv Philips X-ray unit...	5	10 MA	48	5M5
1,000 pkv G.E. X-ray unit, reflected beam.	12×3	3 MA	72	40/5AA5/40 40/5M5/40
1,000 pkv G.E. X-ray unit, transmitted beam........	12	3 MA	72	40/5AA5/40 40/5M5/40
Cesium-137, 0.63 Mev....	10×10	63 curies	36	40AA5/250 40M5/250
Cobalt-60, 1.1 and 1.3 Mev..	7.8	17.1 curies	36	40AA5/250 40M5/250
Betatron, 22 Mev.........	0.25×0.12	100 roentgens per minute per meter	144	120AAO 120MO

Figure 9 gives the exposure techniques for plutonium having a thickness of from 0.25 to 1.5 inches. In this case Kodak Type AA film was used. Otherwise, all other factors remained unchanged. Note particularly the shaded area in Figure 9, obtained when cesium-137 was used as the source of radiation. This area is determined by the theoretical exposure curve (*dashed*) and the experimentally obtained curve calling for a shorter exposure time because of the influence of the autoradiation of the plutonium.

It is of interest to know what radiographic sensitivity or radiographic resolution can be expected. By the term "per cent radiographic sensitivity" (3) one means the size of the smallest radiographically distinguishable cavity, expressed in per cent of the total thickness of the specimen. It has been found necessary to abandon this nomenclature for heavy metals such as plutonium and uranium because of the ultimate use of the parts inspected. Since, in most of the cases known to the authors, the fabricators, as well as possible users of radiographically inspected plutonium parts, have been mainly interested in the absolute dimensions of the voids, the term "radiographic resolution" has been introduced to mean the size of the smallest radiographically detectable defect in mils or millimeters. Penetrameters with cylindrical holes having diameters equal to their heights are used for this purpose. Normally, such penetrameters are manufactured from the same material as the specimen under investigation. In the case of plutonium this is impractical because of the health hazard connected with the metal.

Many experiments were performed to try to find a substitute material. A very satisfactory answer has been found in the metal tantalum, which demonstrates absorption characteristics similar to this type of plutonium. For these investigations, tantalum penetrameters were placed on the surface facing the source of radiation, in order to evaluate the radiographic resolution of the various exposure techniques.

Table 3 lists these values for Type M film and thicknesses of plutonium ranging from $\frac{1}{32}$ to $\frac{1}{2}$ inch, whereas Table 4 lists them for Type AA film and a thickness

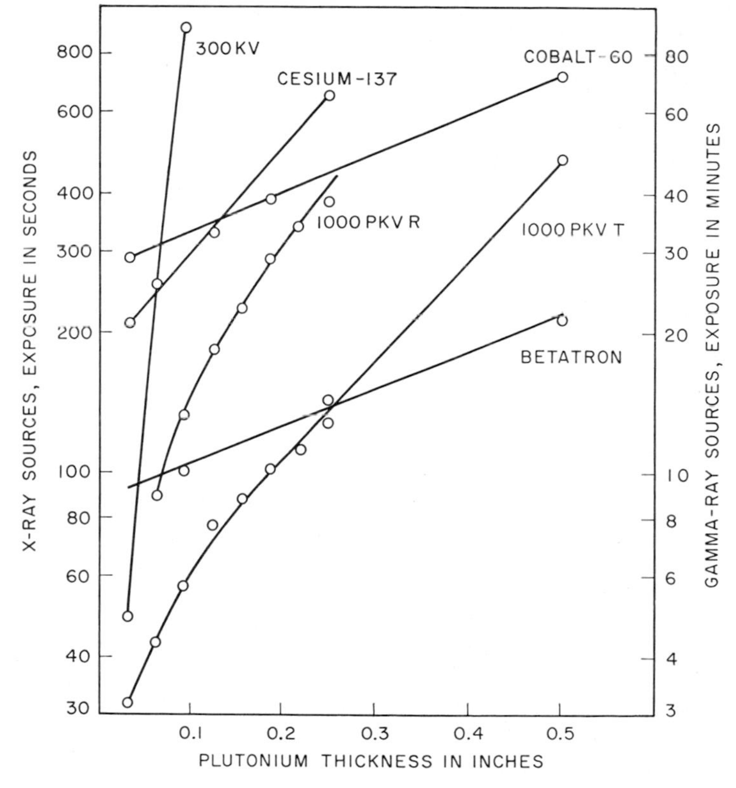

FIG. 8.—Exposure times versus plutonium thickness up to 0.5 inch for various X-ray and gamma-ray sources. Kodak Type M film. Screens appropriate to source. Average film density 1.5. Developed 6 minutes at 68° F in EK liquid developer. Target-to-film distance as listed in Table 2.

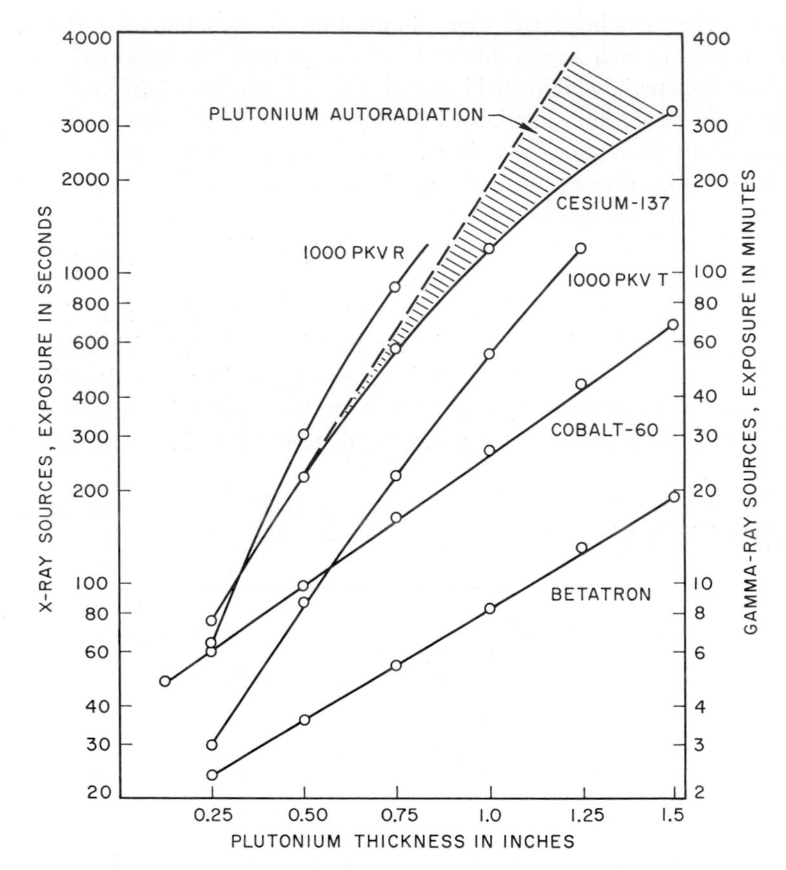

FIG. 9.—Exposure times versus plutonium thickness up to 1.5 inch for various X-ray and gamma-ray sources. Kodak Type AA film. Screens appropriate to source. Average film density 1.5. Developed 6 minutes at 68° F in EK liquid developer. Target-to-film distance as listed in Table 2.

TABLE 3*

RADIOGRAPHIC RESOLUTION FOR VARIOUS THICKNESSES OF PLUTONIUM RANGING FROM $\frac{1}{32}$ INCH TO $\frac{1}{2}$ INCH, WITH VARIOUS SOURCES OF RADIATION

Thickness (Inch)	300 KV	1,000 pkv Reflected	1,000 pkv Transmitted	Cs-137	Co-60	Betatron	Betatron Enlarged 2×
0.03125	0.004	0.010	0.010	0.010	0.020	0.020	0.010
.0625	0.006	.010	.010	.010	.020	.020	.010
.125	....	.010	.010	.010	.020	.020	.020
.25	....	.020	.020	.020	.020	.020	0.020
0.5	....	0.020	0.020	0.020	0.030	0.020	....

* Kodak Type M film. For exposure conditions, see Table 2.

range from $\frac{1}{4}$ to $1\frac{1}{2}$ inches. When the most satisfactory source of radiation is chosen, 0.020-inch cavities can be detected in this thickness range most of the time.

Figure 10 is a radiograph of the plutonium-aluminum alloy ingot, which shows that the alloy was not homogeneous or free of cavities. It also shows 0.020-inch penetrameters of iron, aluminum, tatalum, and titanium, thereby illustrating the radiographic contrast caused by the different absorption coefficients of these materials.

Figure 11 is a good example of the detection of radioactive impurities present in the surface of the ingot by autoradiographic means. In photo *b* the plastic container in which the specimen is placed is in direct contact with the film; in photo *c* the object-to-film distance is $\frac{1}{4}$ inch, great enough to absorb all the radiation coming from the radioactive impurities; and in photo *a*, 0.001-inch nickel is placed between the specimen and the film. It seems that this amount of nickel absorbs about as much as does $\frac{1}{4}$ inch of air.

TABLE 4*

RADIOGRAPHIC RESOLUTION FOR VARIOUS THICKNESSES OF PLUTONIUM RANGING FROM $\frac{1}{4}$ INCH TO $1\frac{1}{2}$ INCHES, WITH VARIOUS SOURCES OF RADIATION

Thickness (Inches)	1,000 pkv Reflected	1,000 pkv Transmitted	Cs-137	Co-60	Betatron
0.25	0.020	0.020	0.020	0.020	0.030
0.5	.020	.020	.020	.020	.030
0.75	0.020	.020	.020	.030	.030
1	....	.030	.030	.030	.030
1.25	....	0.030	0.030	.030	.030
1.5	....	....	....	0.030	0.030

* Kodak Type AA film. For exposure conditions, see Table 2.

CONCLUSION

By selecting the right energy of radiation, satisfactory radiographic inspection of plutonium can be performed, thereby giving valuable information about the internal physical condition of the material. When familiar with the type of radiation originating in this metal, radiographers can undertake preventive measures to avoid its damaging influence on the radiographic image. At the same time, this radiation can be utilized to study the quality of the coatings surrounding the material. Such study is absolutely necessary to protect the objects under investigation, as well as the lives of workers handling this material.

The authors wish to express their appreciation to Messrs. W. J. Maraman and J. W. Anderson, who were responsible for the manufacture of the step-wedges; to Mr. W. C. Herrmann for the photographic illustrations; and to Mr. Thomas L. Hamblet for the drawings.

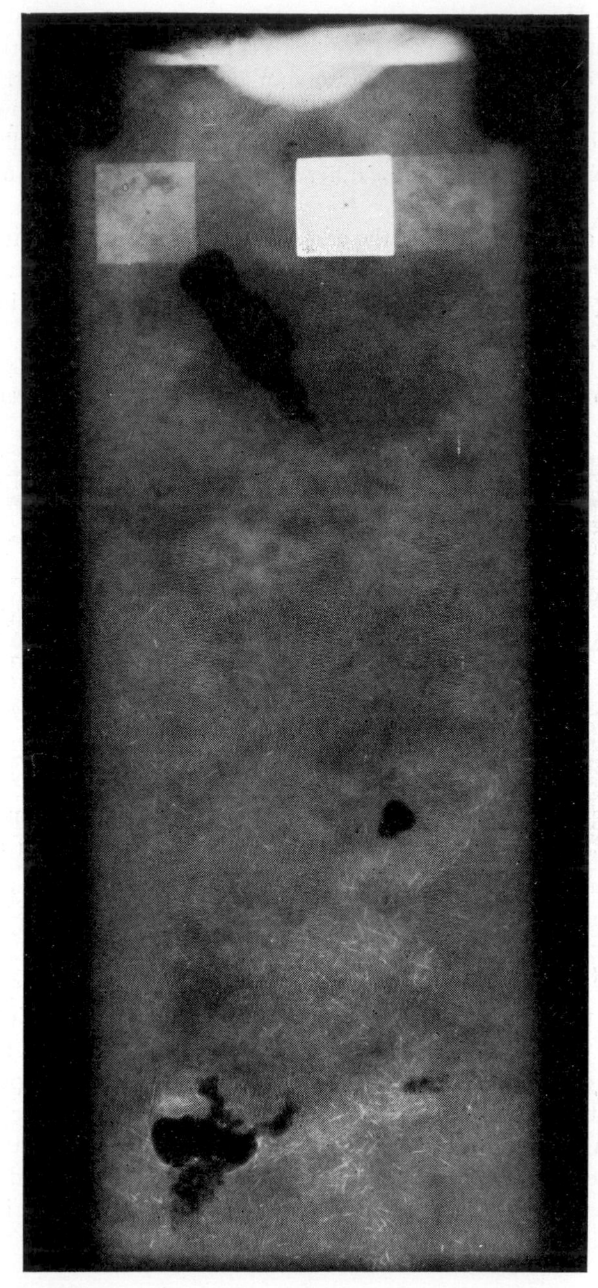

FIG. 10.—Radiograph of Pu-Al ingot, $\frac{3}{4}$ inch thick. 0.020-inch penetrameters from left to right Fe, Al, Ta, and Ti. Radiographic techniques: 300 ckv; 400 MA sec; 48-inch target-to-film distance; 5M5.

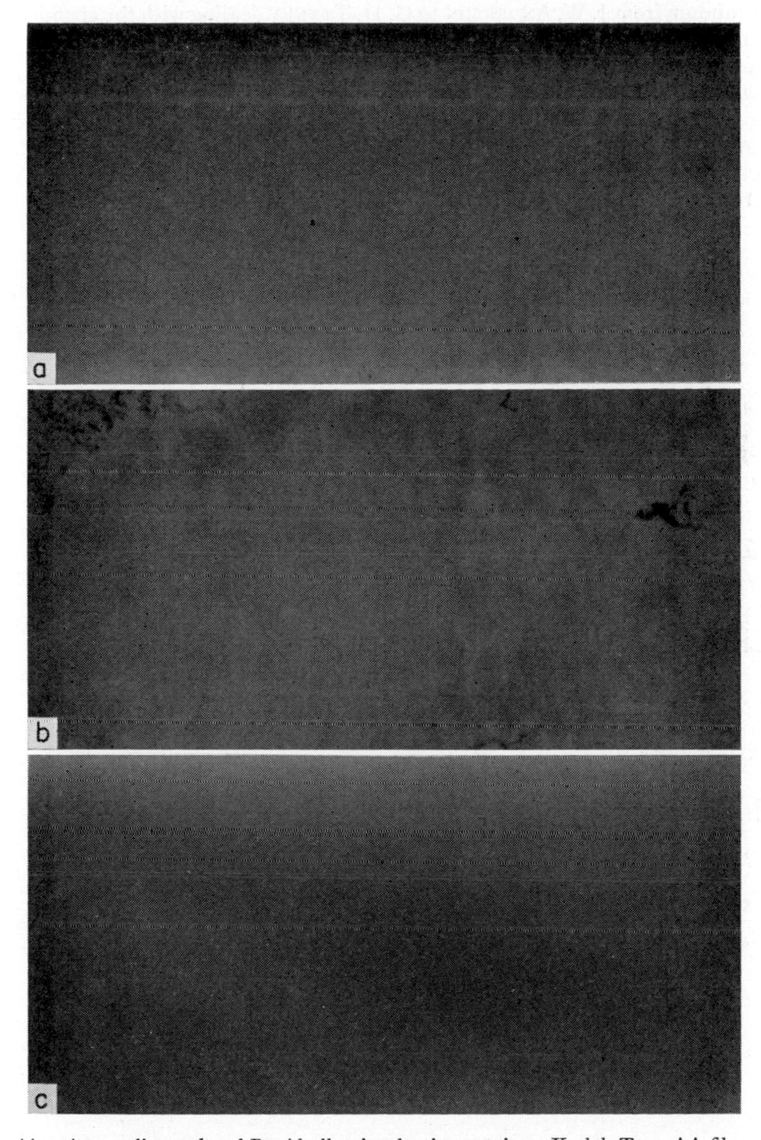

FIG. 11.—Autoradiographs of Pu-Al alloy in plastic container. Kodak Type AA film. *a*, 0.001-inch nickel between object and film; 75-minute exposure; obtained film density 1.16. *b*, Object-to-film distance 0 inch; 30-minute exposure; obtained film density 1.36. *c*, Object-to-film distance $\frac{1}{4}$ inch; 30-minute exposure; obtained film density 1.06.

REFERENCES

1. Memorandum from J. W. ANDERSON to G. H. TENNEY dealing with the chemical purity of two step-wedges, September 26, 1957.
2. Memorandum from J. F. KEPHART and E. H. PLASSMANN to G. H. TENNEY dealing with gamma and X-radiation from plutonium, containing data extracted from HOLLANDER, PERLMAN, and SEABORG, *Review of Modern Physics*, **25**:469, 1953, and from laboratory notes, dated October 2, 1957.
3. TENNEY, GEROLD H. *Nondestructive Testing of Uranium.* (A.S.T.M. STP No. 213, 1957.)

CHAPTER XXI

PHASE DIAGRAMS OF PLUTONIUM ALLOYS STUDIED AT HARWELL

M. B. Waldron

Studies of the constitution of plutonium-alloy systems have been undertaken at the Atomic Energy Research Establishment from two points of view. First, there are those systems that are of outstanding technological importance either to the reactor or to the weapons program; second, there are those systems that are of interest because of their possible contribution to our knowledge of the alloying behavior of the transuranic series of elements. However, as in the case of uranium alloys, certain systems studied for fundamental reasons have proved to be of considerable practical importance. Because of the large amount of work that has to be done with a relatively small scientific effort and the fact that, until recently, a very major part of the available effort was absorbed in building up facilities, few complete systems of a declassifiable nature have been established. A larger number of systems have been studied in specific composition ranges. In most cases this has only established information that is described in more detail in the other contributions to this book from the A.E.R.E. However, a few details which supplement these contributions will also be described.

The general features of the glove boxes used at A.E.R.E. have been described by Lee and Mardon in chapter xiv, and many of the constitutional studies have used the same equipment. However, more recently a suite of boxes has been commissioned by Williamson, Poole, and Marples (1) for alloy studies in which the keynote is versatility. Two typical boxes are shown in Figures 1 and 2. Standard boxes using rubber sealing strip of a kind made for automobile windows are employed, and the equipment inside is made up of parts that are sufficiently small to pass through the port used for our standard transfer bags. If possible, commercial products are employed; for example, an automobile windscreen wiper and an electric drill are combined to form a metallographic polishing machine. The vacuum furnace in the second photograph can be used for dilatometry, thermal analysis, or halide reductions, or it can quickly be interchanged with an argon-arc furnace. Two of the phase diagrams described later were determined in a suite containing five boxes of the type shown in Figures 1 and 2.

M. B. Waldron is with the Atomic Energy Research Establishment, Harwell, Berkshire, United Kingdom.

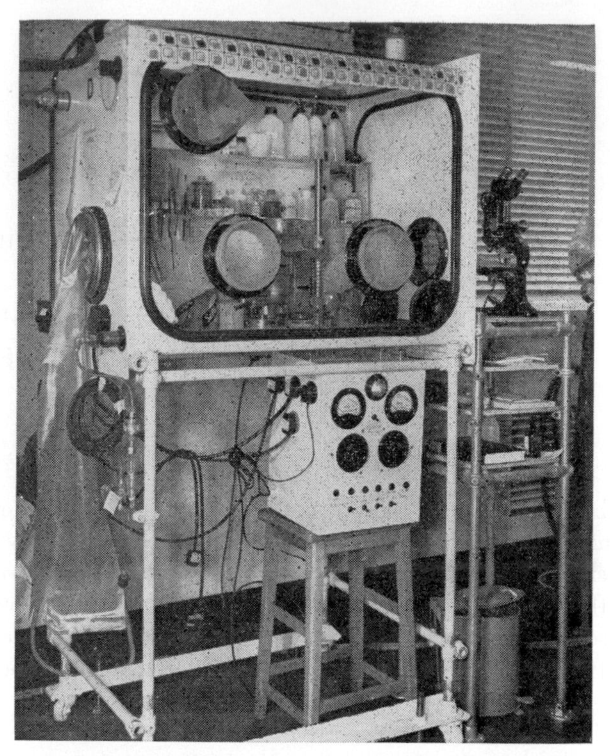

FIG. 1.—A glove box equipped for the metallographic polishing of plutonium

FIG. 2.—A glove box equipped with a vacuum arc-melting furnace

Apart from the use of glove boxes, the techniques employed were typical of any investigation of phase diagrams of the rarer metals. Alloys were usually prepared by arc melting. Samples weighed up to 10 gm, though usually considerably less than 1 gm; thermal analysis, dilatometry, metallography, and X-ray crystallography were all employed. Details of these techniques have been reported in published papers (1–3).

THE PLUTONIUM-IRON SYSTEM (2)

The form of the diagram of the plutonium-iron system is shown in Figure 3. It is closely similar to the diagram published by Konobeevsky (4) when our investigation was well in hand. We completed the study for two reasons: (1) to obtain a comparison between results found by us by conventional methods and those obtained by the Russians using microtechniques and (2) to determine the phase equilibria involving δ- and ϵ-plutonium, which were not shown in the Russian diagram.

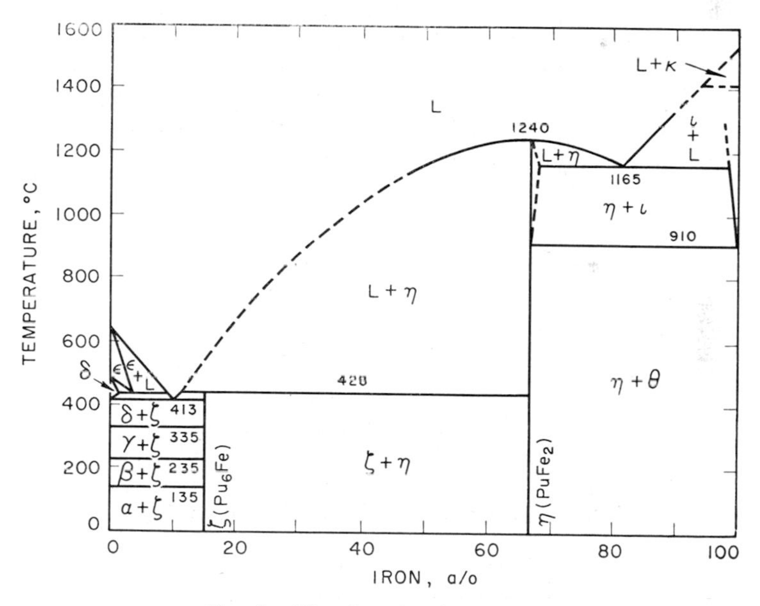

FIG. 3.—The plutonium-iron system

The agreement with the Russian work is good, the major discrepancies being that $PuFe_2$ was found to melt at 1,240° ± 5° C instead of 1,230° C, and the $PuFe_2$-iron eutectic temperature was determined as 1,165° ± 5° C instead of 1,180° C. The plutonium-rich end of the system is shown in Figure 4, where an unusual form of eutectoid reaction, from ϵ to δ + liquid, occurs. This partial melting is confirmed by the record of two dilatometer runs shown in Figure 5, which indicate that an abrupt collapse of the specimen occurred in the region of 405°–425° C on both heating and cooling. The series of photomicrographs, Figures 6–9, illustrates the results of quenching a 1.5 a/o iron alloy from temperatures within this field.

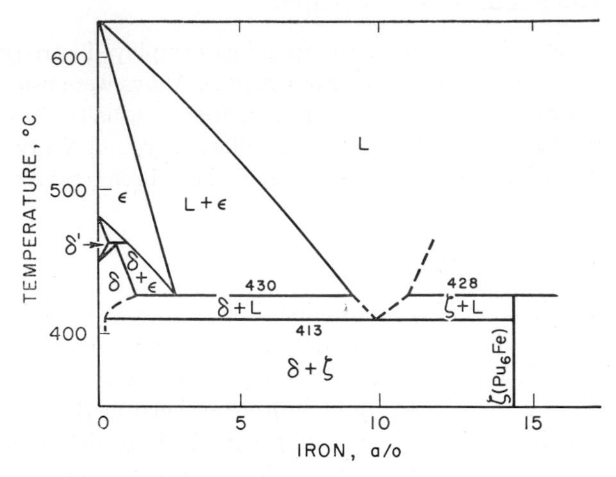

FIG. 4.—Enlargement of the plutonium-rich end of the plutonium-iron system

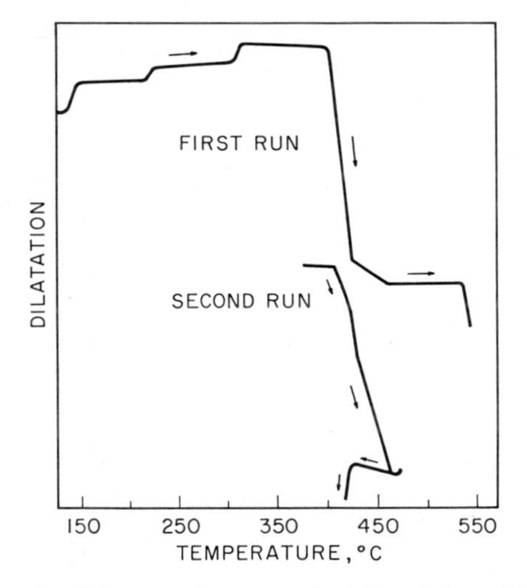

FIG. 5.—Dilatometric curves of a 1.08 a/o iron alloy

TABLE 1

PHASE	STRUCTURE	UNIT-CELL DIMENSIONS (A)			
		Harwell	Russian	U.S.	Canadian
Pu_6Fe......	Body-centered tetragonal	$a = 5.349 \pm 0.003$ $c = 10.405 \pm .005$	5.347 10.403	5.359 ± 0.004 10.41 ± 0.01	....
$PuFe_2$......	Face-centered cubic, C15	Pu-rich: $a = 7.189 \pm .002$ Fe-rich: $a = 7.177 \pm 0.004$ (1150° C)	7.178	7.191 ± 0.001	7.190 ± 0.005 7.150 ± 0.005

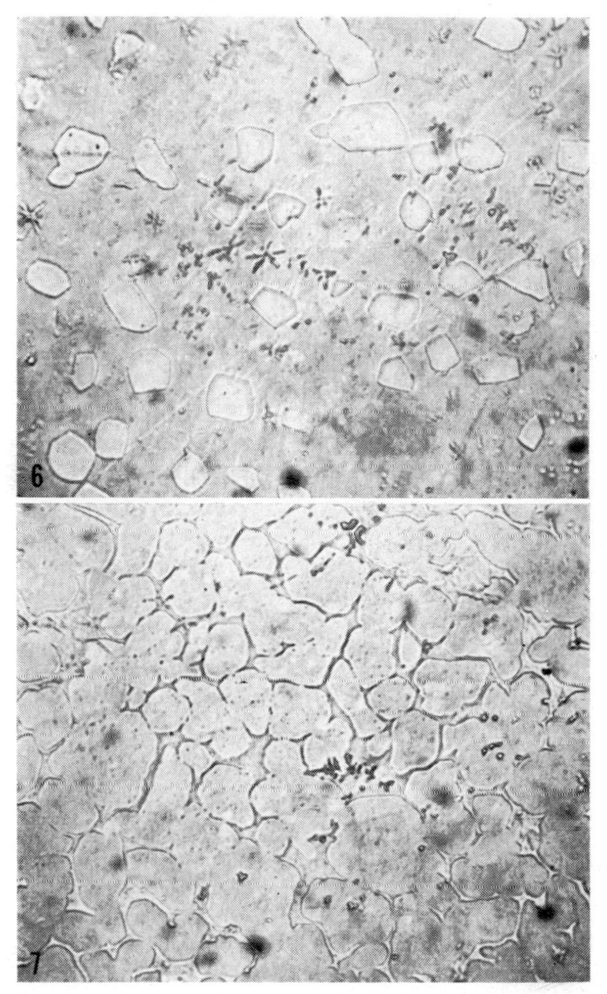

FIGS. 6–7.—Fig. 6: Plutonium–1.5 a/o iron alloy quenched from 420° C. Light-gray globules of liquid in δ-plutonium (transformed to α at room temperature). 1,250×. Fig. 7: Plutonium–1.5 a/o iron alloy quenched from 445° C. δ- and ϵ-plutonium at this temperature. On quenching, the δ-phase has transformed (ultimately to α at room temperature) and the ϵ has undergone retrograde melting, followed by resolidification, thus producing the appearance of an intergranular liquid phase in the microstructure. 1,250×.

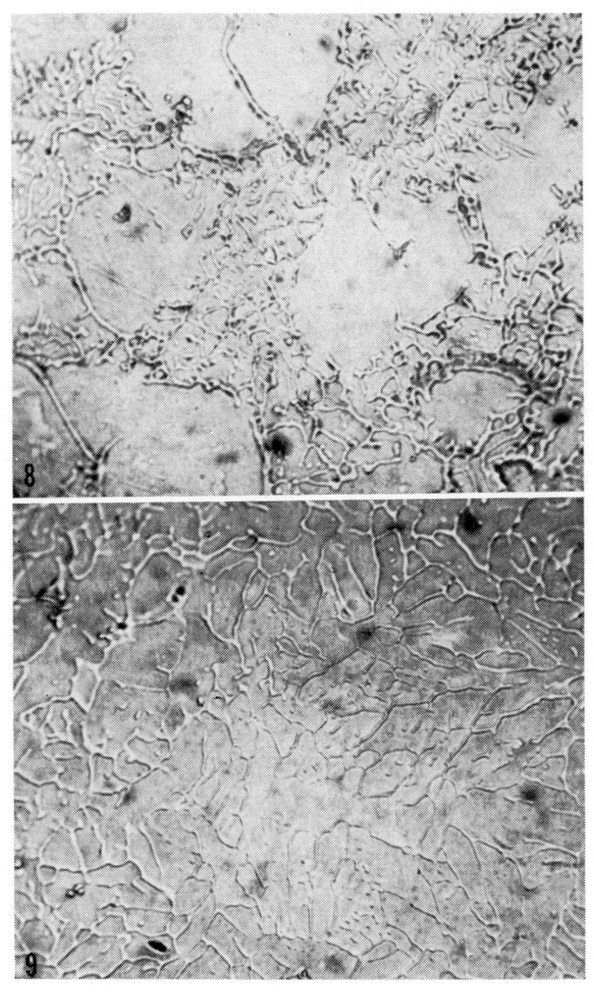

FIGS. 8–9.—Fig. 8: Plutonium–1.5 a/o iron alloy quenched from 465° C. Largely ε-plutonium at this temperature. On quenching, the retrograde melting of ε has again produced the appearance of a network of liquid. 1,250×. Fig. 9: Plutonium–1.5 a/o iron alloy quenched from 490° C. All ε-plutonium at this temperature. The partial retrograde melting that occurred on quenching from this temperature produced a more uniform and continuous network of the liquid phase. 1,250×

The crystal structures of the compounds were confirmed by powder patterns taken with an 11.4-cm Philips camera modified to make it suitable for handling alpha-active specimens, the lid of the camera being replaced by a light-trap fitting close to the film, thus permitting the specimen to be loaded and removed from the camera in a glove box without contaminating the film. Copper K-α radiation was used for specimens lying between plutonium and $PuFe_2$, cobalt K-α being necessary for more iron-rich alloys. The unit-cell dimensions found are given in Table 1.

THE PLUTONIUM-THORIUM SYSTEM (3)

The phase diagram of the plutonium-thorium system is shown in Figure 10. The large solubility of plutonium in α-thorium is noteworthy and was established by metallography, lattice parameter measurements, thermal analysis, and dilatometry. The peritectic formation of the ζ-phase is believed to represent the equilibrium

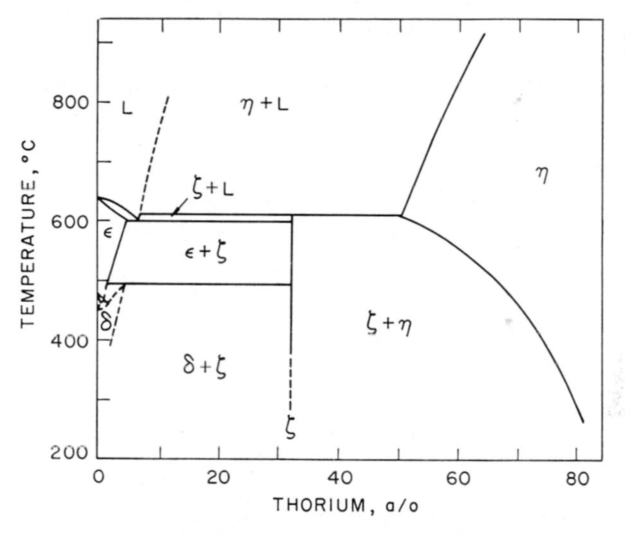

FIG. 10.—The plutonium-thorium system

form of the diagram, and arrests corresponding to this were found on heating. On rapid cooling, formation of the ζ-compound appears to be depressed below the eutectic temperature at 605° C, so that it forms by a non-equilibrium peritectoid reaction at a temperature that depends on the cooling rate. The heating and cooling curves for a 30 a/o thorium alloy illustrate these effects (Fig. 11). Further support for this form of the diagram comes from the microstructure of alloys quenched from within the supposed liquid + ζ field (Fig. 12). This photomicrograph shows that the solid phase which exists in equilibrium with liquid at 602° C is the ζ-compound, consisting of angular particles. Figure 13, on the other hand, shows that the solid phase which coexists with liquid at 614° C is α-thorium, the particles of which are rounded.

The ζ-phase appears to contain between 30 and 33 a/o thorium and has a complex diffraction pattern that can be indexed on the basis of an orthorhombic unit

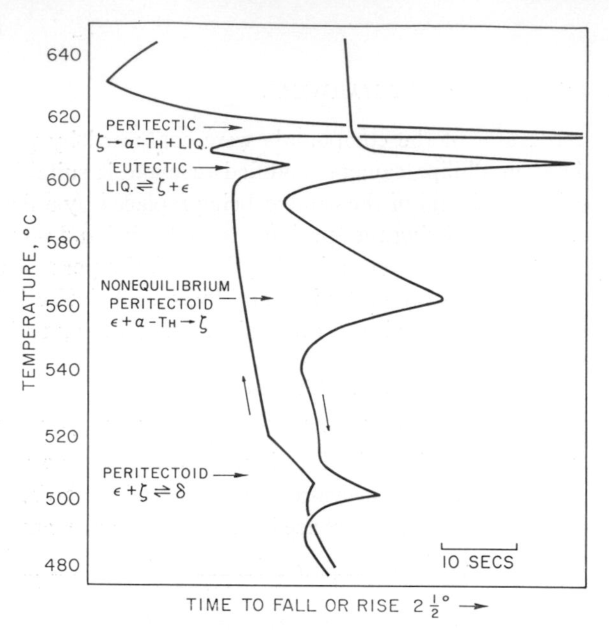

FIG. 11.—Thermal-analysis curves of a 30 a/o thorium alloy

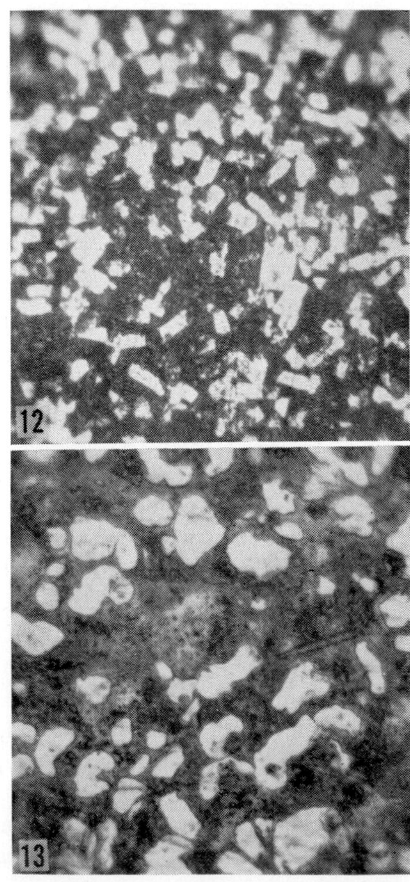

FIGS. 12–13.—Fig. 12: Plutonium–15 a/o thorium alloy annealed at 602° C for $\frac{3}{4}$ hour and quenched. Angular grains of the ζ-compound appear white in a matrix of the quenched liquid. 900×. Fig. 13: Plutonium–15 a/o thorium alloy annealed at 614° C for $\frac{1}{2}$ hour and quenched. Rounded grains of α-thorium appear white in a matrix of the quenched liquid. 900×.

cell with $a = 9.820 \pm 0.005$ A, $b = 8.164 \pm 0.005$ A, and $c = 6.681 \pm 0.005$ A. If this phase has the formula $Pu_{13}Th_6$ (31.6 per cent Th), good agreement is found with measured densities.

The measured densities of the entire series of plutonium-thorium alloys are plotted in Figure 14. It will be seen that plutonium dissolved in α-thorium has an

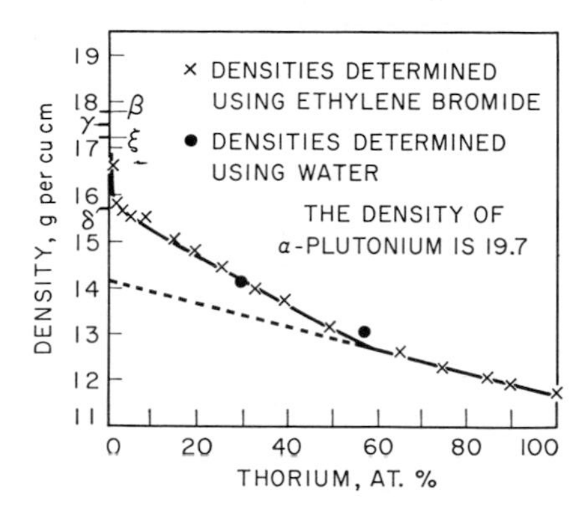

FIG. 14.—The densities of plutonium-thorium alloys as a function of their composition

effective atomic size considerably greater than even that of δ-plutonium. This will be commented on later. Further details of the plutonium-thorium system have been reported by Poole, Williamson, and Marples (3).

THE PLUTONIUM-URANIUM ALLOYS (5)

Investigation of the system of plutonium-uranium alloys had just begun at Harwell when the diagram determined at Los Alamos was made available to us. Because of its great technological and fundamental interest, work on this diagram is continuing, particularly in those sections where tentative results are shown in the American work. The essential features of the diagram are confirmed, as indicated in Figure 15, but there are some differences that can be noted. In particular, we have established the field as indicated in the diagram for the homogeneity range of the body-centered cubic phases, ϵ-plutonium and γ-uranium.

Thermal-analysis results suggest eutectoid decompositions from ϵ to $\delta' + \eta$ and from δ' to $\delta + \eta$, as shown. The lower-temperature transformations have been extensively studied by thermal analysis and dilatometry without entirely confirming the two transitions shown by the Los Alamos diagram at 271° and 279° C.[1] We have found the arrest in alloys between 0 and 4 w/o uranium at approximately 280° C and an arrest at 295° C in alloys between 5 and 25 w/o uranium, but no evidence was found for both arrests, even in the alloys containing 4 and 5 w/o uranium. However, we cannot propose an alternative form of the diagram that fits all our observations.

We have obtained a diffraction pattern of the plutonium-uranium ζ-phase that

[1] The Los Alamos version of the plutonium-uranium phase diagram has recently been revised in this region; see chap. xxii.

can be indexed according to the unit cell proposed by Coffinberry and Ellinger (6) with lattice parameters as given in the accompanying table. These values compare

w/o Plutonium	Unit-Cell Edge	w/o Plutonium	Unit-Cell Edge	w/o Plutonium	Unit-Cell Edge
30..........	10.60±0.02 A	50.........	10.61±0.01 A	65.........	10.65±0.01 A

with a value 10.664 ± 0.005 A given by the Los Alamos workers for the composition UPu (6).

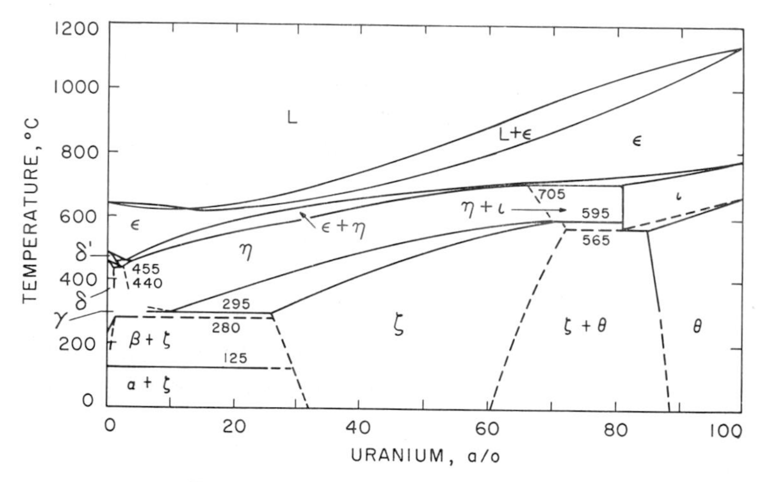

FIG. 15.—The plutonium-uranium system

PLUTONIUM ALLOYS WITH TITANIUM, ZIRCONIUM, AND HAFNIUM (7)

Investigation of the systems of plutonium alloys with titanium, zirconium, and hafnium is still continuing, and, although important features have been established, there is not yet sufficient information with which to propose diagrams.

In the plutonium-zirconium system it has been shown that δ-plutonium dissolves up to approximately 70 a/o zirconium. This fact is established by Figure 16, which shows the lattice parameter values for alloys annealed at 500° C and quenched. Microscopic examination, dilatometry, and thermal analysis have shown that the field of this face-centered cubic phase is continuous from the δ-phase of pure plutonium. It also appears probable that ε-plutonium forms a continuous series of solid solutions with β-zirconium. At lower temperatures, below 330° C, a phase of approximate composition Pu_4Zr has been found by X-ray and metallographic identification, but its structure has not been determined. It was believed originally that there was a compound Pu_xZr with $x \geqq 15$, as reported by Coffinberry and Waldron (8), but this is no longer claimed, all the observed lines on diffraction patterns being explicable in terms of the phase Pu_4Zr and retained phases of pure plutonium.

A second phase, $PuZr_2$, has been found to occur at temperatures below 380° C. This phase has a diffraction pattern that can be indexed on the structure proposed by Silcock (9) for UZr_2; it is hexagonal, space group $P\bar{6}m2$, with 3 atoms per unit cell having $a = 5.055 \pm 0.005$ A and $c = 3.123 \pm 0.005$ A. The pattern closely resembles that of a cubic cell having $a = 4.390$ A and several forbidden reflections. There is good agreement between the observed and calculated intensities. Plutonium and zirconium atoms share positions $\frac{2}{3}, \frac{1}{3}, \frac{1}{2}; \frac{1}{3}, \frac{2}{3}, \frac{1}{2}$; and zirconium atoms occupy 0, 0, 0.

It has also been found that α-zirconium dissolves up to 13 a/o plutonium, the α/β transition temperature being depressed to 610° C. Figure 17 shows the lattice

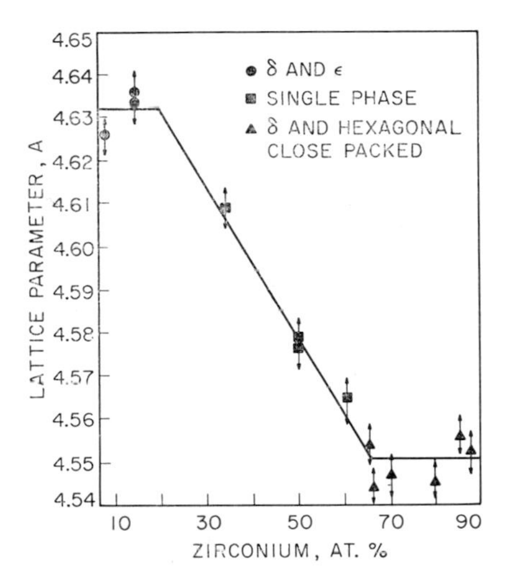

Fig. 16.—The length of the unit-cell edge of the face-centered cubic δ-plutonium solid solutions in the plutonium-zirconium system.

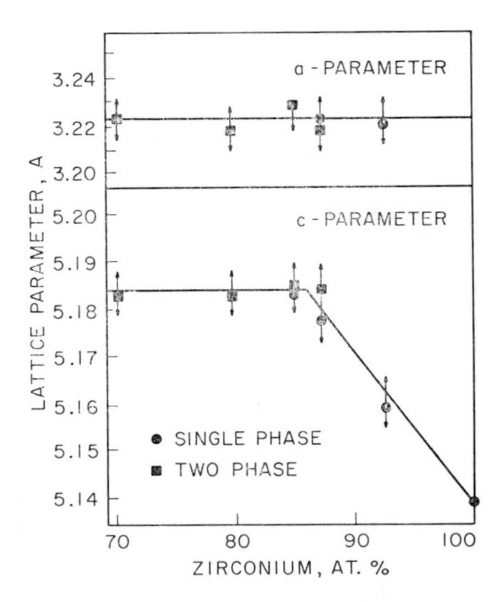

Fig. 17.—The dependence on composition of a_0 and c_0 of the hexagonal α-zirconium solid solutions in the plutonium-zirconium sytsem.

parameter measurements in the α-zirconium field for alloys quenched from 500° C. Figure 18 shows the results of density measurements on alloys across the phase field, from which it can be deduced that plutonium atoms dissolve in α-zirconium with an effective diameter which they possess in δ-plutonium and more nearly equal to that of ε-plutonium.

Preliminary lattice parameter measurements in the plutonium-titanium and plutonium-hafnium systems indicate that ε-plutonium may dissolve approximately 30 a/o of either titanium (a = 4.596 A) or hafnium (a = 4.612 A). Plutonium does

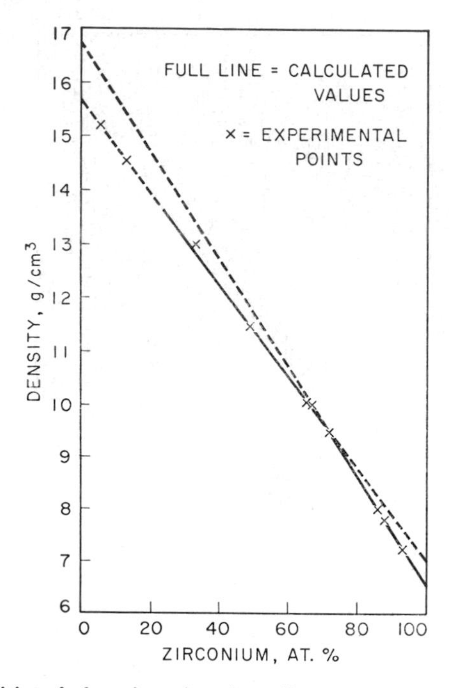

FIG. 18.—The densities of plutonium-zirconium alloys as a function of their composition

not dissolve in α-titanium sufficiently to change the lattice parameter noticeably, but in α-hafnium the lattice parameters change from a = 3.196 A, c = 5.056 A to a = 3.192 A, c = 5.072 A, which is approximately equivalent to a 5 a/o solubility of plutonium.

DISCUSSION

The alloying behavior of plutonium can be considered in the light of the considerable volume of information that is presented in this and other chapters in this book dealing with its physical metallurgy. It can be seen that, while plutonium forms many compounds that are isostructural with the corresponding uranium ones, the forms of the phase diagrams are substantially different, even where analogies between compounds exist. The difference in atomic diameters (7 and 4 per cent, respectively, for δ- and ε-plutonium compared with γ-uranium) accounts for this in many cases and is further evidence that the heavy elements fit in, generally speaking, with the concepts of alloy behavior propounded for simpler metals by Hume-Rothery. Thus thorium dissolves nearly 50 a/o plutonium, whose atomic diameter is well within the range considered favorable for solid solubility; in contrast, uranium and thorium exhibit liquid immiscibility, the difference in atomic diameters being 16 per cent. Other plutonium systems showing measurable solubility in at least one phase of plutonium are listed in Table 2 and demonstrate the validity of the size-factor criterion for solid solubility.

Another factor found to be of considerable significance in the Hume-Rothery approach to alloy formation is the effective valency. Zachariasen (10) has given a scheme relating the metallic radii of this series of elements to their valencies, in a manner analogous to their radius/valence relationships in covalent compounds. We have made use of these values at Harwell in attempting to understand certain

TABLE 2

System	Difference in Atomic Diameter (Per Cent) Relative to		Valency of Solute	Extent of Solubility (a/o)	
	δ-Pu	ε-Pu		δ-Pu	ε-Pu
Pu-Th......	+9	+12	4	0.4 (500° C)	5.5 (600° C)
Pu-U......	−6	−3.5	6(?)	0.1	100
Pu-Zr......	−2	+1	4	0.70 (600° C)	100 (probable)
Pu-Ti......	−11	−8	4	0.30	Not known
Pu-Hf......	−3.5	−1	4	0.30	Not known
Pu-Fe......	−23	−21	?	0.1	3
Pu-Ni......	−25	−23	?	0.25	4
Pu-Os......	−18	−16	?	Nil	6
Pu-Ru......	−19	−17	?	Nil	5
Pu-Al......	−13	−11	3	12	3

features of the alloying behavior of plutonium; other valency schemes proposed by Sarkisov (11), Mott (12), and Thewlis (13) have been examined with less encouraging results.

The existence of six allotropic modifications of plutonium with widely different sizes and presumably related different electronic configurations indicates that the free energies will be closely similar. Hence plutonium can be expected to dissolve in metallic solvents in whichever state minimizes the total free-energy contribution. It seems likely that α-zirconium will dissolve solutes whose diameters lie between 3.17 and 3.23 A—the values for the two closest distances of approach among the atoms in the hexagonal crystal structure of α-zirconium. Although δ-plutonium has a co-ordination similar to that of α-zirconium, its atomic diameter is large, 3.29 A. It is therefore understandable that plutonium appears to adopt the ε-plutonium configuration, atomic diameter 3.20 A. However, since ε-plutonium has the same crystal structure as β-zirconium and again a favorable size factor, it is not surprising that the α/β transition of zirconium is slightly depressed in temperature by the presence of plutonium. It is also consistent with a Zachariasen valency for ε-plutonium of 4.8, since other solutes of higher valency than 4 also depress the α/β transformation of zirconium. Nevertheless, Pu-Zr is the only zirconium system showing substantial solubility in α-zirconium along with a depression of the transition temperature. The extensive solubility of zirconium in δ-plutonium is consistent with the favorable size factor and similar effective valencies (δ-plutonium 4.5, zirconium 4).

The behavior of plutonium in solution in thorium is also interesting, in that it dissolves with an apparent radius of 1.7 A compared with 1.645 A for the δ-phase, the largest observed in the pure metal. On Zachariasen's scheme, this corresponds to a valency of 4.0, that also proposed by him for thorium. Thus it seems that the thorium matrix induces the plutonium atom to expand to the value whose associated valency is ideal for thorium, but no further, even though greater expansion would still further favor solubility on size-factor grounds. Again, solubility of thorium is greater in δ- than in ϵ-plutonium at a given temperature. This is consistent with the size-factor valency effect and with co-ordination, which all favor solubility in δ-plutonium.

The complete intersolubility of ϵ-plutonium and γ-uranium is consistent with the favorable size factor, similar Zachariasen valencies, and similarity of crystal structure. Until it is known whether there is any close relationship between the δ-plutonium structure and the η (Pu_3U) solid solution, comment on the restricted solubility of uranium in δ-plutonium would be premature. The solubility of plutonium in β-uranium is much greater than that known for any other metal in β-uranium; this may be another consequence of the readiness of plutonium to adopt a configuration that suits a particular environment.

REFERENCES

1. Williamson, G. K., Poole, D. M., and Marples, J. A. C. *Journal of the Institute of Metals*, **85**:431–36, 1956–57.
2. Mardon, P. G., Haines, H. R., Pearce, J. H., and Waldron, M. B. *Journal of the Institute of Metals*, **86**:166–71, 1957–58.
3. Poole, D. M., Williamson, G. K., and Marples, J. A. C. *Journal of the Institute of Metals*, **86**:172–76, 1957–58.
4. Konobeevsky, S. T. "Phase Diagrams of Some Plutonium Systems," in *Session of the Division of Chemical Science (of the U.S.S.R. Academy of Sciences), Conference on the Peaceful Uses of Atomic Energy, Moscow, 1955*, pp. 362–75; also pp. 207–14 in English translation by the Consultants Bureau (for sale by Superintendent of Documents, Washington, D.C., price $1.00).
5. Mardon, P. G. Unpublished data.
6. Coffinberry, A. S., and Ellinger, F. H. "The Intermetallic Compounds of Plutonium," *Proceedings of the International Conference on the Peaceful Uses of Atomic Energy*, **9**:138–46. New York: United Nations, 1956.
7. Williamson, G. K., Poole, D. M., Marples, J. A. C., and Waldron, M. B. Unpublished data.
8. Coffinberry, A. S., and Waldron, M. B. "The Physical Metallurgy of Plutonium," chap. 4 of *Progress in Nuclear Energy*, Ser. V, Vol. **1**, pp. 354–410. London: Pergamon Press, Ltd., 1956.
9. Silcock, J. M. *Transactions of the American Institute of Mining, Metallurgical and Petroleum Engineers*, **209**:521, 1957.

10. ZACHARIASEN, W. H. "Crystal Chemistry of the 5f Elements," in *The Actinide Elements*, ed. G. T. SEABORG and J. J. KATZ, chap. 18, pp. 769–96. ("National Nuclear Energy Series," Div. IV, Vol. **14B.**) New York: McGraw-Hill Book Co., Inc., 1954.
11. SARKISOV, E. S. *Doklady Akademii Nauk S.S.S.R.*, **60**:371, 1948.
12. MOTT, B. W. *Research*, **6**:238, 1953.
13. THEWLIS, J. Private communication.

CHAPTER XXII

PLUTONIUM PHASE DIAGRAMS STUDIED AT LOS ALAMOS

F. W. Schonfeld

THE PHASE DIAGRAMS

The plutonium phase diagrams presented in this chapter are based on data obtained at the Los Alamos Scientific Laboratory by conventional methods. The techniques employed were chiefly optical metallography, X-ray metallography, thermal analysis, and dilatometry. To a very limited extent the filtration method

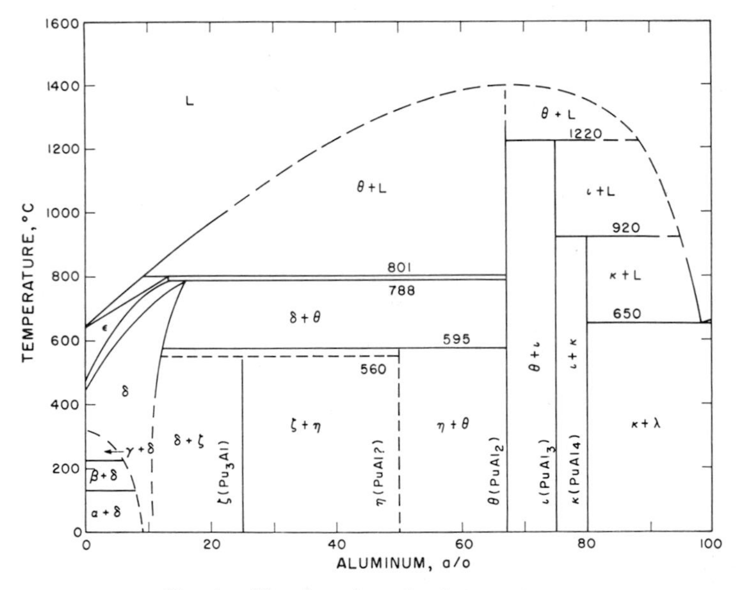

FIG. 1.—The plutonium-aluminum system

has been used in the determination of liquidus curves. In many of the alloy systems the data are incomplete and sufficient only for tentative conclusions; all the diagrams should be considered subject to change as additional experimental information is obtained.

F. W. Schonfeld is at the University of California, Los Alamos Scientific Laboratory, Los Alamos, New Mexico.

In Figures 1–15 are presented the phase diagrams of the plutonium binary systems that had been investigated with sufficient thoroughness to make possible the construction of tentative or partial diagrams previous to November, 1957. Into these diagrams as shown here, however, are incorporated a number of revisions and extensions of experimental results that have been established since that time. Also, three diagrams determined since November, 1957, are included in this chapter as Figures 16, 17, and 18. Crystal-structure data for most of the intermediate phases of the diagrams are reported by Ellinger in chapter xxv, and details of phase-equilibrium relationships involving the delta-prime phase of plutonium are discussed by Elliott and Larson in chapter xxiv for a number of systems.

Because of their very simple form, phase diagrams are not shown for the binary

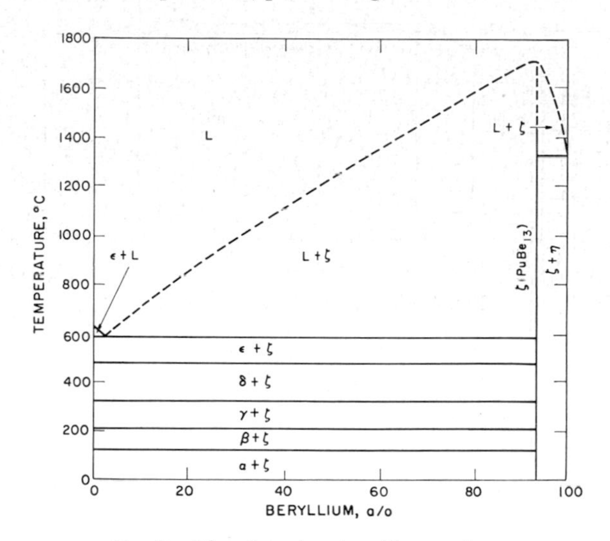

FIG. 2.—The plutonium-beryllium system

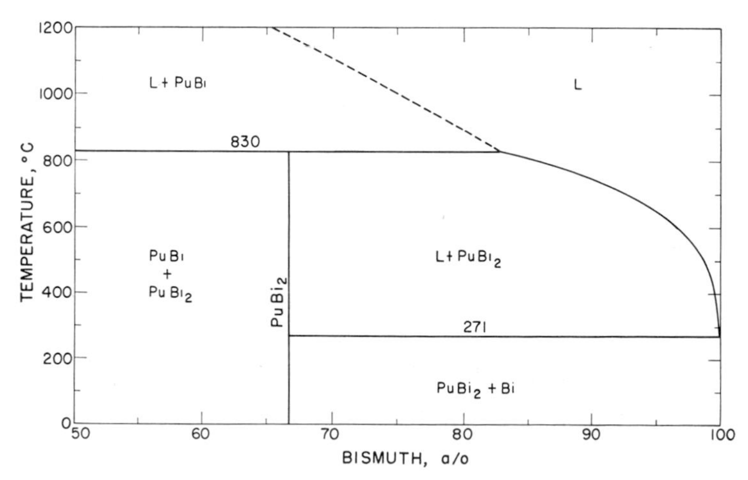

FIG. 3.—The high-bismuth end of the plutonium-bismuth system

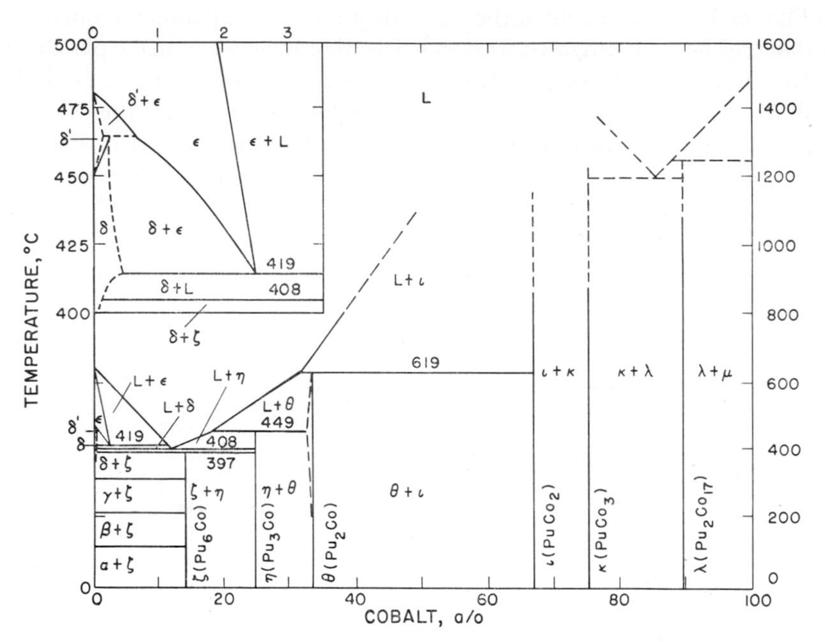

FIG. 4.—The plutonium-cobalt system

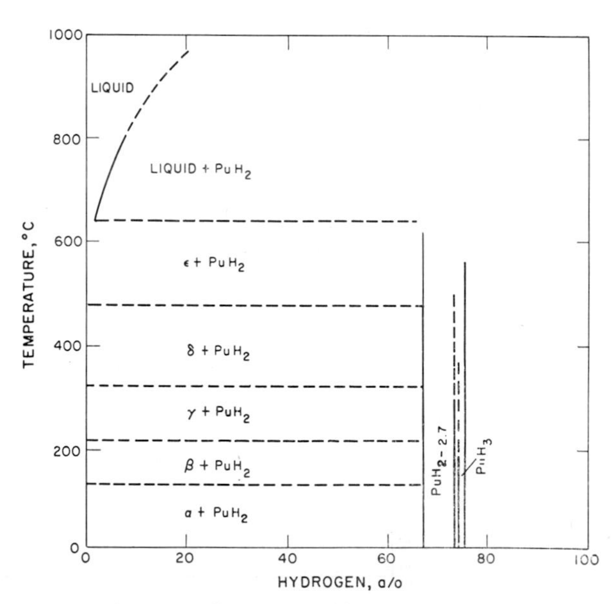

FIG. 5.—The plutonium-hydrogen system

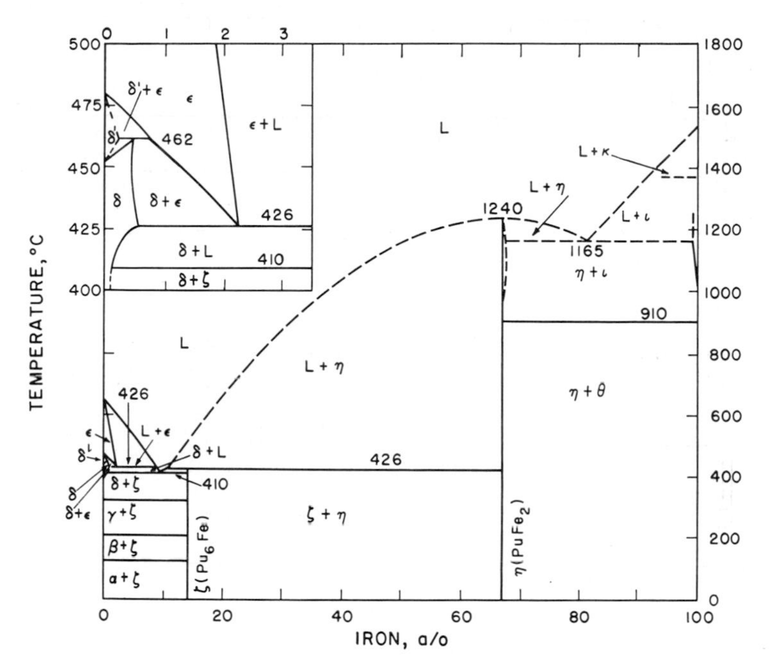

FIG. 6.—The plutonium-iron system

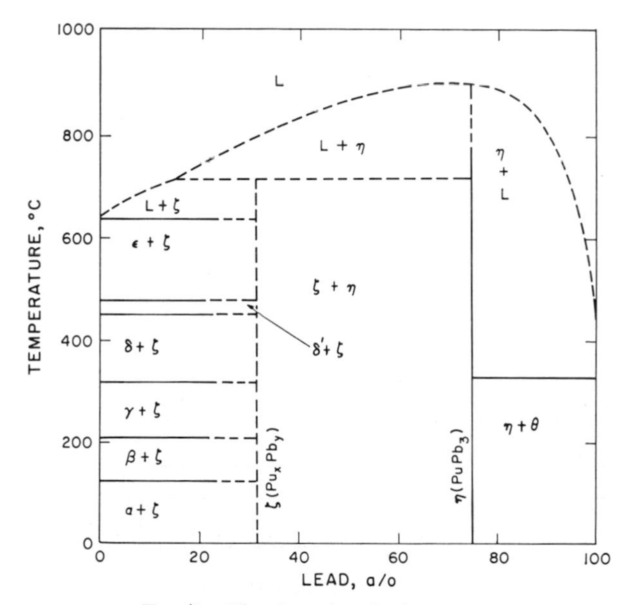

FIG. 7.—The plutonium-lead system

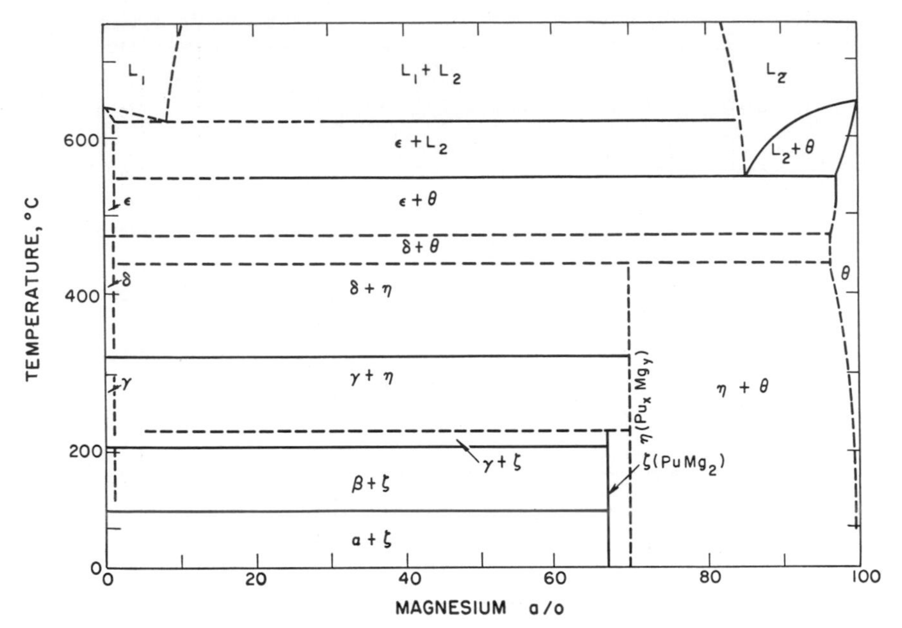

FIG. 8.—The plutonium-magnesium system

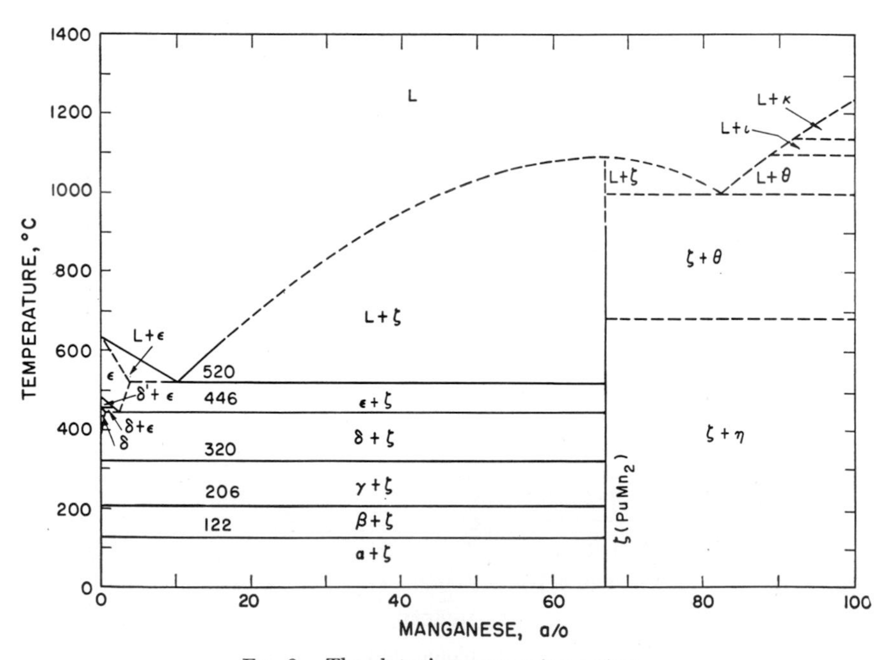

FIG. 9.—The plutonium-manganese system

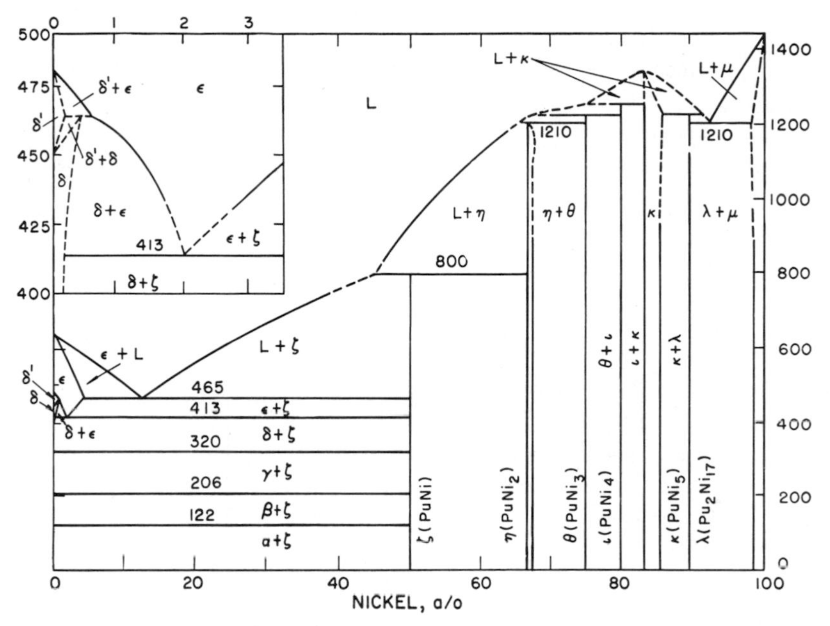

FIG. 10.—The plutonium-nickel system

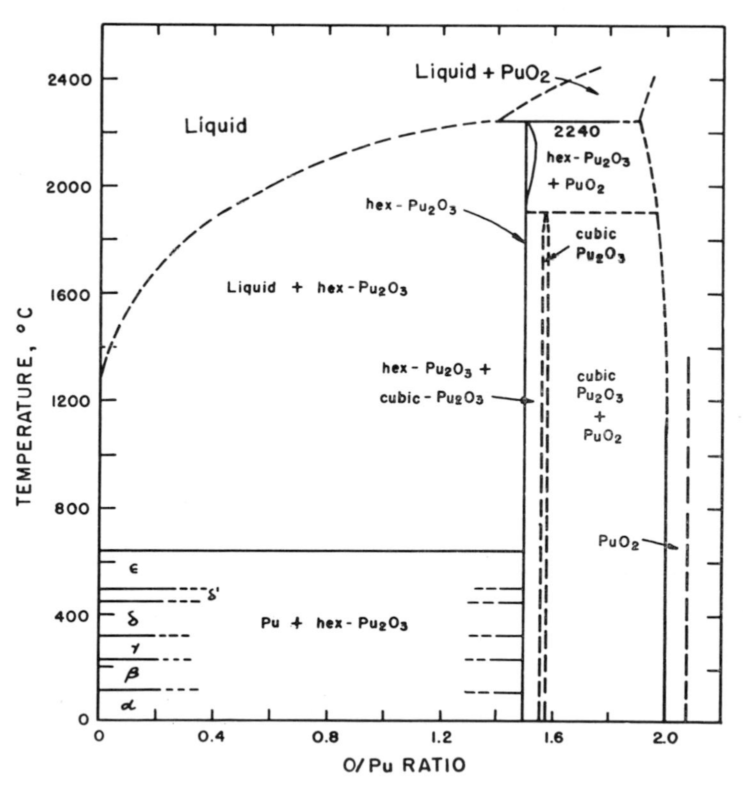

FIG. 11.—The plutonium-oxygen system

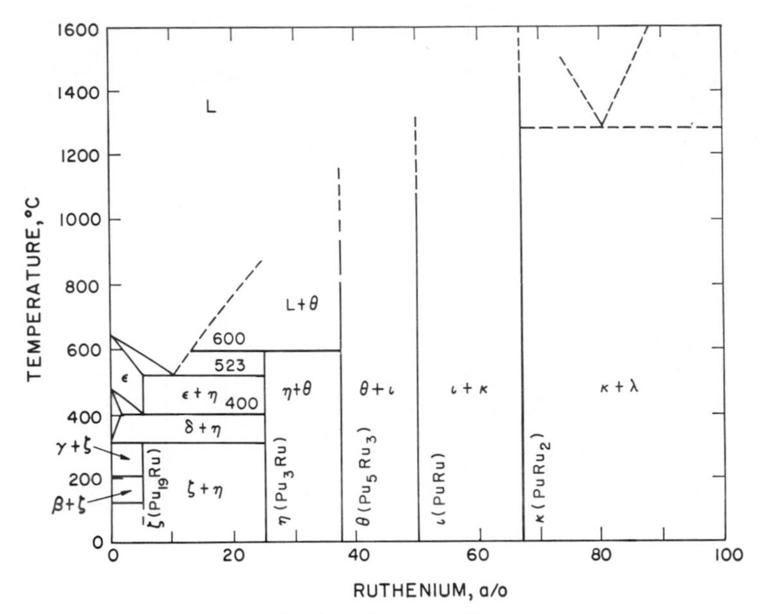

FIG. 12.—The plutonium-ruthenium system

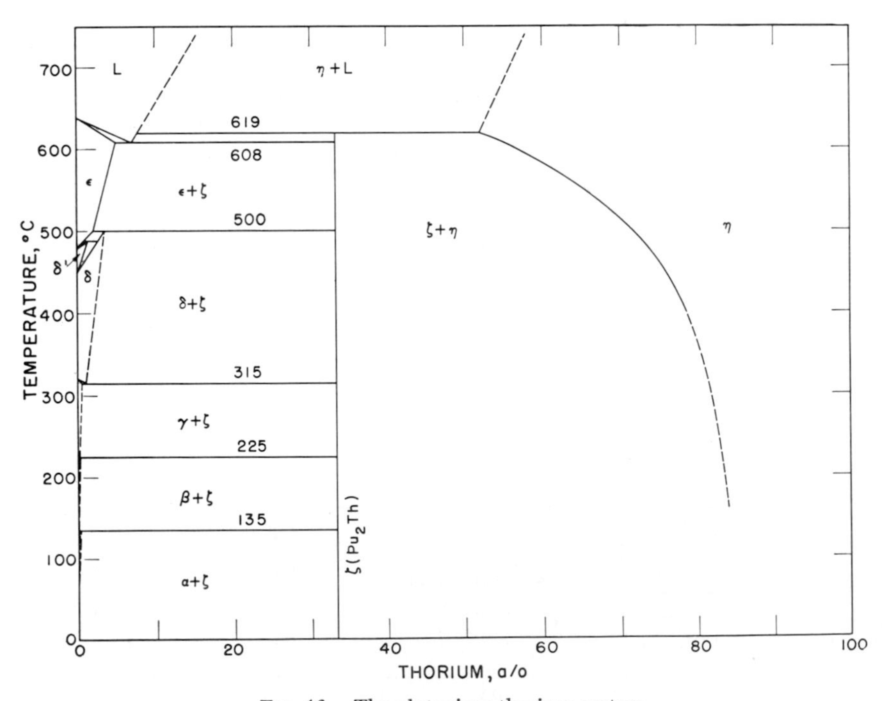

FIG. 13.—The plutonium-thorium system

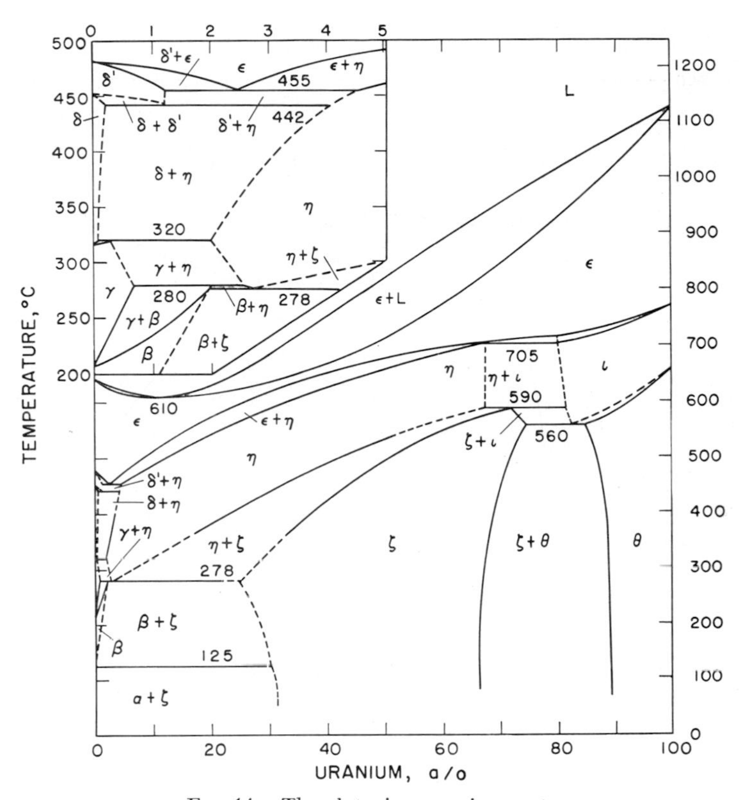

FIG. 14.—The plutonium-uranium system

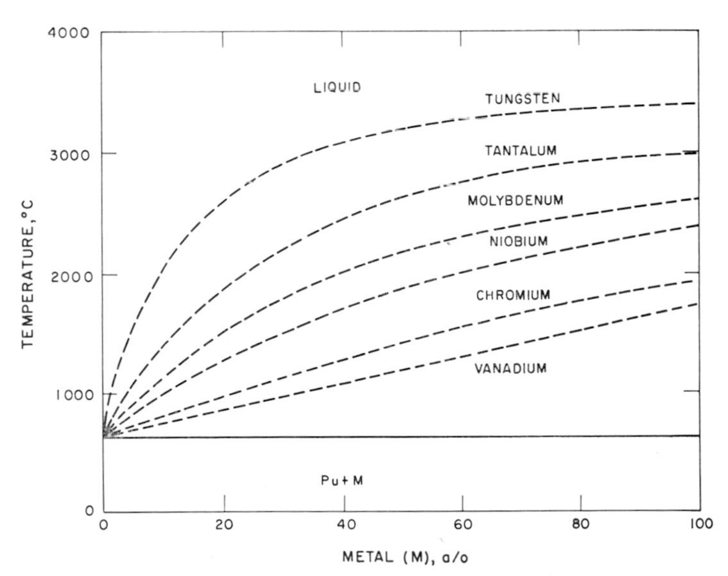

FIG. 15.—The form of the phase diagram and the sequence of liquidus curves for alloys of plutonium with six metals on groups V-A and VI-A.

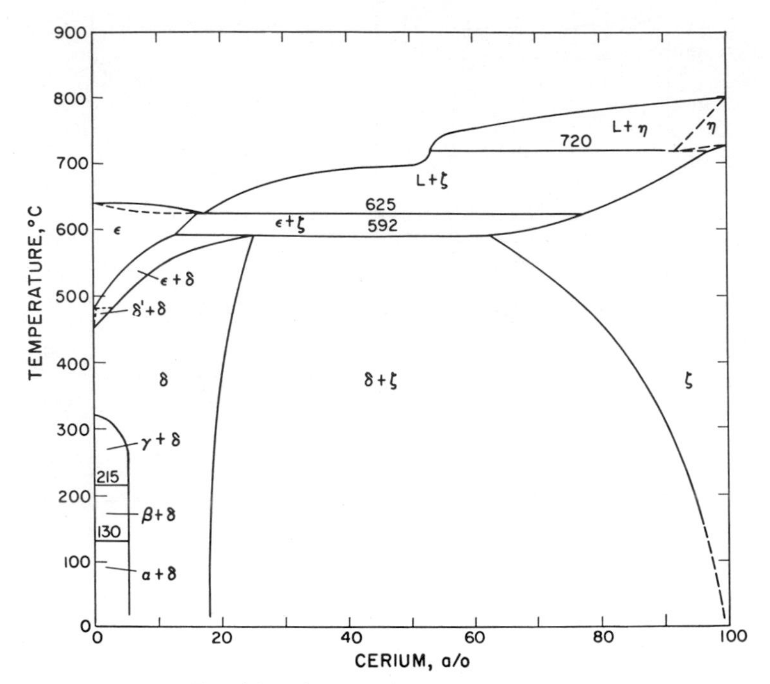

FIG. 16.—The plutonium-cerium system

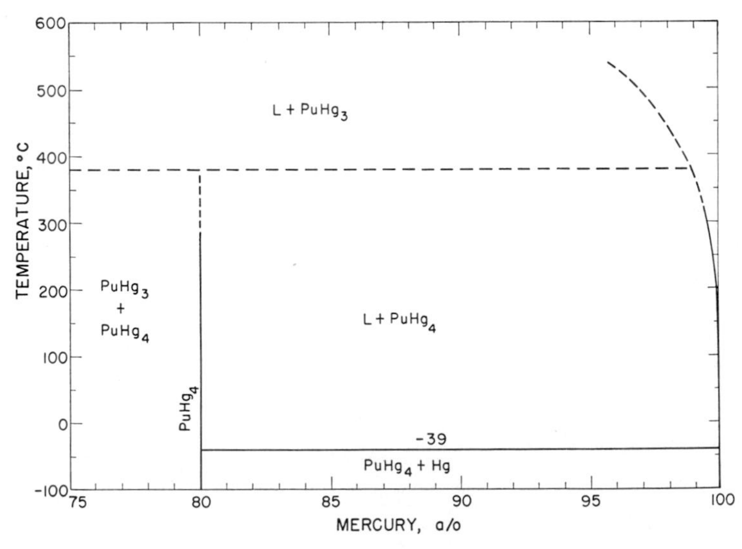

FIG. 17.—The high-mercury end of the plutonium-mercury system

systems involving those metals that appear to be completely immiscible with plutonium in both the liquid and the solid state. Metals for which such failure to form alloys with plutonium has been experimentally established are lithium, sodium, potassium, calcium, strontium, and barium.

The tentative liquidus curves shown in Figure 15 are based on preliminary tests of the solubilities in liquid plutonium of six refractory metals. They are intended to indicate only that the sequence of the solubilities of these metals is the inverse of the sequence of their melting temperatures. At Los Alamos no eutectic microstructures have ever been observed in alloys of any of the six binary systems represented, but it is not certain that some slight lowering, to the extent of a few degrees,

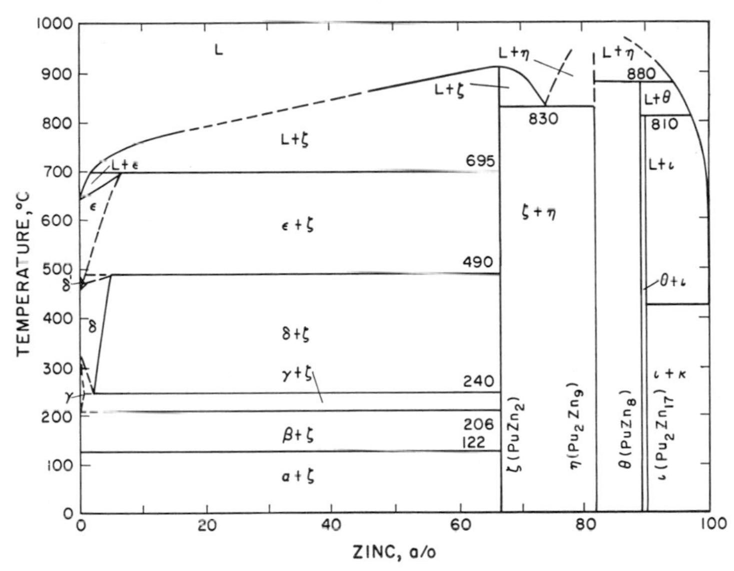

FIG. 18.—The plutonium-zinc system

of the melting temperature of plutonium is not caused by the presence of small amounts of the more soluble of the six metals. (See chap. xxiii for Russian phase diagrams of the plutonium-chromium, plutonium-molybdenum, and plutonium-vanadium systems.)

In addition to the alloying data shown by the phase diagrams, fragmentary information pertaining to the following systems has been determined:

1. *Plutonium-arsenic.*—One compound, PuAs, is known (1). It appears to have a rather high melting temperature.

2. *Plutonium-carbon.*—Two stable compounds of the plutonium-carbon system are PuC and Pu_2C_3. Some evidence suggests that there is a eutectic reaction between PuC and plutonium at a low carbon concentration, and it is believed that PuC forms by peritectic reaction at about 1,200° C.

3. *Plutonium-copper.*—Although it is believed that plutonium forms at least three compounds with copper, the composition of only the most plutonium-rich of these, $PuCu_2$, has been well established in Los Alamos. A eutectic between this

compound and plutonium occurs at 625° C and about 3 a/o copper, and at least one other eutectic composition appears to exist at about 90 a/o copper. Solid solubilities in this system appear to be low.

4. *Plutonium-germanium.*—It is known that five intermetallic compounds exist in the plutonium-germanium system: Pu_3Ge, Pu_3Ge_2(?), Pu_2Ge_3, $PuGe_2$, and $PuGe_3$. A eutectic composition between $PuGe_3$ and germanium has been observed.

5. *Plutonium-gold.*—Plutonium-gold intermetallic phases are known to exist, but they have not yet been identified. There appears to be a eutectic composition between plutonium and the most plutonium-rich compound.

6. *Plutonium-indium.*—The plutonium-indium system exhibits complete liquid miscibility and a significant degree of solid solubility in at least one plutonium allotrope—the delta phase. Several intermetallic compounds are known to exist, but only one, Pu_3In, has been identified at Los Alamos.

7. *Plutonium–rare earths.*—The plutonium-cerium phase diagram is shown in Figure 16. The plutonium-praseodymium and plutonium-neodymium diagrams appear to have a similar form. In the plutonium-lanthanum system there are also terminal solid solubilities, but a miscibility gap exists between liquid phases.

8. *Plutonium-rhenium.*—One intermediate phase, $PuRe_2$, is known in this system. There is a eutectic composition between this compound and plutonium.

9. *Plutonium-silicon.*—Two plutonium-rich compounds in the plutonium-silicon system have tentatively been designated Pu_5Si_3 and Pu_3Si_2. Additional compounds having more firmly established compositions are PuSi, Pu_2Si_3, and $PuSi_2$. Two eutectic reactions have been identified; one occurs between plutonium and Pu_5Si_3(?) at 570° C and 4 a/o silicon, the other between silicon and $PuSi_2$ at 82 a/o silicon. The maximum solid solubility in delta plutonium seems to be about 1 a/o silicon.

10. *Plutonium-silver.*—Silver forms immiscible liquids with plutonium. A preliminary investigation of the plutonium-silver system has indicated that the solubility of silver in liquid plutonium is negligible but that the maximum solubility of plutonium in silver is of the order of 35 a/o. Within the composition range of 0–35 a/o plutonium, silver appears to form one or more compounds with plutonium.

11. *Plutonium-tin.*—Available evidence suggests that in this system there are only two intermetallic compounds, $PuSn_3$ and Pu_4Sn(?). The microstructures of plutonium-rich alloys indicate that there is a eutectic at about 13 a/o tin and that terminal solid solubilities are low.

12. *Plutonium-titanium.*—Preliminary experiments have revealed no intermediate phases in the plutonium-titanium system and suggest that titanium dissolves to an extent greater than 10 a/o in at least one plutonium allotrope.

13. *Plutonium-zirconium.*—The few available data indicate that the liquidus and solidus lines are representative of a continuous series of solid solutions between epsilon plutonium and beta zirconium. The solubility of zirconium in delta plutonium is quite high and makes retention of the delta phase to room temperature easily possible. One plutonium-rich compound has been observed to form peritectoidally at about 250° C, but its composition has not been fixed.

THE ALLOYING BEHAVIOR OF PLUTONIUM

The phase diagrams shown in Figures 1–18 and the supplementary data on plutonium alloys reported above are entirely the results of work done at Los Alamos. In order to discuss the alloying behavior of plutonium in general terms, however, it is appropriate to take account also of the results of phase-equilibrium studies performed on plutonium alloys at other laboratories. In the following brief review of the general alloying characteristics of plutonium, this is done, much of the data discussed having been determined since November, 1957.

Although the alloying behavior of plutonium bears some resemblance to the alloying characteristics of both uranium and the rare-earth elements, in many respects it is markedly different from either. It might be said to occupy a position intermediate between these two, but, if so, the further observation should be made that the similarity of plutonium alloys to those of the rare-earth metals appears to be greater than their resemblance to the alloys of uranium. This behavior may be conveniently summarized by means of a few general statements regarding the reactions of plutonium with the various families of other metals in the periodic table.

The alloying behavior of plutonium with respect to the A subgroups of the periodic table will be considered first. Although only its reactions with lithium, sodium, and potassium have been investigated experimentally, plutonium appears to be completely immiscible with all the alkali metals in both the liquid and the solid state. The same statement can be made with regard to calcium, strontium, and barium among the alkaline earths, but plutonium is soluble to the extent of about 15 atomic per cent in liquid magnesium, and if (like uranium) it forms a liquid miscibility gap with beryllium, this gap is a very narrow one, lying on the high-beryllium side of the compound $PuBe_{13}$. Plutonium forms two compounds with magnesium by solid-state reaction (see Fig. 8).

Among the group III-A earth metals, plutonium has been found to have a large liquid miscibility gap with lanthanum and, by analogy with uranium, is suspected of reacting in a similar manner with scandium and yttrium. It has, however, been found to be completely miscible in the liquid state with all the rare-earth (f-transition) metals that have been studied (Ce, Pr, Nd), and it appears also to have complete liquid miscibility with all the actinide metals that have been investigated (Th, U, Np). In fact, so far as is now known, plutonium is completely miscible in the liquid state with all metals other than the exceptions noted above and one more (Ag, discussed below).

Plutonium has been found to form rather extensive solid solutions (especially delta-phase Pu) with the rare-earth metals, but it appears to form no compounds with them. Among the actinides, however, it is known to form one compound with thorium and two intermediate solid solutions of wide homogeneity range with uranium (Fig. 14). Epsilon plutonium has complete solid solubility with gamma uranium and with gamma neptunium (2). Except for neptunium, additions of either lanthanide or actinide metals have not been found to lower the melting temperature of plutonium more than about 30° C.

The d-transition metals of groups IV-A, V-A, and VI-A form only two known compounds with plutonium and cause very little or no eutectic lowering of its melting temperature. Both the compounds are products of solid-state reaction in the plutonium-zirconium system (2, 3). In common with the rare earths and a number of B-subgroup metals, the group IV-A metals (Ti, Zr and Hf) are extensively soluble in delta plutonium, and the solubilities of titanium and hafnium in the epsilon phase are of the order of 30 atomic per cent (2). Beta zirconium forms a continuous series of solid solutions with epsilon plutonium.

In the binary systems of plutonium with d-transition elements of groups V-A, VI-A, VII-A, and VIII-A, all solid solubilities are quite restricted. As has been noted by Konobeevsky (4), there exists among these systems a trend toward an increasing number of compounds with increasing number of d electrons in the d-transition metal. Thus vanadium, niobium, tantalum, chromium, molybdenum, and tungsten form no compounds with plutonium; manganese and rhenium each form one compound; iron, ruthenium, and osmium form two, five, and four compounds, respectively; and cobalt and nickel each form six compounds. Another fairly systematic trend that seems to prevail among these binary systems is that a decreasingly lower eutectic temperature near the plutonium end of the phase diagram accompanies an increase in the number of d electrons, except that the lowest of all the known eutectic temperatures occurs in the plutonium-cobalt instead of the plutonium-nickel system. Thus, as mentioned above, the group V-A and VI-A metals cause little or no eutectic lowering of the melting temperature of plutonium (Fig. 15); manganese (Fig. 9), ruthenium (Fig. 12), and osmium (chap. xxiii) form eutectics at temperatures from 115° to 145° C lower than the 640° C melting temperature of plutonium; and iron, cobalt, and nickel (Figs. 6, 4, and 10) form eutectics at temperatures from 175° to 232° C lower than 640° C.

In the plutonium-iron and plutonium-cobalt systems there is a striking occurrence of the phenomenon of retrograde melting (see Figs. 6 and 10, and chap. xxiv). Thus a 2 a/o iron alloy that is entirely solid epsilon phase at 440° C becomes about one-fifth liquid (i.e., liquid plus approximately 80 v/o delta phase) on cooling to 420° C, and at 400° C it is again entirely solid (delta plus Pu_6Fe). A similar situation exists in the plutonium-cobalt system, except that temperatures are slightly lower and the phases formed on resolidification are delta and Pu_3Co (instead of delta and Pu_6Co).

The tendency toward a eutectiferous form of phase diagram persists, but in much reduced degree, in the binary systems of plutonium with the group I-B metals. Thus a plutonium-rich eutectic with copper melts at 625° C, and a plutonium-gold eutectic has a composition and melting temperature close to those of pure plutonium. As indicated above, however, silver forms immiscible liquids with plutonium. A preliminary investigation of the plutonium-silver system has indicated that the solubility of silver in liquid plutonium is negligible but that the maximum solubility of plutonium in liquid silver is of the order of 35 a/o. Silver appears to form one or more compounds with plutonium, as do copper and gold.

So far as is known, plutonium forms compounds with all the B-subgroup metals and metalloids. Another generalization that can be made is that most of these compounds have melting temperatures appreciably higher than the melting temperatures of the components in their respective binary systems. Hence liquidus curves tend to rise sharply near the extremities of the phase diagrams, with little or no eutectic lowering of the melting temperatures of the end members. These characteristics appear to be common to the alloying behaviors of all the lanthanide and actinide metals.

Except for solid solutions in delta and epsilon plutonium, solid solubilities are quite restricted among the phases occurring in the binary alloys of plutonium with B-subgroup elements. Solid solubilities to the extent of several atomic per cent in both epsilon and delta phases have, however, been experimentally established for a number of B-subgroup solutes. Most of the delta solid solutions (including those of the rare-earth and group IV-A metals) are either stable at room temperature or can be retained at room temperature by rapid cooling from their high-temperature stability ranges. It has not been found possible, however, to retain any of the epsilon solid solutions at room temperature.

The occurrence of significant solid solubility in the alpha, beta, gamma, and delta-prime phases of plutonium is much rarer than its incidence in the delta and epsilon allotropes. Only neptunium has been reported to have detectable solubility in alpha plutonium (2), and the experimental evidence relating to possible solid solubility in the beta and gamma phases has been difficult to interpret. In many cases of the retention of the beta or gamma phases at room temperature, it has not been clear whether the retained phase represented equilibrium conditions at a higher temperature or a supersaturated solid solution resulting from a martensitic transformation of the delta solid solution. Such retained phases have been observed in alloys of plutonium with aluminum, magnesium, thorium, uranium, titanium, and zirconium. Only titanium is known to be soluble in delta-prime plutonium to an extent greater than about 1.5 atomic per cent (see chap. xxiv).

The number of persons who have participated in plutonium-alloy work at Los Alamos is too large for every name to be mentioned, but the outstanding efforts of the following major contributors to this work are especially deserving of grateful acknowledgment.

R. N. R. Mulford and G. E. Sturdy investigated the plutonium-hydrogen and plutonium-oxygen systems, and R. O. Elliott and E. M. Cramer determined the solubility of plutonium in liquid bismuth.

V. O. Struebing prepared almost all the alloys used in the various investigations; and F. H. Ellinger, R. O. Elliott, M. Gibbs, P. Vigil (X-ray metallography), E. M. Cramer, K. Imlah, C. O. Matthews, C. R. Tipton (optical metallography), W. N. Miner, D. D. Whyte, and G. W. Wensch (thermal analysis and dilatometry) contributed data for many of the results reported here.

The phase diagrams shown in Figures 16 and 18 are the work of F. H. Ellinger, E. M. Cramer, and C. C. Land. The liquidus curve in Figure 15 is based on the solubility data of D. F. Bowersox and J. A. Leary (5).

REFERENCES

1. GORUM, A. E. *Acta Crystallographica*, **10**:144, 1957.
2. WALDRON, M. B., GARSTONE, J., LEE, J. A., MARDON, P. G., MARPLES, J. A. C., POOLE, D. M., and WILLIAMSON, G. K. "The Physical Metallurgy of Plutonium," in *Proceedings of the Second United Nations International Conference on the Peaceful Uses of Atomic Energy*, **6**:162–69. Geneva: United Nations, 1958.
3. BOCHVAR, A. A., KONOBEEVSKY, S. T., KUTAITSEV, V. I., MENSHIKOVA, I. S., and CHEBOTAREV, N. T. "The Interaction between Plutonium and Other Metals in Connection with Their Arrangement in Mendeleev's Periodic Table," in *Proceedings of the Second United Nations International Conference on the Peaceful Uses of Atomic Energy*, **6**:184–93. Geneva: United Nations, 1958.
4. KONOBEEVSKY, S. T. "Phase Diagrams of Some Plutonium Systems," in *Session of the Division of Chemical Science (of the U.S.S.R. Academy of Sciences), Conference on the Peaceful Uses of Atomic Energy, Moscow, 1955*, pp. 362–75; also pp. 207–14 in English translation by Consultants Bureau (for sale by Superintendent of Documents, Washington, D.C., price $1.00).
5. BOWERSOX, D. F., and LEARY, J. A. *Journal of Inorganic and Nuclear Chemistry*, **9**: 108–12, 1959.

CHAPTER XXIII

PLUTONIUM PHASE DIAGRAMS PUBLISHED BY THE RUSSIANS

F. W. Schonfeld

In July, 1955, at the Conference on the Peaceful Uses of Atomic Energy held in Moscow by the Academy of Sciences of the U.S.S.R., S. T. Konobeevsky (1) reported the results of Soviet research on the physical metallurgy of plutonium. Included in his paper were phase diagrams for seven binary systems of plutonium with other metals, and the simple form of the diagram for an eighth system was described in the text. His diagrams were drawn on a weight per cent basis, but in this chapter they are redrawn to the atomic per cent scale.

In Figures 1–7 are shown Konobeevsky's diagrams for the binary systems of plutonium with beryllium, iron, lead, manganese, nickel, osmium, and vanadium. Although it was indicated by Konobeevsky that the plutonium-chromium phase diagram is similar in form to that of the plutonium-vanadium system, a drawing of this diagram was not published until later by Bochvar, Konobeevsky, Kutaitsev, Menshikova, and Chebotarev (2).

In the paper by Bochvar *et al.*, which was presented at the second United Nations International Conference on the Peaceful Uses of Atomic Energy, several of the original diagrams of Konobeevsky were repeated, and eight additional plutonium-alloy diagrams were published. The eight new diagrams are included in this chapter as Figures 8–15. They show the binary systems of plutonium with aluminum, bismuth, chromium, copper, molybdenum, thorium, uranium, and zirconium.

Comparison of the Soviet diagrams with those determined at Los Alamos (chap. xxii) indicates that good agreement exists with respect to the general character of a majority of the systems that have been investigated both in Russia and in the United States. Minor discrepancies between critical temperatures and compositions in most cases probably fall within the range of experimental uncertainty in their determination. Three of the Russian diagrams, those for plutonium-copper, plutonium-osmium, and plutonium-zirconium, represent systems on which little or no work has been done at Los Alamos. In only four or five systems do there appear to be major discrepancies between results obtained in Moscow and at Los Alamos,

F. W. Schonfeld is at the University of California, Los Alamos Scientific Laboratory, Los Alamos, New Mexico.

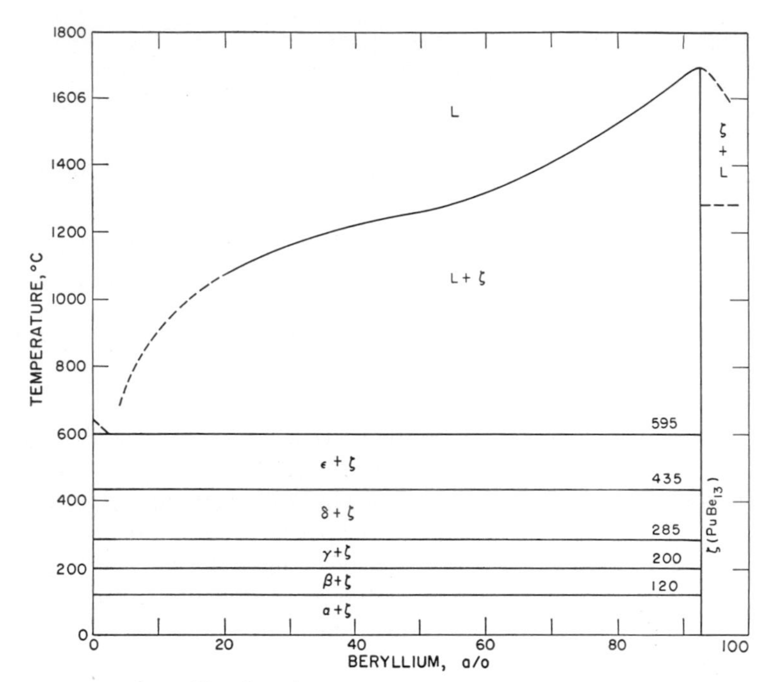

Fig. 1.—The plutonium-beryllium system

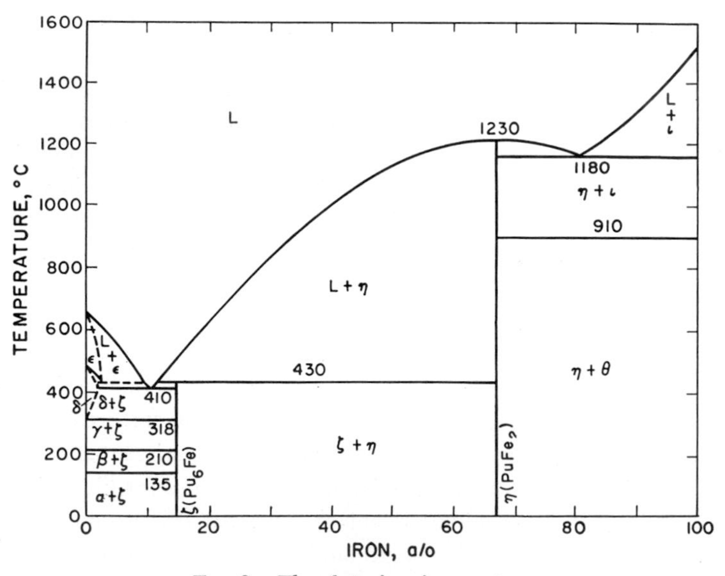

Fig. 2.—The plutonium-iron system

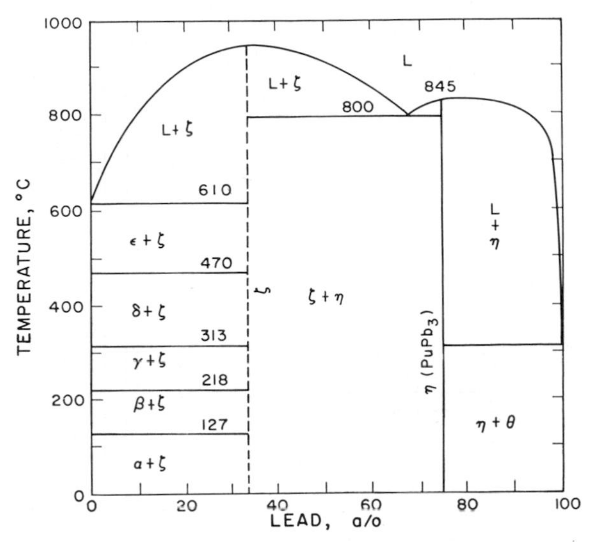

FIG. 3.—The plutonium-lead system

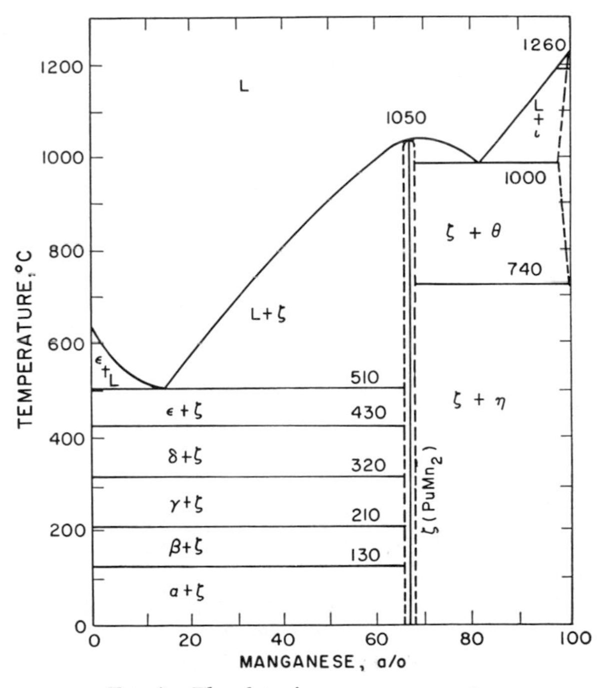

FIG. 4.—The plutonium-manganese system

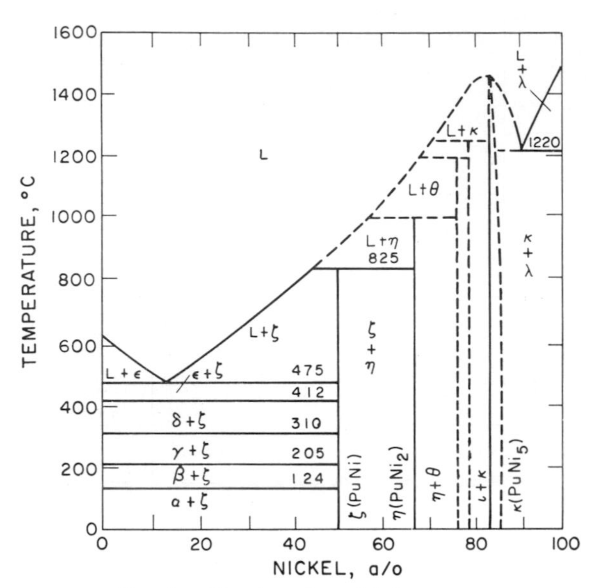

FIG. 5.—The plutonium-nickel system

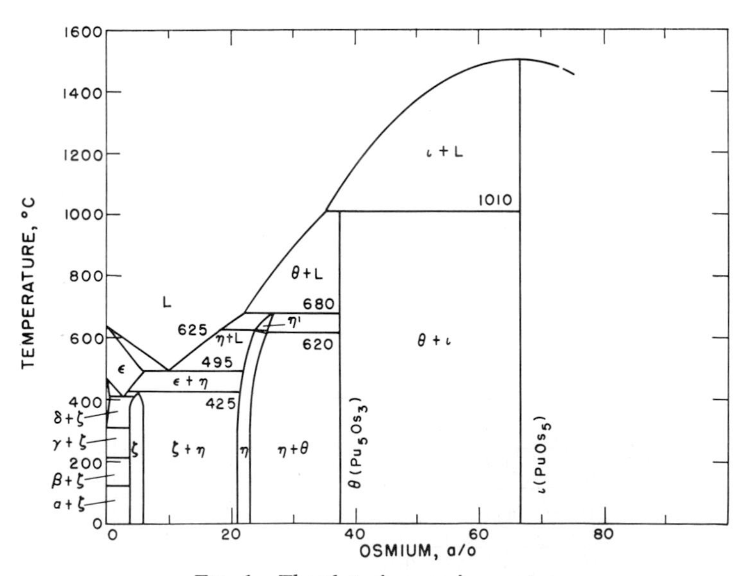

FIG. 6.—The plutonium-osmium system

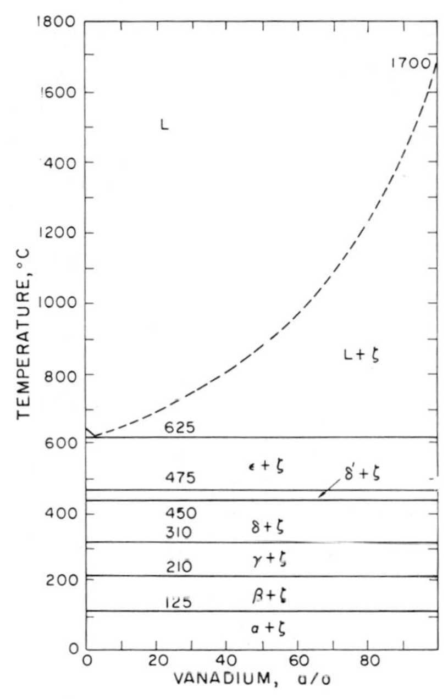

FIG. 7.—The plutonium-vanadium system

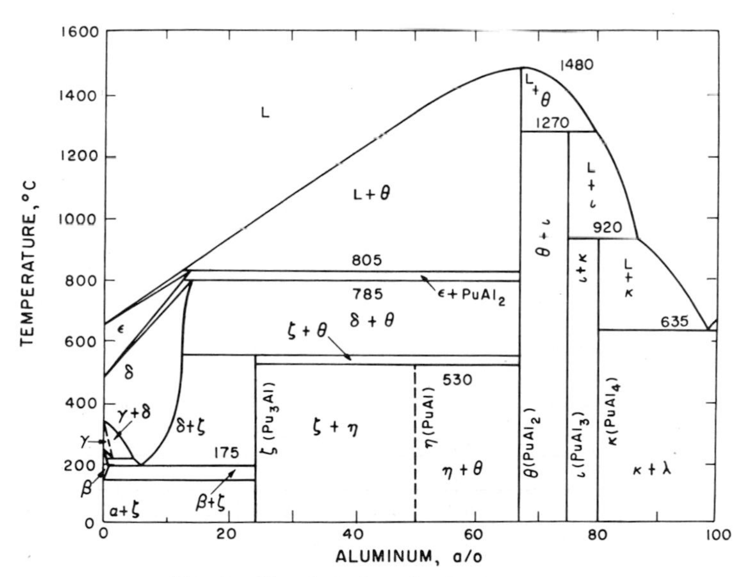

FIG. 8.—The plutonium-aluminum system

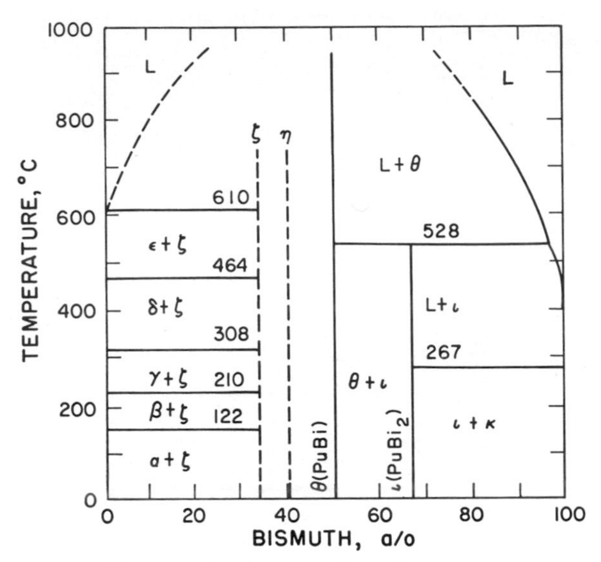

FIG. 9.—The plutonium-bismuth system

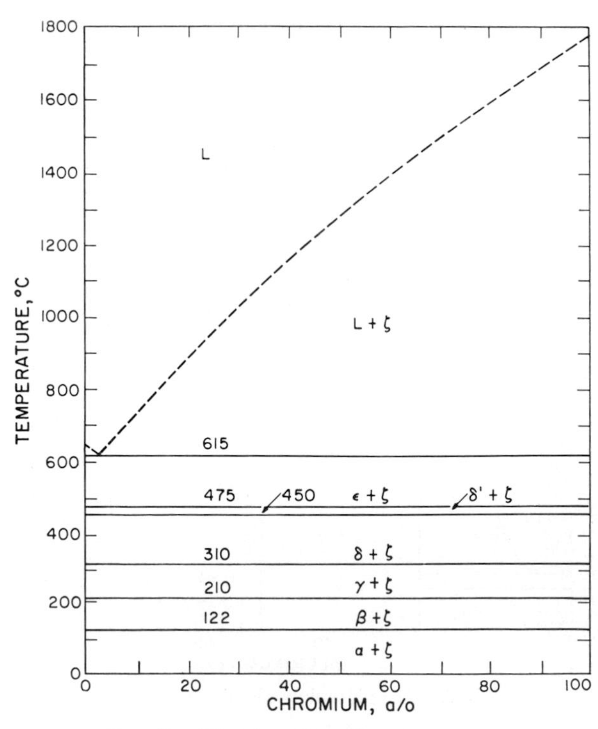

FIG. 10.—The plutonium-chromium system

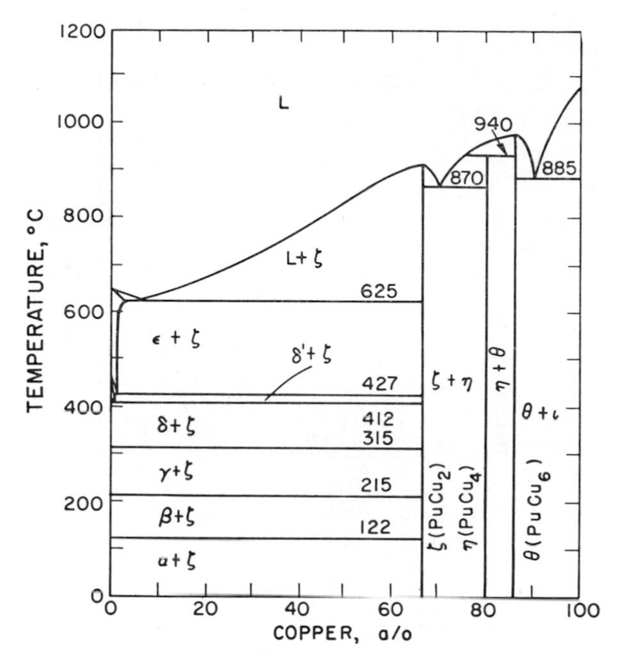

FIG. 11.—The plutonium-copper system

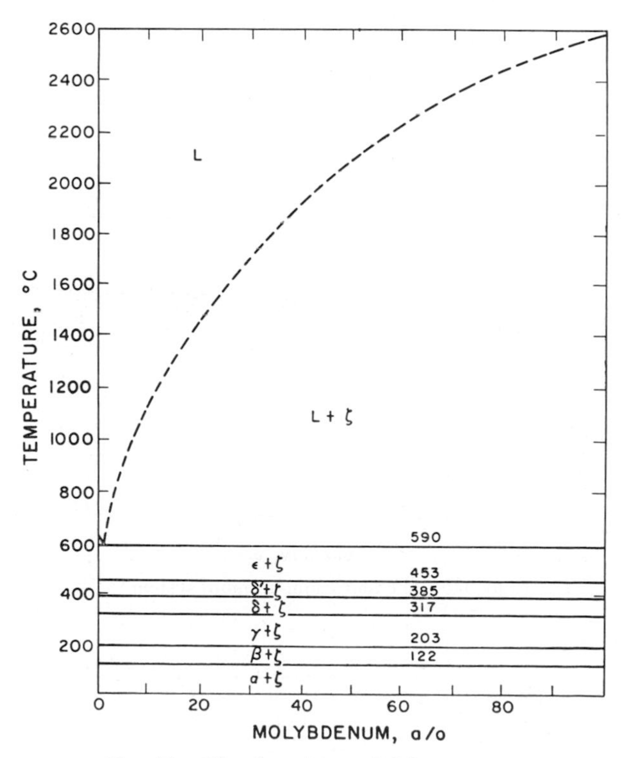

FIG. 12.—The plutonium-molybdenum system

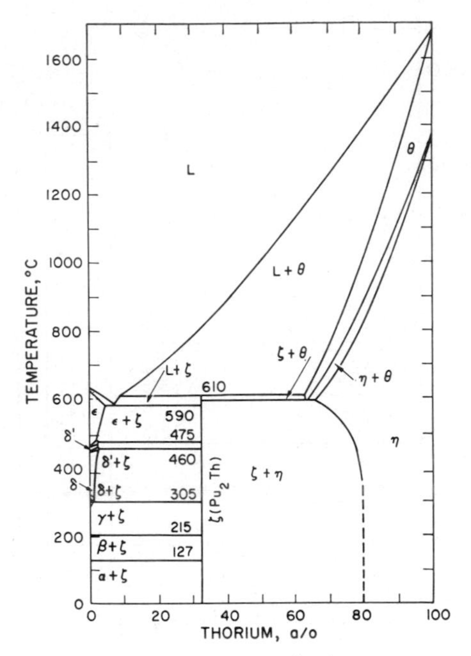

FIG. 13.—The plutonium-thorium system

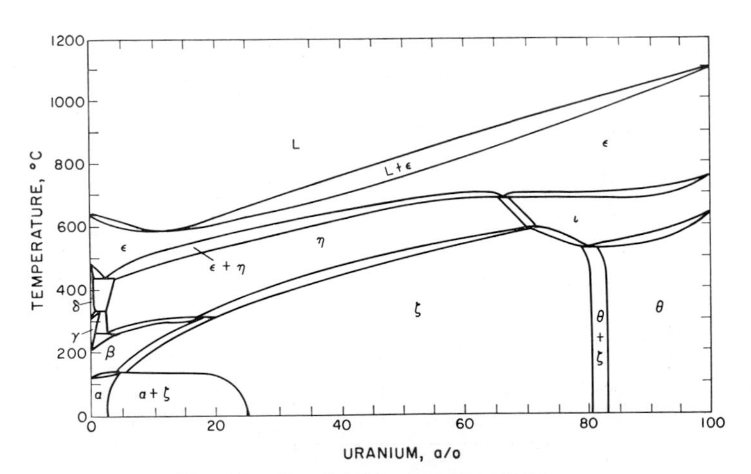

FIG. 14.—The plutonium-uranium system

and, although it is to be expected that these differences will soon be resolved through the performance of further experimental work in both countries, it is difficult at this stage to speculate in any particular case regarding the probable outcome of such further work.

The eutectoid reaction shown in the Russian version of the plutonium-aluminum system (Fig. 8) was arrived at through the use of high-pressure equipment operated over a range of temperature (3), a technique that has not yet been applied at Los Alamos. Also, since the publication of their diagram for the plutonium-uranium system (Fig. 13), the Russians have decided (3) that the high-uranium side of their

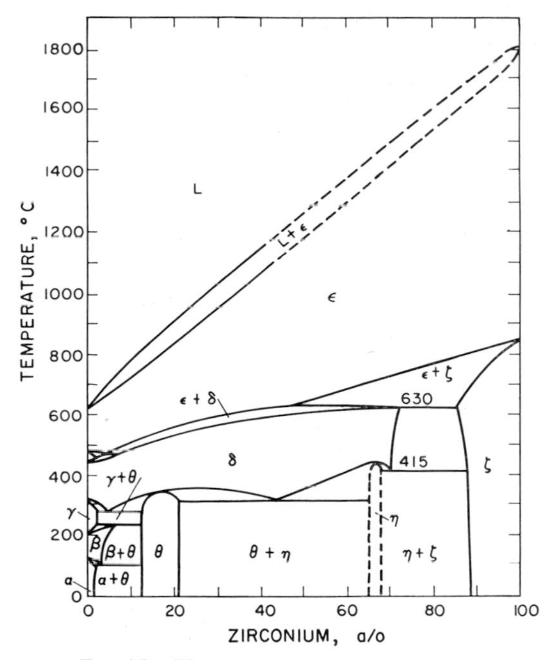

FIG. 15.—The plutonium-zirconium system

diagram should be revised to conform more closely with the American and British versions (chaps. xxii and xxi); but the disagreement in this system regarding the extent of the beta-plutonium field remains unresolved. Moreover, neither the Harwell (4) nor the Los Alamos investigators have been able to confirm the remarkable lowering in temperature of the alpha/beta transformation of thorium that the Russian workers report for solid solutions of plutonium in this metal (Fig. 12).

The difference in melting behavior of the plutonium-rich compound shown in the American and Russian diagrams of the plutonium-lead system (chap. xxii, Fig. 7, and Fig. 3, this chapter) is based on relatively little evidence in the case of the American work; i.e., the Los Alamos interpretation appears to be in need of either confirmation or revision. On the other hand, it appears that the Russians have investigated the plutonium-nickel system somewhat less thoroughly than have the Americans (chap. xxii, Fig. 10, and Fig. 5, this chapter).

REFERENCES

1. Konobeevsky, S. T. "Phase Diagrams of Some Plutonium Systems," in *Session of the Division of Chemical Science (of the U.S.S.R. Academy of Sciences), Conference on the Peaceful Uses of Atomic Energy, Moscow, 1955*, pp. 362–75; also pp. 207–14 in English translation by Consultants Bureau (for sale by Superintendent of Documents, Washington, D.C., price $1.00).
2. Bochvar, A. A., Konobeevsky, S. T., Kutaitsev, V. I., Menshikova, I. S., and Chebotarev, N. T. "The Interaction between Plutonium and Other Metals in Connection with Their Arrangement in Mendeleev's Periodic Table," in *Proceedings of the Second United Nations International Conference on the Peaceful Uses of Atomic Energy*, **6**:184–93. Geneva: United Nations, 1958.
3. Bochvar, A. A. Private communication.
4. Waldron, M. B. Private communication.

CHAPTER XXIV

DELTA-PRIME PLUTONIUM

R. O. Elliott and A. C. Larson

INTRODUCTION

The phase transformations that occur in plutonium between 400° and 500° C have been a source of concern and speculation since the time of the earliest experiments on the allotropy of plutonium (1, 2) (see also chap. iv). A typical early dilatometric curve in this temperature range is shown in the upper portion of Figure 1. This curve shows a steadily increasing magnitude of negative expansion at the high-temperature end of the delta-phase range, starting about 30° C below the transition to the epsilon phase on heating. The interpretation of this curve was based largely on high-temperature X-ray diffraction data, which showed a mixture of delta and epsilon phases, the proportion of epsilon increasing with increasing temperatures (3). This X-ray evidence for a progressive transformation from delta to epsilon was thought to account for the "rounded-off" shape of the dilatometric curves just preceding the transition to the epsilon phase near 480° C on heating.

The lower portion of Figure 1 shows a more recent dilatometric curve, obtained with a high-purity plutonium specimen and an adiabatic, self-heating dilatometric method. The breaks are much sharper at the transformation points, and the curve shows evidence that another plutonium allotrope exists between 451° and 480° C on heating (4, 5) (see also chap. xii). This new phase was designated delta-prime and was found to have a negative linear thermal-expansion coefficient of $-120 \times 10^{-6}/°$ C. The existence of the delta-prime allotrope was confirmed with high-temperature X-ray diffraction powder patterns obtained with high-purity plutonium specimens. The X-rays did not reveal the presence of delta-prime until about 465° C, however, and the delta/delta-prime/epsilon phase transformation behavior resembled that of an alloy rather than that of a pure metal. Also, the average linear thermal expansion of the delta-prime phase was found by the X-ray diffraction powder method to be only $-16 \pm 28 \times 10^{-6}/°$ C (3).

In spite of these unresolved discrepancies between X-ray diffraction and dilatometric results, there was good agreement that the delta-prime phase in plutonium was strongly influenced by relatively small amounts of impurities. It was noted that

The authors are at the University of California, Los Alamos Scientific Laboratory, Los Alamos, New Mexico.

the heat effect accompanying the delta/delta-prime transition is barely detectable and disappears on slight additions of a second component (6). It was possible to obtain a delta-prime diffraction pattern only if the metal was of 99.97 w/o purity, and as little as 0.15 w/o of impurities was enough to cause delta-prime to be eliminated from the metal (3). Actually, evidence for delta-prime has been seen on both dilatometric and thermal-analysis traces of relatively impure plutonium specimens (7), suggesting that, although delta-prime may indeed be very sensitive to small amounts of certain impurities, it is relatively unaffected by others.

The objective of the present chapter is to show how the delta-prime phase is influenced by the usual impurities in plutonium, as well as by small amounts of other alloying elements. This information is intended to aid in understanding the alloy-like behavior of a metal that, by ordinary standards, would be considered quite pure. Hence the chapter provides data relating to plutonium-rich alloys containing all the following elements, commonly found as impurities in plutonium: carbon, aluminum, silicon, chromium, manganese, iron, cobalt, nickel, and uranium. It also includes information pertinent to plutonium-rich alloys containing

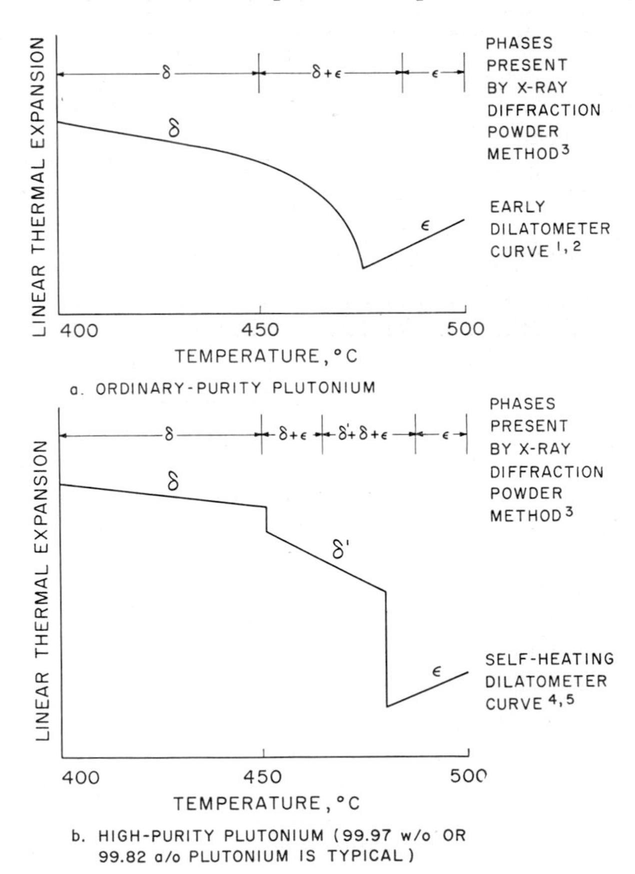

FIG. 1.—The general form of (*a*) early and (*b*) recent dilatometric curves of plutonium in the range of 400°–500° C.

titanium, vanadium, cerium, and thorium but is not concerned with the effects on delta-prime of the alkali metals, the alkaline earths, copper, tin, lead, oxygen, nitrogen, or hydrogen. With plutonium specimens of both ordinary and high purity, the unit-cell dimensions and thermal-expansion coefficients of the delta-prime phase were determined, as well as the temperatures of the delta/delta-prime and delta-prime/epsilon transitions.

EXPERIMENTAL

A High-Temperature, High-Vacuum X-Ray Diffractometer

The principal research tool used in this work was the General Electric XRD-3 X-ray diffractometer, shown in Figure 2. Some features in the basic design of this

Fig. 2.—The high-temperature, high-vacuum X-ray diffractometer

equipment are similar to those previously described elsewhere (8). The furnace chamber is designed for easy removal and replacement because it must be carried from the diffractometer to an open-front hood, where the alpha-active specimens are loaded. Use of respirators, rubber surgical gloves, and other safety precautions for handling plutonium specimens is required. Although some radioactive contamination is tolerated on the inside of the furnace chamber, the outside is kept "cold" or free of alpha contamination. The chamber is made of brass, water-cooled on the sides and in the base. A steel plate, fitting onto the XRD-3 machine in the same position as the standard specimen holder, and a system of adjustment screws are located underneath the furnace chamber. These screws permit rotary or translatory alignment of the specimen at elevated temperatures. An aluminum block containing a rotary vacuum seal with V-type neoprene packings is located above the furnace chamber. The furnace chamber rotates with the goniometer, while the high-vacuum pumping system remains stationary. The seal design allows for $\pm \frac{1}{32}$-inch translatory adjustment of the chamber.

A beryllium X-ray window, 10 mils in thickness, extends a little more than halfway around the furnace chamber and is sealed gas-tight to the chamber with a rectangular-shaped arrangement of a clamp and a neoprene gasket. The gasket has an O-ring cross-section and was specially molded for the purpose. Thermocouple and power leads are brought into the lower part of the furnace chamber through neoprene compression seals. The system can be evacuated to 2×10^{-5} mm of Hg within a few hours.

The heating element is made with two concentric Alundum tubes, slotted to allow passage of the incident and diffracted X-rays. The inner tube, $\frac{3}{4}$ inch I.D. by $3\frac{1}{2}$ inches long, is wound with 6 turns per inch of Nichrome V wire, 0.032 inch in diameter. The inner and outer tubes are cemented together with Alundum cement, and a reflecting sheet of 0.0002-inch-thick aluminum foil is wrapped around the outside of the element. The furnace requires 30 watts at 500° C.

The specimen and furnace are supported on a ceramic insulating pedestal made of machined and fired Lavite. The specimen—a square wafer about $\frac{5}{8}$ by $\frac{5}{8}$ by $\frac{1}{10}$ inch thick—is backed against a silver plate to insure temperature uniformity and is clamped into a tantalum holder. Specimen temperature is measured by means of a chromel-alumel thermocouple, which has its hot junction located within the specimen and about $\frac{1}{32}$ inch beneath the surface being examined.

It was found that the high-temperature diffractometer could also be used for detecting thermal arrests. Simultaneous use of these two techniques—i.e., X-ray diffraction and thermal analysis—revealed that, at the delta-prime/epsilon transition, the specimen surface is about 6° C cooler than the thermocouple, and hence the following temperature corrections were used: $-7.5°$ at 500° C; $-6.5°$ at 485° C; $-5°$ at 460° C; $-4°$ at 445° C; $-3°$ at 430° C; $-2°$ at 415° C; $-1°$ at 400° C. These corrections brought our delta/delta-prime and delta-prime/epsilon transformation temperatures for plutonium into close agreement with the average of published dilatometric and thermal-analysis results, and reasonable agreement was obtained for the Pu-Fe and Pu-Co eutectic temperatures near 410° and 405° C, respectively. The fact that the specimen's surface was cooler than its interior is thought to be a result of the X-ray furnace design, which necessitated having an open space in the furnace tube directly in front of the specimen.

Specimen Preparation

The required amounts of plutonium and alloying element were placed in a square-bottom magnesia crucible about 1 inch deep by $\frac{5}{8}$ inch square on the inside, heated in an induction furnace under vacuum, held at the desired temperature until well mixed, and then rapidly cooled to room temperature. The alloy casting obtained in this way was then broken out of the crucible, machined to final dimensions, and stored in ethyl alcohol. When small additions of iron or nickel to pure plutonium were desired, homogeneous plutonium-iron and plutonium-nickel eutectic master alloys were first prepared. Portions of these master alloys were then realloyed with plutonium to make plutonium-rich binary iron or nickel alloys. The castings weighed about 25 gm before machining.

Three different plutonium melting stocks of ordinary purity were used to make alloys. These were selected for best over-all purity from a number of stocks available, and their analyses are listed in Table 1. The plutonium-rich binary alloys contained from 0.125 to 10.0 a/o alloying addition, and all compositions of these alloys designated below are nominal.

TABLE 1

CHEMICAL ANALYSES OF PLUTONIUM METALS

(In ppm by Weight)

Impurity	Ordinary-Purity			High-Purity
	J-1352-53	J-3015	J-3037	KJ-329
Li	< 0.2	< 0.2	< 0.2	< 0.2
Be	< 0.2	< 0.2	< 0.2	< 0.2
Na	<10	<10	<10	<10
Mg	< 5	< 5	< 5	< 5
Ca	< 5	< 5	< 5	38*
Al	25	15	20	< 5
La	<10	<10	<10	<10
Si	65	140	140	12
Pb	< 1	1	< 1	< 1
Cu	15	< 1	3	2
Cr	<20	20	30	<30
C	30	60	130	23
O_2	90	50	70	70–240†
Fe	70	15	70	27
Ni	70	70	270	10
F	< 2	< 1	....	....
B	< 0.5	< 0.5	< 0.5	< 0.5
Co	< 5	< 5	< 5	< 5
Mn	7	5	7	20
S	....	....	....	<10
Mo	<30	....	....	....
Ag	< 0.1	10	< 0.5	< 1
Zn	<10	<10	<10	<10
Sn	1	< 1	< 1	< 1
Bi	....	< 1	< 1	....
Th	....	....	....	....
Total impurities (less O_2)	<382	<386	<719	<221

* Probably reduced to <5 during subsequent vacuum melting.
† Samples were not uniform.

PROCEDURE

Before being loaded, each specimen was electropolished at 6.5 volts for 5 minutes in the following solution: 8 parts (by volume) orthophosphoric acid, 5 parts ethyl alcohol, 5 parts glycerine. Following the electropolishing, the specimen was washed in distilled water and ethyl alcohol, dried, and finally loaded into the furnace chamber. During heating, the vacuum pressure first increased from 2×10^{-5} to 5×10^{-5} mm of Hg and then dropped to less than 5×10^{-6} mm of Hg on holding at a constant elevated temperature. The best vacuums were obtained when the plutonium specimens were at elevated temperatures—evidently the result of plutonium gettering.

The goniometer was oscillated at 2° per minute (a fast scanning rate) over the critical diffraction region as the specimen was heated slowly, usually from 0.1° to 0.5° C per minute. It was possible to detect the beginning and the end of a solid-state transition simply by observing the changes that occurred on the diffraction trace. It was not necessary to record the entire diffraction trace; a small diffraction interval, often only 5° or 10° of two-theta, was sufficient for the identification of phases during a transition. When lattice constants were desired, the traces were taken at 0.2° per minute (a slow scanning speed), and temperatures were recorded before and after each peak. The diffraction-line positions were taken as the center of a peak at about half-height. The line positions of the first eight delta-prime reflections were measured at a number of temperatures between 450° and 479° C, using the highest resolution possible. The data were processed on the Maniac I, an electronic digital computer, using the method of least squares to fit the following equation:

$$\sin^2 \theta_c = \frac{(h^2 + k^2)\lambda^2}{4a_t^2(1+\bar{a}_a t)^2} + \frac{l^2\lambda^2}{4c_t^2(1+\bar{a}_c t)^2} - 2d \sin \theta_0 \cos^2 \theta_0 - d^2 \cos^4 \theta_0 ,$$

where the terms containing d correct for a missetting of the specimen with respect to the axis of the diffractometer (9); θ_c is the calculated diffraction angle; h, k, l are Miller indices; λ is the wavelength of the X-radiation used; a_t and c_t are lattice constants of the tetragonal unit cell at temperature t; $\bar{a}_a$ and $\bar{a}_c$ are the average linear coefficients of thermal expansion along the a- and c-axes, respectively, for the temperature range of 450°–479° C; θ_0 is the observed diffraction angle; and d is the displacement of the specimen either before or behind the axis of the diffractometer.

Most of the alloys were cast, machined, and then examined by means of X-ray diffraction. The transformation data were obtained on heating. Some of the alloys were given homogenizing heat treatments at 500° C for about 1 week and then re-examined. Heat treatments were not given any of the plutonium-iron or plutonium-cobalt alloys because retrograde melting in these systems resulted in alteration of the specimen surfaces at temperatures very little higher than 400° C. After X-ray examination at high temperatures, all specimens were examined metallographically for phase identification, homogeneity, impurity inclusions, surface effects, etc.

RESULTS AND DISCUSSION

Plutonium-Carbon, Plutonium-Vanadium, and Plutonium-Chromium

The phase-equilibrium relationships, between 400° and 500° C, of the plutonium-rich binary alloys containing carbon, vanadium, or chromium are so simple that no diagram is needed to explain them. In plutonium alloys containing 5 a/o of either carbon, vanadium, or chromium, the delta/delta-prime transition occurs at 452° ±

5° C, and the delta-prime/epsilon transition occurs at 480° ± 5° C. The second phases in equilibrium with the plutonium phases in this temperature range are PuC, pure vanadium, and pure chromium, respectively. These elements are believed to be virtually insoluble in delta, delta-prime, and epsilon plutonium because they were not found to have any significant influence on the delta/delta-prime and delta-prime/epsilon transition temperatures of pure plutonium.

These results for the vanadium and chromium alloys are the same as data previously reported by the Soviet workers (6), but the result for carbon is new and somewhat surprising. The high-purity plutonium used at Los Alamos for the initial

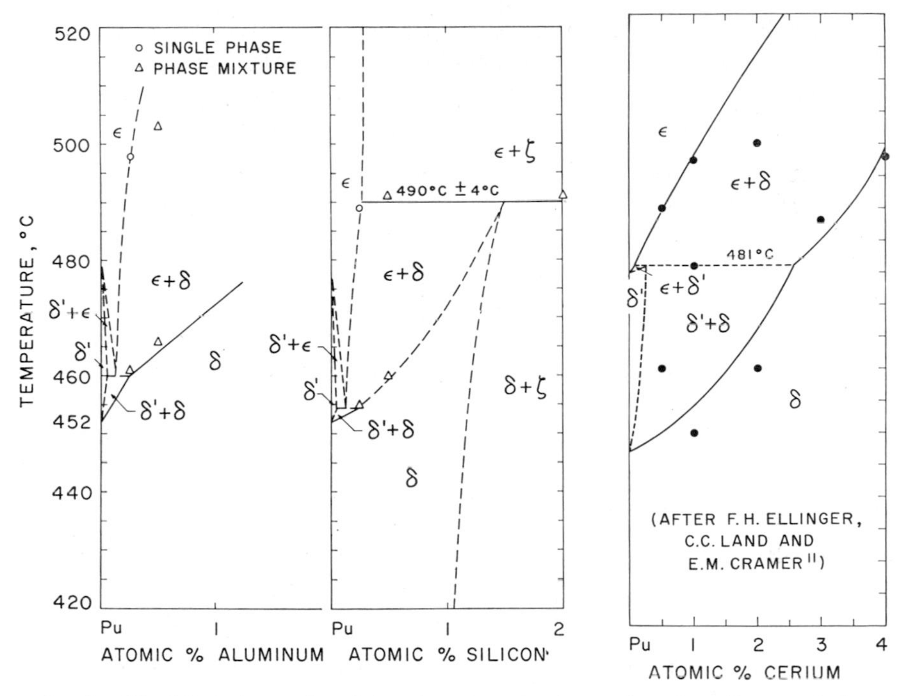

FIG. 3.—The delta-prime region of the plutonium-aluminum, plutonium-silicon, and plutonium-cerium phase diagrams.

detection and X-ray identification of the delta-prime phase was of much improved purity, largely because of a decrease in its carbon content (4). It was also known that appreciable amounts of carbon are soluble in face-centered cubic alpha thorium (10), and some solubility of carbon in face-centered cubic delta plutonium was believed probable. To learn that the reduction of carbon impurity in plutonium was not of major importance in the discovery of the delta-prime phase was, therefore, quite surprising.

PLUTONIUM-ALUMINUM, PLUTONIUM-SILICON, AND PLUTONIUM-CERIUM

The plutonium-rich end of the plutonium-cerium phase diagram and the plutonium-rich portions of the plutonium-aluminum and plutonium-silicon systems are shown in Figure 3. These alloying additions to plutonium show greater solubilities in the delta phase than in either epsilon or delta-prime plutonium. The delta-prime phase is eliminated in plutonium-rich binary alloys containing less than 0.5 a/o aluminum or silicon, or less than 2 a/o cerium (11). It is probable that the high-plutonium end of the plutonium-zinc phase diagram is similar to the high-plutonium regions in the plutonium-aluminum and plutonium-silicon diagrams (no delta-prime phase has been observed in alloys containing 0.5 a/o zinc) (12). The plutonium-aluminum and plutonium-silicon diagrams are shown schematically because of the following difficulties in determining the delta-prime phase boundaries with the high-temperature X-ray diffractometer: (1) surface enrichment either in the alloy addition or in impurities due to the reactivity of the alloying element with residual gases in the diffractometer vacuum chamber (see the plutonium-titanium discussion below) and (2) the early appearance of the epsilon phase on heating (see discussion below of the delta/delta-prime and delta-prime/epsilon transformations).

Plutonium-Manganese, Plutonium-Iron, Plutonium-Cobalt, and Plutonium-Nickel

The phase diagrams for the plutonium-rich binary alloys containing manganese, iron, cobalt, and nickel are shown in Figure 4. These alloying additions to plutonium also show greater solubilities in the epsilon phase than in either the delta or the delta-prime phase. The delta-prime phase is eliminated in plutonium-rich binary alloys containing less than 2 a/o manganese or less than 1 a/o of either iron, cobalt, or nickel.

An outstanding feature of the plutonium-iron and plutonium-cobalt alloys is their retrograde melting behavior, i.e., certain alloys in these systems become partly molten on reaching their respective eutectic temperatures and then solidify again as the temperature is increased. The British workers (13) first found this behavior in plutonium-iron alloys by means of dilatometric and thermal-analysis methods. The similar finding for plutonium-cobalt alloys is new, and the X-ray data of the present investigation confirm the validity of the plutonium-iron and plutonium-cobalt phase-diagram constructions showing retrograde melting.

The plutonium-rich portions of the phase diagrams for the binary alloys containing iron, cobalt, and nickel are quite similar, as may be seen in Figure 4. An exception is the retrograde melting behavior exhibited by plutonium-iron and plutonium-cobalt alloys but not by the nickel alloys. The similarities can be seen by visualizing the changes that would occur in the plutonium-nickel diagram if its eutectic temperature were below 413° C, instead of at 465° C. The fact that the plutonium-nickel eutectic temperature is at 465° C accounts for a greater solid solubility of nickel in the epsilon phase than of either iron or cobalt in this phase, in their respective systems where the eutectic temperatures are 410° and 405° C. Note that the eutectoid composition of the epsilon phase occurs in the vicinity of 2 a/o in all three of these systems, as well as in the plutonium-manganese system. These similarities may be related to the similarity in atomic radius of the manganese, iron, cobalt, and nickel atoms.

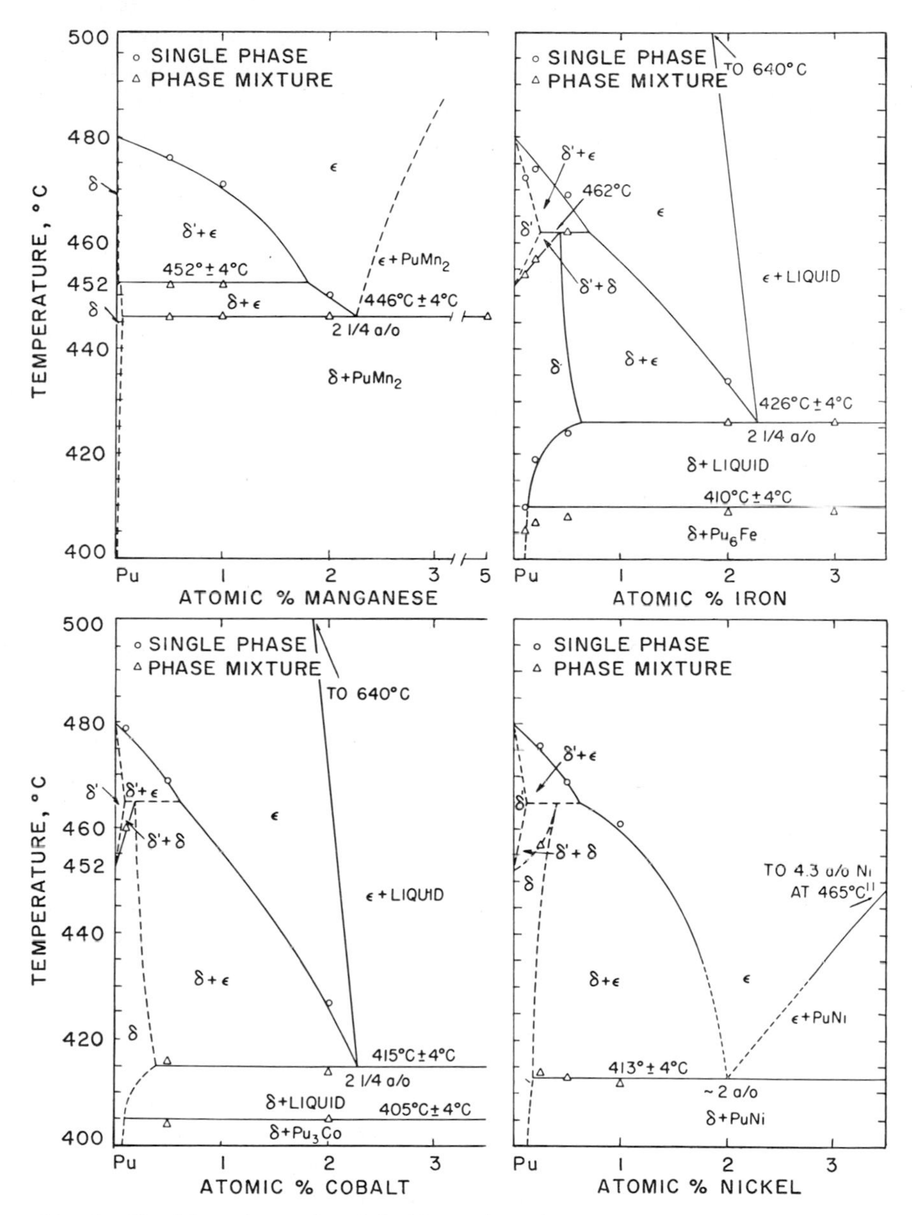

FIG. 4.—The delta-prime region of the binary phase diagrams of plutonium with manganese, iron, cobalt, and nickel.

Plutonium-Titanium, Plutonium-Thorium, and Plutonium-Uranium

It is evident from the preceding discussion that the transformation behavior of plutonium between 400° and 500° C is influenced by small amounts of many of its usual impurity elements. A number of these impurities are more soluble in either the delta or the epsilon phase than in the delta-prime phase, thus causing a mixture of the delta and epsilon phases to replace delta-prime in this temperature range. Undoubtedly this fact accounts for the failure, for nine years, of plutonium metallurgists to discover the delta-prime phase. It also serves to emphasize the unusualness of the situations found in the plutonium-titanium, plutonium-thorium, and plutonium-uranium systems where appreciable quantities of these alloy additions were found to be soluble in the delta-prime phase. The plutonium-rich portions of these systems are shown in Figure 5 for the temperature range between 400° and 500° C. About 5 a/o titanium, 1–1.5 a/o thorium, and 1–1.5 a/o uranium are contained in delta-prime solid solutions at 446° C, approximately 490° C, and 442° C, respectively. As the titanium content is increased in the tetragonal delta-prime solid solution, the a_0 lattice constant also increases, but the c_0 lattice constant decreases.

Of special interest in connection with the plutonium-titanium alloys was the observation of a "self-plating" effect, in which the excess alpha titanium formed a well-bonded layer at the outer surfaces of the specimen. The formation of this layer was accompanied by the disappearance of plutonium oxide phases both within the metal and at its surface. The layer was detectable by both high-temperature X-ray diffractometry and by subsequent metallographic examination. For example, a layer of alpha titanium about 1 μ in thickness was found on a plutonium alloy containing 10 a/o titanium after it had been heated under vacuum in the X-ray diffractometer. After being sealed in an evacuated silica capsule and held for 1 month at 465° C, the thickness of the layer increased to about 10 μ, and the concentration of titanium at the surface resulted in a visible depletion of the alpha-titanium phase in the interior of the metal. The interior was later found by chemical analysis to contain only 8 a/o titanium. The surface layer was appreciably harder than pure alpha titanium and had hexagonal lattice constants of $a_0 = 2.954$ A and $c_0 = 4.715$ A. Another alloy containing 7 a/o titanium was sealed in a silica capsule under 1 atm of oxygen pressure and held for 192 hours at 465° C. Metallographic examination revealed a very broad, irregular, discontinuous surface layer of alpha titanium beneath a thick coating of plutonium oxide phases, again with a visible depletion of the alpha-titanium phase in the interior of the metal.

This "self-plating" effect is believed to be a phenomenon not previously observed in alloy systems. It appears that a third component, oxygen, is involved and provides the driving force for the migration of excess titanium to the specimen surface. It was noted earlier in this chapter that the vacuum in the X-ray diffractometer furnace improved after the plutonium or plutonium-alloy specimen was heated. (Oxide was observed to form continuously on the specimen surfaces while at elevated temperatures.) The high hardness of the alpha-titanium surface layer is consistent with interstitial solid solutions of either oxygen or nitrogen in alpha titanium. The hexagonal lattice constants of pure alpha titanium are $a_0 = 2.9504 \pm 0.0003$ A and $c_0 = 4.6833 \pm 0.0003$ A, and it will dissolve up to 34 a/o of oxygen in solid solution, resulting in an increase in the value of c_0 while a_0 changes only slightly (14, 15). The lattice constants obtained for the alpha-titanium surface layer correspond to those of alpha titanium with about 10 a/o oxygen in solid solution. It is believed that this solid solution is thermodynamically more stable than plutonium oxide, causing the observed disappearance of the plutonium oxide

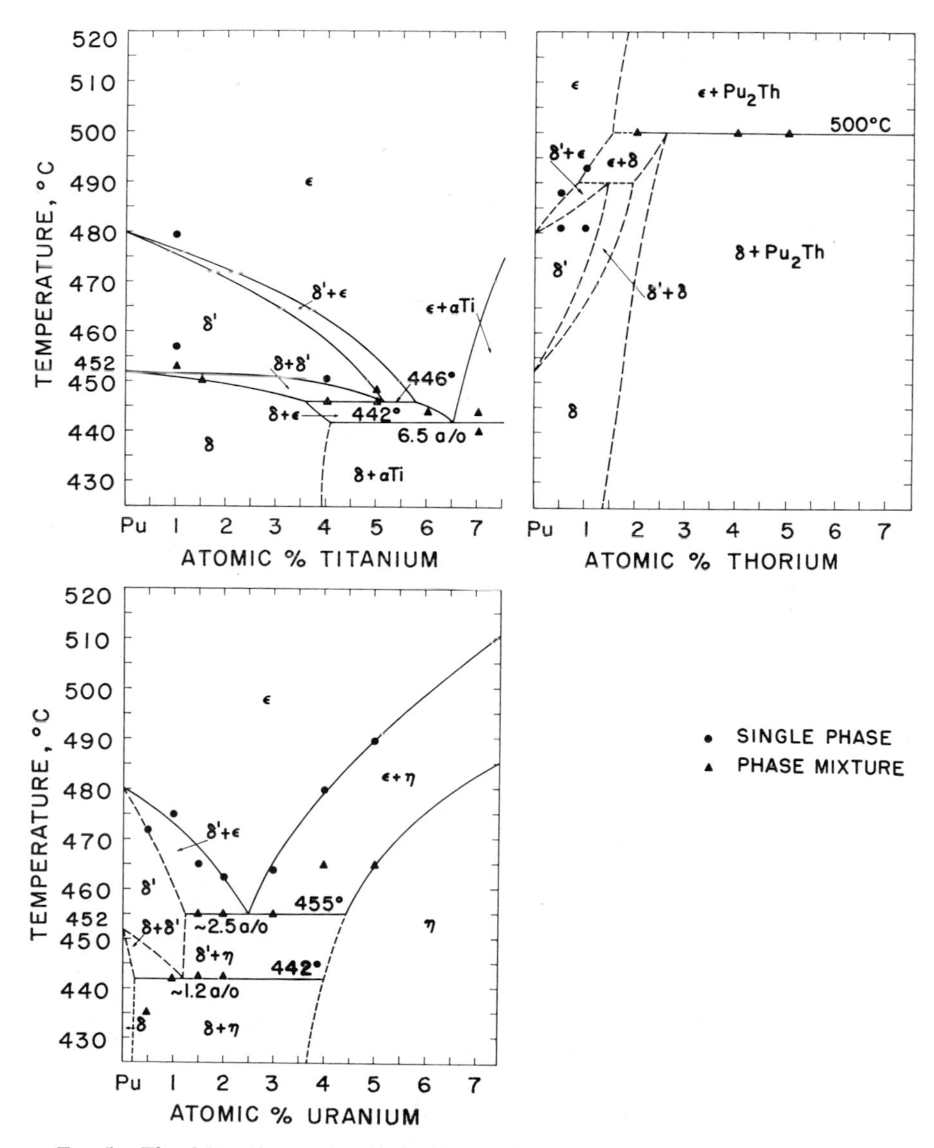

FIG. 5.—The delta-prime region of the binary phase diagrams of plutonium with titanium, thorium, and uranium.

phases when the alpha-titanium surface layer was formed. Nitrogen also forms an interstitial solid solution in titanium, similar to that formed by oxygen in titanium. It is possible that nitrogen is also partly responsible for the "self-plating" observed with plutonium-titanium alloys, but no experiments were conducted to establish this.

The Metallic Size of Plutonium Atoms

The average volume per atom in pure delta plutonium at room temperature (11) is about 25.0 A^3. This value was obtained by extrapolating unit-cell data for delta-phase alloys to zero alloy content. The average volume per atom in pure epsilon

TABLE 2

Volumes per Atom at 24° C (in A^3)

(After Larson and Cromer Ref. 16)

Element	Volumes in Elemental States	Volumes in Intermetallic Compounds (A^3)	
	Volume, A^3	Alloying Element	Plutonium
Carbon	5.6	5.8	24.9
Aluminum	16.5	17.8	24.7
Silicon	19.9	19.9*	23.8
Titanium	Does not form intermetallic compounds with plutonium		
Vanadium	Does not form intermetallic compounds with plutonium		
Chromium	Does not form intermetallic compounds with plutonium		
Manganese	13.0	12.7	22.0
Iron	11.7	12.2	22.2
Cobalt	11.1	11.0	22.4
Nickel	11.0	12.3	21.0
Zinc	15.1	17.6	23.2
Cerium	Does not form intermetallic compounds with plutonium		
Thorium	32.7	32.7*	26.0

* Elemental volumes were used.

plutonium at room temperature is about 22.8 A^3, and this value was obtained by extrapolating the high-temperature diffraction data to room temperature.

The average volumes per plutonium atom were calculated in various intermetallic compounds of as many as possible of the binary systems studied, and the results are given in Table 2. Equations were written for each compound, relating the volume per formula unit to the sum of the volumes of the individual atoms which comprise the unit. Two such equations were then solved simultaneously for the average volume per atom. The data used in making these calculations are summarized in chapter xxv of this volume. When more than two compounds were available, those were selected for which the data were believed to be the most reliable.

In Table 2 note that the volumes per plutonium atom in intermetallic compounds of plutonium with aluminum, silicon, zinc, or thorium are 24.7, 23.8, 23.2, or

26.0 A^3, respectively. On the other hand, note that the volumes per plutonium atom in compounds of plutonium with manganese, iron, cobalt, or nickel are 22.0, 22.2, 22.4, or 21.0 A^3, respectively. The volume per plutonium atom is larger in intermetallic compounds with aluminum, silicon, zinc, or thorium than in compounds with manganese, iron, cobalt, or nickel. In addition, the series of phase diagrams given in this chapter shows that the former group favors delta-phase (25.0 A^3) solid solutions, while the latter group favors epsilon-phase (22.8 A^3) solid solutions. In both these groups, relatively small amounts of the alloying elements were found to close off the delta-prime phase region, thus eliminating it in their plutonium-rich alloys or in plutonium metal containing sufficient quantities of these elements as impurities.

Delta/Delta-Prime and Delta-Prime/Epsilon Transition Temperatures

The delta/delta-prime and delta-prime/epsilon transition temperatures on heating are given in Table 3 for three plutonium specimens of ordinary purity and one

TABLE 3

DELTA/DELTA-PRIME AND DELTA-PRIME/EPSILON TRANSITION TEMPERATURES OF PLUTONIUM ON HEATING (IN ° C)

Transformation	Plutonium Stock							
	Ordinary-Purity						High-Purity KJ-329	
	J-1352-53		J-3015		J-3037			
	Start	End	Start	End	Start	End	Start	End
$\delta \rightarrow \epsilon$	437	....	437	....	425	....	....	....
$\delta \rightarrow \delta'$	455	460	454	?	455	459	452±2	452±2
$\delta' \rightarrow \epsilon$	....	<482	....	476	....	479	474	480±2

of high purity. Chemical analyses of these specimens are given in Table 1. A small proportion of epsilon phase formed in the ordinary-purity metals at temperatures 20–30° C below the delta/delta-prime transition temperature. This epsilon coexisted with delta up to the delta/delta-prime transition, after which it coexisted with delta-prime. The amount of epsilon appeared to increase slightly as the temperature was raised, but the main, rapid delta-prime/epsilon transition did not occur much below 480° C.

In high-purity plutonium, delta transformed isothermally to delta-prime at 452° ± 2° C. With continued heating, a small amount of delta-prime was next observed to change to epsilon near 474° C, and this phase mixture remained unchanged to 480° ± 2° C, where the delta-prime/epsilon transition took place rapidly.

The most striking difference found between the phase-transformation behaviors of high-purity and ordinary-purity plutonium was with regard to the early appearance of small amounts of the epsilon phase during heating. The early appearance of epsilon was observed consistently in ordinary-purity metals and in plutonium-rich alloys (except those containing titanium) made with ordinary-purity metals. It is believed that this resulted from impurities of the manganese, iron, cobalt, and nickel type. The microstructures of ordinary-purity plutonium melting stocks contain numerous metallic and non-metallic impurity inclusions, some of which resemble in appearance the phases found in the binary plutonium-rich alloys containing iron, cobalt, and nickel. The plutonium matrix surrounding such inclusions would be expected to change from delta to epsilon at temperatures below that for the delta/delta-prime transition, in accordance with the phase diagrams shown in Figure 4. The delta/delta-prime transition occurred at slightly higher temperatures, while the delta-prime/epsilon occurred at slightly lower temperatures, in the ordinary-purity metals than they did in the high-purity metal. The phase diagrams in this chapter show that small amounts of a number of the usual impurities in plutonium, by acting collectively in a multicomponent system, raise the delta/delta-prime transition temperature, while other impurities lower the delta-prime/epsilon transition temperature.

The Unit-Cell Dimensions of Delta-Prime Plutonium

The dimensions and volumes of delta-prime unit cells at 450° C are listed in Table 4 for plutonium specimens of ordinary and high purity, together with values

TABLE 4

UNIT-CELL DIMENSIONS AND VOLUMES OF DELTA-PRIME PLUTONIUM AT 450° C

Units	This Study (X-Ray Diffractometer)		Ellinger (3) (X-Ray Powder Method)
	High-Purity KJ-329	Ordinary-Purity J-3015	
BCT (a_0) (A)......	3.3261±0.0008	3.3283±0.0010	3.3119±0.0021
BCT (c_0) (A)......	4.4630± .0014	4.4512± .0016	4.5265± .0056
Unit-cell volume (A^3)...........	49.374 ±0.023	49.308 ±0.027	49.650 ±0.076

obtained previously by an X-ray diffraction powder method. The uncertainties in the numbers are standard deviations derived from least-squares calculations. The diffractometer values for ordinary-purity metal show a slight increase in the a_0 lattice constant and a decrease in the c_0 lattice constant compared with the values obtained with high-purity metal. This is the same variation in unit-cell dimensions as was observed with increasing amounts of titanium in solid solution. No correlation is evident between the diffractometer and powder-method results, but even small errors in temperature measurement will cause large discrepancies in the delta-prime lattice constants, especially in c_0 because of the exceptionally high thermal expansion in the direction of the c-axis. Note that the unit-cell volumes are in much closer agreement than are the lengths of the unit-cell edges.

THE THERMAL EXPANSION OF DELTA-PRIME PLUTONIUM

The mean linear coefficients of thermal expansion of delta-prime plutonium along the a- and c-axes of its body-centered tetragonal crystal lattice are listed in Table 5 for plutonium specimens of ordinary and high purity, together with the values from previous studies. The uncertainties in the numbers are standard deviations derived from the least-squares calculations. The mean linear coefficient of thermal expansion, $\bar{a}_p$, for the bulk, randomly oriented polycrystalline material was

TABLE 5

LINEAR THERMAL EXPANSION OF DELTA-PRIME PLUTONIUM (PER ° C)

Coefficient	This Study (X-Ray Diffractometer)		Ellinger (3) (X-Ray Powder Method)	Jette (4, 5) (Dilatometric Method)
	High-Purity, KJ-329	Ordinary-Purity, J-3015		
$10^6\,\bar{a}_a$	444.8±12.1	329.9±15.1	305±35	....
$10^6\,\bar{a}_c$	−1063.5±18.2	−828.7±19.1	−659±67	....
$10^6\,\bar{a}_p$	− 65.6±10.1	− 60.8±11.9	− 16±28	−120

calculated from the single-crystal coefficients, using the relation $(1 + \bar{a}_p\Delta t)^3 = (1 + \bar{a}_a\Delta t)^2\,(1 + \bar{a}_c\Delta t)$. The following results were obtained for high-purity plutonium KJ-329:

$$a_t = 3.3261 \pm 0.0008 + (14.80 \pm 0.40) \times 10^{-4}\,(t - 450)\,,$$

$$c_t = 4.4630 \pm 0.0014 - (47.46 \pm 0.81) \times 10^{-4}\,(t - 450)\,,$$

where t is the temperature in ° C and the uncertainties are standard deviations obtained from the least-square calculations.

The diffractometer values for ordinary-purity metal show a reduction in the thermal expansion along the a-axis and an increase, in the positive direction, along the c-axis, compared with the values obtained with high-purity metal. There is no significant difference between the bulk linear thermal expansion coefficients ($\bar{a}_p$) of delta-prime in these two grades of metal, and the bulk values are intermediate between the bulk value previously obtained by the diffraction-powder method and that obtained with a linear dilatometer.

The authors are grateful to Dr. A. S. Coffinberry and Professor E. R. Parker for their joint supervision of the investigation; to Messrs. F. W. Schonfeld, F. H. Ellinger, and E. M. Cramer for their assistance and suggestions; and to the other members of the Los Alamos plutonium chemistry and metallurgy groups whose services contributed to this work.

REFERENCES

1. SMITH, C. S. *Metal Progress*, **65**:No. 5, 81–89, 1954.
2. LORD, W. B. H. *Nature*, **173**:534–35, 1954.

3. ELLINGER, F. H. *Transactions of the American Institute of Mining, Metallurgical and Petroleum Engineers*, **206**:1256–59, 1956.
4. JETTE, E. R. *Journal of Chemical Physics*, **23**:365–68, 1955.
5. ———. "Plutonium Metal," in *Nuclear Metallurgy*, pp. 29–37. ("Institute of Metals Division Special Report Series," No. 1.) New York: American Institute of Mining and Metallurgical Engineers, 1955.
6. KONOBEEVSKY, S. T. "Phase Diagrams of Some Plutonium Systems," in *Session of the Division of Chemical Science (of the U.S.S.R. Academy of Science), Conference on the Peaceful Uses of Atomic Energy, Moscow, 1955*, pp. 362–75; also pp. 207–14 in English translation by the Consultants Bureau (for sale by Superintendent of Documents, Washington, D.C., price $1.00).
7. COFFINBERRY, A. S., and WALDRON, M. B. "The Physical Metallurgy of Plutonium," chap. 4 of *Progress in Nuclear Energy*, Ser. V, Vol. **1**, pp. 354–410. London: Pergamon Press, Ltd., 1956.
8. SUMSION, H. T., and SOWMAN, H. G. *A High-Temperature X-Ray Furnace for the General Electric X-Ray Spectrometer*. (USAEC Rept. KAPL-1303, 1955.)
9. SMAKULA, A., and KALMAJS, J. *Physical Review*, **99**:1737–43, 1955.
10. ROUGH, F. A., and BAUER, A. A. *Constitutional Diagrams of Uranium and Thorium Alloys*. Reading, Mass.: Addison-Wesley Publishing Co., Inc., 1958.
11. ELLINGER, F. H., LAND, C. C., and CRAMER, E. M. "The Plutonium-Cerium Phase Diagram," in *Extractive and Physical Metallurgy of Plutonium and Its Alloys*, ed. W. D. WILKINSON, pp. 149–67. New York: Interscience Publishers, 1960.
12. CRAMER, E. M., ELLINGER, F. H., and LAND, C. C. "The Plutonium-Zinc Phase Diagram," in *Extractive and Physical Metallurgy of Plutonium and Its Alloys*, ed. W. D. WILKINSON, pp. 169–80. New York: Interscience Publishers, 1960.
13. MARDON, P. G., HAINES, H. R., PEARCE, J. H., and WALDRON, M. B. *Journal of the Institute of Metals*, **86**:166–71, 1957–58.
14. CLARK, H. T. *Transactions of the American Institute of Mining and Metallurgical Engineers*, **185**:588–89, 1949.
15. BUMPS, E. S., KESSLER, H. D., and HANSEN, M. *Transactions of the American Society for Metals*, **45**:1008–28, 1953.
16. LARSON, A. C., and CROMER, D. T. Unpublished data.

CHAPTER XXV

A REVIEW OF THE INTERMETALLIC COMPOUNDS OF PLUTONIUM

F. H. Ellinger

INTRODUCTION

In 1955 a review of the intermetallic compounds of plutonium was prepared for the First International Conference on the Peaceful Uses of Atomic Energy (1). It covered briefly the crystallographic data of the intermetallic compounds and contained some discussion of certain compounds of special interest. It included all

TABLE 1

DISTRIBUTION OF PLUTONIUM INTERMETALLIC COMPOUNDS OVER THE PERIODIC TABLE

IA	IIA	IIIA	IVA	VA	VIA	VIIA	VIII			IB	IIB	IIIB	IVB	VB	VIB
H 2															
Li 0	Be 1											B	C 2	N 1	O 4
Na 0	Mg 2											Al 5	Si 5	P (1)	S 2
K 0	Ca 0	Sc	Ti 0	V 0	Cr 0	Mn 1	Fe 2	Co 6	Ni 6	Cu 3	Zn 3	Ga	Ge 5	As (1)	Se
Rb	Sr	Y	Zr 2	Nb 0	Mo 0	Tc	Ru 5	Rh	Pd	Ag (1)	Cd	In (1)	Sn (2)	Sb	Te (1)
Cs	Ba 0	La 0	Hf	Ta 0	W 0	Re 1	Os 5	Ir	Pt	Au	Hg (2)	Tl	Pb 2	Bi (2)	Po
Fr	Ra	Ac													

Th 1	Pa	U 2	Np	Pu	Am	Cm			

the known results to 1955 that had been obtained at the Los Alamos Scientific Laboratory and the wartime Manhattan Project in the United States and at the Chalk River Project in Canada. The present chapter contains more extensive

F. H. Ellinger is at the University of California, Los Alamos Scientific Laboratory, Los Alamos, New Mexico.

TABLE 2

Compounds Mentioned in This Chapter and Reference Numbers

Compound	Ref. No.
$LaAl_2$	50
$LaAs$	89
$LaBi$	100
La_2C_3	23
LaH_2	7
$LaHg_3$	98
$LaHg_4$	98
LaN	24
$LaNi_2$	79
$LaNi_5$	83
La_2O_3	32, 33
LaP	64
$LaPb_3$	92
LaS	69
La_2S_3	71
$LaSi_2$	62
$LaSn_3$	92
$LaTe$	96
$CeAl_2$	51
$CeAs$	89
$CeBe_{13}$	14
$CeBi$	100
Ce_2C_3	23
$CeFe_2$	73
CeH_2	7
CeN	24
$CeNi_2$	80, 81
$CeNi_5$	83
Ce_2O_3	32, 33
CeO_2	42
CeP	64
$CePb_3$	93
CeS	68, 69
Ce_2S_3	71
$CeSi_2$	58, 62
$CeSn_3$	93
$CeTe$	96
$PrAs$	89
$PrBi$	100
Pr_2C_3	23
PrH_2	7
PrN	24
$PrNi_2$	79
$PrNi_5$	84
Pr_2O_3	32, 33
PrO_2	43
PrP	64
$PrPb_3$	94
PrS	69
$PrSi_2$	62
$PrSn_3$	94
$PrTe$	96
Nd_2C_3	23
NdH_2	7
NdN	25
NdO	28
Nd_2O_3	32, 33
NdP	65
NdS	69
$NdSi_2$	62
$NdTe$	96
Sm_2C_3	23
SmH_2	7
SmH_3	9
SmH	26
SmO	29
Sm_2O_3	37, 38, 39
$SmSi_2$	62
Eu_2O_3	37, 38, 39
Gd_2C_3	23
$GdFe_2$	74
GdH_2	8
GdH_3	8
$GdMn_2$	74
$GdNi_5$	85
Gd_2O_3	37, 38, 39
Tb_2C_3	23
Tb_2O_3	37, 38, 39
TbO_2	44
Dy_2C_3	23
DyH_3	9
Dy_2O_3	37, 38, 39
Ho_2C_3	23
Ho_2O_3	37, 38, 39
ErH_2	9
ErH_3	9
Er_2O_3	37, 38, 39
Tm_2O_3	37, 38, 39
Yb_2O_3	37, 38, 39
Lu_2O_3	37, 38, 39
Actinide Isotypes	
Ac_2O_3	34
Ac_2S_3	71
$ThBe_{13}$	14
ThC	18
ThN	27
$ThNi_5$	86
Th_2Ni_{17}	78
ThO	27
ThO_2	45
ThP	66
Th_3P_4	66
ThS	68, 70
$ThSi$	57
Th_2Si_3	57
$ThSi_2$	61
PaO	30
$PaO_{2.2}$	41
UAl_2	52, 72
UAl_4	55
UAs	90
UBe_{13}	14
UBi	101, 102
UC	19
U_2C_3	22
U_6Co	72
UCo_2	72
U_6Fe	72
UFe_2	72
UGe_3	88
UHg_3	97
UHg_4	97
U_6Mn	72
UMn_2	72
UN	19
UO	19
UO_2	46, 47
UP	67
UPb_3	88, 89
US	68
USi	58
β-USi_2	58
USi_2	58
USn_3	52
UTe	95
UZn_9	111
U_2Zn_{17}	112
$NpAl_2$	104
$NpAl_4$	104
$NpBe_{13}$	15
NpC	20, 31
NpN	16
NpO	16
NpO_2	41
$NpSi_2$	58
$PuAg_3$	1, 11
Pu_3Al	1
$PuAl$	1
$PuAl_2$	1, 11
$PuAl_3$	1, 11, 53
$PuAl_4$	1, 11, 54
$PuAs$	1, 63
$PuBe_{13}$	1, 2, 11
$PuBi$	1
$PuBi_2$	1, 107
PuC	1, 16
Pu_2C_3	1, 21
Pu_6Co	....
Pu_3Co	1, 110
Pu_2Co	1
$PuCo_2$	1, 11
$PuCo_3$	....
Pu_2Co_{17}	....
$PuCu_2$	107
$PuCu_3$(?)	1, 107
$PuCu_7$(?)	1, 107
Pu_6Fe	1, 2
$PuFe_2$	1, 2, 11
Pu_3Ge	....
Pu_3Ge_2	....
Pu_2Ge_3	1
$PuGe_2$	1
$PuGe_3$	1
PuH_2	1, 5, 6
PuH_3	1, 6
$PuHg_3$	1
$PuHg_4$	1
Pu_3In	1
$PuMg_2$	....
$PuMg_x$	1
$PuMn_2$	1, 2, 11
PuN	16

TABLE 2—*Continued*

Compound	Ref. No.	Compound	Ref. No.	Compound	Ref. No.
ACTINIDE ISOTYPES—*Continued*		Pu_3Ru	....	AmO_2	40, 41
PuNi	1, 105	Pu_5Ru_3	....	Am_2S_3	71
$PuNi_2$	1, 2, 11	PuRu	....	Cm_2O_3	40
$PuNi_3$	1, 105	$PuRu_2$	....	CmO_2	40
$PuNi_4$	1	PuS	68		
$PuNi_5$	1, 11	Pu_2S_3	71	MISCELLANEOUS	
Pu_2Ni_{17}	1, 11	Pu_2Si(?)	1	AlB_2	60
PuO	1, 16	Pu_3Si_2(?)	1	$AuCu_3$	87
Pu_2O_3	1, 31	PuSi	1	CaF_2	10
Pu_2O_3–Pu_4O_7	1, 36	Pu_2Si_3	1, 59	Cu_2Mg	49
PuO_2	1, 41	$PuSi_2$	1, 58	$CaZn_5$	82
$Pu_{19}Os$	2	$PuSn_3$	1	CsCl	91
Pu_3Os	2	PuTe	1, 63	FeB	56
Pu_5Os_3	2	Pu_2Th	103	$(Fe, Mn)_2O_3$	38
$PuOs_2$	2	Zeta Pu	1	Fe_2P	75, 76, 77
PuP	1, 63	Eta Pu	1	$MgZn_2$	108
Pu_xPb	2, 107	$Pu_{19}Zr$	1	Mn_2O_3	37, 38
$PuPb_3$	1, 2	$PuZr_2$	106	NaCl	17
$PuRe_2$	....	AmO	16	$NaZn_{13}$	12, 13
$Pu_{19}Ru$	....	Am_2O_3	35	$SrPb_3$	48

crystallographic data on the compounds described in 1955 and includes the additional information gained since then by the above-mentioned sources and also by the Atomic Energy Research Establishment at Harwell, England, and by the Russians as reported by Konobeevsky (2). Although some of the compounds described in this review are not strictly intermetallic compounds, they have been included for completeness.

The distribution of plutonium intermetallic compounds over the periodic table and the number of compounds that plutonium forms with each element is summarized in Table 1. This survey is not complete (parentheses around a number indicate uncertainty regarding that number), but it is readily apparent that the transition metals of group VIII and the B-subgroup elements tend to form the most intermetallic compounds with plutonium.

The crystallographic nomenclature in this chapter follows that used in *Structure Reports* (3) and in the *International Tables for X-Ray Crystallography* (4). The symbol Z stands for the number of formula units in the unit cell, and D_x stands for the density (gm/cm^3) calculated from the unit-cell volume and contents. All distances are expressed in angstrom units (A). The isotypes listed for each plutonium compound have been limited to the corresponding compounds in which plutonium is replaced by any elements of the lanthanide or actinide series.

The compounds are described in alphabetical order of the alloying component. Table 2 lists all the plutonium compounds and isotypes mentioned in this chapter and gives the corresponding references. Where no references are listed, the information is found only in this chapter.

PLUTONIUM-ALUMINUM

Pu_3Al

Tetragonal, partially ordered $SrPb_3$ type (tetragonally deformed $AuCu_3$ type)

$a = 4.499 \pm 0.002$, $c = 4.538 \pm 0.002$ A

$Z = 1$, $D_x = 13.45$

Space group.—P4/mmm (?) (Int. Tab. 123).

Atomic positions

1Al in (a): 0,0,0

1Pu in (c): $\frac{1}{2},\frac{1}{2},0$

2Pu in (e): $0,\frac{1}{2},\frac{1}{2}$; $\frac{1}{2},0,\frac{1}{2}$

Interatomic distances

Pu-8Pu = 3.195	Al-8Pu = 3.195
Pu-4Al = 3.181	Al-4Pu = 3.181

PuAl (Composition Uncertain)

The intermetallic is formed by a peritectoid reaction between $PuAl_2$ and delta plutonium, and its diffraction pattern has never been obtained free from the superimposed patterns of these other phases. Its diffraction pattern does, however, suggest a distorted CsCl type of structure, thus this phase has tentatively been assigned the composition PuAl.

$PuAl_2$

Cubic face-centered, Cu_2Mg type

Pu-rich: $a = 7.838 \pm 0.001$ A

Al-rich: $a = 7.848 \pm 0.001$

$a = 7.831 \pm 0.005$ (Runnalls)

$a = 7.832$ (Harwell)

$Z = 8$, $D_x = 8.06$

Space group.—Fd3m (Int. Tab. 227).

Atomic positions

$(0,0,0; 0,\frac{1}{2},\frac{1}{2}; \frac{1}{2},0,\frac{1}{2}; \frac{1}{2},\frac{1}{2},0) +$

8Pu in (a): 0,0,0; $\frac{1}{4},\frac{1}{4},\frac{1}{4}$

16Al in (d): $\frac{5}{8},\frac{5}{8},\frac{5}{8}$; $\frac{5}{8},\frac{7}{8},\frac{7}{8}$; $\frac{7}{8},\frac{5}{8},\frac{7}{8}$; $\frac{7}{8},\frac{7}{8},\frac{5}{8}$

Interatomic distances (based on a=7.848 A)

Pu-4Pu = 3.398	Al-6Al = 2.775
-12Al = 3.254	-6Pu = 3.254

Isotypes.—$LaAl_2$, $CeAl_2$, UAl_2, $NpAl_2$.

$PuAl_3$

Hexagonal, $PuAl_3$ type (new structure type)

$a = 6.10 \pm 0.02$ A (Larson *et al.*)

$c = 14.47 \pm 0.02$

$a = 6.08 \pm 0.01$ (Runnalls)

$c = 14.40 \pm 0.03$

$Z = 6$, $D_x = 6.67$

Space group.—$P6_3/mmc$ (Int. Tab. 194).

Atomic positions

$2Pu_I$ in (b): $\pm(0,0,\frac{1}{4})$

$4Pu_{II}$ in (f): $\pm(\frac{1}{3},\frac{2}{3},z;\ \frac{1}{3},\frac{2}{3},\frac{1}{2}-z)$ with $z = 0.0892 \pm 0.0001$

$6Al_I$ in (h): $\pm(x,2x,\frac{1}{4};\ 2\bar{x},\bar{x},\frac{1}{4};x,\bar{x},\frac{1}{4})$ with $x = 0.5160 \pm 0.0030$

$12Al_{II}$ in (k): $\pm(x,2x,z;\ 2\bar{x},\bar{x},z;\ x,\bar{x},z;\ \bar{x},2\bar{x},\frac{1}{2}+z;\ 2x,x,\frac{1}{2}+z;\ x,x,\frac{1}{2}+z)$ with $x = 0.8337 \pm 0.0040$, $z = 0.0815 \pm 0.0007$

Interatomic distances

Pu_I-6Al	= 3.01	Al_I-2Pu	= 3.02
-6Al	= 3.06	-2Pu	= 3.06
Pu_{II}-3Al	= 3.02	-2Al	= 2.76
-3Al	= 3.03	-4Al	= 2.96
-6Al	= 3.05	-2Al	= 3.34
		Al_{II}-2Pu	= 3.01
		-2Pu	= 3.05
		-2Al	= 2.94
		-2Al	= 2.96
		-4Al	= 3.05

Mean Pu-12Al = 3.04

$PuAl_4$

Orthorhombic body-centered, UAl_4 type

$a = 4.41$ A (Singer)
$b = 6.29$
$c = 13.79$
$a = 4.42 \pm 0.02$ (Runnalls)
$b = 6.26 \pm 0.02$
$c = 13.66 \pm 0.03$
$Z = 4,\ D_x = 6.02$

Space group.—Imma (Int. Tab. 74).

Atomic positions

$(0,0,0;\ \frac{1}{2},\frac{1}{2},\frac{1}{2})+$

4Pu in (e): $\pm(0,\frac{1}{4},z)$ with $z = 0.113$

$4Al_I$ in (e): $\pm(0,\frac{1}{4},z)$ with $z = -0.113$

$4Al_{II}$ in (b): $0,0,\frac{1}{2};\ 0,\frac{1}{2},\frac{1}{2}$

$8Al_{III}$ in (h): $\pm(0,y,z);\ \pm(0,\frac{1}{2}-y,z)$ with $y = 0.033$, $z = 0.314$

Interatomic distances (based on $a = 4.41$, $b = 6.29$, $c = 13.79$)

Pu-4Al	= 3.01	Al_I-2Al	= 3.09
-5Al	= 3.12	-4Al	= 3.12
-2Al	= 3.15	Al_{II}-4Pu	= 3.12
-2Al	= 3.29	-2Al	= 2.57
Al_I-1Pu	= 3.12	-4Al	= 3.12
-2Pu	= 3.15	-2Al	= 3.15
-4Al	= 2.78	Al_{III}-2Pu	= 3.01

Al_{III}-1Pu	= 3.29	Al_{III}-2Al	= 2.82
-1Al	= 2.57	-1Al	= 3.09
-1Al	= 2.73	-1Al	= 3.56
-2Al	= 2.78		

Isotypes.—UAl_4,$NpAl_4$.

Remarks.—The atomic parameters are those of Borie, but Singer modified the *z* parameter for the plutonium atoms from 0.111 to 0.113.

PLUTONIUM-ARSENIC

PuAs

Cubic face-centered, NaCl type

$a = 5.855 \pm 0.004$ A (Gorum)

$Z = 4$, $D_x = 10.39$

Space group.—Fm3m (Int. Tab. 225).

Atomic positions.—See PuC.

Interatomic distances

Pu-6As = 2.93 As-6Pu = 2.93

Isotypes.—LaAs, CeAs, PrAs, UAs.

PLUTONIUM-BERYLLIUM

$PuBe_{13}$

Cubic face-centered, $NaZn_{13}$ type

Be-rich: $a = 10.282 \pm 0.001$ A

Be-rich: $a = 10.284 \pm 0.001$ (Runnalls)

Pu-rich: $a = 10.278 \pm 0.001$ (Runnalls)

$Z = 8$, $D_x = 4.35$

Space group.—Fm3c (Int. Tab. 226).

Atomic positions

$(0,0,0;\ 0,\frac{1}{2},\frac{1}{2};\ \frac{1}{2},0,\frac{1}{2};\ \frac{1}{2},\frac{1}{2},0)\ +$

8 Pu in (*a*): $\pm(\frac{1}{4},\frac{1}{4},\frac{1}{4})$

8 Be in (*b*): $0,0,0;\ \frac{1}{2},\frac{1}{2},\frac{1}{2}$

96 Be in (*i*): $\pm(0,y,z)\circlearrowleft,\ \pm(\frac{1}{2},z,y)\circlearrowleft,\ \pm(0,\bar{y},z)\circlearrowleft,\ \pm(\frac{1}{2},\bar{z},y)\circlearrowleft$

with $y = 0.1806 \pm 0.0003$, $z = 0.1192 \pm 0.0003$

Interatomic distances (based on a = 10.283 A)

Pu-24Be = 2.988

Be_I-12Be = 2.225

Be_{II}-2Be = 2.153

-1Be = 2.225

-2Be = 2.245

-4Be = 2.313

-1Be = 2.451

-2Pu = 2.988

Isotypes.—$CeBe_{13}$, $ThBe_{13}$, UBe_{13}, $NpBe_{13}$.

Remarks.—The atomic parameters used for calculating the interatomic distances are those of $NaZn_{13}$ taken from the work of D. P. Shoemaker, R. E. Marsh, F. J. Ewing, and L. Pauling.

PLUTONIUM-BISMUTH

PuBi

Cubic face-centered, NaCl type

$a = 6.350 \pm 0.001$ A

$z = 4, D_x = 11.62$

Space group.—Fm3m (Int. Tab. 225).

Atomic positions.—See PuC.

Interatomic distances

Pu-6Bi = 3.18 Bi-6Pu = 3.18

Isotypes.—LaBi, CeBi, PrBi, UBi (?).

Remarks.—Although PuBi is pyrophoric, it was prepared successfully by melting a mixture of plutonium and bismuth filings in an evacuated clear silica capillary which was then used for the X-ray specimen.

$PuBi_2$

$PuBi_2$ is pyrophoric and a satisfactory diffraction pattern of it has not been obtained. Its composition is based upon metallographic and thermal-analysis observations.

PLUTONIUM-CARBON

PuC

Cubic face-centered, NaCl type

$a = 4.97 \pm 0.01$ A

$Z = 4, D_x = 13.6$

Space group.—Fm3m (Int. Tab. 225).

Atomic positions

$(0,0,0; 0,\frac{1}{2},\frac{1}{2}; \frac{1}{2},0,\frac{1}{2}; \frac{1}{2},\frac{1}{2},0) +$

4 Pu in (a): 0,0,0

4 C in (b): $\frac{1}{2},\frac{1}{2},\frac{1}{2}$

Interatomic distances

Pu-6C = 2.48 C-6Pu = 2.48

Isotypes.—ThC, UC, NpC.

Pu_2C_3

Cubic body-centered, Pu_2C_3 type

$a = 8.129 \pm 0.001$ A (Zachariasen)

$Z = 8, D_x = 12.70$

Space group.—$I\bar{4}3d$ (Int. Tab. 220).

Atomic positions

$(0,0,0; \frac{1}{2},\frac{1}{2},\frac{1}{2}) +$

16Pu in (c): $x,x,x; \frac{1}{4}+x, \frac{1}{4}+x, \frac{1}{4}+x; \frac{1}{2}+x, \frac{1}{2}-x,\bar{x},\bar{x}$;

$\frac{3}{4}+x, \frac{1}{4}-x, \frac{3}{4}-x$ↄ
with $x = 0.050 \pm 0.003$

24C in (*d*): $x,0,\frac{1}{4}$ↄ; $\bar{x},\frac{1}{2},\frac{1}{4}$ↄ; $\frac{1}{4}+x,\frac{1}{2},\frac{1}{4}$ↄ; $\frac{3}{4}-x,0,\frac{1}{4}$ↄ
with $x = 0.280$

Interatomic distances

Pu-3Pu = 3.35	C-1C = 1.54
-2Pu = 3.52	-2Pu = 2.48
-6Pu = 3.70	-2Pu = 2.51
-3C = 2.48	-2Pu = 2.84
-3C = 2.51	
-3C = 2.84	

Isotypes.—U_2C_3, La_2C_3, Ce_2C_3, Pr_2C_3, Nd_2C_3, Sm_2C_3, Gd_2C_3, Tb_2C_3, Dy_2C_3, Ho_2C_3.

PLUTONIUM-COBALT

Pu_6Co

Tetragonal body-centered, U_6Mn type

$a = 10.46 \pm 0.02$ A
$c = 5.33 \pm 0.01$
$Z = 4$, $D_x = 17.00$

Space group.—I4/mcm (Int. Tab. 140).

Atomic positions.—See Pu_6Fe.

Interatomic distances (*based on* $a = 10.46$, $c = 5.33$ A)

Pu_I-$1Pu_I$ = 2.75	Co-2Co = 2.66
$-2Pu_{II}$ = 2.86	$-8Pu_{II}$ = 2.81
$-4Pu_I$ = 3.30	
$-4Pu_{II}$ = 3.34	
$-2Pu_{II}$ = 3.43	
Pu_{II}-$1Pu_{II}$ = 2.73	
$-1Pu_I$ = 2.86	
$-2Pu_{II}$ = 3.14	
$-2Pu_I$ = 3.34	
$-2Pu_{II}$ = 3.41	
$-1Pu_I$ = 3.44	
$-2Pu_{II}$ = 3.51	
-2Co = 2.81	

Average Pu-Pu distance = 3.24

Isotypes.—U_6Co.

Pu_3Co

Orthorhombic

$a = 3.470$ A (Elliott and Larsen)
$b = 10.939$
$c = 9.196$
$Z = 4$, $D_x = 14.76$

Space group.—Probably Cmcm (Int. Tab. 63).

Pu_2Co

Hexagonal, Fe_2P type

Co-rich: $a = 7.763 \pm 0.002$ A $c/a = 0.470$

$c = 3.648 \pm 0.003$

Pu-rich: $a = 7.902 \pm 0.004$ $c/a = 0.449$

$c = 3.549 \pm 0.002$

$Z = 3$, $D_x = 14.0$

Space group.—P321 (Int. Tab. 150).

Atomic positions

$1Co_I$ in (b): $0,0,\frac{1}{2}$

$2Co_{II}$ in (d): $\frac{1}{3},\frac{2}{3},z$; $\frac{2}{3},\frac{1}{3},\bar{z}$ with $z = \frac{1}{8}$

$3Pu_I$ in (e): $x,0,0$; $0,x,0$; $\bar{x},\bar{x},0$ with $x = -0.26$

$3Pu_{II}$ in (f): $x,0,\frac{1}{2}$; $0,x,\frac{1}{2}$; $\bar{x},\bar{x},\frac{1}{2}$ with $x = 0.40$

Interatomic distances (*based on* $a = 7.763$, $c = 3.648$ A)

Pu_I-$2Pu_{II}$	= 3.21	Co_I-$6Pu_I$	= 2.72
$4Pu_{II}$	= 3.28	-$3Pu_I$	= 3.11
-$2Pu_I$	= 3.50	Co_{II}-$3Pu_I$	= 2.95
-$2Co_I$	= 2.72	-$3Pu_{II}$	= 2.74
-$2Co_{II}$	= 2.95		
Pu_{II}-$2Pu_I$	= 3.21		
-$4Pu_I$	= 3.28		
-$1Co_I$	= 3.11		
-$2Co_{II}$	= 2.74		

$PuCo_2$

Cubic face-centered, Cu_2Mg type

Pu-rich: $a = 7.081 \pm 0.001$

$a = 7.075 \pm 0.005$ (Runnalls)

$Z = 8$, $D_x = 13.35$

Space group.—Fd3m (Int. Tab. 227).

Atomic positions.—See $PuAl_2$.

Interatomic distances

Pu-4Pu	= 3.066	Co-6Co	= 2.503
-12Co	= 2.936	-6Pu	= 2.936

Isotypes.—UCo_2, $CeCo_2$.

$PuCo_3$

Rhombohedral, $PuNi_3$ type

Pu_2Co_{17}

Hexagonal, Th_2Ni_{17} type

$a = 8.325 \pm 0.002$ A

$c = 8.104 \pm 0.003$

$Z = 2$, $D_x = 10.10$

Space group.—$P6_3/mmc$ (Int. Tab. 194).

Atomic positions

$2Pu_I$ in (b): $\pm (0,0,\frac{1}{4})$

$2Pu_{II}$ in (d): $\pm$ ($\frac{1}{3}$,$\frac{2}{3}$,$\frac{3}{4}$)
$6Co_{I}$ in (g): $\frac{1}{2}$,0,0; 0,$\frac{1}{2}$,0; $\frac{1}{2}$,$\frac{1}{2}$,0; $\frac{1}{2}$,0,$\frac{1}{2}$;
0,$\frac{1}{2}$,$\frac{1}{2}$; $\frac{1}{2}$,$\frac{1}{2}$,$\frac{1}{2}$
$12Co_{II}$ in (j): $\pm$ (x,0,$\frac{1}{4}$); $\pm$ (0,x,$\frac{1}{4}$); $\pm$ (x,x,$\frac{1}{4}$); $\pm$ ($\bar{x}$,0,$\frac{1}{4}$);
$\pm$ (0,$\bar{x}$,$\frac{1}{4}$); $\pm$ ($\bar{x}$,$\bar{x}$,$\frac{1}{4}$) with $x = \frac{1}{3}$
$12Co_{III}$ in (k): $\pm$ (x,2x,0); $\pm$ ($\bar{x}$,x,0); $\pm$ (2x,x,0); $\pm$ (x,2x,$\frac{1}{2}$);
$\pm$ ($\bar{x}$,x,$\frac{1}{2}$); $\pm$ (x,2x,$\frac{1}{2}$) with $x = \frac{1}{6}$
$4Co_{IV}$ in (f): $\pm$ ($\frac{1}{3}$,$\frac{2}{3}$,z); $\pm$ ($\frac{1}{3}$,$\frac{2}{3}$,$\frac{1}{2}-z$) with z = 0.11

Remarks.—The atomic parameters are those of Th_2Ni_{17} as determined by Florio *et al.*

PLUTONIUM-COPPER

Based mainly on metallographic evidence, there are believed to be three copper-rich intermetallic phases. The existence of $PuCu_2$ is quite firmly established, but the compositions of the remaining two are uncertain.

PLUTONIUM-GERMANIUM

Pu_3Ge AND Pu_3Ge_2

These two plutonium-rich phases have been observed by means of X-ray and metallographic methods, but their crystal structures are not known. The composition Pu_3Ge is fairly certain, but the composition Pu_3Ge_2 has not been proved.

The other Pu-Ge intermetallics are as follows:

Pu_2Ge_3

Hexagonal, pseudo-AlB_2 type
a = 3.975 $\pm$ 0.002 A
c = 4.198 $\pm$ 0.002
$Z = \frac{1}{2}$, D_x = 10.06

Space group.—P6/mmm (Int. Tab. 191).

Atomic positions
1Pu in (a): 0,0,0
1.5Ge in (d): $\frac{1}{3}$,$\frac{2}{3}$,$\frac{1}{2}$; $\frac{2}{3}$,$\frac{1}{3}$,$\frac{1}{2}$

Interatomic distances

Pu-12Ge = 3.11	Ge-3Ge = 2.29
-6Pu = 3.98	-6Pu = 3.11

Remarks.—See Pu_2Si_3.

$PuGe_2$

Tetragonal body-centered, $ThSi_2$ type
Ge-rich: a = 4.102 $\pm$ 0.002 A
c = 13.81 $\pm$ 0.01
Z = 4, D_x = 10.98

Space group.—I4/amd (Int. Tab. 141).

Atomic positions.—See $PuSi_2$.

Interatomic distances

Pu-12Ge = 3.12   Ge-3Ge = 2.36

-6Pu = 3.12

$PuGe_3$

Cubic primitive, $AuCu_3$ type

$a = 4.223 \pm 0.001$ A

$Z = 1, D_x = 10.07$

Space group.—Pm3m (Int. Tab. 221).

Atomic positions

1Pu in (*a*): 0,0,0

3Ge in (*c*): $0,\frac{1}{2},\frac{1}{2}$; $\frac{1}{2},0,\frac{1}{2}$; $\frac{1}{2},\frac{1}{2},0$

Interatomic distances

Pu-12Ge = 2.99   Ge-8Ge = 2.99

-4Pu = 2.99

Isotypes.—UGe_3.

PLUTONIUM-GOLD

Plutonium-gold intermetallic phases are known to exist, but they have not yet been identified.

PLUTONIUM-HYDROGEN

$PuH_{2.0}$

Cubic face-centered, CaF_2 type

$PuH_{2.0}$: $a = 5.359 \pm 0.001$ A

$PuH_{2.5}$: $a = 5.34 \pm 0.01$

$Z = 4, D_x = 10.40$

Space group.—Fm3m (Int. Tab. 225).

Atomic positions

$(0,0,0; 0,\frac{1}{2},\frac{1}{2}; \frac{1}{2},0,\frac{1}{2}; \frac{1}{2},\frac{1}{2},0) +$

4 Pu in (*a*): 0,0,0

8 H in (*c*): $\frac{1}{4},\frac{1}{4},\frac{1}{4}$; $\frac{3}{4},\frac{3}{4},\frac{3}{4}$

Interatomic distances

Pu-8H = 2.32   H-4Pu = 2.32

Isotypes.—LaH_2, CeH_2, PrH_2, NdH_2, SmH_2, GdH_2, ErH_2.

Remarks.—This hydride exists over a homogeneity range from $PuH_{2.0}$ to $PuH_{2.7}$, and it is interesting to note that the unit-cell dimension decreases with increase in hydrogen content. The hydrogen atoms in excess of those required for the fluorite-type structure were found by means of neutron diffraction on the isostructural cerium hydride and deuteride to be located in the octohedral interstices, i.e., the positions $\frac{1}{2},0,0$; $0,\frac{1}{2},0$; $0,0,\frac{1}{2}$; $\frac{1}{2},\frac{1}{2},\frac{1}{2}$.

PuH_3

Hexagonal, PuH_3 type

$a = 3.78 \pm 0.01$ A

$c = 6.76 \pm 0.01$

$Z = 2, D_x = 9.61$

Space group.—$P6_3/mmc$ (Int. Tab. 194).

Atomic positions

2Pu in (*c*): $\pm(\frac{1}{3},\frac{2}{3},\frac{1}{4})$

$2H_I$ in (*b*): $\pm(0,0,\frac{1}{4})$

$4H_{II}$ in (*f*): $\pm(\frac{1}{3},\frac{2}{3},z)$, $\pm(\frac{2}{3},\frac{1}{3},\frac{1}{2}+z)$ with $z = 0.607$

Interatomic distances

Pu-$3H_I$ = 2.18	Pu-6Pu = 3.78
-$6H_{II}$ = 2.39	-6Pu = 4.02
-$2H_{II}$ = 2.41	
H_I-3Pu = 2.18	
-$6H_{II}$ = 2.39	
H_{II}-1Pu = 2.41	
-3Pu = 2.39	
-$1H_{II}$ = 1.94	
-$3H_I$ = 2.39	

Isotypes.—SmH_3, GdH_3, DyH_3, ErH_3.

Remarks.—The agreement between the observed and calculated intensities of the powder pattern of PuH_3 confirmed the close-packed arrangement of the plutonium atoms; the positions of the hydrogen atoms were suggested by W. H. Zachariasen.

PLUTONIUM-INDIUM

Pu_3In

Cubic primitive, $AuCu_3$ type

$a = 4.703 \pm 0.002$ A

$Z = 1$, $D_x = 13.3$

Space group.—Pm3m (Int. Tab. 221).

Atomic positions.—See $PuGe_3$.

Interatomic distances

Pu-8Pu = 3.32	In-12Pu = 3.32
-4In = 3.32	

Remarks.—Although Pu_3In is described above as having the $AuCu_3$ type of structure, i.e., fully ordered, it has not been observed with complete ordering of the atoms. The powder patterns of the alloys that were examined showed weaker superlattice lines than would be required by complete ordering. It is believed, however, that proper thermal treatment of Pu_3In would result in a fully ordered structure.

PLUTONIUM-IRON

Pu_6Fe

Tetragonal body-centered, U_6Mn type

$a = 10.404 \pm 0.004$ A

$c = 5.355 \pm 0.002$

$a = 10.40 \pm 0.02$ (Marden *et al.*)

$c = 5.345 \pm 0.005$

a = 10.403 (Konobeevsky)
c = 5.347
$Z = 4$, $D_x = 17.07$

Space group.—I4/mcm (Int. Tab. 140).

Atomic positions

$(0,0,0; \frac{1}{2},\frac{1}{2},\frac{1}{2})+$
4 Fe in (a): $\pm\ (0,0,\frac{1}{4})$
8 Pu_I in (h): $\pm\ (x,\frac{1}{2}+x,0)$; $\pm\ (\frac{1}{2}-x,x,0)$ with $x = 0.407$
16 Pu_{II} in (k): $\pm\ (x,y,0)$; $\pm\ (y,\bar{x},0)$; $\pm\ (x,\bar{y},\frac{1}{2})$; $\pm\ (y,x,\frac{1}{2})$
with $x = 0.214$, $y = 0.102$

Interatomic distances (based on a = 10.404, *c* = 5.355)

Pu_I-1Pu_I	= 2.74	Fe-2Fe	= 2.68
-2Pu_{II}	= 2.85	-8Pu_{II}	= 2.81
-4Pu_I	= 3.30		
-4Pu_{II}	= 3.35		
2Pu_{II}	= 3.41		
Pu_{II}-1Pu_{II}	= 2.71		
-1Pu_I	= 2.85		
-2Pu_{II}	= 3.14		
-2Pu_I	= 3.35		
-1Pu_I	= 3.41		
-2Pu_{II}	= 3.42		
-2Pu_{II}	= 3.49		
-2Fe	= 2.81		

Average Pu-Pu distance = 3.23

Isotypes.—U_6Fe.

Remarks.—The atomic parameters given for Pu_6Fe are those determined by Baenziger *et al.* for U_6Mn.

Thermal expansion.—The average coefficients of linear expansion of Pu_6Fe parallel to the axes of the unit cell between room temperature and 400° C have been found to be

$$\bar{\alpha}_a = (73 \pm 2) \times 10^{-6}/^\circ \text{C}$$

$$\bar{\alpha}_c = (-10 \pm 1) \times 10^{-6}/^\circ \text{C}$$

Pu_6Fe-U_6Fe Solid Solutions

Pu_6Fe and U_6Fe were found to be miscible in all proportions. Although it is difficult to prepare single-phase alloys of the ternary composition $(Pu,U)_6Fe$, the powder photographs do show essentially a single phase, and there is a uniform change in the unit-cell dimensions between Pu_6Fe and U_6Fe, as shown in Figure 1.

$PuFe_2$

Cubic face-centered, Cu_2Mg type
Pu-rich: $a = 7.191 \pm 0.001$ A
$a = 7.150 \pm 0.005$ (Runnalls)
$a = 7.190 \pm 0.005$ (Runnalls)

Pu-rich: $a = 7.18 \pm 0.01$ (Mardon *et al.*)
$a = 7.178$ (Konobeevsky)
$z = 4$, $D_x = 12.53$

Space group.—Fd3m (Int. Tab. 227).

Atomic positions.—See $PuAl_2$.

Interatomic distances (*based on* $a = 7.191$)

Pu-4Pu = 3.114	Fe-6Fe = 2.542
-12Fe = 2.981	-6Pu = 2.981

Isotypes.—$CeFe_2$, $GdFe_2$, UFe_2.

$PuFe_2$-UFe_2 Solid Solutions

$PuFe_2$ and UFe_2 are mutually soluble in any proportion. The variation in the unit-cell dimension from $PuFe_2$ to UFe_2 is shown in Figure 2.

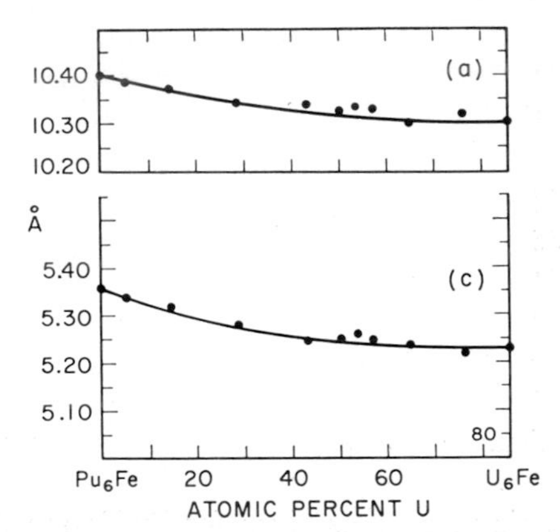

Fig. 1.—Unit-cell dimensions versus composition of Pu_6Fe-U_6Fe solid solution

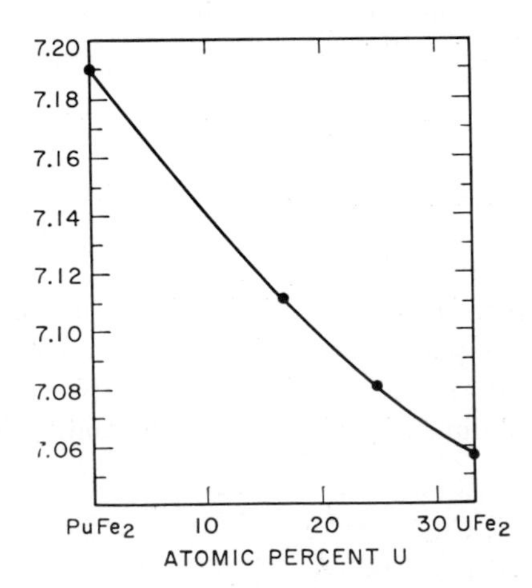

Fig. 2.—Unit-cell dimension versus composition of $PuFe_2$-UFe_2 solid solution

PLUTONIUM-LEAD

$PuPb_3$

Cubic, probably ordered $AuCu_3$ type

$a = 4.808 \pm 0.001$ A

$a = 4.81$ (Konobeevsky)

$Z = 1, D_x = 12.86$

Space group.—Pm3m (Int. Tab. 221).

Atomic positions.—See $PuGe_3$.

Interatomic distances

Pu-12Pb = 3.40 Pb-8Pb = 3.40

-4Pu = 3.40

Isotypes.—$LaPb_3$, $CePb_3$, $PrPb_3$, UPb_3. A plutonium-rich, plutonium-lead compound is known to exist, but its composition is uncertain.

PLUTONIUM-MAGNESIUM

$PuMg_2$

Cubic face-centered, CaF_2 type

$a = 7.34 \pm 0.01$ A

$z = 4, D_x = 4.83$

Space group.—Fm3m (Int. Tab. 225).

Atomic positions.—See PuH_2.

Interatomic distances

Pu-8Mg = 3.18 Mg-4Pu = 3.18

Remarks.—$PuMg_2$ is formed by a reaction in the solid state, and its diffraction pattern has never been obtained free from the superimposed patterns of other phases. Earlier it was reported (1) as possibly being Pu_2Mg, but re-evaluation of the data indicates that the composition $PuMg_2$ is more probably correct.

$PuMg_x$ (where x Is Approximately 2)

Hexagonal

$a = 13.8 \pm 0.1$ A

$c = 9.7 \pm 0.1$

Like $PuMg_2$, this intermetallic is formed by a peritectoid reaction, and its approximate composition was determined by microstructural studies.

PLUTONIUM-MANGANESE

$PuMn_2$

Cubic face-centered, Cu_2Mg type

Pu-rich: $a = 7.292 \pm 0.001$ A

$a = 7.290 \pm 0.005$ (Runnalls)

Pu-rich: $a = 7.29$ (Konobeevsky)

Mn-rich: $a = 7.26$ (Konobeevsky)

$Z = 8, D_x = 11.95$

Space group.—Fd3m (Int. Tab. 227).
Atomic positions.—See $PuAl_2$.
Interatomic distances (based on a = 7.292 A)

Pu-4Pu	= 3.157	Mn-6Mn	= 2.578
-12Mn	= 3.023	-6Pu	= 3.023

Isotypes.—$GdMn_2$, UMn_2.

PLUTONIUM-MERCURY

$PuHg_3$

Apparently this intermetallic is isostructural with hexagonal UHg_3, but a satisfactory diffraction pattern of $PuHg_3$ has never been obtained.

$PuHg_4$

This intermetallic is isostructural with UHg_4. The dimension of the pseudo-cube is $a = 3.61 \pm 0.01$ A.

PLUTONIUM-NICKEL

PuNi

Study of this compound by D. T. Cromer, now in progress, has revealed an orthorhombic symmetry.

$PuNi_2$

Cubic face-centered, Cu_2Mg type
Pu-rich: $a = 7.141 \pm 0.001$ A
Ni-rich: $a = 7.115 \pm 0.001$
$a = 7.16 \pm 0.01$ (Runnalls)
$a = 7.14$ (Konobeevsky)
$Z = 8$, $D_x = 13.1$
Space group.—Fd3m (Int. Tab. 227).
Atomic positions.—See $PuAl_2$.
Interatomic distances (based on a = 7.115 A)

Pu-4Pu	= 3.081	Ni-6Ni	= 2.516
-12Ni	= 2.950	-6Pu	= 2.950

Isotypes.—$LaNi_2$, $CeNi_2$, $PrNi_2$.

$PuNi_3$

This compound has been found to have a rhombohedral unit cell. Its structure is being determined by D. T. Cromer.

$PuNi_4$

The crystal structure of $PuNi_4$ is not known, but its diffraction pattern suggests a close relationship to the $PuNi_5$ structure.

$PuNi_5$

Hexagonal, $CaZn_5$ type
Pu-rich: $a = 4.872 \pm 0.002$ A
$c = 3.980 \pm 0.001$

Ni-rich: $a = 4.861 \pm 0.002$
$c = 3.982 \pm 0.001$
$a = 4.875 \pm 0.005$ (Runnalls)
$c = 3.970 \pm 0.005$
$Z = 1, D_x = 10.8$

Space group.—P6/mmm (Int. Tab. 191).

Atomic positions

1Pu in (*a*): 0,0,0
$2Ni_I$ in (*c*): $\frac{1}{3},\frac{2}{3},0; \frac{2}{3},\frac{1}{3},0$
$3Ni_{II}$ in (*g*): $\frac{1}{2},0,\frac{1}{2}; 0,\frac{1}{2},\frac{1}{2}; \frac{1}{2},\frac{1}{2},\frac{1}{2}$

Interatomic distances (based on a = 4.861, *c* = 3.982 A)

$Pu\text{-}6Ni_I = 2.81$	$Ni_I\text{-}3Pu = 2.81$
$\text{-}12Ni_{II} = 3.14$	$\text{-}6Ni_{II} = 2.44$
	$\text{-}3Ni_I = 2.81$
	$Ni_{II}\text{-}4Pu = 3.14$
	$\text{-}4Ni_I = 2.44$
	$\text{-}4Ni_{II} = 2.44$

Isotypes.—$LaNi_5$, $CeNi_5$, $PrNi_5$, $GdNi_5$, $ThNi_5$.

Pu_2Ni_{17}

Hexagonal, Th_2Ni_{17} type
$a = 8.29 \pm 0.02$ A
$c = 8.01 \pm 0.02$
$Z = 2, D_x = 10.3$

Space group.—$P6_3/mmc$ (Int. Tab. 194).

Atomic positions.—See Pu_2Co_{17}.

PLUTONIUM-NITROGEN

PuN

Cubic face-centered, NaCl type
$a = 4.908 \pm 0.001$ A
$a = 4.905 \pm 0.001$ (Zachariasen)
$Z = 4, D_x = 14.22$

Space group.—Fm3m (Int. Tab. 225).

Atomic positions.—See PuC.

Interatomic distances

Pu-6N = 2.45
N-6Pu = 2.45

Isotypes.—LaN, CeN, PrN, NdN, SmN, ThN, UN, NpN.

Remarks.—Attempts to prepare higher nitrides of plutonium have not been successful.

PLUTONIUM-OSMIUM

According to Konobeevsky (2) there are five Pu-Os intermetallic phases. The crystal structures of $Pu_{19}Os$ and high- and low-temperature modifications of Pu_3Os, and Pu_5Os_3 were not determined. Crystal data given for $PuOs_2$ are as follows:

$PuOs_2$

Hexagonal, $MgZn_2$ type

$a = 5.337$ A

$c = 8.683$

$Z = 4, D_x = 19.2$

Space group.—$P6_3/mmc$ (Int. Tab. 194).

PLUTONIUM-OXYGEN

PuO

Cubic face-centered, NaCl type

$a = 4.96 \pm 0.01$ A

$Z = 4, D_x = 13.9$

Space group.—Fm3m (Int. Tab. 225).

Atomic positions.—See PuC.

Interatomic distances

Pu-6O = 2.48

O-6Pu = 2.48

Isotypes.—NdO, SmO, ThO, PaO, UO, NpO, AmO.

Remarks.—The limits of error of the lattice constant, as with PuC, refer to the range of lattice constants found for different preparations of varying purity rather than to the precision of measurement.

Pu_2O_3

Hexagonal, La_2O_3 type

$a = 3.841 \pm 0.006$ A

$c = 5.958 \pm 0.005$

$a = 3.840 \pm 0.004$ (Templeton and Dauben)

$c = 5.957 \pm 0.006$

$Z = 1, D_x = 11.47$

Space group.—$P\bar{3}m1$ (Int. Tab. 164).

Atomic positions

2Pu in (*d*): $\frac{1}{3},\frac{2}{3},z; \frac{2}{3},\frac{1}{3},\bar{z}$ with $z = 0.235$

$1O_I$ in (*a*): 0,0,0

$2O_{II}$ in (*d*): $\frac{1}{3},\frac{2}{3},z; \frac{2}{3},\frac{1}{3},\bar{z}$ with $z = 0.63$

Interatomic distances

Pu-3Pu = 3.57	O_I-6Pu = 2.62
-6Pu = 3.84	
-3Pu = 3.86	O_{II}-4Pu = 2.36
$-4O_{II}$ = 2.36	
$-3O_I$ = 2.62	

Isotypes.—La_2O_3, Ce_2O_3, Pr_2O_3, Nd_2O_3, Ac_2O_3, Am_2O_3.

Pu_2O_3-Pu_4O_7

Cubic body-centered, Mn_2O_3 type

Pu-rich: $a = 11.04 \pm 0.02$ A

O-rich: $a = 11.00 \pm 0.02$

$Z = 16$, $D_x = 10.2$

Space group.—Ia3 (Int. Tab. 206).

Atomic positions

$(0,0,0; \frac{1}{2},\frac{1}{2},\frac{1}{2}) +$

8Pu in (*b*): $\frac{1}{4},\frac{1}{4},\frac{1}{4}; \frac{1}{4},\frac{3}{4},\frac{3}{4}; \frac{3}{4},\frac{1}{4},\frac{3}{4}; \frac{3}{4},\frac{3}{4},\frac{1}{4}$

24Pu in (*d*): $\pm (x,0,\frac{1}{4})\circlearrowleft; \pm (x,\frac{1}{2},\frac{3}{4})\circlearrowleft$ with $x = -0.030$

48O in (*e*): $\pm (x,y,z)\circlearrowleft; \pm (x,\bar{y},\frac{1}{2}-z)\circlearrowleft; \pm (\frac{1}{2}-x,y,\bar{z})\circlearrowleft;$
$\pm (\bar{x},\frac{1}{2}-y,z)\circlearrowleft.$

Isotypes.—Sm_2O_3 through Yb_2O_3, Am_2O_3, Cm_2O_3.

Remarks.—The atomic parameters of the oxygen atoms are approximately $x = 0.385$, $y = 0.145$, and $z = 0.380$, as given by Pauling for $(Fe,Mn)_2O_3$.

PuO_2

Cubic face-centered, CaF_2 type

$a = 5.3960 \pm 0.0003$ A

$Z = 4$, $D_x = 11.46$

Space group.—Fm3m (Int. Tab. 225).

Atomic positions.—See PuH_2.

Interatomic distances

Pu-8O = 2.336

O-4Pu = 2.336

Isotypes.—CeO_2, PrO_2, TbO_2, ThO_2, $PaO_{2.2}$, UO_2, NpO_2, AmO_2, CmO_2.

PLUTONIUM-PHOSPHORUS

PuP

Cubic face-centered, NaCl type

$a = 5.664 \pm 0.004$ A (Gorum)

$Z = 4$, $D_x = 9.87$

Space group.—Fm3m (Int. Tab. 225).

Atomic positions.—See PuC.

Interatomic distances

Pu-6P = 2.832

P-6Pu = 2.832

Isotypes.—LaP, CeP, PrP, NdP, ThP, UP.

PLUTONIUM-RHENIUM

$PuRe_2$

Hexagonal, $MgZn_2$ type

$a = 5.396 \pm 0.001$ A

$c = 8.729 \pm 0.001$
$Z = 4, D_x = 18.45$

Space group.—$P6_3/mmc$ (Int. Tab. 194).

Atomic positions

4Pu in (f): $\pm (\frac{1}{3},\frac{2}{3},z)$; $\pm (\frac{1}{3},\frac{2}{3},\frac{1}{2}-z)$ with $z = 0.062$
$2Re_I$ in (a): 0,0,0; 0,0,$\frac{1}{2}$
$6Re_{II}$ in (h): $\pm (x,2x,\frac{1}{4})$; $\pm (2\bar{x},\bar{x},\frac{1}{4})$; $\pm (x,\bar{x},\frac{1}{4})$ with $x = -\frac{1}{6}$

Interatomic distances

Pu-$3Re_{II}$ = 3.14	Re_I-$6Re_{II}$ = 2.68
-$6Re_{II}$ = 3.16	-$2Re_I$ = 4.36
-$3Re_I$ = 3.16	-6Pu = 3.16
-1Pu = 3.28	Re_{II}-$2Re_I$ = 2.68
-3Pu = 3.30	-$4Re_{II}$ = 2.70
	-2Pu = 3.14
	-4Pu = 3.16

PLUTONIUM-RUTHENIUM

$PU_{19}RU$, PU_3RU, AND PU_5RU_3

The identities of these three plutonium-rich intermetallic phases have been well established by metallography and X-ray diffraction, but their crystal structures are not yet known. The other two plutonium-ruthenium intermetallics are as follows:

PuRu

Cubic primitive, CsCl type
$a = 3.363 \pm 0.001$ A
$Z = 1, D_x = 14.87$

Space group.—Pm3m (Int. Tab. 221).

Atomic positions

1Pu in (a): 0,0,0
1Ru in (b): $\frac{1}{2},\frac{1}{2},\frac{1}{2}$

Interatomic distances

Pu-8Ru = 2.91	Ru-8Pu = 2.91

$PuRu_2$

Cubic face-centered, Cu_2Mg type
Pu-rich: $a = 7.476 \pm 0.001$ A
$Z = 8, D_x = 14.06$

Space group.—Fd3m (Int. Tab. 227).

Atomic positions.—See $PuAl_2$.

Interatomic distances

Pu-4Pu = 3.24	Ru-6Ru = 2.64
-12Ru = 3.10	-6Pu = 3.10

PLUTONIUM-SILICON

The existence of two plutonium-rich intermetallics in this system has been established by X-ray and metallographic observations, but the crystal structures are not known. Their compositions fall in the range of 30–40 atomic per cent silicon. Other plutonium-silicon intermetallics are as follows:

PuSi

Orthorhombic, FeB type

$a = 5.727 \pm 0.005$ A
$b = 7.933 \pm 0.003$
$c = 3.847 \pm 0.001$
$Z = 4, D_x = 10.15$

Space group.—Pbnm (Int. Tab. 62).

Atomic positions

4Pu in (c): $\pm (x_1,y_1,\frac{1}{4}); \pm (\frac{1}{2}-x_1,y_1+\frac{1}{2},\frac{1}{4})$
with $x_1 = 0.125$, $y_1 = 0.180$

4Si in (c): $\pm (x_2,y_2,\frac{1}{4}); \pm (\frac{1}{2}-x_2,y_2+\frac{1}{2},\frac{1}{4})$
with $x_2 = 0.611$, $y_2 = 0.028$

Interatomic distances

Pu-2Si = 2.95	Si-2Si = 2.35
-2Si = 3.01	-2Pu = 2.95
-1Si = 3.03	-2Pu = 3.01
-1Si = 3.07	-1Pu = 3.03
-1Si = 3.18	-1Pu = 3.07
-4Pu = 3.62	-1Pu = 3.18
-2Pu = 3.73	

Isotypes.—ThSi, USi.

Remarks.—The atomic parameters are those of USi, taken from Zachariasen.

Pu_2Si_3

Hexagonal, pseudo-AlB_2 type

$a = 3.876 \pm 0.002$
$c = 4.090 \pm 0.002$
$Z = \frac{1}{2}, D_x = 8.77$

Space group.—P6/mmm (Int. Tab. 191).

Atomic positions

1Pu in (a): 0,0,0
1.5Si in (d): $\frac{1}{3},\frac{2}{3},\frac{1}{2}; \frac{2}{3},\frac{1}{3},\frac{1}{2}$

Interatomic distances

Pu-12Si = 3.03	Si-6Pu = 3.03
-6Pu = 3.88	-3Si = 2.24
-2Pu = 4.09	

Isotypes.—β-USi_2, Th_2Si_3.

Remarks.—The crystal structure of the uranium analogue of this compound was first determined by W. H. Zachariasen, who reported it to be hexagonal of the AlB_2 type. He designated it as β-USi_2, to distinguish it from tetragonal USi_2. However, chemical analysis of the plutonium compound gave values close to 60 rather than the 67 a/o silicon required for $PuSi_2$. Likewise, Jacobson and co-workers found that the silicon content of β-$ThSi_2$ was close to 60 a/o silicon. To account for the low value of the silicon content in this apparently AlB_2 type of structure, the presence of silicon vacancies is postulated.

$PuSi_2$

Tetragonal body-centered, $ThSi_2$ type

Si-rich: $a = 3.967 \pm 0.001$

$c = 13.72 \pm 0.03$

$a = 3.98 \pm 0.01$ (Zachariasen)

$c = 13.58 \pm 0.05$

$Z = 4, D_x = 9.08$

Space group.—$I4_1/amd$ (Int. Tab. 141).

Atomic positions

$(0,0,0; \frac{1}{2},\frac{1}{2},\frac{1}{2}) +$

4Pu in (a): $0,0,0; 0,\frac{1}{2},\frac{1}{4}$

8Si in (e): $0,0,z; 0,0,\bar{z}; 0,\frac{1}{2},\frac{1}{4}+z; 0,\frac{1}{2},\frac{1}{4}-z$

with $z = 0.417$

Interatomic distances

Pu-12Si	= 3.03	Si-6Pu	= 3.03
-8Pu	= 3.96	-3Si	= 2.29

Isotypes.—$LaSi_2$, $CeSi_2$, $PrSi_2$ $NdSi_2$, $SmSi_2$, $ThSi_2$, USi_2, $NpSi_2$.

PLUTONIUM-SILVER

Runnalls prepared the intermetallic phase $PuAg_3$ and has reported the following crystal data:

Hexagonal

$a = 12.730 \pm 0.003$ A

$c = 9.402 \pm 0.005$

$Z = 16, D_x = 11.33$

Space group.—$P6_3$ (Int. Tab. 173), $P6_3/m$ (Int. Tab. 176), or $P6_322$ (Int. Tab. 182).

The atomic positions have not been determined.

PLUTONIUM-SULFUR

PuS

Cubic face-centered, NaCl type

$a = 5.536 \pm 0.001$ A (Zachariasen)

$Z = 4, D_x = 10.60$

Space group.—Fm3m (Int. Tab. 225).

Atomic positions.—See PuC.

Interatomic distances

Pu-6S	= 2.768	S-6Pu	= 2.768

Isotypes.—LaS, CeS, PrS, NdS, ThS, US.

Pu_2S_3-Pu_3S_4

Cubic body-centered, Th_3P_4 type

$a = 8.4543 \pm 0.0005$ A (Zachariasen)

$Z = 10\frac{2}{3}$ Pu atoms, 16 S atoms, $D_x = 8.41$

Space group.—I$\bar{4}$3d (Int. Tab. 220).
Atomic positions
(0,0,0; $\frac{1}{2},\frac{1}{2},\frac{1}{2}$) +
10$\frac{2}{3}$Pu in (*a*): ($\frac{3}{8}$,0,$\frac{1}{4}$)↻; ($\frac{1}{8}$,0,$\frac{3}{4}$)↻;
16S in (*c*): (x,x,x); ($\frac{1}{2}+x,\frac{1}{2}-x,\bar{x}$)↻; ($\frac{1}{4}+x,\frac{1}{4}+x,\frac{1}{4}+x$);
($\frac{3}{4}+x,\frac{1}{4}-x,\frac{3}{4}-x$)↻; with $x = 0.083 \pm 0.015$
Interatomic distances.—Pu-8S = 2.92; shortest, S-S = 3.14.
Isotypes.—Ac_2S_3, Am_2S_3, La_2S_3, Ce_2S_3-Ce_3S_4.

PLUTONIUM-TELLURIUM

PuTe

Cubic face-centered, NaCl type
$a = 6.183 \pm 0.004$ A (Gorum)
$Z = 4$, $D_x = 10.33$
Space group.—Fm3m (Int. Tab. 225).
Atomic positions.—See PuC.
Interatomic distances
Pu-6Te = 3.09 Te-6Pu = 3.09
Isotypes.—LaTe, CeTe, PrTe, NdTe, UTe.

PLUTONIUM-THORIUM

Pu_2Th is the only intermediate phase in the plutonium-thorium system. Its crystal structure is not known, but the Harwell group have tentatively reported an orthorhombic unit cell containing six formula units and having the dimensions
$a = 9.820$ A
$b = 8.164$
$c = 6.681$

PLUTONIUM-TIN

$PuSn_3$

Cubic primitive, $AuCu_3$ type
$a = 4.630 \pm 0.001$ A
$Z = 1$, $D_x = 9.96$
Space group.—Pm3m (Int. Tab. 221).
Atomic positions.—See $PuGe_3$.
Interatomic distances
Pu-12Sn = 3.27 Sn-4Sn = 3.27
-8Pu = 3.27
Isotypes.—$LaSn_3$, $CeSn_3$, $PrSn_3$, USn_3.

PLUTONIUM-URANIUM

Eta Pu-U

This phase has a homogeneity range that, at its maximum width, extends from about 2 to about 70 a/o uranium. It is not stable at room temperature. Its crystal lattice has been tentatively described as tetragonal, with unit-cell dimensions of

$a = 10.57 \pm 0.05$, $c = 10.76 \pm 0.05$ A for the composition Pu_3U. With 52 atoms in the unit cell, the calculated density would be 17.2 gm/cm^3.

ZETA PU-U

This phase is stable at room temperature, where its composition limits lie between about 33 and 65 a/o uranium. The limits of the homogeneity range at its maximum width extend from about 25 to about 74 a/o uranium. The powder patterns taken of this phase at room temperature can be indexed on the basis of a primitive cubic unit cell with $a = 10.664 \pm 0.005$ A for the composition PuU. The measured density indicates that this cell contains 58 atoms; the calculated density would be 18.95 gm/cm^3. The apparently cubic unit cell was found to expand anisotropically, however. The lines of the powder pattern split according to tetragonal symmetry. Hence it appears that zeta is actually tetragonal with an axial ratio of unity at room temperature.

PLUTONIUM-ZINC

$PuZn_2$

Cubic face-centered, Cu_2Mg type

Pu-rich: $a = 7.760 \pm 0.001$ A

Zn-rich: $a = 7.747 \pm 0.001$

$Z = 8$, $D_x = 10.54$

Space group.—Fd3m (Int. Tab. 227).

Atomic positions.—See $PuAl_2$.

Interatomic distances (based on $a = 7.747$)

Pu-4Pu	= 3.354	Zn-6Zn	= 2.739
-12Zn	= 3.212	-6Pu	= 3.212

$PuZn_x$

There is metallographic and X-ray evidence indicating the existence of an intermetallic compound intermediate in composition between 67 and 90 atomic per cent zinc.

$PuZn_{8.5 \text{ or } 9}$

This compound has been found to be hexagonal and evidently isostructural with the corresponding U-Zn compound. However, the uranium compound has been reported by Chiotti *et al.* to be UZn_9 with an axial ratio of 0.999, whereas Makarov *et al.* report it to be U_2Zn_{17} with an axial ratio of 2.93, the a_0 dimension (8.99A) being the same in both structures. As yet, it has not been conclusively shown to which of these two structures the plutonium compound corresponds.

PLUTONIUM-ZIRCONIUM

Two intermetallic phases have been observed in plutonium-zirconium alloys. The plutonium-rich phase forms by a peritectoid reaction at some temperature between 250° and 300° C and gives a complex diffraction pattern. It has a high content of plutonium; a 95 atomic per cent plutonium alloy was found to consist essentially of this phase; hence it is believed to correspond stoichiometrically to $Pu_{19}Zr$. The existence of this phase has been confirmed at Harwell, but there its composition is believed to correspond to the formula Pu_4Zr.

Harwell also reports the existence of $PuZr_2$, isostructural with UZr_2 (delta phase).

DISCUSSION OF INTERATOMIC DISTANCES

It is interesting to compare the Pu-Pu interatomic distances in the intermetallic compounds with the distances in the pure plutonium allotropes. This is best done with the plutonium-rich compounds wherein there are many Pu-Pu contacts, but our present knowledge of the crystal structures of such compounds is limited to Pu_6Fe, Pu_6Co, Pu_3Al, and Pu_3In.

For Pu_6Fe and Pu_6Co the average Pu-Pu distances (co-ordination number 13) are 3.23 and 3.24 A, respectively. Correction of this distance to standard co-ordination number 12 gives a value of about 3.21 A, hence a plutonium atom radius of

TABLE 3

INTERATOMIC DISTANCES IN Cu_2Mg-TYPE PLUTONIUM COMPOUNDS

Co-ordination	M Atom						
	Al	Ru	Mn	Fe	Co	Ni	Zn
Pu-4Pu........	3.394	3.237	3.157	3.114	3.066	3.092	3.354
Pu-12M.......	3.249	3.099	3.023	2.981	2.936	2.960	3.212
M-6M........	2.771	2.643	2.578	2.542	2.503	2.525	2.739
M-6Pu........	3.249	3.099	3.023	2.981	2.936	2.960	3.212
r_{Pu}/r_M........	1.10	1.18	1.25	1.25	1.26	1.27	1.18

1.61 A. This is intermediate between 1.58 for alpha-plutonium and 1.64 for delta-plutonium and, according to Zachariasen's (109) interpretation of interatomic distances, corresponds to a valency of 4.5–5 electrons per atom.

In Pu_3In and Pu_3Al ($AuCu_3$ type and distorted $AuCu_3$ type, respectively) each plutonium atom is surrounded by 8 plutonium atoms and 4 alloying atoms. In the case of Pu_3Al, the Pu-Pu distance is 3.20 compared with 3.28 A for delta plutonium. The shorter distance in the compound would be expected, in view of the smaller size of the aluminum atom. In the case of Pu_3In, the Pu-Pu distance is 3.32 A. Evidently the indium atoms expand the lattice, even though the metal radius of the indium atom is the same as that of alpha plutonium (1.58 A) and smaller than that of delta plutonium (1.64 A).

In Table 3 the interatomic distances of the plutonium compounds of the Cu_2Mg type may be compared. In this type of structure the plutonium atoms are considered to be in contact if the ratio of the radius of the plutonium atom to that of the alloying atom is greater than 1.225. This is the case for the manganese, iron, cobalt, and nickel compounds but not for the aluminum, ruthenium, and zinc compounds, and it is found that the Pu-Pu distances in the latter are large compared with those in the former. The apparent size of the plutonium atom does vary in a systematic manner with the radius ratio (r_{Pu}/r_M) in these compounds, showing that plutonium behaves similarly to other metals that form this type of structure wherein the size factor is of primary importance.

The preparation of most of the intermetallic compounds by V. O. Struebing and the measurement of powder diffraction patterns by Marian Gibbs are gratefully acknowledged. The author wishes to thank those who contributed information to this report, credit for which is acknowledged in the references.

REFERENCES

1. Coffinberry, A. S., and Ellinger, F. H. *Proceedings of the International Conference on the Peaceful Uses of Atomic Energy*, **9**:138–46. United Nations, 1956.
2. Konobeevsky, S. T. 1955. "Phase Diagrams of Some Plutonium Systems," in *Session of the Division of Chemical Sciences (of the U.S.S.R. Academy of Sciences), Conference on the Peaceful Uses of Atomic Energy, Moscow, 1955*, pp. 362–75; also pp. 207–14 in English translation by the Consultants Bureau (for sale by Superintendent of Documents, Washington, D.C., price $1.00).
3. Published by N. V. A. Oosthoek's Uitgevers MIJ, Utrecht, Holland.
4. Published by the Kynoch Press, Birmingham, England.
5. Mulford, R. N. R., and Sturdy, G. E. *Journal of the American Chemical Society*, **77**:3449, 1955.
6. ———. *Ibid.*, **78**:3897, 1956.
7. Holley, C. E., Mulford, R. N. R., Ellinger, F. H., Koehler, W. C., and Zachariasen, W. H. *Journal of Physical Chemistry*, **59**:1226, 1955.
8. Sturdy, G. E., and Mulford, R. N. R. *Journal of the American Chemical Society*, **78**:1083, 1956.
9. Mulford, R. N. R., and Ellinger, F. H. Unpublished work.
10. Bragg, W. L. *Strukturbericht*, **1**:148, 1914.
11. Runnalls, O. J. C. *Canadian Journal of Chemistry*, **34**:133, 1956.
12. Zintl, E., and Haucke, W. *Strukturbericht*, **6**:8, 1938.
13. Shoemaker, D. P., Marsh, R. E., Ewing, F. J., and Pauling, L. *Acta Crystallographica*, **5**:637, 1952.
14. Baenziger, N. C., and Rundle, R. E. *Structure Rept.*, **12**:28, 1949.
15. Runnalls, O. J. C. *Acta Crystallographica*, **7**:222, 1954.
16. Zachariasen, W. H. *Structure Rept.*, **12**:48, 1949.
17. Bragg, W. H., and Bragg, W. L. *Strukturbericht*, **1**:72, 1913.
18. Wilhelm, H. A., and Chiotti, P. *Structure Rept.*, **13**:68, 1949.
19. Rundle, R. E., Baenziger, N. C., Wilson, A. S., and McDonald, R. A. *Structure Rept.*, **11**:83, 170, 220, 1948.
20. Zachariasen, W. H. Report ANL-4798, 1951.
21. ———. *Acta Crystallographica*, **5**:17, 1952.
22. Mallet, M. W., Gerds, A. F., and Vaughan, D. A. *Transactions of the Electrochemical Society*, **98**:505, 1951.
23. Gschneidner, K. A. Private communication.
24. Iandelli, A., and Botti, R. *Strukturbericht*, **5**:42, 1937.
25. ———. *Ibid.*, p. 43.
26. Eick, H. A., Baenziger, N. C., and Eyring, L. *Journal of the American Chemical Society*, **78**:5987, 1956.
27. Rundle, R. E. *Acta Crystallographica*, **1**:180, 1948.
28. Ellinger, F. H. 1953. AEC Rept. AECU-2593.

29. Ellinger, F. H., and Zachariasen, W. H. *Journal of the American Chemical Society*, **75**:5650, 1953.
30. Zachariasen, W. H. *Acta Crystallographica*, **5**:19, 1952.
31. Templeton, D. H., and Dauben, C. H. AEC Rept. UCRL-1886 (1952).
32. Zachariasen, W. H. *Strukturbericht*, **1**:262, 744, 1926.
33. Pauling, L. *Strukturbericht*, **1**:744, 785, 1928.
34. Zachariasen, W. H. *Structure Rept.*, **12**:172, 1949.
35. Templeton, D. H., and Dauben, C. H. *Journal of the American Chemical Society*, **75**:4560, 1953.
36. Mooney, R. C. L., and Zachariasen, W. H. *The Trans-uranium Elements*, Paper 20.1. ("National Nuclear Energy Series," Div. IV, Vol. **14B**.) New York: McGraw-Hill Book Co., Inc., 1949.
37. Zachariasen, W. H. *Strukturbericht*, **1**:333, and **2**:38, 1928.
38. Pauling, L. *Strukturbericht*, **1**:785, and **2**:38, 1930.
39. Templeton, D. H., and Dauben, C. H. *Journal of the American Chemical Society*, **76**:5237, 1954.
40. Asprey, L. B., and Ellinger, F. H. AEC Rept. AECD-3627 (1953).
41. Zachariasen, W. H. *Structure Rept.*, **12**:148, 1949.
42. Goldschmidt, V. M., and Thomassen, L. *Strukturbericht*, **1**:197, 1923.
43. Goldschmidt, V. M. *Strukturbericht*, **1**:198, 1926.
44. Gruen, D. M., Koehler, W. C., and Katz, J. J. *Structure Rept.*, **15**:187, 1951.
45. Goldschmidt, V. M., and Thomassen, L. *Strukturbericht*, **1**:208, 1923.
46. ———. *Ibid.*, p. 212.
47. Rundle, R. E., Baenziger, N. C., Wilson, A. S., and McDonald, R. A. *Structure Rept.*, **11**:222, 1948.
48. Zintl, E., and Neumayer, S. *Strukturbericht*, **3**:174, 639, 1933.
49. Friauf, J. B. *Strukturbericht*, **1**:490, 531, 1927.
50. Nowotny, H. *Structure Rept.*, **9**:10, 1941.
51. Wallbaum, H. J. *Structure Rept.*, **9**:5, 1940.
52. Rundle, R. E., and Wilson, A. S. *Structure Rept.*, **11**:30, 1948.
53. Larson, A. C., Cromer, D. T., and Stambaugh, C. K. *Acta Crystallographica*, **10**:443, 1957.
54. Singer, J. Unpublished work, 1950.
55. Borie, B. S., Jr. *Journal of Metals, AIME*, **3**:800, 1951.
56. Bjurström, T., and Arnfelt, H. *Strukturbericht*, **2**:795, and **3**:12, 1929.
57. Jacobson, E. L., Freeman, R. D., Thorp, A. G., and Searcy, A. W. *Journal of the American Chemical Society*, **78**:4850, 1956.
58. Zachariasen, W. H. *Structure Rept.*, **12**:123, 1949.
59. Runnalls, O. J. C., and Boucher, R. R. *Acta Crystallographica*, **8**:592, 1955.
60. Hofmann, W., and Jänicke, W. *Strukturbericht*, **3**:311, 1935.
61. Brauer, G., and Mitius, A. *Structure Rept.*, **9**:121, 1942.
62. Brauer, G., and Haag, H. *Structure Rept.*, **13**:131, 1950.
63. Gorum, A. E. *Acta Crystallographica*, **10**:144, 1957.
64. Iandelli, A., and Botti, E. *Strukturbericht*, **5**:43, 1936.
65. ———. *Ibid.*, p. 43, 1937.
66. Meisel, K. *Strukturbericht*, **7**:112, 1939.
67. Zumbusch, M. *Structure Rept.*, **8**:107, 1941.
68. Zachariasen, W. H. *Structure Rept.*, **12**:139, 1949.
69. Iandelli, A. *Gazzetta chimica italiana*, **85**:881, 1955.
70. Strotzer, E. F. *Structure Rept.*, **8**:112, 1941.
71. Zachariasen, W. H. *Structure Rept.*, **12**:179, 1949.
72. Baenziger, N. C., Rundle, R. E., Snow, A. I., and Wilson, A. S. *Structure Rept.*, **13**:93, 1950.

73. JEPSON, J. O., and DUWEZ, P. *Transactions of the American Society for Metals*, **47**: 543, 1955.
74. ENDTER, F., and KLEMM, W. *Structure Rept.*, **9**:81, 1944.
75. HÄGG, G. *Strukturbericht*, **1**:593, 1928.
76. FRIAUF, J. B. *Strukturbericht*, **2**:284, 1930.
77. HENDRICKS, S. B., and KOSTING, P. R. *Strukturbericht*, **2**:285, 1930.
78. FLORIO, J. V., BAENZIGER, N. C., and RUNDLE, R. E. *Acta Crystallographica*, **9**:367, 1956.
79. VOGEL, R. *Structure Rept.*, **11**:87, 1947.
80. NOWOTNY, H. *Structure Rept.*, **11**:87, 1942.
81. FÜLLING, W., MOELLER, K., and VOGEL, R. *Structure Rept.*, **11**:87, 1942.
82. HAUCHKE, W. *Structure Rept.*, **11**:59, 1940.
83. NOWOTNY, H. *Structure Rept.*, **11**:59, 1942.
84. VOGEL, R., and FÜLLING, W. *Structure Rept.*, **11**:59, 1947.
85. ENDTER, F., and KLEMM, W. 1943. *Structure Rept.*, **11**:59, 1943.
86. HEUMANN, T. *Structure Rept.*, **11**:59, 1948.
87. JOHANSSON, C. H., and LINDE, I. O. *Strukturbericht*, **1**:506 and 486, 1925.
88. IANDELLI, A., and FERRO, R. *Annali di chimica* (Rome), **42**:598, 1952.
89. IANDELLI, A., and BOTTI, E. *Strukturbericht*, **5**:44, 1937.
90. IANDELLI, A. *Rendiconti delle reale accademia nazionale dei Lincei*, **13**:138, 1952.
91. DAVEY, W. P. *Strukturbericht*, **1**:74, 107, 1921.
92. ROSSI, A. *Strukturbericht*, **3**:646, 1933.
93. ZINTL, E., and NEUMAYR, S. *Strukturbericht*, **3**:647, 1933.
94. ROSSI, A. *Strukturbericht*, **3**:647, 1934.
95. FERRO, R. *Zeitschrift für anorganische Chemie*, **275**:320, 1954.
96. IANDELLI, A. *Gazzetta chimica italiana*, **85**:881, 1955.
97. RUNDLE, R. E., and WILSON, A. S. *Structure Rept.*, **11**:164, 1948.
98. IANDELLI, A., and FERRO, R. *Structure Rept.*, **15**:88, 1951.
99. TEITEL, R. J. *Journal of Metals, AIME*, **4**:397, 1952.
100. IANDELLI, A., and BOTTI, E. *Strukturbericht*, **5**:45, 1937.
101. BREWER, L., EDWARDS, R. K., and TEMPLETON, D. H. *Structure Rept.*, **12**:29, 1949.
102. TEITEL, R. J. *Journal of Metals, AIME*, **9**:131, 1957.
103. POOLE, D. M., WILLIAMSON, G. K., and MARPLES, J. A. C. *Journal of the Institute of Metals*, **86**:172, 1957.
104. RUNNALLS, O. J. C. *Journal of Metals, AIME*, **5**:1460, 1953.
105. CROMER, D. T. Unpublished work.
106. WALDRON, M. B. Private communication.
107. CRAMER, E. M. Unpublished work.
108. FRIAUF, J. *Strukturbericht*, **1**:180, 228, 1927.
109. ZACHARIASEN, W. H. *Acta Crystallographica*, **5**:660, 1952.
110. ELLIOTT, R. O., and LARSON, A. C. Unpublished work, 1957.
111. CHIOTTI, P., KLEPFER, H. H., and GILL, K. J. *Journal of Metals, AIME*, **9**:51, 1957.
112. MAKAROV, E. S., and VINOGRADOV, S. I. *Kristallografija, SSSR*, **1**:634, 1956.

CHAPTER XXVI

THE PREPARATION OF PLUTONIUM-ALUMINUM AND OTHER PLUTONIUM ALLOYS

O. J. C. Runnalls

INTRODUCTION

Alloys are usually prepared by adding the alloying constituents as metals to the solvent melt. It is often possible, however, to introduce alloying elements as chemical compounds, such as oxides or halides, provided that such compounds are reduced by the solvent metal, M_1, according to the reaction

$$M_1 + M_2X \rightarrow M_2 + M_1X \, .$$

The product metal, M_2, alloys with metal M_1, which is present in excess, while the product compound, M_1X, may be separated from the alloy by dissolution in a flux or by volatilization.

The above technique offers particular advantages when applied to the preparation of plutonium alloys, namely, (*a*) one eliminates the expensive steps of producing and manipulating the chemically reactive plutonium metal; (*b*) an inert-gas atmosphere within the glove-box system may not be required if, as is often the case, the alloys produced are not pyrophoric; (*c*) when plutonium metal is added to a light metal, such as beryllium, the plutonium may segregate before alloying because of its low melting point, 640° C, and high density, 15.9–19.8 gm/cm³ (1). This problem may be significantly reduced by adding the plutonium as plutonium trifluoride, for example, since plutonium trifluoride has a higher melting point, 1,425° C (2) and a lower density, 9.32 gm/cm³ (3).

THE USE OF THERMODYNAMIC DATA

Thermodynamic data were used to determine the feasibility of reducing a plutonium compound with a given metal. The plutonium compounds considered were the dioxide, PuO_2, and the fluorides, PuF_3 and PuF_4. Other halides could also be used, but they are more difficult to prepare and have undesirable hygroscopic properties.

The heats of formation at 298° K and the free energies of formation at 298°,

O. J. C. Runnalls is with the Atomic Energy of Canada Limited, Chalk River, Ontario.

500°, 1,000°, and 1,500° K for oxides and halides of those elements which might be used to reduce PuO_2, PuF_3, and PuF_4 are shown in Tables 1 and 2. Values for the plutonium compounds are included.

The stability of compounds of plutonium relative to those of other metals is conveniently depicted, as in Figures 1 and 2, by plotting the standard free energy of formation against temperature. Figure 2 indicates, for example, that alkaline-earth and rare-earth fluorides are more stable than plutonium trifluoride at temperatures up to 1,500° K. Thus the reduction of plutonium trifluoride with an alkaline-earth or rare-earth metal should proceed spontaneously.

TABLE 1

HEATS AND FREE ENERGIES OF FORMATION FOR SEVERAL REACTIVE METAL OXIDES*

Compound	Heat of Formation at 298° K (Kcal/Mole)	Free Energy of Formation (Kcal/Mole)			
		298° K	500° K	1,000° K	1,500° K
Al_2O_3	−400	−380	−366	−336	−306
BaO	−133	−126	−121	−111	− 95
BeO	−147	−140	−135	−123	−112
CaO	−152	−144	−140	−131	−114
Ce_2O_3	−435	−414	−400	−361	−321
CeO_2	−245	−232	−223	−199	−175
HfO_2	−266	−252	−243	−221	−201
La_2O_3	−435	−416	−403	−370	−336
Li_2O	−143	−134	−128	−112	− 96
MgO	−144	−136	−131	−119	−103
PuO_2	−251	−238	−230	−210	−190
Sc_2O_3	−411	−391	−378	−345	−311
SiO_2	−208	−195	−187	−165	−144
SrO	−141	−134	−129	−116	−102
ThO_2	−293	−280	−272	−251	−230
TiO_2	−226	−212	−204	−182	−161
UO_2	−259	−247	−238	−218	−198
ZrO_2	−262	−248	−239	−216	−194

* Extracted from the survey by A. Glassner, Rept. ANL-5107 (August, 1953), except for the data on PuO_2, which were taken from L. Brewer, L. Bromley, P. W. Gilles, and N. L. Lofgren, Rept. AECD-2661 (February, 1949).

The following illustration shows how the data were used to indicate possible alloying reactions when the plutonium compound was more stable than the compound of the reducing metal.

For the reaction

$$PuF_3(s) + Al(l) \rightarrow Pu(l) + AlF_3(l)\,, \qquad (1)$$

$$\Delta F^{\circ}_{1,500^{\circ}\,K} = -15\ \text{kcal} = RT \ln K\,,$$

where K is the thermodynamic equilibrium constant for the reaction. Thus $K = 6.6 \times 10^{-3}$.

If, as an approximation, it is assumed that aluminum and plutonium are present as pure metals, then

$$\begin{aligned} K^{\circ}_{1{,}500^{\circ}\,\mathrm{K}} &= 6.6 \times 10^{-3} \\ &= \text{Partial pressure of } AlF_3 \text{ gas, } P_{AlF3} \\ &\div \text{ vapour pressure of } AlF_3 \text{ at } 1{,}500^{\circ}\,\mathrm{K}. \\ P_{AlF3} &= 6.6 \times 10^{-3} \times 0.5 \text{ atm} \\ &= 2.5 \text{ mm Hg.} \end{aligned}$$

The equilibrium should be shifted farther to the right in reaction (1) if excess aluminum is present, because of the added contribution of the free energy of solution of plutonium in aluminum metal. It might be expected, therefore, that if plutonium trifluoride and aluminum were heated in vacuum, so that the volatile aluminum trifluoride was continuously removed, a fluoride-free plutonium-aluminum alloy would be formed.

TABLE 2

HEATS AND FREE ENERGIES OF FORMATION FOR SEVERAL REACTIVE METAL FLUORIDES*

Compound	Heat of Formation at 298° K (Kcal/Mole)	Free Energy of Formation (Kcal/Mole)			
		298° K	500° K	1,000° K	1,500° K
AlF	−102	− 97	− 94	− 87	− 81
AlF_3†	−356	−339	−328	−301	−273
AsF_3	−198	−191	−188	...	...
BaF_2	−287	−274	−266	−284	−229
BeF_2	−227	−217	−211	−195	−184
CaF_2	−290	−278	−269	−250	−230
CeF_3	−416	−399	−387	−360	−334
CsF	−132	−124	−119	−106	− 87
GaF	− 56	− 51	− 47	...	...
GaF_3	−255	−239	−228	−202	−187
HfF_4	−435	−413	−399	−363	...
InF_3	−250	−234	−224	−199	−178
KF	−135	−127	−122	−109	− 93
LaF_3	−421	−404	−392	−366	−340
LiF	−146	−139	−134	−122	−112
MgF_2	−264	−250	−242	−221	−199
NaF	−136	−129	−124	−112	− 95
PuF_3	−375	−357	−345	−316	−288
PuF_4	−424	−402	−387	−354	−325
RbF	−133	−125	−121	−108	− 90
ScF_3	−367	−350	−338	−311	−287
SiF_4	−371	−361	−354	−338	−321
SrF_2	−289	−277	−269	−250	−232
ThF_4	−477	−454	−438	−403	−373
TiF_4	−370	−351	−338	−326	−316
TlF_3	−175	−159	−149	−128	−110
UF_3	−340	−322	−310	−281	−252
UF_4	−443	−421	−407	−373	−339
ZrF_4	−445	−424	−415	−378	−360

* Extracted from the survey by A. Glassner, Rept. ANL-5107 (August, 1953), except for the data on PuF_3 and PuF_4, which were taken from L. Brewer, L. Bromley, P. W. Giles, and N. L. Lofgren, Rept. UCRL-633 (March, 1950).

† Corrected by using the value for ΔH°_{298} of −356 kcal/mole reported by P. Gross, C. Hayman, and D. L. Levi, *Trans. Faraday Soc.*, **477**:50, 1954.

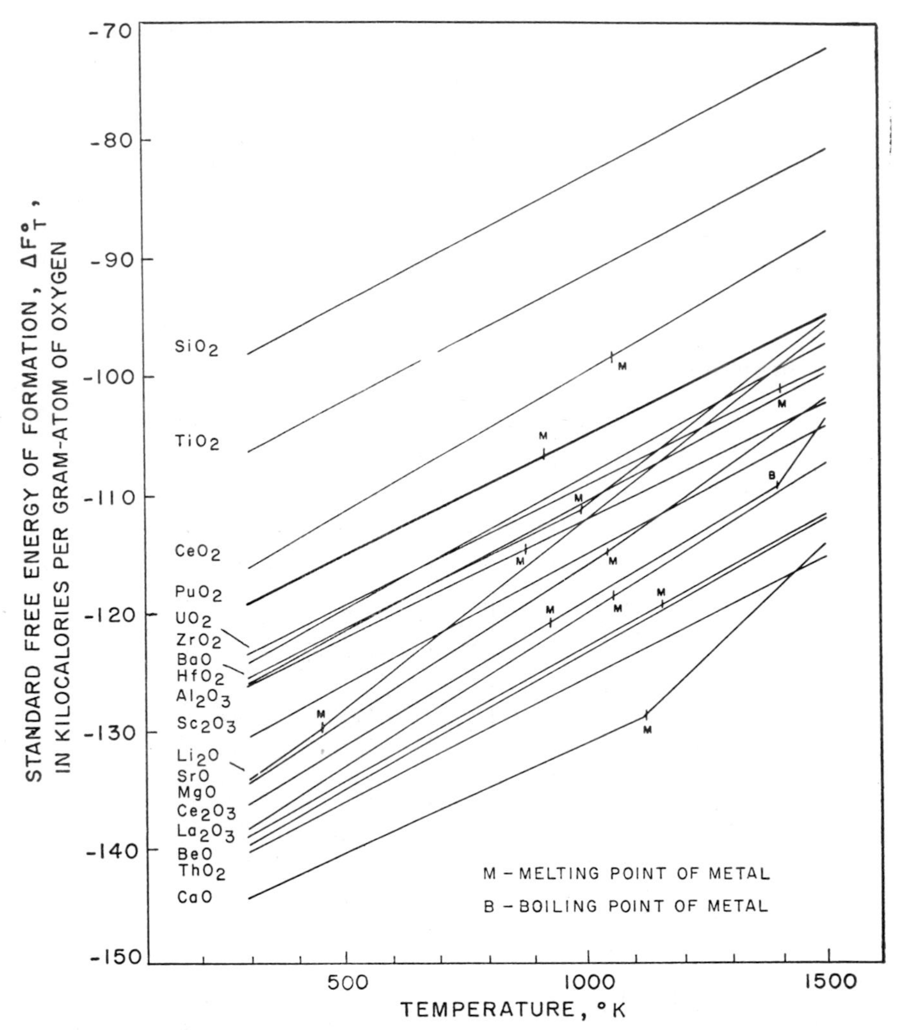

Fig. 1.—Standard free energy of formation versus temperature for a number of reactive metal oxides.

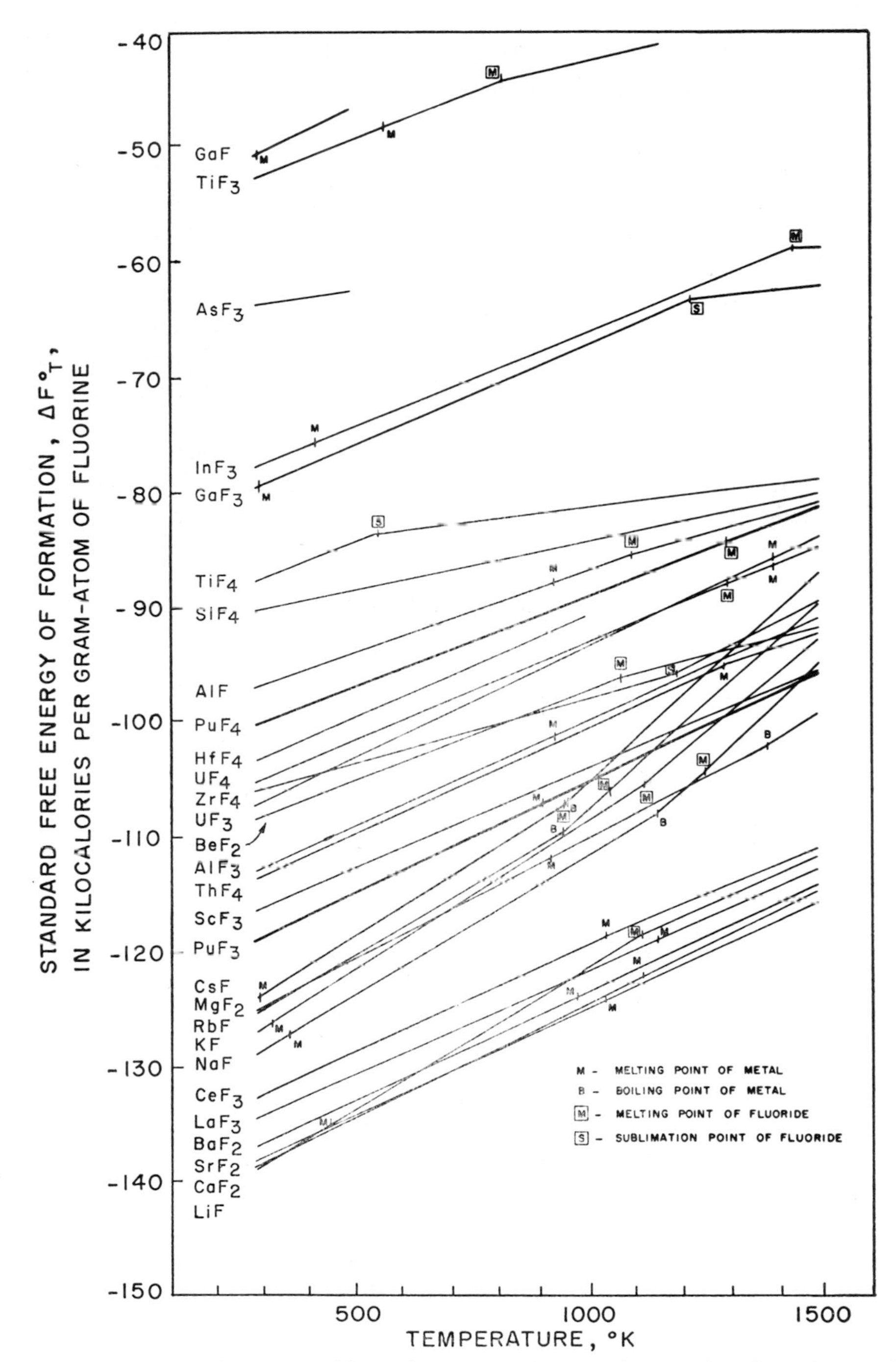

FIG. 2.—Standard free energy of formation versus temperature for a number of reactive metal fluorides.

PLUTONIUM-ALUMINUM ALLOY PREPARATION

Aluminum Reduction of Plutonium Trifluoride

SMALL-SCALE EXPERIMENTS

The first experiments on the aluminum reduction of plutonium trifluoride were performed with a plutonium trifluoride charge varying from 100 to 1,000 mg. The plutonium trifluoride, 99.5 wt. per cent pure, had been prepared by fluorinating thermally decomposed plutonium oxalate, i.e., plutonium dioxide, with a mixture of dry hydrogen fluoride and hydrogen at 650° C in a platinum tray. An aluminum ingot weighing 2–3 gm was placed on top of a weighed plutonium charge in a graphite crucible, which was then heated in the vacuum furnace illustrated in Figure 3. The furnace temperature attained for a fixed power input was known from a previous calibration with thermocouples. The pressure in the system was maintained at, or below, 10^{-4} mm mercury by a 4-inch oil-diffusion pump coupled to a mechanical pump.

The reduction of the fluoride proceeded slowly when the furnace temperature was 1,000° C. At 1,100° C the reduction rate was more rapid, as illustrated in

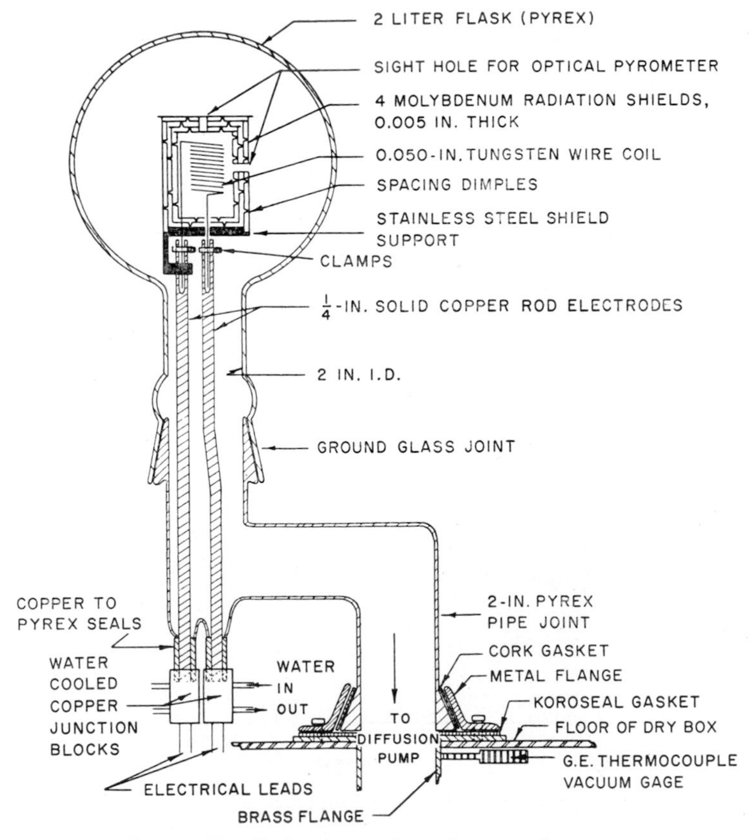

FIG. 3.—Detail of resistance-heated vacuum furnace

Table 3. Reduction was carried out in two stages in experiments 5, 6, and 7, since a violent evolution of gas and liquid metal had occurred with plutonium trifluoride charges greater than 200 mg when the furnace temperature was raised rapidly to 1,100° C. Apparently, the violent reaction could be avoided by maintaining the charge at 900° C for 1–2 hours initially. The explanation for the latter phenomenon was obtained in the larger-scale experiments reported below.

TABLE 3

RESULTS OF THE GRAM-SCALE REDUCTION OF PLUTONIUM TRIFLUORIDE WITH EXCESS ALUMINUM

Expt. No.	Plutonium Trifluoride Charge (Mg)	Reduction Temperature (° C)	Time at Temperature (Hours)	Plutonium in Plutonium-Aluminum Alloy (Wt. Per Cent)*	Plutonium Yield (Wt. Per Cent)
1	110	1,100	2	3.4	98
2	113	1,000	0.5	0.9	17
3	91	1,050	1	2.8	69
4	138	1,100	2	3.4	85
5	265	{900 1,100	1 3	9.8	87
6	533	{900 1,100	2 3	21.5	93
7	1,014	{900 1,100	2 5	80.0	91

* The plutonium content of the alloy was determined by alpha-counting an aliquot from a sample which had been dissolved in dilute hydrochloric acid.

LARGER-SCALE EXPERIMENTS

Purpose.—The preceding experiments were carried out primarily to indicate a method that might be used to convert 350 gm of plutonium into an alloy suitable for irradiation in the Canadian Heavy Water Reactor, NRX, at Chalk River. The fabrication of a fuel rod containing the 350 gm of plutonium was required for reactor enrichment and to provide eventually a large source of plutonium rich in the higher isotopes.

Apparatus.—The plutonium nitrate solutions were evaporated in an Inconel "boat" on oxalic acid crystals. A Nichrome heating coil, protected and supported by a Vycor tube, was suspended above the liquid surface inside a stainless-steel evaporator. The effluent gases passed through a calcium carbonate scrubbing tower in the exhaust line.

After evaporation, the dry residue was calcined and fluorinated in a Monel tube heated by a horizontal-tube furnace. The ends of the Monel tube were wrapped with copper cooling coils and were sealed with Teflon plugs. Anhydrous HF was admitted to the tube through a series of Monel valves and pipes. Attempts were made to meter the HF flow by using a platinum hot-wire gauge, but with only fair success. Tank hydrogen was purified by passage over hot palladized asbestos, silica gel, Ascarite, and magnesium perchlorate. The exit gases were scrubbed free of HF in towers filled with calcined dolomite, CaO·MgO.

The alloys were prepared in the silica-tube induction furnace illustrated in Figure 4. The water-cooled induction coil was powered by a 15-kva spark-gap generator. Two concentric tubes of zircon and Alundum surrounding the graphite crucible provided adequate thermal insulation. The apparent temperature was measured by sighting a disappearing-filament type of optical pyrometer on the crucible interior through an observation window in the top brass plate. The relation between the

Fig. 4.—Silica-tube induction furnace inside glove box

true and the observed temperatures was determined at the melting points of silver, 960° C, and copper, 1,083° C. The furnace was evacuated by a 2-inch oil-diffusion pump, backed by a mechanical pump. The pressure was measured with a thermocouple gauge.

Procedure.—The plutonium was received in shipping flasks, each containing from 60 to 95 gm of plutonium as plutonium nitrate in 200 ml of 3–5 N nitric acid. The solution, together with a 200-ml rinse of 2 N nitric acid, was transferred to the Inconel evaporation boat. The boat contained oxalic acid crystals, 25 per cent in excess of the weight required to convert the plutonium to plutonium (IV) oxalate.

The solution could be evaporated to dryness in 16–18 hours, with negligible spattering losses. The dried product was heated to 650° C in 2 hours in an air stream supplied at 600 ml/min. The furnace temperature was maintained at 650° C for 2 hours to complete the conversion of plutonium nitrate and oxalate to plutonium dioxide. The air flow was then replaced with a mixture of anhydrous hydrogen fluoride at 500 ml/min and purified hydrogen at 100 ml/min. The rate of conversion to plutonium trifluoride was determined by observing the fast neutrons emitted from the plutonium trifluoride due to the nuclear reaction F^{19} $(\alpha,n)Na^{22}$. After 5–6 hours at 650° C, the furnace chamber was cooled to 100° C in a hydrogen fluoride/hydrogen atmosphere and then to room temperature in hydrogen alone. The plutonium trifluoride was weighed into an aluminum can, which was then placed on top of a weighed aluminum ingot in a graphite crucible. A graphite cylinder was seated on the crucible to act as a condenser for the aluminum fluoride vapor. The assembly was lowered into the furnace on a rod inserted through a hole in the bottom plate.

When the furnace had been evacuated to a pressure of 10^{-4} mm mercury, the crucible was heated to the reduction temperature in about 30 minutes. Initially, the reduction temperature used was 1,125° C. In later experiments, however, a temperature of 900° C was found to be more satisfactory. Once the rate of reduction had been determined in initial experiments, the heating time required could be estimated with reasonable certainty for a given charge. When the alloy had cooled in vacuum, the crucible was removed from the vacuum furnace, and the alloy was remelted in air in a second induction furnace. The molten alloy was then poured manually into a water-cooled mold of the desired shape.

Results.—The fast-neutron intensity measurements recorded near the fluorination furnace indicated that the conversion of plutonium dioxide to plutonium trifluoride for a 250-gm plutonium charge was complete after 6 hours at 650° C. The resulting fluoride was removed easily from the Inconel reaction vessel as a finely divided powder. The temperature used initially for the reduction of 30-gm charges of plutonium trifluoride was 1,075°–1,125° C. During the heating-up period some bubbling of the melt was observed at 950° C. At 1,000° C, however, the liquid surface became quiescent. At 1,125° C a white condensate of aluminum fluoride deposited rapidly on the graphite sleeve above the hot zone of the furnace. When the condensate had been removed later from the graphite and weighed and, once the plutonium content of the alloy had been determined by volumetric chemical analysis, the respective masses indicated that the reduction at 1,125° C had proceeded as follows:

$$PuF_3 + x\mathrm{Al} \rightarrow PuAl_{x-1} + AlF_3. \qquad (2)$$

When the weight of the plutonium trifluoride charge was increased to 50 gm, a more violent evolution of gas was observed on heating through the temperature range from 950° to 1,000° C. The reaction was of sufficient violence, in fact, to eject large globules of liquid aluminum from the crucible.

Since some reaction accompanied by the evolution of fumes had been observed at temperatures as low as 860° C, an experiment was conducted at 900° C. A massive condensate resulted and was easily removed from the condenser as a compact cylinder. The weight of the condensate and the weight of plutonium in the alloy indicated that the reduction had proceeded by a new mechanism, whereby gaseous aluminum monofluoride was evolved, i.e.,

$$PuF_3 + xAl \rightarrow PuAl_{x-3} + 3AlF. \quad (3)$$

The rate of reduction of plutonium trifluoride was approximately five times more rapid at 900° C than at 1,125° C, as shown in Table 4. In addition, crucible attack at the lower temperature was insignificant. Thus all subsequent reductions were carried out at 900° C.

TABLE 4

TYPICAL RESULTS FOR LARGER-SCALE ALUMINUM REDUCTIONS OF PLUTONIUM TRIFLUORIDE

Expt. No.	Plutonium Charge (Gm Pu in PuF_3)	Aluminum Charge (Gm)	Reduction Temperature (° C)	Reduction Time (Hours)	Plutonium Alloyed (Gm)	Plutonium Yield (Wt. Per Cent)
1	27.4	257	1,075	2	15.8	58
2	27.9	264	1,125	2	23.7	85
3	48.8	272	1,125	5	44.0	90
4	62.0	275	900	1	42.5	69
5	64.5	270	900	2	62.8	97
6	62.0	275	900	1.5	59.7	96

In the final stages of development the plutonium yield for the conversion of nitrate solution to alloy ingot averaged 98 per cent.

Discussion.—Klemm and Voss (4) reported that aluminum monofluoride volatilized from a mixture of aluminum trifluoride and aluminum heated above 700° C in vacuum and was believed to disproportionate on condensation, according to the reaction

$$3AlF \rightarrow AlF_3 + 2Al. \quad (4)$$

It is interesting to note that the condensates from the aluminum reductions of plutonium trifluoride at 900° C, which had an aluminum-to-fluorine ratio of 1 to 1, contained small beads of aluminum metal. The aluminum spheres decreased in diameter as the distance from the reaction zone increased. An X-ray powder diffraction pattern obtained from a sample of the condensate showed only aluminum and aluminum trifluoride lines. Thus, in agreement with the observations of Klemm and Voss, the gaseous aluminum monofluoride had disproportionated on condensation. In addition, the aluminum had coalesced into the largest spheres at the hottest end of the condenser, as would be expected.

The rate of reduction of plutonium trifluoride is markedly higher at 900° C, where aluminum monofluoride is evolved, than at 1,000°–1,125° C, where the effluent is aluminum trifluoride (see Tables 3 and 4). At 1,000° C and above, either gaseous aluminum monofluoride does not form in the system, or it disproportionates in the gas phase. The above observations suggest that the volatility of aluminum monofluoride must be considerably higher than that of aluminum trifluoride.

The boiling points for aluminum monofluoride and aluminum trifluoride have been tabulated by Brewer (5) as 1,650° and 1,545° K, respectively. The former figure was given in parentheses as an estimated value only and would appear to require correction downward.

Aluminum Reduction of Plutonium Dioxide

The aluminum reduction of plutonium dioxide should proceed spontaneously, since, for the reaction

$$3PuO_2 + 4Al \rightarrow 3Pu + 2Al_2O_3 \,, \qquad (5)$$

$$-\Delta F_{1,500^\circ \mathrm{K}} = 42 \text{ kcal} \,.$$

If excess aluminum is used, the equilibrium should be shifted farther to the right because of the energy contribution from the alloying reaction.

Small-Scale Experiments

Plutonium dioxide was prepared by evaporating a plutonium nitrate solution on oxalic acid crystals and calcining the dry residue at 600° C in air. Cylindrical graphite crucibles, 1 by 5 cm, were used as containers. The normal charge was a 2–3-gm aluminum ingot covered with 100 mg of plutonium dioxide.

Initially, the charged crucibles were heated in a Nichrome-wound furnace at 1,000° C in air. Since little reaction occurred, however, even after 3 hours at 1,000° C, later experiments were carried out in a vacuum furnace similar to that shown in Figure 3. Alloys containing 2.5 wt. per cent plutonium were produced by heating plutonium dioxide and aluminum for 3 hours at 1,100° C in the latter furnace. The best reduction efficiency obtained was 60 per cent. The unreacted plutonium dioxide was present in pockets which were separated from the alloy by a barrier layer of aluminum oxide.

The foregoing experiments indicated that undesirably long heating times would be required to reduce the plutonium dioxide completely, unless the barrier film of aluminum oxide could be removed either mechanically or chemically during the alloying reaction. A suitable method for the chemical removal of the aluminum oxide film was suggested by Saller's (6) experiments on the preparation of uranium-aluminum alloys. When Saller added a briquetted mixture of uranium oxide and cryolite, $3NaF \cdot AlF_3$, to liquid aluminum at 1,100°–1,300° C, the alloy formed rapidly, with an average uranium yield of 80 per cent. The product aluminum oxide had dissolved in the liquid cryolite, which floated to the surface of the melt as a slag layer.

Saller's oxide-reduction technique would provide an even simpler flow sheet for the preparation of plutonium-aluminum alloys than the aluminum reduction of plutonium trifluoride, since the hydrofluorination and vacuum reduction stages would not be required. Initial experiments to test the method with plutonium dioxide were carried out, using 10–60-gm plutonium charges as described later. More recent experiments have indicated that the method is applicable to small-scale reductions, also. In a typical experiment, a charge consisting of 217 mg of plutonium as plutonium dioxide, 1 gm of cryolite, and a 1.5-gm aluminum ingot was heated in a graphite crucible in air for 20 minutes at 1,200° C. The resulting alloy button contained 208 mg of plutonium, a 96 per cent yield.

Larger-Scale Experiments

Apparatus.—The larger-scale experiments were carried out in a glove-box line (Fig. 5) which had been specially designed to permit the preparation, casting, and machining of plutonium-aluminum alloys. The evaporation and calcination techniques described previously were modified somewhat in the new glove-box line.

Precipitates of plutonium (IV) oxalate were prepared by reacting plutonium nitrate solution with oxalic acid crystals in a 1-liter glass beaker. The precipitate

Fig. 5.—Glove-box line for the preparation, casting, and machining of plutonium-aluminum alloys. Duplication of chap. vii, Fig. 1.

was dried in the beaker inside a glass bell jar under infrared heating bulbs. The effluent gases were passed through a water-cooled condenser before being discharged to the glove-box exhaust line. The dried oxalate was calcined to plutonium dioxide in a stainless-steel boat, $2\frac{1}{2}$ inches square and 8 inches long, inside a 4-inch-diameter stainless-steel tube heated externally.

The alloys were prepared in air in graphite crucibles, which were heated in a small induction furnace powered by the 15 kva spark-gap generator. The crucibles were 8 inches long and 3 inches in outside diameter, with a wall thickness of $\frac{3}{8}$ inch.

Procedure.—A plutonium nitrate solution containing about 100 gm of plutonium in 600 ml of 4 N nitric acid was stirred with a 25 per cent excess of oxalic acid crystals to precipitate plutonium (IV) oxalate in the glass beaker. The precipitate was allowed to settle overnight, and the supernatant liquid was decanted off. The usual volume of liquid removed was 500 ml, and it contained approximately 5 mg of plutonium. The wet sludge was dried under the infrared lights in 3–4 hours. Occasionally, when rather impure plutonium solutions were obtained, which formed gels on reaction with oxalic acid rather than granular precipitates, all the liquid was removed by evaporation. The heating time required then was 16–20 hours.

The dry residues from two evaporations, i.e., 200 gm of plutonium, were transferred to the stainless-steel boat and were heated in air in the tube furnace for 2 hours at 650° C. The finely divided plutonium dioxide product was removed easily from the boat and was weighed prior to the reduction step.

Approximately 100 alloy billets containing a total of 6 kg of plutonium were prepared in the glove-box line shown in Figure 5, using the following reduction procedure. The plutonium dioxide powder was mixed with powdered cryolite in the ratio of 1 gm plutonium dioxide to 2 gm cryolite. Deep-drawn aluminum cans, $2\frac{1}{2}$ inches long and $\frac{1}{2}$ inch in diameter, were filled with the dry mixture and were placed in a restraining metal die. The powder was compressed to one-quarter its original volume by using a stainless-steel piston tapped with a hammer. Each can contained about 2 gm of plutonium. A weighed aluminum ingot was melted under a cryolite layer in the graphite crucible. Early reductions were made at 1,100° C. Later, the temperature was raised to 1,200° C. The cans containing the pressed plutonium dioxide/cryolite mixture were dropped into the melt at the approximate rate of one every 10 seconds. Thus a 60-gm plutonium addition was made in 5 minutes. The melt was held at a temperature of 1,200° C for a further 10 minutes before the induction furnace was turned off.

This reduction technique has been modified recently to eliminate the pelletizing step. The present practice, as employed in the routine production of alloy ingots containing 30 gm of plutonium, is as follows. The powdered plutonium dioxide/cryolite, mixed in the ratio of 1 gm plutonium dioxide to 2 gm cryolite, is placed on top of cryolite powder in a cold graphite crucible. A layer of cryolite powder is added, followed by the aluminum ingot and a covering layer of cryolite. The total weight of powder in the three cryolite layers equals that which is mixed initially with the plutonium dioxide, i.e., the ratio of cryolite to plutonium dioxide in the crucible is 4 to 1. The crucible is heated in air to 1,200° C in 15 minutes, held at temperature for 15 minutes, then cooled in the furnace.

The cooling rate is rapid. Within 10 minutes, both the cryolite and the alloy layers solidify. When the crucible is cold, it is broken by a hammer blow, and the alloy ingot is removed, slag-free. The ingot may then be remelted and cast into the desired shape.

Results.—Data on the plutonium yield from eight of the first reductions of plutonium dioxide/cryolite mixtures with liquid aluminum at 1,100° C are given in Table 5. The plutonium contents of the starting solutions and of samples from the solid alloy ingots were determined by chemical analysis in the Chemical Control Branch of the Chalk River laboratories.

As Table 5 indicates, the plutonium yield in the trial reductions exceeded 90 per cent. The oxide-reduction method was adopted, therefore, as the standard technique for the preparation of plutonium-aluminum alloys. Since kilogram quantities of plutonium were later converted to alloys, the opportunity arose to examine the process in more detail.

Initially, the plutonium content of each alloy was determined from a spectrophotometric analysis on a small sample which had been dissolved in HCl. The analytical results often indicated that the plutonium fraction varied as much as 10 per cent in adjacent turnings from a chill-casting or on halved samples dipped from the melt. Since the analytical method was accurate to ± 1 per cent, the variation in the results was attributed to segregation within the alloy, though no gross segregation was visible in X-radiographs of chill-cast cylinders, 1-inch-diameter.

Several slags were dissolved, and several graphite crucibles were leached in acids to determine the plutonium loss in typical reduction experiments. The maximum loss was 0.3 per cent of the plutonium added initially, in those reductions where the

TABLE 5

PLUTONIUM ALLOYING YIELDS FROM REDUCTIONS OF PLUTONIUM DIOXIDE/CRYOLITE MIXTURES WITH ALUMINUM AT 1,100° C

Expt. No.	Plutonium Charge (Gm)	Aluminum Charge (Gm)	Plutonium Alloyed (Gm)	Plutonium Yield (Per Cent)
1	20.0	228	18.2	91
2	16.5	200	16.3	99
3	36.3	200	36.0	99
4	44.1	259	42.9	97
5	55.4	263	53.3	96
6	44.1	240	43.6	99
7	62.5	160	61.4	98
8	12.5	281	12.0	96

plutonium dioxide/cryolite mixture had been pressed and added to the aluminum at 1,200° C. In a typical assay, a 200-gm slag layer was separated from above a 300-gm alloy containing 60 gm of plutonium. When the slag had been crushed, small, shiny particles of alloy were seen, which dissolved in dilute hydrochloric acid. The plutonium recovered was 100 mg. When the cryolite had been completely dissolved in hot sulfuric acid, a further 20 mg of plutonium was present in solution. An additional loss of 50 mg of plutonium had occurred during the pelletizing step. The estimated total loss was, therefore, 170 mg of plutonium, or 0.3 per cent.

In the zero-energy reactor ZEEP at Chalk River, Hone (7) made individual reactivity measurements on thirty-four 20 wt. per cent plutonium-aluminum chill-cast cylinders, each containing 60 gm of plutonium. The measurements were reproducible to ± 0.3 per cent. Thus the relative plutonium contents of the cylinders could be estimated with corresponding precision. One of the alloys, PR-8, was used as a standard, and all other measurements were normalized to it. In Table 6 several typical results have been listed, together with the plutonium content determined from chemical analyses and from an estimated 99.7 per cent conversion of weighed plutonium dioxide to alloy.

As a result of the good agreement between the plutonium contents obtained from the reactivity measurements and from the plutonium dioxide weighed in, the latter method has been preferred over chemical analyses for accounting purposes.

Recently, plutonium yields in the plutonium dioxide/aluminum reaction have been re-examined because of the change introduced in the reduction procedure, i.e., where the pelletizing step had been eliminated. The slags from four reductions were dissolved and assayed for plutonium. These initial assays indicated that the present reduction procedure resulted in somewhat higher plutonium losses, 0.8 per

TABLE 6

PLUTONIUM CONTENTS OF 20 WT. PER CENT PLUTONIUM-ALUMINUM ALLOYS OBTAINED USING THREE DIFFERENT ACCOUNTING METHODS

(Plutonium Content in Grams)

Alloy No.	From Chemical Analysis		Estimated from the Plutonium Dioxide Weighed In	From Reactivity Measurements (Normalized to PR-8)
	1	2		
PR-8......	59.5	59.6	59.6	59.5
-9......	61.1	....	59.7	59.5
-10.....	61.5	....	60.0	60.6
-11.....	60.9	....	59.8	59.3
-12.....	59.5	....	59.7	60.3
-13.....	60.8	....	59.2	59.1
-20.....	55.6	61.9	59.8	59.5
-21.....	52.6	50.5	59.7	59.1
-22.....	62.8	65.5	59.8	59.3
-23.....	57.2	56.6	59.6	59.5

cent. Further experiments on the effect of the reaction time and temperature on the reduction efficiency are in progress. The purpose of the experiments is to establish the procedure required to increase the plutonium yield to the 99.7 per cent attained with the earlier reduction method.

Discussion.—When the plutonium-aluminum billet is first removed from the reduction crucible, it is covered with an adherent yellow film. After the alloy stands in air for 1–2 days, the film changes to a fluffy white powder, which can be brushed off. X-ray analyses have shown that the yellow compound is sodium peroxide, Na_2O_2, and that the resulting white powder is sodium hydroxide. The sodium peroxide is believed to have formed by the oxidation of a sodium metal layer which is precipitated at the surface of the alloy billet when the alloy is cooled from 1,200° C to the solidification temperature of 645° C. Presumably, the sodium is introduced to the liquid alloy by a reaction between the sodium fluoride (in $3NaF \cdot AlF_3$) and water vapor at 1,200° C, producing sodium oxide, Na_2O, which is then reduced to sodium metal by the liquid aluminum. Little sodium remains in the finished casting, however, since spectrographic analyses on samples of remelted and cast billets have shown that the sodium content is below 0.005 wt. per cent.

Alloys containing up to 25 wt. per cent plutonium have been prepared thus far by the reaction of aluminum with plutonium dioxide/cryolite mixtures in air. Although no attempt has been made to prepare alloys richer in plutonium, it should be possible to increase the plutonium concentration to 50–60 wt. per cent before the rising temperature of the liquid becomes a limitation. The chemical reactivity of such alloys should not be a problem, since compositions up to $PuAl_2$, 81 wt. per cent plutonium, are resistant to oxidation.

The plutonium yield in converting plutonium dioxide to plutonium-aluminum alloy appears to be consistently higher than for the similar aluminum reduction of uranium oxides. A number of uranium-aluminum alloys containing U^{233} and U^{235} were made in the equipment described above by adding a pressed oxide-cryolite mixture to aluminum at 1,200° C. The uranium contents have been determined precisely by means of both reactivity measurements on cast cylinders and chemical analyses from large pieces of each casting, i.e., 30–40 per cent of the total. The uranium yields have varied from 93 to 98 per cent. It is interesting to note that in Saller's (6) experiments the uranium yield varied also, usually within the range of 75–98 per cent. One possible explanation for the difference in behavior is that the thermodynamics of the plutonium dioxide reaction are more favorable because the free energy of formation of uranium dioxide exceeds that of plutonium dioxide by about 8 kcal. It is possible also, of course, that the solubility of uranium oxide in aluminum-saturated liquid cryolite exceeds the plutonium dioxide solubility limit of 0.01 wt. per cent.

PREPARATION OF OTHER PLUTONIUM ALLOYS

A number of other plutonium alloys have been prepared by the reduction of plutonium fluorides and oxide with metals such as beryllium, gallium, magnesium, silicon, thorium, and uranium. The preparative techniques used are described in more detail below.

The Reaction between Plutonium Fluorides and Other Group IIIA Metals

Reaction with Gallium

The low free energies of formation for the gallium fluorides given in Table 2 indicate that it would not be possible to reduce plutonium trifluoride with gallium metal. The calculated equilibrium vapor pressure of gallium trifluoride above an equimolar plutonium trifluoride/gallium mixture at 1,500° K is only 10^{-14} atm, for example. In view of the unexpected behavior of plutonium trifluoride/aluminum mixtures, however, it seemed worthwhile to examine the plutonium trifluoride/gallium system experimentally.

The plutonium trifluoride charges used in the experiments weighed from 100 to 300 mg and, once mixed with a similar weight of gallium, were heated in the vacuum furnace illustrated in Figure 3 in beryllium oxide crucibles. No reaction took place up to 1,050° C, however, and the cool glass bulb surrounding the furnace assembly rapidly became coated with a dark-red deposit of gallium fluoride. The reduction of the plutonium trifluoride was complete after 2 hours. Part of the resulting plutonium-gallium alloy ingot was kept for X-ray diffraction studies, and part was dissolved for a plutonium assay. Several typical results are given in Table 7.

Although the calculated ratios of fluorine to gallium in the gaseous effluents were close to unity, it is probable that the compound evolved was gallium trifluoride, since an appreciable amount of gallium metal would volatilize during the reduction period. When, for example, pure gallium metal weighing 676 mg was heated in the system, 457 mg had distilled off after 2 hours at 1,150° C. A 60 per cent plutonium-gallium alloy of similar weight lost a smaller, but still significant, amount, 40 mg, after the same heat treatment. In a reduction experiment the gallium loss should be intermediate between the above two extremes, since pure gallium is present only at the start and becomes richer in plutonium as the reduction proceeds. It is interesting to note that the weight lost from the charge in experiment 4, Table 7, was 221 mg. If all the fluorine had been evolved as gallium trifluoride, the weight loss would have been 115 mg. The difference, 106 mg, would appear to be a reasonable gallium distillation loss.

TABLE 7

CALCULATED FLUORINE:GALLIUM RATIOS IN GASEOUS EFFLUENT FROM GALLIUM REDUCTION OF PLUTONIUM TRIFLUORIDE

No.	Reactants in Charge		Reaction Time at 1,150° C (Hours)	Wt. Per Cent Plutonium in Resulting Plutonium-Gallium Alloy	Calculated Fluorine: Gallium Ratio in Effluent
	Plutonium Trifluoride (Mg)	Gallium (Mg)			
1......	130.2	176.6	2.0	56.0	0.98
2......	259.1	219.1	3.0	70.5	1.40
3......	267.3	223.3	3.0	72.3	1.34
4......	267.1	226.2	3.0	72.3	1.31
5......	194.0	242.2	2.3	60.3	0.99
6......	194.9	243.2	2.8	56.1	1.15
7......	207.1	307.7	2.3	47.7	1.18

Plutonium-gallium alloys have also been prepared by the gallium reduction of plutonium tetrafluoride. The plutonium tetrafluoride reduction rate is about twice that of plutonium trifluoride.

Indium and Thallium

Several attempts were made to prepare plutonium-indium and plutonium-thallium alloys by the indium and thallium reductions of plutonium trifluoride or tetrafluoride in vacuum at temperatures up to 1,100° C, without success. Little of the thallium charge could be retained in the crucible because of its high volatility. Some of the indium was lost by distillation, also. There was no significant alloying reaction in either case. Some reaction had been expected between indium and the plutonium fluorides because of the similarity in the thermodynamic properties of indium trifluoride and gallium trifluoride, as shown in Table 2. Since no reaction took place with indium, however, one might have predicted the results of subsequent experiments, which showed that plutonium fluorides were not reduced by the even less reactive liquid thallium.

Plutonium-Beryllium Alloys

Beryllium Reduction of Plutonium Trifluoride

The preparation of plutonium-beryllium alloys by the beryllium reduction of plutonium trifluoride in vacuum according to the reaction

$$2PuF_3 + xBe \rightarrow Pu_2Be_{x-3} + BeF_2 \qquad (6)$$

has been described by Runnalls and Boucher (8). The alloys were prepared in the tungsten-coil furnace (Fig. 3) in beryllium oxide crucibles. The alloy formation depended on the removal of the beryllium fluoride product which distilled from the hot zone and condensed on the cool glass bulb as a black deposit. Since plutonium-beryllium alloys emit neutrons as a result of the (α,n) reaction with beryllium, the chemical reaction between plutonium trifluoride and beryllium was conveniently examined by positioning a neutron counter near the reduction furnace. When the mixture of plutonium trifluoride and beryllium powders reacted to form the plutonium-beryllium alloy, the neutron yield increased markedly. Thus a slow increase in the neutron yield from a plutonium trifluoride/beryllium mixture at 900° C indicated that the reaction was proceeding at a measurable rate at that temperature. The reduction rate at 1,125° C was rapid. For a 400-mg plutonium trifluoride charge, the conversion to the plutonium-beryllium alloy was complete in 25 minutes. The plutonium reduction yield on the latter scale averaged 99 per cent.

Although plutonium-beryllium alloys of widely different atom ratios—i.e., from 1:10 to 1:300—were prepared by the beryllium reduction of plutonium trifluoride, it was sometimes desirable to adjust the plutonium-beryllium ratio of the alloy, particularly for neutron-yield studies. This was most easily accomplished by evaporating beryllium from a beryllium-rich alloy at 1,200° C, where the sublimation rate was high.

Beryllium Reduction of Plutonium Dioxide

Beryllium metal should reduce plutonium dioxide spontaneously, since, for the reaction

$$2Be + PuO_2 \rightarrow Pu + 2BeO\,, \qquad (7)$$

$$-\Delta F_{298^\circ K} = 47 \text{ kcal}\,.$$

Recent experiments by Michaud and Boucher (9) at Chalk River have indicated that the reduction proceeds rapidly, even when no flux is present to dissolve the product beryllium oxide. After a plutonium dioxide/beryllium mixture had been heated in vacuum for 30 minutes at 850° C, for example, most of the plutonium had been converted to the intermetallic compound $PuBe_{13}$, as determined by X-ray diffraction analyses.

Several sintered alloys were prepared in argon, also, by heating a plutonium dioxide/beryllium mixture in the presence of a cryolite flux, in which 7 wt. per cent beryllium oxide should dissolve at 1,000° C, according to Khazanov (10). The charges were heated from 20° to 1,100° C in 30 minutes, then cooled to room temperature in 15 minutes. The only plutonium phase which was detected in the products was $PuBe_{13}$, indicating that the reduction efficiency was high. Confirmation of the high efficiency was obtained by measuring the number of neutrons emitted by the alloys. A typical product had a neutron yield of 58.6 neutrons per million alpha-particles, very near the thick-target neutron yield of 60.0. The high neutron yields indicated also that the diameter of the $PuBe_{13}$ crystallites in the alloys was below 2 μ (8).

Plutonium-Silicon Alloys

A plutonium-silicon alloy near the composition $PuSi_2$ was prepared by the reduction of plutonium trifluoride with calcium disilicide at 1,550° C by Westrum (11). No reaction had been observed when the mixture of plutonium trifluoride and calcium disilicide was heated to 1,300° C. The reactants had been mixed according to the equation

$$10PuF_3 + 3CaSi_2 \rightarrow 10Pu + 3CaF_2 + 6SiF_4 . \qquad (8)$$

If the heat of formation of calcium disilicide is as high as Westrum has indicated, i.e., 200 kcal, then the direct reduction

$$4PuF_3 + 3Si \rightarrow 4Pu + 3SiF_4 \qquad (9)$$

is a more favorable reaction, thermodynamically, since for reaction (8) $-\Delta H_{298} = -131$ kcal/mole Pu and for reaction (9) $-\Delta H_{298} = -97$ kcal/mole Pu.

The above conclusion has been confirmed experimentally, in fact, by heating plutonium trifluoride with excess silicon in beryllium oxide crucibles to 1,300° C in vacuum. A complete reduction of the plutonium trifluoride occurred in 15 minutes at 1,300° C. The resulting fluoride-free plutonium silicides were hard and brittle, with a silvery metallic luster. The composition of the alloys prepared ranged from 70 to 95 wt. per cent plutonium.

Plutonium-Thorium Alloys

It may be seen from Figure 2 that thorium tetrafluoride is more stable than plutonium tetrafluoride but less stable than plutonium trifluoride. Thus the reduction of plutonium tetrafluoride should proceed spontaneously in the presence of a thorium excess, forming plutonium trifluoride rather than plutonium metal. A thorium reaction with plutonium trifluoride could occur, but only under reduced pressure, where the gaseous thorium tetrafluoride product was continuously removed.

When a mixture of plutonium trifluoride powder and thorium metal turnings was heated to 1,250° C for 1 hour in vacuum, however, no reaction occurred. A mixture of plutonium tetrafluoride and thorium, on the other hand, reacted exothermically at 650° C, and the temperature rose quickly to above 1,100° C. Sixty per cent of the initial charge of 200 mg of plutonium tetrafluoride reacted to form a plutonium-thorium alloy. The remainder of the plutonium tetrafluoride had been reduced to plutonium trifluoride. Further experiments on the reaction between plutonium tetrafluoride and thorium are in progress.

Plutonium-Uranium Alloys

The possibility of preparing plutonium-uranium alloys by the molten uranium reduction of plutonium trifluoride was considered, i.e.,

$$4PuF_3 + xU \rightarrow Pu_4U_{x-3} + 3UF_4\,.$$

The calculated uranium tetrafluoride pressure over the system at 1,500° K is only 10^{-9} atm. No allowance has been made, however, for the energy contribution from the alloying of the plutonium with the uranium.

A few preliminary experiments have indicated that the reaction proceeds slowly at 1,200° C in vacuum. A plutonium-uranium alloy containing 2.5 wt. per cent plutonium was prepared, for instance, by heating a mixture of uranium and plutonium trifluoride at 1,200° C for 15 minutes. This represented a 20 per cent plutonium yield.

The Use of Calcium and Magnesium as Reducing Metals

Reaction with Plutonium Trifluoride

The fluorides of plutonium should be spontaneously reduced by calcium and magnesium, since both calcium and magnesium fluorides are more stable fluorides, as indicated in Figure 2.

Plutonium metal was prepared in 1-gm amounts by reducing plutonium trifluoride in calcium vapor in small iron bombs. The best reductions were obtained when the plutonium trifluoride was contained in a tantalum crucible, with the bomb heated to 1,000° C, i.e., where the calcium vapor pressure was $\sim$ 10 mm mercury.

Plutonium-magnesium alloys containing 3 wt. per cent plutonium were prepared by reacting liquid magnesium with plutonium trifluoride in helium for 15 minutes at 800° C in graphite crucibles. The plutonium recovery was only 30 per cent, however, and magnesium distillation losses were high.

Possible Reactions Using Plutonium Dioxide

The production of uranium metal by the calcium and magnesium reduction of uranium dioxide has been described by Buddery (12). Because of the similarity in the thermodynamic properties of the dioxides of plutonium and uranium, it should be possible to prepare plutonium metal by a similar technique. By analogy with the uranium reductions, one would expect the plutonium product to consist of small beads of metal dispersed in a sintered matrix of calcium or magnesium oxide, since insufficient heat is liberated during the bomb reaction to melt the product oxide of calcium or magnesium. Thus the metal particles are prevented from coalescing into a single billet.The addition of a fluoride flux which dissolves the calcium or magnesium oxide should promote the formation of an agglomerated metal product, since, effectively, the melting points of the oxides are lowered. There are several fluxes which might be satisfactory solvents. According to Berent (13), for example, a mixture of 10 per cent magnesium oxide in cryolite melts at 890° C. Also, Thomson and Kaye (14) reported that a melt of calcium fluoride and sodium fluoride in the ratio of 2 to 1 at 900° C would dissolve 23.5 per cent calcium oxide, and Khazanov (10) found that 14 per cent calcium oxide dissolved in molten cryolite at 1,000° C.

Preliminary reduction experiments were carried out using 250-mg charges of plutonium dioxide mixed with calcium chips which were heated to 1,000° C for 15 minutes in iron bombs. Microscopic and X-ray examination of the products showed that the reduction had gone to completion and that, as expected, the plutonium metal was present as small spherical particles. Further experiments are under way to investigate the effect of flux additions on the particle size of the product plutonium metal.

SUMMARY

Several binary plutonium alloys have been prepared by reducing plutonium fluorides or oxide with an excess of the alloying metal. The reactions of aluminum with plutonium trifluoride and with plutonium dioxide were examined in the most detail. The aluminum reduction of plutonium trifluoride was dependent on the removal of gaseous aluminum fluorides from the reaction zone. The rate of reduction was about five times more rapid at 900° C than at 1,125° C because of the evolution of the highly volatile aluminum monofluoride at the former temperature. At 1,000° C and above, the gas evolved was aluminum trifluoride, indicating that gaseous aluminum monofluoride was no longer stable in the system. In experiments using 60-gm plutonium charges, the plutonium yields averaged 98 per cent.

Several kilograms of plutonium dioxide have been converted to plutonium-aluminum alloys by the reduction of plutonium dioxide with aluminum in the presence of cryolite at 1,200° C. The reduction proceeded readily, owing to the dissolution of the product aluminum oxide in the liquid cryolite. The reduction yields were consistently above 99 per cent.

The reaction between plutonium trifluoride and other group IIIA metals has been examined. Charges of plutonium trifluoride weighing 200 mg were completely reduced when heated with an excess of gallium metal in vacuum for 2 hours at 1,150° C, forming plutonium-gallium alloys. Plutonium trifluoride was not reduced by either indium or thallium, however.

Plutonium-beryllium alloys have been prepared by reducing both plutonium trifluoride and plutonium dioxide with beryllium in vacuum. The plutonium trifluoride/beryllium reaction was complete after 25 minutes at 1,125° C, and produced a fluoride-free alloy in high yield. The plutonium dioxide/beryllium reaction was essentially complete after 30 minutes at 850° C.

The silicon reduction of plutonium trifluoride proceeded readily at 1,300° C in vacuum, resulting in the formation of fluoride-free plutonium-silicon alloys in 15 minutes.

A limited number of experiments have also been carried out on the preparation of plutonium alloys of thorium, uranium, and magnesium by the direct reduction of plutonium trifluoride or tetrafluoride. Some experiments on the calcium and magnesium reduction of plutonium dioxide are in progress.

REFERENCES

1. Jette, E. R. *Journal of Chemical Physics*, **23**:365–68, 1955.
2. Westrum, E. F., and Wallman, J. C. *Journal of the American Chemical Society*, **73**: 3530, 1951.
3. Zachariasen, W. H. "X-Ray Diffraction Studies of Fluorides of Plutonium and Neptunium; Chemical Identity and Crystal Structure," in *The Transuranium Elements*, ed. G. T. Seaborg, J. J. Katz, and W. M. Manning, paper 20.5, pp. 1462–72. ("National Nuclear Energy Series," Div. IV, Vol. **14B.**) New York: McGraw-Hill Book Co., Inc., 1949.
4. Klemm, W., and Voss, E. *Zeitschrift für anorganische Chemie*, **251**:233, 1943.
5. Brewer, L. "The Fusion and Vaporization Data of the Halides," in *Chemistry and Metallurgy of Miscellaneous Materials: Thermodynamics*, ed. L. L. Quill, paper 7, pp. 193–275. ("National Nuclear Energy Series," Div. IV, Vol. **19B.**) New York: McGraw-Hill Book Co., Inc., 1950.
6. Saller, H. A. "Preparation, Properties and Cladding of Aluminum-Uranium Alloys," *Proceedings of the International Conference on the Peaceful Uses of Atomic Energy*, **9**:214–20. New York: United Nations, 1956.
7. Hone, D. W. AECL Rept. CRR-23, p. 19, 1954.
8. Runnalls, O. J. C., and Boucher, R. R. *Canadian Journal of Physics*, **34**:949, 1956.
9. Michaud, G. G., and Boucher, R. R. Unpublished data, 1957.
10. Khazanov, E. I. *Legkie Metally*, **5**:16, 1956.
11. Westrum, E. F. "Preparation and Properties of Plutonium Silicides," in *The Transuranium Elements*, ed. G. T. Seaborg, J. J. Katz, and W. M. Manning, paper 6.5, pp. 729–30. ("National Nuclear Energy Series," Div. IV, Vol. **14B.**) New York: McGraw-Hill Book Co., Inc., 1949.
12. Buddery, J. H. "Production of Uranium Metal by the Reduction of Uranium Tetrafluoride," in *Progress in Nuclear Energy*, Ser. V, Vol. **1**, chaps. 1–3, pp. 24–32. London: Pergamon Press, Ltd., 1956.
13. Berent, Y. K. *Kimicheskiĭ Referativnyĭ Zhurnal*, **4**:85, 1941.
14. Thomson, D., and Kaye, L. *Transactions of the Electrochemical Society*, **67**:169, 1935.

PART III

PLUTONIUM IN NUCLEAR REACTORS

CHAPTER XXVII

THE ROLE OF PLUTONIUM IN NUCLEAR POWER

R. M. Kiehn

Plutonium has been the forgotten reactor fuel. Although the word "ignored" might be a better choice than the word "forgotten," one still may ask the question, why? A brief survey indicates that there are two dominant reasons why plutonium reactor technology has not advanced as rapidly as uranium reactor technology. The first reason is associated with the fact that plutonium is a man-made element and has not been generally available for non-military application in large quantities; the second reason is associated with the extreme health hazard presented by this alpha-active, bone-seeking element, which makes laboratory handling costly and difficult.

Plutonium is formed as the result of a neutron capture by U^{238} and the subsequent two-stage beta decay of the intermediate U^{239} and Np^{239}. The element does not occur significantly in nature because it disintegrates with a half-life of about 2.43×10^4 years. Although exact figures are unavailable, a simple computation indicates that the present world supply of plutonium is probably measured in amounts of a few thousand kilograms, i.e., a few tons. If this amount of plutonium were fissioned to produce electricity, it could supply the present electrical energy demand of the United States for only a few days. I present these figures in order to stress the fact that the future of plutonium not only depends on the development of techniques to use the fuel in nuclear reactors but is also contingent on the development of methods of producing, extracting, and metallurgically processing large amounts of the material. Most of the plutonium now available in metallic form has been produced by aqueous chemical processing of uranium metal irradiated in the plutonium-production reactors, such as those at Hanford and Savannah River.

Although the nuclear properties of plutonium, which will be discussed briefly, have been known for a number of years to be attractive from a fission reactor standpoint, out of the 200–250 reactors currently built or contemplated (1), there has been only one plutonium reactor constructed (now dismantled) (2) and only two or three others seriously considered (3, 4). A change in reactor-design thinking seems to be necessary to correct this disproportionate emphasis, a change that has been recently stimulated by the Atomic Energy Commission.

R. M. Kiehn is at the University of California, Los Alamos Scientific Laboratory, Los Alamos, New Mexico.

The three important nuclear parameters of plutonium are (1) the energy-dependent values of ν, the average number of neutrons produced per fission; (2) σ_f, the fission cross-section; and (3) α, the ratio of the parasitic absorption cross-section to the fission cross-section. Experimental values and estimates for σ_f and α of Pu^{239} are presented as functions of neutron energy in Figure 1 (5). A few significant differences are noticed if these values are compared with those measured or estimated for U^{235}. The value of α for Pu^{239} is larger than that for U^{235} in the thermal-energy region and perhaps lower than that for U^{235} in the Mev region. The value of $\eta = \nu/(1 + \alpha)$—the number of neutrons produced per neutron absorbed—is about the same for the two isotopes in the thermal neutron region but significantly larger for Pu^{239} in the Mev region.

By use of these parameters, a rough measure of a reactor's nuclear design potential may be made in terms of three quantities which may be called (1) the fuel

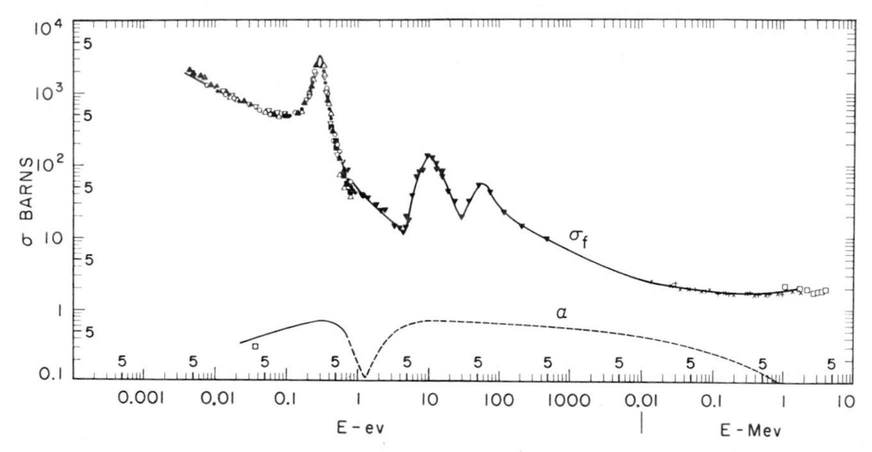

FIG. 1.—The fission cross-section and α of Pu^{239} as functions of neutron energy

reactivity contribution, $\nu/(1 + \alpha)$; (2) the fuel-burning efficiency, $1/(1 + \alpha)$; and (3) the fuel-breeding ratio capability, $(\nu - 1 - \alpha)/(1 + \alpha)$. Because more than one neutron is emitted per fission, it is possible to design reactors in such a way that the excess neutrons are used to create new nuclear fuel by means of nuclear transmutations. (The process of forming Pu^{239} by neutron bombardment of U^{238} has been described above.) In fact, it is possible in breeder reactors to produce more fuel than the amount used. Experimentally, a plutonium critical assembly has been constructed in England which produced 2.2 new atoms of Pu^{239} for every atom of Pu^{239} destroyed (6)—a direct verification of the feasibility of nuclear breeding.

The three quantities mentioned above have been estimated in Table 1 for several neutron energies. From the table it is seen that, from a nuclear standpoint, Pu^{239} is a much better fast (high-energy) reactor fuel than is U^{235}, and only a slightly poorer thermal reactor fuel. However, the large breeding capability of Pu^{239} allows relatively large practical losses to be sustained in a plutonium-breeder reactor and still enables the system to produce more fuel than it consumes.

The table also points out other features of the higher neutron economy obtainable with fast-reactor systems. It is interesting to note that, in fast-reactor systems, α for plutonium is further decreased because Pu^{240} readily fissions above 300 kev; stating it more simply, the burning efficiency of plutonium may be larger than is denoted in Table 1. U^{236}, on the other hand, does not readily fission below 1 Mev and hence will gradually build up in a U^{235} reactor, thereby contaminating the fuel. Only at the expense of further nuclear transformations, or by isotopic separation means, may this undesirable higher isotope be processed.

The basis for recent increased interest in plutonium reactor fuels consists of two general arguments. The first and more speculative argument reasons that the energy content of the estimated uranium reserves in the form of U^{235} will not appreciably extend the energy resources of the world beyond the limit of fossil fuels. Utilization of all the uranium (140 times the amount of U^{235}) by converting the U^{238} to plutonium and then burning the plutonium in reactors would, however, appreciably extend the world's fuel reserves. The second argument recognizes that, as uranium-fueled reactors are brought into operation, plutonium supplies will gradually be built up. Efficient and more economical utilization of nuclear fuels demands a use for this material in the long-range future.

TABLE 1

REACTOR DESIGN POTENTIALITIES OF Pu^{239} AND U^{235}

Neutron Energy	Reactor Design Function	Pu^{239}	U^{235}	
Thermal......	$\nu/(1+\alpha)$	2.03	2.08	Reactivity
	$1/(1+\alpha)$	0.704	0.844	Burning efficiency
	$(\nu-1-\alpha)/(1+\alpha)$	1.027	1.078	Breeding ratio
100 kev.......	$\nu/(1+\alpha)$	2.40	1.98	
	$1/(1+\alpha)$	0.833	0.806	
	$(\nu-1-\alpha)/(1+\alpha)$	1.40	0.988	
1 Mev........	$\nu/(1+\alpha)$	2.78	2.26	
	$1/(1+\alpha)$	0.937	0.893	
	$(\nu-1-\alpha)/(1+\alpha)$	1.806	1.262	

The first argument may be dramatically demonstrated by means of a simple graph (see Fig. 2). Estimates have been made as follows: (1) World power demands are now at a level of about 10^{17} Btu/yr and are increasing at a rate between 3 and 5 per cent per year (7–9). (2) "Economically recoverable" fossil fuel reserves are estimated to be equivalent to 4×10^{19} to 1×10^{20} Btu = 40–100 Q (7). (3) Although the present estimated "economically recoverable" reserves of uranium are equal to only about 0.15×10^6 tons in the United States, 0.2×10^6 tons in Canada, and 0.32×10^6 tons in South Africa (which are the world's major producers at present), world reserve estimates have been placed at a figure of 25×10^6 tons of uranium.

For purposes of discussion, let us speculate that fossil reserves are equivalent to 100 Q and place uranium reserves at a higher, 10^8 tons, figure. Although in 10^8 tons of uranium there is a potential energy of 7,200 Q (72 times fossil reserves), there are only 51 Q available in the form of U^{235}. At a 3 per cent annual increase in power demand, the fossil fuels would be depleted in about 150 years, and U^{235} reactors would be capable of meeting subsequent world power requirements for only an additional decade or two. Reactors which produced and consumed plutonium would significantly extend the fuel reserves by possibly several hundred years.

The second argument is premised by the statement that there is a need for nuclear power in the immediate future, if not now. Fuel reserves are not favorably located within the boundaries of some major industrial population centers and countries; economic nuclear power appears to be a panacea for the fuel-import problem. If nuclear power is not to be dependent on the somewhat expensive and limited isotope separation facilities, the reactor systems employed must rely on natural uranium or nuclear transformation processes as the source of new fuel. Both schemes imply a use for plutonium.

Natural uranium reactors must burn a relatively large proportion of the total fuel if nuclear fuel costs (neglecting fabrication costs) are to compete with fossil

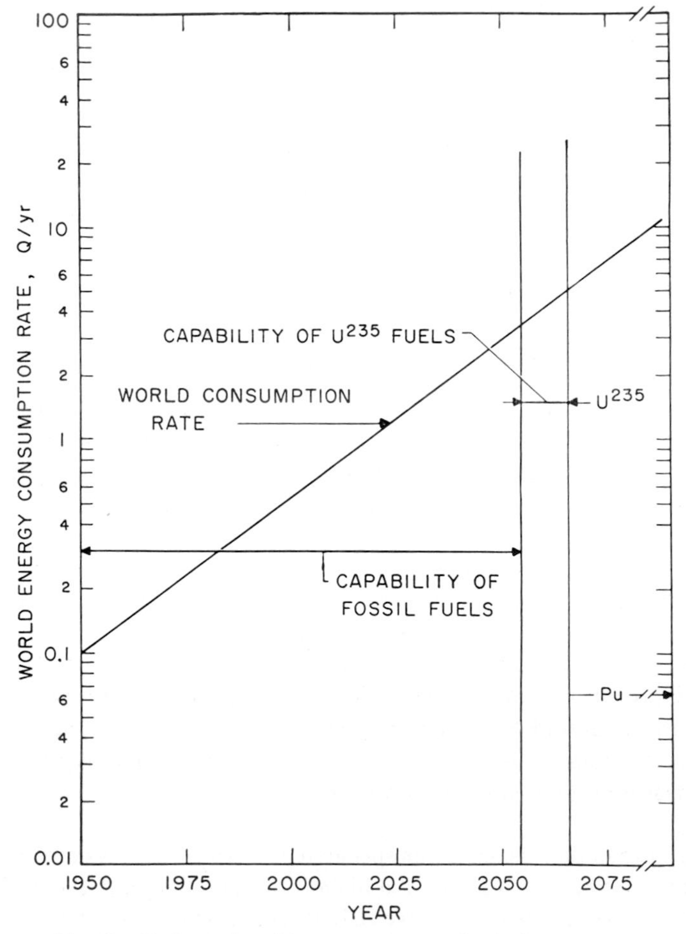

FIG. 2.—Estimated world energy consumption in future years

fuel costs. Irradiations of the order of 10,000 Mwd/ton are necessary to obtain fuel costs of the order of 1 mil/kw-hr at present-day uranium prices. Unfortunately, natural uranium reactor systems begin to lose significant amounts of the reactivity necessary to maintain a chain reaction when the fuel is irradiated for more than 3,000 Mwd/ton. Enrichment of the fuel with produced plutonium by means of the so-called plutonium recycle system will enable a greater fraction of the uranium to be consumed, hopefully lowering the cost per kilowatt-hour assigned to the nuclear fuel. Economic utilization of nuclear fuel demands an increase in plutonium technology.

From a composition standpoint, there are two classes of plutonium compounds and alloys pertinent to reactor design: those containing up to about 2 atomic per cent plutonium, which take on the metallurgical characteristics of the host elements, and those alloys containing 10–50 atomic per cent plutonium, in which the metallurgical properties are dominated, if not determined, by the plutonium content. The very dense plutonium alloys will probably not be of interest for reactors because power-density limitations will not allow efficient extraction of the produced heat.

What properties must these plutonium fuels possess? First, they must have good corrosion resistance to, or be capable of being protected from, attack by high-temperature coolants, such as water, "inert" gases, and liquid metals. Second, the physical properties of a fuel alloy must allow successful heat extraction of energy generated at rates greater than 500 w/gm of plutonium. The material must have adequate heat conductivity and high-temperature strength to survive thermal stress and shock and, throughout its operating temperature range, must be free from discontinuous geometrical changes, such as those induced by phase transformations. In addition to meeting these and other familiar high-temperature requirements, the material must be capable of withstanding radiation damage, a problem peculiar to nuclear reactors. Just how much irradiation burnup can be sustained will depend on the individual reactor design, but a desirable goal would be of the order of a few per cent of the total atoms contained in the fuel.

Instigated by the possibilities of breeding, an interest in compositions containing mixtures of plutonium and uranium will always be present. Reactor interest will center around mixtures containing between 5 and 10 uranium atoms per plutonium atom and mixtures containing 50 or more uranium atoms per plutonium atom. The choice of elements to be compounded with plutonium will also be dependent on the specific reactor design. A good rule to guide metallurgists would be to keep the macroscopic neutron-absorption cross-section of the diluent materials to less than 10 per cent of the Pu^{239} macroscopic fission cross-section, if the absorber is not U^{238}. Thermal cross-sections should be used for thermal reactor designs, and fast-neutron cross-sections compared for fast-neutron reactors.

A few combinations of plutonium and a diluent element have been considered, and a short review will indicate some of the possibilities. Plutonium-uranium metal combinations are, of course, obvious. Plutonium-aluminum alloys have been utilized but have the limitation of a low melting temperature. Solid solutions of UO_2 and PuO_2 have also been considered. A characteristic of oxide fuels is that physical and nuclear properties have been sacrificed in order to gain the advantages of high burnup that are associated with them. The British have proposed the use of plutonium carbide to extend their Calder Hall type of system.

Not only solid-state compositions are of interest; reactor proposals utilizing molten metal fuels, such as plutonium with bismuth, magnesium, iron, cobalt, or nickel, have been made, but their success depends primarily on a suitable container material capable of being fabricated and withstanding high temperatures and compatible not only with the fuel but also with the coolant.

At Los Alamos, consideration is being given to the use as a nuclear reactor fuel of a molten plutonium alloy containing about 10 atomic per cent iron or cobalt. These alloys melt at about 410°–420° C and have been contained in tantalum at temperatures as high as 650° C for thousands of hours. Extensive evaluation of the fuel and container materials under mock conditions of heat flux and irradiation will be carried out before an actual reactor is constructed. Such a reactor system, as contemplated by the Los Alamos Scientific Laboratory, will serve as a pilot plant yielding information and experience with the operation, neutron stability, handling and charging problems, and the design merits of a molten plutonium-fueled reactor. The reactor fuel will be quite dense and not directly suited to power-plant design, as heat-extraction limitations do not allow a high specific power for such dense fuels. Fortunately, ternary alloys containing plutonium can be made which have about the same melting point (410° C) and allow concentrations of plutonium to be varied from 0 to 90 atomic per cent; a wide range of core configurations is thus possible, and the power-density problem is somewhat alleviated. The Los Alamos Molten Plutonium Reactor Experiment will attempt to study the problems of a design consisting of a molten plutonium-alloy core, contained in tantalum, cooled by molten sodium, and operated at temperatures in the vicinity of 500°–600° C. A dense fuel will be used to limit the critical mass inventory of plutonium.

Although the nuclear and reactor-physics aspects of plutonium-fueled reactors are not completely understood, physicists do have enough information to make speculations and design comparisons on paper.They are not so sure, however, that they have suitable materials with which to construct their reactors. The progress of plutonium power-reactor technology now depends on the effort and ingenuity of metallurgists and engineers to devise suitable high-temperature radiation-resistant materials, fabrication technques, handling schemes, and, most importantly, processing methods. The metallurgist should realize that he is now the kingpin in the plutonium power-reactor field; the physicist is his assistant. In conclusion, I wish to emphasize that nuclear engineers can, and will, make considerable sacrifices in their guarded breeding ratios and electrical conversion efficiencies, if they can be assured of material reliability.

REFERENCES

1. Lane, J. A. "Where Reactor Development Stands Today," *Nucleonics*, **14**:30 (August, 1956).
2. Glasstone, S. S., *Principles of Nuclear Reactor Engineering*, p. 832. New York: D. Van Nostrand Co., Inc., 1955.

3. Kiehn, R. M. *LAMPRE: A Molten Plutonium Fueled Reactor Concept* (Los Alamos Scientific Laboratory Rept. LA-2112, 1957).
4. Grebe, J. J. "Why Is Dow-Detroit Edison Working on a Fast Breeder Reactor for Power?" *Nucleonics*, **12**:13 (February, 1954).
5. Hughes, D. J. *Neutron Cross Sections* (USAEC Rept. BNL-325, 1956).
6. Holmes, J. E. R., *et al.* "Experimental Studies on Fast Neutron Reactors at A.E.R.E.," *Proceedings of the International Conference on the Peaceful Uses of Atomic Energy*, **5**:331–41. New York: United Nations, 1956.
7. Putnam, P. C. *Energy in the Future.* New York: D. Van Nostrand Co., Inc., 1953.
8. United Nations. "World Energy Requirements in 1975 and 2000," *Proceedings of the International Conference on the Peaceful Uses of Atomic Energy*, **1**:3–33. New York: United Nations, 1956.
9. ———. "Contribution of Nuclear Energy to Future World Power Needs," *ibid.*, pp. 85–102.
10. *Twenty-first Semi-annual Report of the Atomic Energy Commission*, p. 7, 1957.
11. Hurst, D. G. "Experiments on Some Characteristics of the NRX Reactor. Part I. Methods and Prolonged Fuel Irradiation," *Proceedings of the International Conference on the Peaceful Uses of Atomic Energy*, **5**:111–18. New York: United Nations, 1956.

CHAPTER XXVIII

THE PREPARATION OF PLUTONIUM-ALUMINUM ALLOY FUEL ELEMENTS FOR THE NRX REACTOR

K. L. Wauchope

INTRODUCTION

Plutonium-aluminum alloys have been irradiated in the NRX reactor since 1952, to provide enrichment and, eventually, a source of plutonium rich in the higher isotopes. The fuel elements originally used were prepared by casting the alloy directly into 10-inch-long aluminum cans which were threaded at each end (1). A pure aluminum plug was cast on top of the alloy to form a seal, and the cans were decontaminated and removed from the glove-box line. The required number

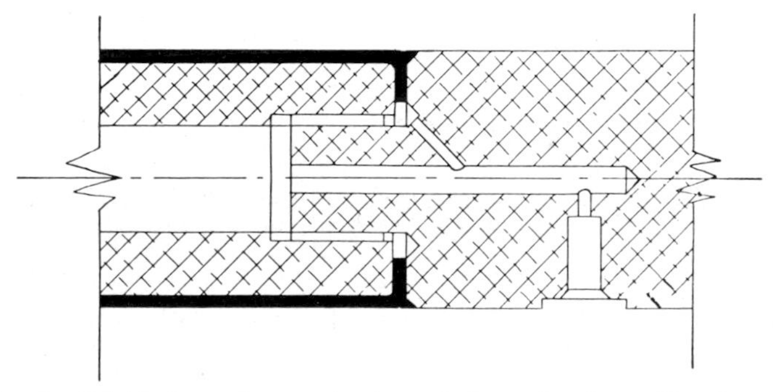

FIG. 1.—Joint between inner and outer end pieces, showing weld and vents

were then screwed together, a finned aluminum sheath was drawn over the assembly, and the ends were sealed by welding to aluminum end pieces.

Use of this method simplified the fabrication of the element but resulted in the presence of two unbonded interfaces between the alloy and the coolant. In 1955, one of the elements failed in the NRX reactor. The failure has been attributed to inadequate cooling following entry of water into the finned sheath and the buildup of a steam pressure sufficiently high to lift the sheath out of contact with the casting can (2). Subsequently, the alloy temperature rose, and eventually the molten alloy penetrated the casting can, the inner and outer sheaths, and the reactor vessel wall. As a result of this accident, the following changes were made in the fuel element design.

K. L. Wauchope is with the Atomic Energy of Canada, Limited, Chalk River, Ontario.

a) The aluminum-aluminum interface between the casting can and the finned sheath was removed. Machined alloy slugs are now loaded directly into the sheath.

b) The diameter of the fuel slug was increased to 1.360 inches, the same as standard NRX uranium rods.

c) A limit of 600,000 Btu/ft^2/hr was placed on the heat flux at the coolant-sheath interface, which imposes a limit of 0.92 gm/cm of fuel length, about 3.7 per cent by weight, on the plutonium concentration of the alloy.

The element in use at present consists of a stack of twelve plutonium-aluminum alloy slugs, 9.022 $\pm$ 0.005 inches long and 1.360 $\pm$ 0.002 inches in diameter, with aluminum end pieces encased in a finned aluminum sheath, 0.080 inch thick. The

FIG. 2.—Plutonium-aluminum alloy rod for NRX reactor

end pieces are drilled deeper than the threading to provide a reservoir for helium gas for leak testing, and the outer end fittings are vented, as shown in Figure 1. A complete element is shown in Figure 2.

Fabrication of such a rod involved the development of new techniques of casting, machining, assembly, and testing, since bare alloy slugs, rather than slugs contained in protective casting cans, were to be handled. The methods which have been developed are described below.

PREPARATION OF ALLOY SLUGS

PREPARATION OF ALLOY

The alloy is prepared by the method developed by Runnalls (3), i.e., plutonium received as nitric acid solution is precipitated as oxalate, the precipitate dried and calcined at 650° C to form oxide, and the oxide reduced with aluminum in the presence of cryolite and sufficient excess aluminum to produce an alloy of the required composition.

The first alloy ingots were prepared by adding a compressed mixture of plutonium dioxide and cryolite in small aluminum cans to molten aluminum at 1,100°–1,200° C. Later alloys have been prepared by adding a powdered mixture of plutonium dioxide and cryolite to a cold graphite crucible, followed by an aluminum slug and a covering layer of cryolite, then heating for 15 minutes at 1,200° C. The tedious, time-consuming steps of preparing compacts and adding them to molten metal at high temperature are thus avoided. A series of chemical analyses of dissolved slags, acid-leached crucibles, and alloy samples indicates that the plutonium yield in the preparation of 3.7 wt. per cent plutonium-aluminum alloy ingots containing approximately 30 gm of plutonium exceeds 99 per cent and is the same regardless of which method is used.

The precipitation and calcination steps are carried out in batches containing up to 200 gm of plutonium. A separate reduction is done for each alloy slug required. To produce enough 3.7 wt. per cent alloy for one slug, 28 gm of plutonium as oxide and 758 gm of high-purity aluminum are used. The reduction billets weigh about 780 gm and contain approximately 27.8 gm of plutonium. The remaining 200 mg are lost to the slag and crucible.

Melting and Casting

The reduction billets are induction-melted in graphite crucibles in air. No flux cover is used.

The molten alloy is heated to 850° C, stirred thoroughly, and poured into a heavy-walled (4 inches O.D. × 1.43 inches I.D. × 14 inches deep) graphite mold tilted about 30° from the vertical. The use of tilted molds avoids blow holes due to gases entrapped by splashing metal during the early stages of pouring. Rejections for this defect have now been reduced almost to zero.

A sample is poured into a small graphite mold immediately before the casting is poured. This sampling method yields a representative sample and has minimized analytical inaccuracies caused by segregation when samples are cut from the solid castings.

The castings are about 12 inches long and 1.41 inches in diameter, with about 2.5 inches of pipe. They weigh about 765 gm and contain approximately 27.2 gm of plutonium. Loss as oxide during melting is, therefore, about 2 per cent. Longitudinal segregation is responsible for a variation of about 5 per cent in the plutonium concentration from the top to the bottom of the casting. Some inverse segregation of plutonium at the surface of the castings has also been noted. The reasons for this type of segregation and methods of preventing it are now being studied.

Finishing

The cast slugs are machined to a diameter of 1.360 ± 0.002 inches. The resulting surface is free of visible porosity. The piped ends are then cut off, and the slugs finished to the length of 9.022 ± 0.005 inches, using a slitting saw set up in the chuck of the lathe and a special jig to hold the slug in position.

Initially, carbon tetrachloride was used as a coolant in the machining operations. However, after some difficulty due to reaction of carbon tetrachloride with small, hot, aluminum-alloy chips, its use was discontinued. Satisfactory machining techniques have now been developed by using a jet of compressed air as coolant and a lubricating coating of graphite on the slitting saw.

The finished slugs weigh 595.0 ± 1.0 gm and contain about 21.7 gm of plutonium. About 3.5 gm of plutonium are contained in the piped-end cap cut from the casting, while another 2 gm are in the machine cuttings.

The finished slugs are cleaned with carbon tetrachloride and removed from the glove box into sealed polythene envelopes. Each slug is subjected to radiographic examination before use, and any showing porosity or internal piping are remelted.

The piped ends are later remelted. Lathe cuttings and melting drosses are, however, chemically processed to recover their plutonium content.

Plutonium Recovery

The over-all plutonium recovery, from nitrate solution to finished slug, has averaged about 77 per cent. Approximately 13 per cent is contained in the end cap and is recovered by melting. Hence the plutonium recovery without chemical processing is about 90 per cent. Approximately 7 per cent contained in lathe cuttings and 2 per cent in melting drosses are recovered by chemical processing. The remaining 1 per cent, representing dusting losses and losses to the crucible and slag during reduction, is not economically recoverable.

ASSEMBLY OF FUEL ELEMENTS

Preparation of Sheaths

Finned aluminum NRX sheaths, 1.410 inches in inner diameter with walls 0.080 inch thick, are used. The sheaths are received from the manufacturer with one end drawn down for easy gripping in the draw-bench carriage. For use with plutonium alloys, this end is drawn farther down, flattened, and sealed by welding. The integrity of the sheath and end weld is checked by a helium-leak test before use.

Loading of Sheaths

The finished slugs, in polythene envelopes, and the inner end pieces are placed in a fume hood having a filtered exhaust. The open end of the sheath is introduced into the fume hood through a hole cut in one end, with the sheath supported on a long bench, as shown in Figure 3.

The bottom end piece is first placed in the sheath. The polythene envelopes are then cut open, and the twelve alloy slugs are inserted, followed by the top end piece. The inside of the open end of the sheath is decontaminated, and into it is inserted a one-hole rubber stopper containing a piece of copper tubing connected to a vacuum valve.

During these operations, continuous air samples are taken. To date, no evidence of contamination has been discovered. Contamination of the tools used in loading and of the operator's gloves has also been very slight, indicating almost complete absence of loose particles on the surface of the alloy slugs.

FIG. 3.—Loading alloy slugs into a finned sheath

VACUUM BAKING

The loaded sheath (fuel element) is removed from the fume hood to an oven, and the vacuum valve is connected to a roughing pump with filtered exhaust. The fuel element is then evacuated, heated to 160° C, held at this temperature for 4 hours, and allowed to cool to room temperature under vacuum. This operation insures the removal of absorbed water from the inner surfaces of the element.

Air samples taken during this operation have failed to show any contamination, nor has any been found on the exhaust filter of the vacuum pump.

DRAWING

The assembly is removed from the oven and returned to the fume hood, where the vacuum connection is removed and a felt plug inserted. The plug consists of three disks of $\frac{3}{8}$-inch-thick felt, slightly larger than the inner diameter of the sheath and held together by a $\frac{1}{2}$-inch bolt and nut. It acts as a wiper as the sheath elongates during drawing and prevents escape of contamination.

The assembly is then taken to a draw bench in an open shop and pulled through a die, which draws the sheath down into close contact with the slugs, i.e., to an inner diameter of 1.360 inches.

During drawing, it is necessary to exert pressure on the stack of slugs in order to prevent gaps from forming between them as they pass through the die. For this purpose an effort was made to use a hydraulic jack running on a track mounted behind the die. The jack was attached to the trailing end of the sheath, so as to follow during drawing. However, it was found that, with both ends of the sheath held rigidly and pressure applied to the stack of slugs, even very slight curvature of the fins, or misalignment of the fins with the die apertures, caused serious damage to the sheath. In three cases out of eight, damage was sufficient to necessitate desheathing and reassembly of the element.

While this difficulty could probably have been overcome by modification of the equipment, it was found to be simpler to have pressure exerted by a strong man against a stout stick, e.g., a broom handle, inserted into the end of the sheath and

FIG. 4.—The drawing operation

bearing against the felt plug mentioned above (Fig. 4). This method, although primitive, has been successful.

During drawing, the sheath elongates by about 18 inches, while the wall thickness remains essentially unchanged. Continuous air samples are taken during the operation, but neither these nor swipes taken on the equipment used have revealed any contamination.

During all operations conducted in open shops a plastic spray in pressure containers, which could be sprayed on exposed alloy surfaces in the event of a sheath failure, is kept at hand. To date, its use has not been necessary.

Finishing and Testing

After drawing, the element is returned to the fume hood and the ends of the sheath are cut off, leaving about $\frac{1}{2}$ inch of metal protruding beyond each end piece. The insides of the ends are decontaminated, and the element is removed to an open machine shop. The element is placed in a lathe, the ends of the sheath are faced off, leaving $\frac{3}{16}$ inch protruding beyond each end piece, and this metal is spun down over the end pieces. The outer end fittings are then screwed into place and welded (Fig. 1). Integrity of the welds is checked by taping a polythene sleeve securely over the element covering both welds, pressurizing the sleeve with helium to about $\frac{1}{2}$ pound above atmospheric pressure, and evacuating through the vents shown in Figure 1 and through a mass-spectrometer leak detector. The vents are then sealed by welding.

The element is subjected to a further leak test before being accepted for irradiation in the reactor. This test is carried out by suspending the element in ethylene glycol within a 4-inch-diameter Pyrex column, which is then evacuated. Any leaks are indicated by bubbles rising through the liquid.

SUMMARY

Plutonium-aluminum alloy fuel elements for the NRX reactor are prepared from cast and machined alloy slugs. The slugs, with the necessary aluminum end pieces, are inserted into finned aluminum sheaths which are drawn down into close contact with the alloy slugs. The ends are sealed by welding to aluminum outer fittings. The integrity of the sheaths and welds is checked by helium and ethylene glycol leak tests before the elements are accepted for irradiation in the reactor.

The alloy preparation, casting, and machining steps are carried out in glove boxes. It has proved possible to carry out all subsequent operations in fume hoods or open shops without spread of contamination, thus materially reducing the difficulty and cost of preparing plutonium-alloy fuel elements.

The methods described have been developed to a point where it is felt that they can, if necessary, serve for routine production of cylindrical, aluminum-sheathed, plutonium-aluminum alloy fuel elements in sizes of about $\frac{1}{4}$ inch in diameter or larger. With some modification, they could probably be adapted for use with Zircaloy sheathed elements.

REFERENCES

1. Runnalls, O. J. C., *et al.*, *The Preparation of an Aluminum-Plutonium Fuel Rod* (AECL Rept. CRC-503, May, 1952).
2. ———. AECL Rept. UK/C-4/114 (October, 1955).
3. ———. *The Preparation of Plutonium-Aluminum Alloys* (AECL-458, November, 1957).

CHAPTER XXIX

THE FABRICATION OF BILLETS CONTAINING PLUTONIUM FOR MTR FUEL ELEMENTS

R. E. Tate

INTRODUCTION

Two properties of plutonium make its incorporation into MTR fuel elements desirable. First, because plutonium is fissionable, it provides neutrons and energy for reactor operation. Plutonium fuel elements permit the study of reactor variables with this fuel. Since plutonium is produced in uranium-fueled reactors and since it is potentially more than 100 times more abundant than U^{235}, knowledge of its performance as a fuel is very important. Second, neutrons are captured by some of the plutonium nuclei, and these plutonium atoms in the fuel elements thereby become basic building blocks for the manufacture of transplutonic elements. The properties of these transplutonic elements are of interest to physicists. These properties are of considerable importance in long fuel cycles and high fuel burnups.

PROBLEMS WITH PLUTONIUM

Plutonium has other properties that make its handling more difficult than that of most materials. Among these problems are (1) a health hazard, (2) criticality considerations, and (3) fabrication problems stemming from the unusual crystallography of plutonium.

The commonest plutonium isotope, Pu^{239}, is primarily an alpha-emitter with a half-life of 24,300 years. The alpha particles are relatively easily stopped by an obstruction no thicker than a surgeon's rubber glove. The primary health hazard involved is the intake of plutonium into the body, where the heavy alpha particles cause extensive tissue damage. Plutonium is only slowly excreted from the body and is a "bone-seeker." The body-burden tolerance has been set at 0.5 μg, and it is from this figure that the maximum permissible air-borne concentration has been set at 7×10^{-5} μg/cubic meter or 8.8 disintegrations per minute per cubic meter of air. It may seem unbelievable that at a number of establishments kilogram lots of this metal are being handled daily without liberating into the atmosphere more than a fraction of a microgram of plutonium per day.

In order to control the spread of air-borne plutonium, the following features

R. E. Tate is at the University of California, Los Alamos Scientific Laboratory, Los Alamos, New Mexico.

should characterize the equipment and methods employed in the fabrication operation: (1) complete inclosure is recommended for most operations; (2) high air velocities are essential for open-front hood operations; (3) surfaces absolutely free from loose contaminated particles will minimize the air-borne particle hazard; (4) skilled operators who can cope with contamination problems as they arise are highly desirable; and (5) low oxidation rates of some alloys may permit less stringent protective measures. In the fabrication of clad fuel-element billets, which is the subject of this chapter, all these considerations have an important bearing on the production process.

The possibility of inadvertently assembling a critical mass is probably greater with plutonium than with other fissionable metals because the critical mass is smaller. Each operation presents a different problem, and such factors as geometry, the presence of diluents such as aluminum, and the presence of moderators like water-cooling jackets are important considerations. Because of the criticality considerations, processes involving materials with high plutonium contents can never be large-batch operations. A competent nuclear physicist experienced in this field should review the procedures contemplated for handling plutonium.

Some of the properties of plutonium have been reported by Smith (1) and by Jette (2) in this country; by Lord (3) and by Ball and co-workers (4) in England; and by Konobeevsky (5) in the U.S.S.R. A comprehensive review of the physical metallurgy of plutonium has been prepared by Coffinberry and Waldron (6). Even if plutonium were not radioactive, it would still be unattractive for fabrication from the standpoint of its physical properties. Bruce Old (7) summed up the problem rather appropriately when he stated in an article on "New Developments in Metallurgy": "We shall ignore plutonium other than to say that its properties rhyme with its chemical designation."

Plutonium has six allotropic forms, more than has any other element. Each of these six phases exists over a comparatively short temperature range, since plutonium melts at 640° C. The thermal-expansion behavior of plutonium is discussed in chapters xii, xiii, and xiv. The three abrupt expansions and two contractions (accompanying the phase changes) pose a number of problems for the metallurgist who would fabricate this metal to close dimensional tolerances and who would clad it in a jacket that would not rupture if the assembly were to be cycled over a temperature range. Although there are other reasons, this expansion behavior alone would seem to preclude the use of unalloyed plutonium in any power reactor.

The room-temperature form of plutonium is monoclinic. It has the properties that this structure of low symmetry suggests. If one raises the temperature high enough to operate in the range where a more favorable structure for fabrication exists, he is confronted with rapid oxidation rates, which lead to loss of material and to the extreme hazard of producing fine plutonium-containing dusts.

PLUTONIUM-ALUMINUM ALLOY

One way to circumvent some of these difficulties is to alloy plutonium with another metal, and it is apparent that aluminum might be a good choice of an alloying element because it is available in excellent purity, it is relatively cheap, it is fabricated easily, and, perhaps most important, it has a neutron cross-section acceptable to reactor designers. Among its disadvantages is, of course, the fact that it does not have a high melting temperature. It was with these considerations in mind that the aluminum-plutonium binary diagram was included among the first few alloy systems selected for study when enough metal and time became available at Los Alamos for phase-diagram investigations of plutonium. The aluminum-rich portion of the diagram is shown in Figure 1. This region is very similar to the corresponding region of the aluminum-uranium system. Pertinent values for the two systems are given in Table 1. It is apparent that the compounds MAl_3 and MAl_4, the eutectic

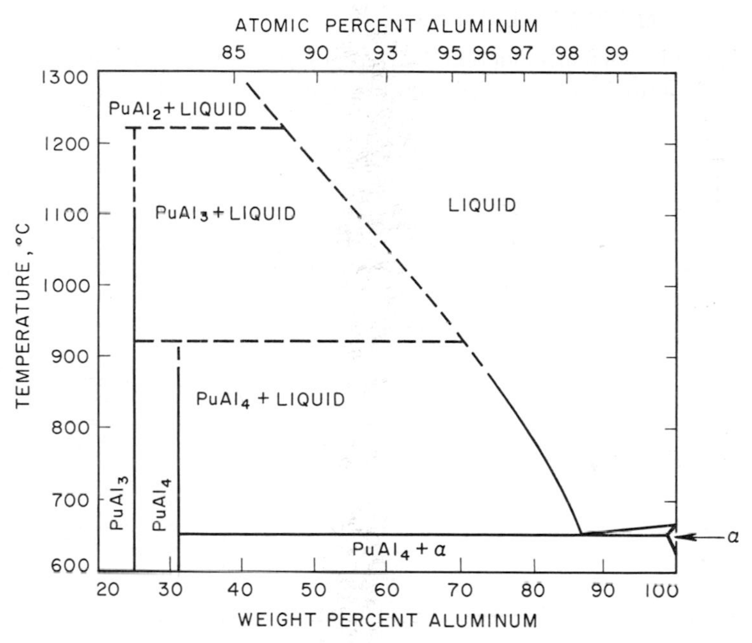

FIG. 1.—Aluminum-rich end of the plutonium-aluminum phase diagram

TABLE 1

COMPARISON OF VALUES FOR THE SYSTEMS URANIUM-ALUMINUM AND PLUTONIUM-ALUMINUM

	Uranium	Plutonium
Eutectic composition (w/o M)	13.0	13.3
Eutectic temperature (° C)	640	650±5
Density Al (gm/cm^3)	2.70	2.70
MAl_4	UAl_4	$PuAl_4$
Peritectic decomposition temperature of MAl_4 (° C)	730	920±5
Density MAl_4 (gm/cm^3)	6.06	6.02
MAl_3	UAl_3	$PuAl_3$
Peritectic decomposition temperature of MAl_3 (° C)	1,350	1,220±10

compositions, and the densities of the eutectic components are all very nearly the same in the two systems. Larger differences exist with respect to the temperature horizontals. In the plutonium system, the eutectic temperature is about 10° higher, the peritectic decomposition temperature of MAl_4 is 190° higher, but the peritectic decomposition temperature of MAl_3 is 130° lower than in the uranium system. Uranium is sometimes used as a metallurgical stand-in for plutonium when it is desired to try out a new process or technique, but one has to be cautious in basing the behavior of plutonium on results obtained with uranium because frequently the differences in behavior of the two metals are large enough to be very significant.

DESCRIPTION OF FUEL ELEMENT

Following this brief introduction to some of the limitations imposed by the properties of plutonium, the fabrication of fuel-element billets containing plu-

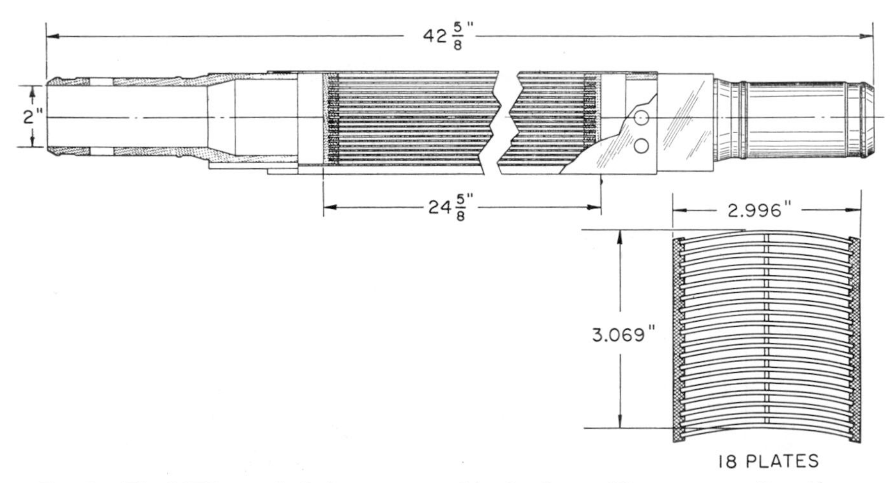

FIG. 2.—The MTR-type fuel element as used in the Omega-West reactor at Los Alamos

tonium can be discussed with a better appreciation of the problems involved. Some time ago the Los Alamos Scientific Laboratory was asked to prepare plutonium-containing fuel plates for MTR-type (Materials Testing Reactor) fuel elements. This type of fuel element was developed at Oak Ridge National Laboratory and was described at the first International Conference on the Peaceful Uses of Atomic Energy by Cunningham and Boyle (8). Since the construction of the Materials Testing Reactor, a number of research reactors in the United States have employed this aluminum-plate fuel element. It is a successful and popular fuel element. Figure 2 shows the fuel element as used in the Omega West Reactor at Los Alamos. The plates are like a sandwich, with the "meat" being a uranium-aluminum alloy and the "bread" being commercially pure aluminum. Using plutonium in these plates does not alter the fundamental fabrication steps employed in making the element, but some differences in the casting technique are necessary, and the techniques for handling the bare alloy are considerably more exacting.

A diagram of the billet assembly supplied by Oak Ridge National Laboratory is shown in Figure 3. Because experience in fabricating the aluminum alloy containing 10 w/o plutonium had already been obtained at Los Alamos, it was decided to use this composition for the alloy cores. The plutonium-alloy core was prepared at Los Alamos by melting, casting, rolling, and blanking. Each billet was assembled, welded, and sealed off under high vacuum before shipment to Oak Ridge. These billets were then rolled to fuel plates and assembled as fuel elements in the usual manner at Oak Ridge National Laboratory.

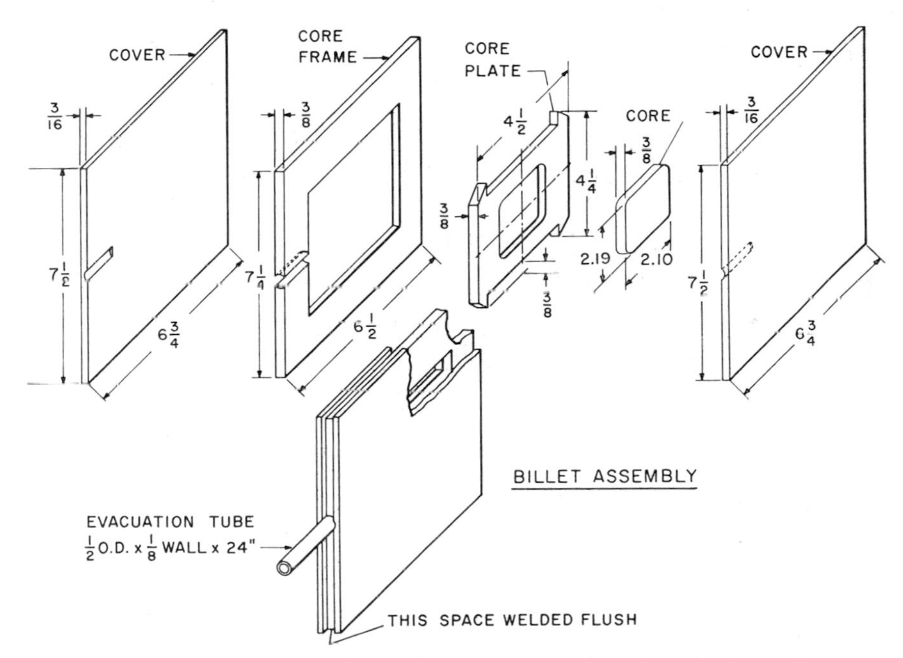

FIG. 3.—Exploded view of the fuel billets containing plutonium-aluminum alloy

ALLOYING AND CASTING

Plutonium reacts readily with air at elevated temperatures, and for this reason it is usually melted under a good vacuum. A schematic view of the furnace used is shown in Figure 4. This furnace was copied from a design developed at the Argonne National Laboratory. By pumping from both above and below the molten metal, very good vacuums for casting are obtained. Melting and alloying are carried out in a type V-2 magnesia crucible supplied by the Norton Company. A charge weighing nominally 650 gm is prepared from high-purity aluminum rods $\frac{1}{4}$ inch in diameter and a short length of freshly machined plutonium rod. Using freshly machined plutonium minimizes the spread of contamination and avoids introducing oxide impurities into the melt. Prior to being loaded into the crucible, the aluminum used in alloying is pickled in an acid bath composed of 60 parts water, 6 parts 70 per cent nitric acid, and 3 parts 50 per cent hydrofluoric acid. Scrap from the blanking operation is also used as feed material, and it is degreased with carbon tetrachloride and cleaned with a rotary wire brush in a glove box prior to being charged. The melt is bottom-poured, a hollow magnesia stopper rod being lifted by a pneumatic control operated from outside the inclosure in which the furnace is located. A rectangular ingot having the dimensions $\frac{3}{4} \times 2\frac{1}{2} \times 6$ inches and weighing 550–600 gm is formed in the mold.

The production of aluminum-alloy castings containing 10 w/o plutonium poses three principal metallurgical problems: (1) dissolution of the plutonium and formation of a homogeneous melt, (2) freezing the ingot in such a manner that gravity segregation does not occur, and (3) freezing the ingot in such a manner that

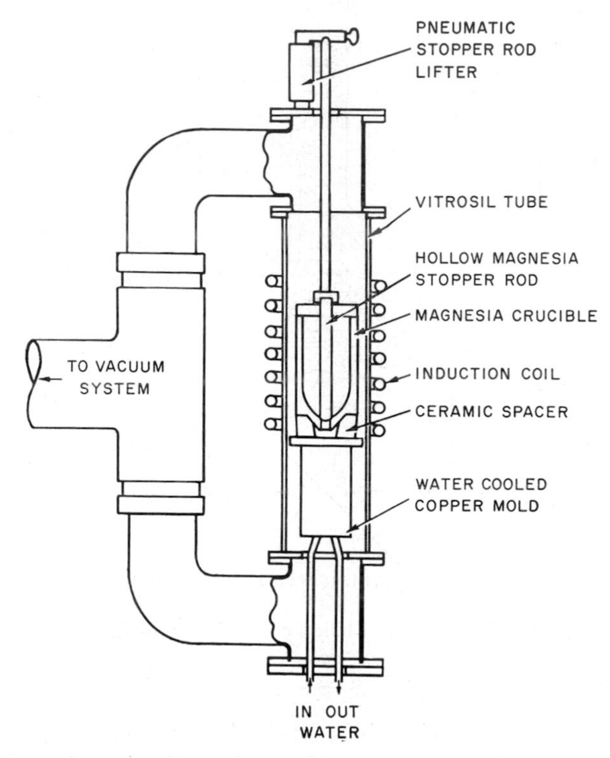

Fig. 4.—Schematic diagram of the vacuum chill-casting furnace

solidification shrinks and similar casting defects are minimized. During preliminary experiments to test the equipment, some aluminum ingots containing 10 w/o uranium were prepared and, after holding the melt for 30 minutes at 750° C, were found to be homogeneous. However, for a plutonium alloy of equivalent composition, 120 minutes at 940° C was barely sufficient to produce a homogeneous ingot with the same equipment.

The furnace is powered by a 20-kw vacuum-tube generator operating at a frequency of 450 kilocycles. Very little electromagnetic stirring of the melt occurs at this frequency, and it seems likely that the use of low-frequency induction heating would speed the formation of a homogeneous melt. At 940° C magnesium vapor is produced by reduction of the crucible by the melt. This vapor causes a glow discharge in the furnace chamber, and after 60 minutes the discharges are so frequent and violent as to make continuation of the run less than prudent. To obtain the required length of time at temperature for a homogeneous ingot, each virgin ingot is cut up and recast after the melt is again held at 940° C for 60 minutes.

As shown in Table 1, the densities of aluminum and $PuAl_4$ are 2.70 and 6.02 gm/cm³, respectively. If an ingot is allowed to solidify slowly, considerable gravity

FIG. 5.—The vacuum chill-casting equipment

segregation occurs, and much of the plutonium is found in the bottom of the ingot. By using a water-cooled copper mold and by controlling the rate of pour, both the gravity segregation problem and the solidification shrinkage problem are minimized. The pour rate is controlled by the size of the pour hole, by the hydrostatic head of the melt, and by the viscosity of the melt, which is in turn controlled by the temperature. For the particular set of conditions that exist in the furnace, a pour-hole diameter of $\frac{3}{32}$ inch and a pouring temperature of 685° C, as indicated by a chromel-alumel thermocouple at the bottom of the hollow stopper rod, have yielded the optimum results. Induction heating of a tantalum cone surrounding the conical crucible bottom prevents premature clogging of the pour hole due to freezing during the several minutes that are required for the melt to pour. Superheating may take place in some of the last metal to pour, and this may be the cause of some solidification shrinkage that occurs in varying degrees near the top of the ingot. Ideally, one wishes to add both metal and heat to the ingot at a rate commensurate with the rate of heat removal from the lower part of the ingot and in this manner to maintain only a relatively thin molten layer in the mold cavity at all times during the pour.

FIG. 6.—The vacuum chill-casting furnace

Two views of the furnace are shown in Figures 5 and 6. The furnace is located in an inclosure, and when the inclosure door is opened for loading the furnace, a strong draft of air is drawn inward through the opening. It is important that the inside of the inclosure be kept scrupulously free from loose contaminated material. During the loading and unloading operations the operator is protected by a respirator, coveralls, cap, and obstetrician's long rubber gloves covering surgeon's short rubber gloves worn next to the hands. Because of constant vigilance and attention to detail, no contamination has been detected outside the furnace inclosure during the production of more than 150 castings.

INSPECTION OF CASTINGS

Two inspection procedures are performed on each ingot before the metallurgical processing is carried further. Each ingot is sealed into a heavy-gauge polyvinyl chloride plastic bag, and a radiograph is made. The plastic bag permits the use of standard radiographic equipment. If the radiograph does not reveal the presence of gross segregation in the form of discrete acicular particles of $PuAl_4$, the sprue of the ingot is sawed off, and a 10-gm sample is cut from both the top and the bottom of the ingot. Chemical analysis of these samples is employed to determine whether the composition of the ingot is within the tolerated limits of 9.5–10.5 weight per cent plutonium.

ROLLING

Satisfactory ingots are rolled to the desired thickness, and reject ingots are sawed up and remelted. Figure 7 shows a train of glove boxes completely inclosing

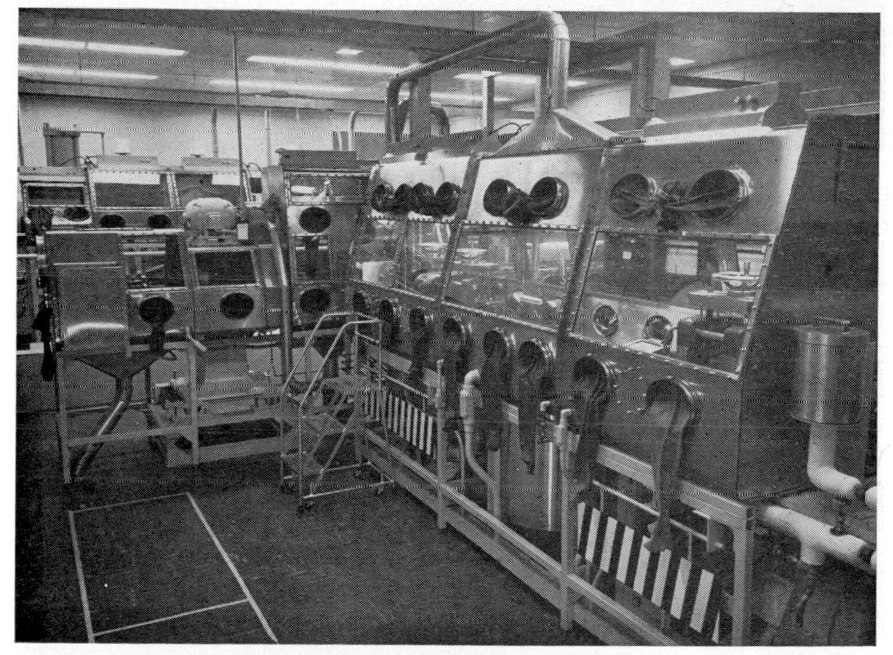

FIG. 7.—The equipment for rolling and swaging plutonium alloys

two rolling mills, one a hot mill for breakdown (near the right side of the picture) and the other a cold mill for finishing (just to the right of the portable stepladder). The aluminum alloy is quite resistant to oxidation and can be heated and rolled in the air atmosphere of the glove box. The ingots are hot-rolled from an air furnace operating at 550° C, and the hollow rolls of the mill are heated to 250° C by circulating a hot fluid through them.

The rolling operation is much like any laboratory rolling experiment except that

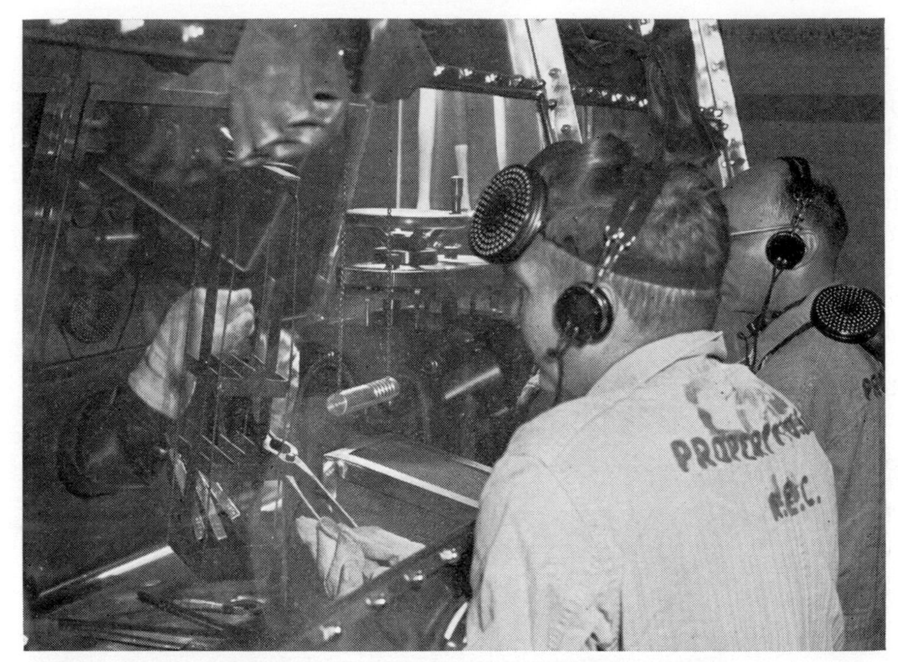

FIG. 8.—The handling of hot plutonium-alloy ingots in a dry box

FIG. 9.—The hot-rolling of plutonium-alloy ingots

it is done in an inclosure with manual manipulation through glove ports. Figures 8 and 9 give some idea of the frustrations encountered when working in an inclosure, especially if the operator is trying to hurry so that the ingots do not cool too much. Each operator wears four pairs of gloves, put on in the following sequence: (1) mortician's gloves to absorb the perspiration, (2) surgeon's gloves to avoid contaminating the hands if a glove-box gauntlet develops a pinhole leak, (3) glove-box gauntlets which allow the hands to work inside the inclosure, and (4) asbestos gloves to minimize the possibility of burning a hole in the glove-box gauntlets. It is obvious that an appreciable loss of dexterity accompanies this arrangement. Throat microphones and headsets are used for communication between members of the three-man team, who are located on different sides of the inclosure. Adequate communication is very important while handling hot metal in a glove box.

Ingots are hot-rolled from a thickness of 0.75 inch to approximately 0.45 inch, with reductions of about 50 mils per pass, and are then cold-rolled to a finished thickness of 0.370 inch, with reductions of 5–10 mils per pass. Between the rolling operations the ingots are cleaned in a glove box by wire-brushing, and they are straightened if necessary with a small inclosed press. The cold-rolling operation is similar to the hot-rolling, but the mortician's gloves and asbestos gloves are not used, and the pace is more relaxed. After rolling, the ingots are again placed in heavy plastic bags and are radiographed, in order to locate any flaws that have not been completely eliminated by the rolling.

BLANKING

Cores measuring 2.10 by 2.19 inches are blanked from the rolled stock, using a conventional punch and die in a standard die set. Cores are blanked only from areas shown by radiography to be flawless. Available equipment does not include a punch press, so that this operation is done on a hydraulic press, as shown in Figure 10. A special inclosure was built of plastic and aluminum to safeguard against any possible health hazard in this operation, and the inclosure moves up and down with the press stroke. Note the plastic bag through which items are removed from the inclosure. By a sealing and cutting-off technique similar to that described by Kelman *et al.* (9), items are removed through a hole in the inclosure without the hole ever being open to the laboratory atmosphere.

QUALITY CONTROL

Scrap from the blanking operation is returned to the casting line for remelting; the cores are forwarded to another glove box for degreasing in carbon tetrachloride. Non-flammable solvents are used in glove boxes in order to preclude the formation of explosive atmospheres. The cores are then weighed to a tenth of a milligram on the balance shown in Figure 11. At this stage the density of the core is determined by weighing it in distilled water and in air. The large difference between the density of aluminum and that of $PuAl_4$, which gave trouble in casting, is now turned to advantage. Using a density value of 6.02 for $PuAl_4$ (10) and the analysis of Aronin

FIG. 10.—Die set and inclosure mounted in a hydraulic press

FIG. 11.—A completely inclosed analytical balance used for weighing plutonium alloys

and Klein (11), the following relationship between x, the weight fraction of plutonium, and ρ, the density, has been derived:

$$x = 1.24924 - 3.37296\,\frac{1}{\rho}.$$

Thus a density measurement of the rolled core permits a reliable and non-destructive determination of the plutonium content of each core to be made. This operation completes the preparation of the core.

BILLET ASSEMBLY

The billets shown in Figure 3 are prepared at a rate of four each day. This rate has been determined by the capacity of the furnace in which they are subsequently baked out and by the rate at which one welder working half-time can turn out consistently good results. The aluminum evacuation tubes are pickled in an acid bath composed of 60 parts water, 6 parts 70 per cent nitric acid, and 3 parts 50 per cent hydrofluoric acid to insure a clean bore that will seal vacuum-tight when the tubes are subsequently pinched off. All interior surfaces of the core plates, core frames, and cover plates are cleaned by brushing with a rotary stainless-steel wire brush. Handling of these parts is done with clean cotton gloves, in order to keep all surfaces clean and to insure that the surfaces will roll-bond without blister formation when the billets are finally rolled to a 60-mil thickness. While these parts are being prepared, the plutonium-alloy cores are pickled in a nitric–hydrofluoric acid bath of the composition previously mentioned, rinsed in boiling water, and air-dried. This procedure insures cleanliness for subsequent roll-bonding and also permits the assembly of the billets in an open-front hood without fear of spreading contamination because of loose material on the surface of the cores.

The billets are assembled in an open-front chemical fume hood. Stages of this operation are shown in Figures 12 and 13. The core is pressed into the mating core plate by using the press shown in the figures, and, while being extremely careful to remember which gloved hand is contaminated, an operator moves the filled core plate into the remainder of the billet assembly, as shown in Figure 12. Removing his contaminated gloves, the operator dons a new pair, and clamps the assembly together, as shown in Figure 13. After being tested for contamination with an alpha-counter, the assembled billet is ready to be welded. Exterior surfaces free of contamination are attained largely because the alloy core is free from loose contamination, because the operators are extremely careful to think before each movement of the hand, and because air flows through the hood at a high velocity.

WELDING

Welding the periphery of the billet assembly to form a vacuum-tight container is accomplished by the inert-arc process, using commercially pure aluminum filler rod. As shown in Figure 14, the assembly is preheated with a gas torch, so that the temperature of the billet is approximately 300° C during the welding (Fig. 15). Note the air sampler beside the welding operator and the fact that he wears a respirator at all times while in the room.

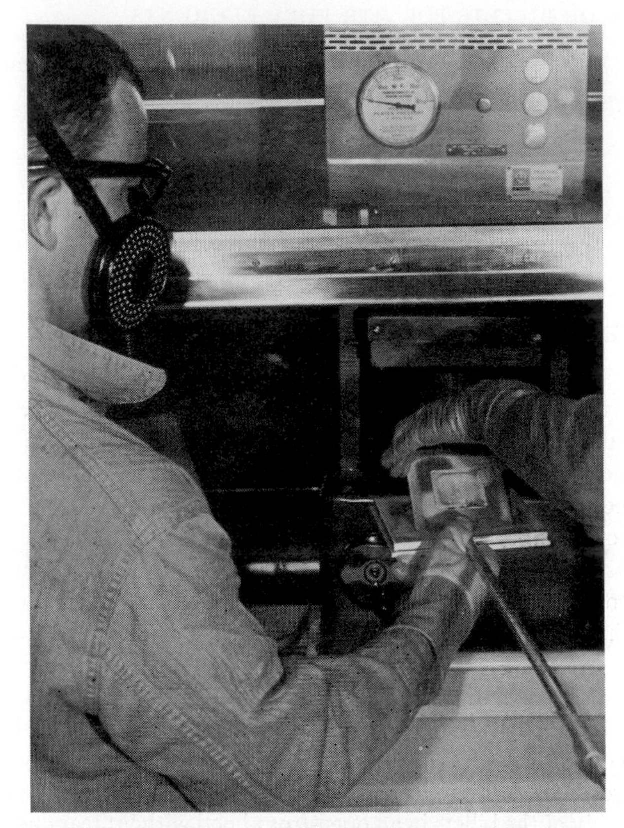

Fig. 12.—Assembly of the billet containing plutonium alloy

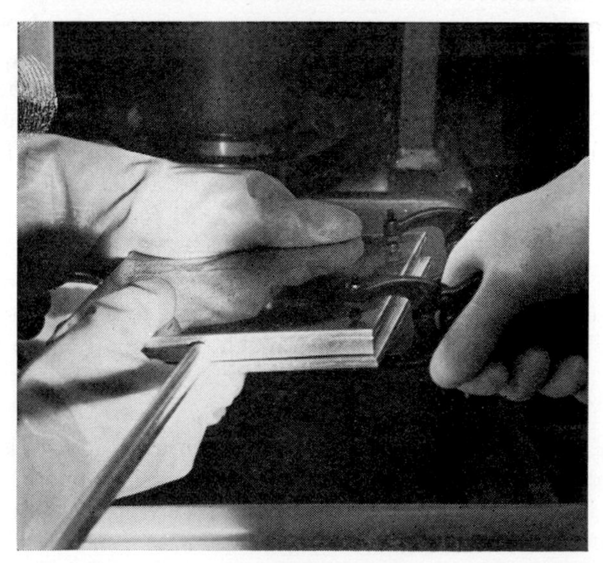

Fig. 13.—The plutonium-containing fuel billet clamped for welding

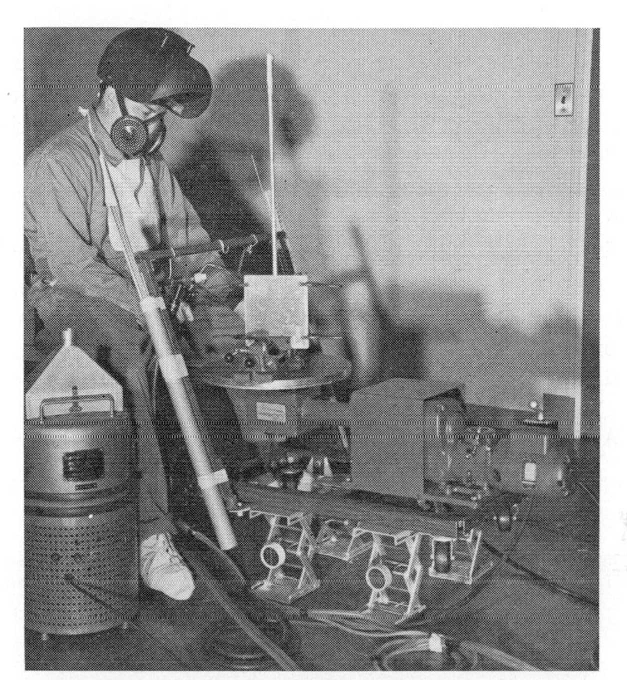

FIG. 14.—The fuel billet being preheated prior to welding

FIG. 15.—The plutonium-containing fuel billet being welded

The performing of preheating and welding steps without the use of inclosures, as shown in Figures 14 and 15, may appear from the health-hazard standpoint to be contrary to good practice in handling plutonium. To be sure, if the cores were a plutonium alloy that oxided readily, welding the billets in this manner would be quite hazardous. However, the oxidation rate of the aluminum-plutonium alloy is extremely low, 3×10^{-4} inch per week at 400° C in pure oxygen (12), and no air-borne oxide is formed. Before the operations shown were attempted, a careful study was made of the contamination problem involved. These operations have been performed more than 200 times, but the air sampler shown in the photographs has never detected a degree of air contamination that even approached the tolerance limit.

BILLET EVACUATION AND SEALING

Immediately after welding, the billet assemblies are transferred to the vacuum manifold shown in Figure 16. Here they are evacuated overnight. Then a plastic

FIG. 16.—The equipment used for leak-checking fuel billets

bag is placed over each billet in turn and the bag inflated with helium. During this time, all the gas that is pumped out of the billet assemblies is conducted to the helium-sensitive leak detector. This is an extremely sensitive test for verifying the tightness of the welds. Only absolutely tight billets are acceptable. The rejection rate was less than 1 per cent on first tests, and no billet has had to be rewelded more than once.

The next operation is to place the billets in the furnace shown in Figure 17 and to evacuate them at the temperature at which they will be rolled. The evacuation is continued overnight. The furnace is then brought up to 540° C and held at that temperature for 4 hours, after which the tubes are pressure-welded shut in two places by crimping them to a thickness of 0.095 inch with the modified bolt cutter, as shown in Figure 17. The tube is heated by a gas torch to 300° C as measured by a temperature-sensitive crayon just prior to crimping, and it is crimped at this temperature. A water-cooled block which protects the aluminum-to-neoprene vacuum joint is not shown in Figure 17. The billets are sealed off at a pressure of less than

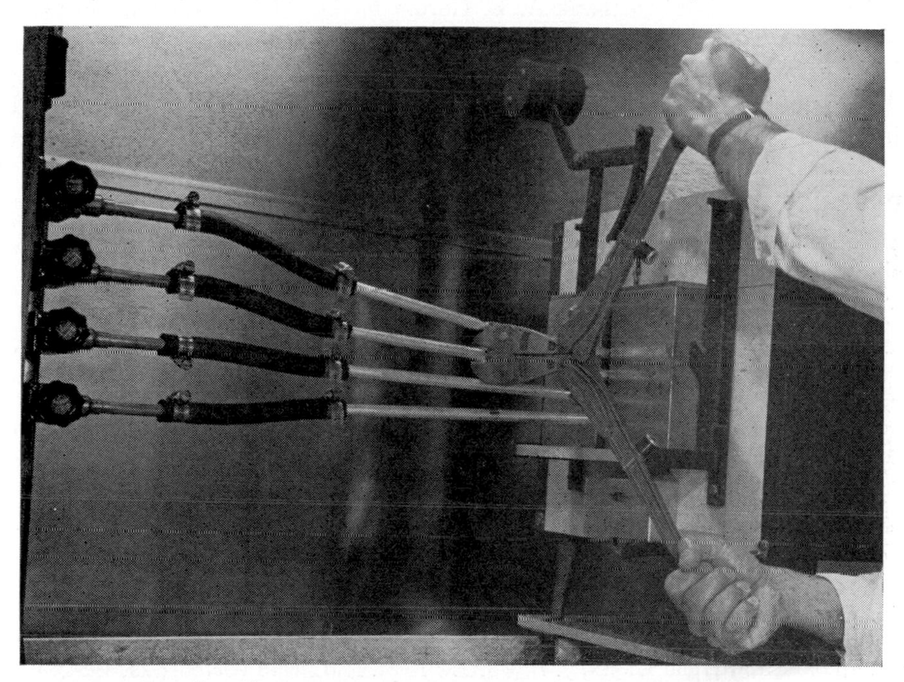

FIG. 17.—The fuel billets being sealed off after evacuation at 540° C

$\frac{1}{2}$ μ of mercury as indicated by a thermocouple gauge on the vacuum manifold. Subsequent rolling of the billets has shown that very few blisters are formed as a result of entrapped gas in the evacuated billet.

The final operation before the billets are shipped to Oak Ridge National Laboratory to be fabricated into fuel plates and fuel elements is to flatten the open end of the tube beyond the crimp and to weld the flattened end shut as a third barrier against gas entering the billet via the tube.

SUMMARY

A moderately large number of billets for the manufacture of MTR fuel-element plates has been produced from the aluminum alloy containing 10 weight per cent plutonium. By common metallurgical techniques—i.e., induction melting, chill-casting, hot-rolling, cold-rolling, and blanking—a straightforward procedure has been developed for fabrication of the plutonium-alloy cores. Cores are press-fitted into mating aluminum core plates, and a vacuum-tight aluminum jacket is assembled around the core and core plate by means of inert-arc welding. To circumvent the health hazards originating from the nature of plutonium, special attention has been paid to inclosing most of the processing equipment and to techniques for minimizing the air-borne radioactivity hazard. Throughout the entire processing, health monitoring instruments have indicated radioactivity values well below tolerance levels.

The author wishes to thank A. E. Gorum for invaluable assistance in many phases of the work and especially for the development of the casting technique; D. E. Grimm for the radiographic work; A. Zerwekh for the chemical analyses; and A. R. Romero for the welding. The work of J. H. Erwin and R. J. Beaver at ORNL in establishing the billet geometry and technique is also gratefully acknowledged.

REFERENCES

1. SMITH, C. S. *Metal Progress*, **65**:81–89, May, 1954.
2. JETTE, E. R. *Journal of Chemical Physics*, **23**:365–68, 1955.
3. LORD, W. B. H. *Nature*, **173**:534, 1954.
4. BALL, J. G., *et al. Nature*, **173**:534, 1954.
5. KONOBEEVSKY, S. T. *Session of the Division of Chemical Sciences (of the U.S.S.R. Academy of Sciences), Conference on the Peaceful Uses of Atomic Energy, Moscow, 1955*, pp. 362–75.
6. COFFINBERRY, A. S., and WALDRON, M. B. *Progress in Nuclear Energy*, Ser. V, Vol. **1**, pp. 354–410. London: Pergamon Press, Ltd., 1956.
7. OLD, B. S. *Metal Progress*, **68**:105–11, August, 1955.
8. CUNNINGHAM, J E., and BOYLE, E. J. *Proceedings of the International Conference on the Peaceful Uses of Atomic Energy*, **9**:203–7. New York: United Nations, 1956;
9. KELMAN, L. R., WILKINSON, W. D., SHUCK, A. B., and GOERTZ, R. C. *Nucleonics*, **14**:77–82, May, 1956.
10. COFFINBERRY, A. S., and ELLINGER, F. H. *Proceedings of the International Conference on the Peaceful Uses of Atomic Energy*, **9**:138–46. New York: United Nations, 1956.
11. ARONIN, L. R., and KLEIN, J. L. USAEC Rept. NMI-1118 (October 29, 1954).
12. WABER, J. T. Private communication.

CHAPTER XXX

THE DEVELOPMENT OF PLUTONIUM-CONTAINING FUELS AT THE ARGONNE NATIONAL LABORATORY

L. R. Kelman and R. J. Dunworth

PLUTONIUM AS A FAST-REACTOR FUEL

Though Pu^{239} is the favored fuel in fast-neutron power-breeder reactors, the unattractive properties of the pure metal make it extremely unlikely that unalloyed plutonium will find use as a solid fuel. It melts at a relatively low temperature (640° C) and undergoes five solid-state transformations on heating from room temperature to its melting point. The unusually large volume changes associated with these transformations (some are negative), the large thermal-expansion coefficients of the allotropes (some also negative), along with the anisotropic behavior of some of these allotropes, result in distortion of plutonium on heating and cooling. Heat-transfer considerations also preclude the use of solid plutonium (or any other fissionable fuel) in highly concentrated form in a practical fast power reactor. The resulting high power-densities would require large cooling areas and thin sections not readily achieved by normal methods of fabrication. A more practical solution is to subdivide or dilute the fissionable material by alloying. A fertile isotope, such as U^{238}, can be used as the diluent, in which case the reactor is an "internal breeder." An inert diluent, such as aluminum, can also be used, in which case all the breeding is accomplished in a surrounding blanket of a fertile isotope.

The degree of dilution of plutonium necessary to make it useful as a fast power-breeder fuel results in its being the minor element, so that the undesirable properties of unalloyed plutonium are no longer of concern. Instead, the metallurgist is concerned about the alloying effect of plutonium on the properties of the diluent. However, the plutonium that is present is sufficient to make it necessary to use the special handling methods that have been and are being developed for working safely with high-intensity alpha-emitters. Methods of dealing with this health-hazard problem were reviewed recently at a symposium held at Harwell and have been published (1). However, recent developments in pyrometallurgical reprocessing of fast reactor-fuel elements necessitate remote handling because short cooling-times and incomplete fission-product removal are involved. The resulting high gamma activity, coupled with the alpha activity, dictates the need for remote handling, whether the fuel is plutonium or U^{235}.

L. R. Kelman and R. J. Dunworth are at the Argonne National Laboratory, Argonne, Illinois.

The first of ANL's fast-neutron power-breeder reactors—Experimental Breeder Reactor No. 1 (EBR-I)—was designed as an experimental facility to further the understanding of the physics of such systems and to study the engineering feasibility of a NaK-cooled reactor. It was described at the 1955 Geneva Conference by Lichtenberger *et al.* (2). The second of ANL's fast reactors—EBR-II—is a much larger engineering facility designed to obtain the information that will permit scaling up to a central power station of competitive size. It also was described at the Geneva Conference by Barnes *et al.* (3) and more recently by Koch and coworkers (4).

In spite of the advantages of Pu^{239}, the fuel that has been used in EBR-I and that will be used for the first loading of EBR-II is U^{235}. This can be attributed in the earlier case to availability and more recently to the relatively greater know-how in uranium fuel-element metallurgy. EBR-II is, however, designed to receive plutonium fuel at the earliest possible loading, and plutonium is the logical fuel for a full-size power reactor of this type. A promising fuel element based on plutonium is in process of development and appears to satisfy EBR-II requirements. However, it is desirable to test this fuel element in a fast-neutron flux simulating the eventual operating conditions of EBR-II. Because no such facility now exists, the U^{235}-fueled EBR-II will serve as the final proving grounds for the fuel element based on plutonium.

The phase diagrams of the plutonium-alloy systems have been studied at Los Alamos, at Harwell, and in Moscow (see chaps. xxi, xxii, xxiii, and xxiv). These studies have shown that three elements—uranium, zirconium, and thorium—can dissolve plutonium in appreciable quantities in the solid state. Uranium is the solute of interest at ANL and will be discussed in greater detail later. The low-plutonium binary zirconium-plutonium and thorium-plutonium alloys in the composition range of reactor-fuel interest are relatively soft and ductile and are fabricated with ease. For example, the addition of 5 w/o plutonium does not change the ductility of zirconium crystal bar—we have achieved 97.5 per cent reduction in cold-rolling this alloy. ANL studies of the thorium-plutonium alloys show that they are stable to thermal cycling 200 times between room temperature and 500° C. Irradiation-damage studies on these alloys are in progress at ANL.

FUEL-ALLOY REQUIREMENTS

Reactor Requirements

The alloys that have been developed for the EBR-II fuel elements result from a combination of reactor physics, reactor-engineering, reprocessing, and metallurgical requirements. U^{238} is used to dilute the fissionable material. Not only does this help the heat-transfer problem, but it also results in considerable internal breeding. The enriched uranium fuel core that will be used at first, pending final testing of a plutonium fuel element, will consist of a total of 363 kg of 5 w/o fissium-uranium alloy containing 170 kg of U^{235}, or 49 per cent enrichment of the uranium (4). The plutonium loading requires somewhat less fissionable material but, because of the metallurgical limitations that will be discussed next, an alloy of uranium–20 w/o plutonium–10 w/o fissium is now being considered with an estimated 15–20 w/o U^{235} added (5). A full-scale power-breeder reactor would use a lower concentration of fissionable isotope because of the much larger core size; an alloy containing less than 20 w/o Pu^{239} and no U^{235} would be appropriate (5). ("Fissium" means the mixture of metallic elements found in fission products, among which molybdenum and ruthenium are prominently represented.)

Refining Requirements

The fission-product elements not removed by the metallurgical refining process determine to a considerable degree the fuel alloy composition that will be used in EBR-II. The anticipated equilibrium alloy resulting from uranium fuel loading in EBR-II will lie in the composition range shown in Table 1. The fission-yield

TABLE 1

Range of Fissium Composition for U^{235} Loading

Element	w/o	Element	w/o	Element	w/o
Zirconium.....	0.1–0.1	Technetium ..	0.5–1.0	Rhodium.....	0.2–0.5
Molybdenum..	1.6–3.4	Ruthenium...	1.2–2.6	Palladium....	0.1–0.3

TABLE 2

Typical Reference Fissium Alloys

Element	U^{235} Fissium (w/o)	Pu^{239} Fissium (w/o)	Element	U^{235} Fissium (w/o)	Pu^{239} Fissium (w/o)
Uranium.........	95.0	69.2	Ruthenium......	1.5	4.3
Plutonium.......	....	20.0	Rhodium........	0.3	0.7
Zirconium........	0.2	0.5	Palladium.......	0.5	2.5
Molybdenum.....	2.5	2.8			

curve for Pu^{239} is, however, slightly different from that for U^{235} because of the extra mass units. Consequently, this changes the relative concentrations of the fission-product elements not removed in the refining process. It is intended that an artificial equilibrium fissium alloy be used to start the reactor, in order to avoid the changing alloy properties that would result if the composition changed on successive recycles.

Since an infinite number of such alloys are possible, a few reference alloys have been made the standards for test purposes. Typical reference alloys for which test data are available are shown in Table 2 (4). The first of these, a uranium–5 w/o fissium alloy, is the one that will be used in the first U^{235} loading. For the plutonium loading, which contains about equal amounts of Pu^{239} and U^{235}, the alloy composition will be somewhere between the two compositions shown in Table 2. In this chapter uranium-fissium alloys are discussed and compared with uranium-plutonium and uranium-plutonium-fissium alloys.

METALLURGICAL REQUIREMENTS

The main metallurgical requirement of the fuel pin in ANL's fast power-breeder reactors is that it be dimensionally stable in the reactor. The alloys must, of course, have metallurgical properties that permit fabrication and handling, sometimes under unusually restrictive conditions, as will be shown later. Also, for safety reasons, it is desirable that the thermal expansion of the alloys be positive and reliable. The thermal conductivity is of great interest because the central metal temperature of the fuel is dependent on it. In addition, the fuel-element alloys must resist attack by NaK or sodium.

Dimensional stability under irradiation is of major concern to the fuel-element metallurgist, since relatively small changes can have an important bearing on the nuclear behavior of the reactor, its efficiency as a power producer, and its safe operation. In order to increase the reactor lifetime of a fuel before it must be reprocessed, we are aiming at a degree of dimensional stability that has not as yet been demonstrated in a working reactor. In EBR-II we have set our sights on at least 2 per cent burnup of the total atoms before we must reprocess the fuel. As an aid to understanding some of the data that are given later in this paper, we shall review the methods used at ANL for most irradiation stability studies (6) and the terminology used in reporting such data.

Small specimens (approximately $\frac{1}{8}$ to $\frac{3}{16}$ inch in diameter by 1 inch long) were irradiated in the high thermal flux of the Materials Testing Reactor in lieu of a fast-reactor test facility. The EBR-I and EBR-II fuel-element designs consist of fuel pins in a loose jacket, so that they are free from physical restraint. Also, a thermal bond of molten NaK or sodium fills the annulus between the core and the jacket. The capsule design for irradiation studies (6) simulates these conditions, as shown in Figure 1. The preferred capsule material is arc-melted crystal-bar hafnium-free zirconium because it provides the best combination of desired properties. The degree of burnup for relatively low exposures is calculated by counting the Co^{60} gamma activity in cobalt-containing, thermal flux monitors that accompany each capsule. For higher burnups, Cs^{137} determinations are made. Changes in the mass abundance of the uranium isotopes have also been used to obtain total atom burnups. Since many of these studies were essentially screening tests wherein many cylindrical specimens were used to study numerous variables, conventional heat-transfer calculations were used to determine temperature, and no external heaters were used. It must be recognized that these calculations are based on conditions at the beginning of the irradiation test, so that the accuracy is greatly influenced by changes in conditions as the test progresses. In more recent capsule tests some variables are being studied more closely. Therefore, thermocouples imbedded in the specimens are being used to supplement the calculated temperatures, and external heating is being used to achieve and control desired temperatures. In addition to these capsule irradiation studies, full-length fuel elements are being irradiated at controlled elevated temperatures. Some of the results of these studies are included in this chapter.

Lengthwise growth of specimens is expressed as microinches per inch per fission per 10^6 total atoms. For small length changes the percentage length change divided by the per cent of atoms burned up is approximately equal to the average instantaneous growth, G_i. However, for larger length changes the changing length of the specimen cannot be disregarded, since the amount of growth at any level of the burnup is dependent on the instantaneous length of the specimen at that time. Thus,

$$G_i = \frac{1}{L}\frac{dL}{dN} = \frac{\ln L/L_0}{\text{fraction of total atoms fissioned}} = \frac{\text{microinch/inch}}{\text{fissions}/10^6\ \text{total atoms}},$$

where L_0 = initial length, L = final length, and N = unit of atom burnup.

Another property of importance is that the fuel be dimensionally stable to the numerous small thermal fluctuations during operation of the reactor, as well as to

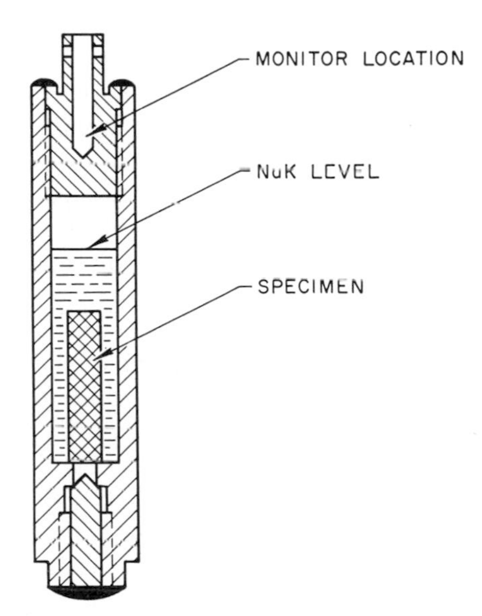

FIG. 1.—The assembled irradiation capsule and specimen

the relatively smaller number of large thermal cycles due to startup and shutdown of the reactor (7). The thermal-cycling growth rate, G_t, is expressed as microinches per inch per thermal cycle and is an average instantaneous growth analogous to the irradiation growth. Thus

$$G_t = \frac{1}{L}\frac{dL}{dN} = \frac{\ln L/L_0}{N} = \text{microinches per inch per thermal cycle},$$

where L_0 = original length, L = final length, and N = number of thermal cycles.

FUEL-ALLOY STUDIES

URANIUM-PLUTONIUM BASE ALLOYS

The concentration of the uranium diluent in the EBR-II fuel will be several times the concentration of plutonium. Therefore, the metallurgical properties can be expected to resemble more closely those of uranium than those of plutonium. It is desirable to examine the binary uranium-plutonium alloys, since they are basic to the EBR-II fuel development. The phase diagram (chap. xxii) was developed by Ellinger and co-workers at Los Alamos, and several features have been confirmed at ANL. The fact that plutonium lowers the melting point, as well as the transformation temperatures, of uranium is pertinent to this development. The amount of lowering of the solidus by 30 w/o plutonium is sufficient to limit severely the maximum fuel temperature in a reactor like EBR-II. Alpha uranium can dissolve up to 15 w/o plutonium at 560° C, with the solubility decreasing to about 10 or 11 w/o at room temperature. Beta uranium can dissolve a maximum of about 18 w/o plutonium at 600° C. Increasing the plutonium concentration beyond the solubility limit in alpha uranium introduces the zeta phase. Alloys containing even small amounts of this compound have been found, in early investigations, to have extremely poor metallurgical properties—the castings contain microcracks that make them so fragile that they can hardly be removed from the mold without cracking, and they tend to fall apart on thermal cycling; the alloys are pyrophoric; and, worst of all, contrary to early test indications, the low-plutonium binary alloys proved to be extremely susceptible to irradiation damage. Recent work at ANL suggests that the poor properties exhibited by the binary uranium-plutonium alloys may not be inherent characteristics but may result from the method of preparation of the alloys.

Fortunately, the major fission-product elements not removed during the refining process—molybdenum and ruthenium—are desirable alloying elements, in that they improve the irradiation stability of uranium and especially of the uranium-plutonium alloys. Molybdenum has long been known to be soluble in uranium and to lower the temperature range of stability of gamma uranium, which has a body-centered cubic structure. Work at ANL indicates that molybdenum acts similarly in the uranium-plutonium alloys and also suppresses formation of the undesirable zeta phase. Recently Dwight (8) has shown that ruthenium is soluble in uranium and has the same tendency as molybdenum to stabilize gamma uranium. Zirconium, which is fortunately a minor fissium element, has been shown by Los Alamos investigators to have a deleterious effect, in that, when present in minor amount (about 5 at. per cent), it stabilizes the undesirable zeta phase.

Irradiation Stability

In EBR-II the degree of radiation damage that can be accommodated will determine the reactor lifetime of the fuel and therefore the frequency of reprocessing. We desire to achieve a minimum of 2 per cent burnup of all atoms. Unlike many other types of reactors, corrosion of the fuel by the coolant is not a limitation, and the fast reactor is insensitive to fission-product poisons.

The amount of volume swelling and linear dimensional changes that can be tolerated in the EBR-II fuel-element design is dependent on factors that will be resolved only when the full-size fuel assemblies are exposed to EBR-II operating conditions. One of the attractive features of the design is that the loose fit of the fuel pin in the jacket will accommodate some dimensional changes. Also, Smith and co-workers (9) have shown that restraint by the jacket can be expected to inhibit some of these changes if filling of the sodium annulus does occur.

Weber has estimated that a volume expansion of about 3.4 per cent must be expected for 1 per cent burnup of all the atoms, due to lattice distortion by the newly formed fission-product elements (10). Thus, for the 2 per cent burnup that is planned for EBR-II, almost half of the 6-mil annulus will be involved, regardless of how well the fuel is stabilized against dimensional changes from other causes. It is expected that this much expansion can be tolerated with ease. Irradiation at moderate temperatures of the uranium-plutonium-fissium alloys of interest to EBR-II resulted in very low volume expansion (see Table 3). The uranium-fissium alloys showed some tendency to expand when irradiated (see Table 4), but they are considerably more stable than most existing uranium-base fuel alloys. It remains to be seen how these alloys will react to the higher temperatures and burnups of EBR-II operating conditions.

TABLE 3

EFFECTS OF IRRADIATION AT MODERATE TEMPERATURES* ON CAST URANIUM-PLUTONIUM-FISSIUM ALLOYS

Alloy (w/o)	Range of Burnup (a/o)	G_i† Av.	V_i‡ Av.	No. of Specimens
70U–20Pu–10.8Fs	0.81–0.92	1.8	2.5	2
75U–20Pu–5.4Fs	.33– .42	1.8	4.2	3
75U–20Pu–5Mo	0.15–0.43	3.2	5.8	7

* The maximum irradiation temperature is estimated not to have exceeded 350° C.
† G_i = Per cent length change/a/o burnup.
‡ V_i = Per cent volume change/a/o burnup.

The uranium-plutonium-fissium and related alloys that we have studied show practically no irradiation-damage effects at moderate temperatures (11). Three cast uranium–20 w/o plutonium base alloys were studied, including 5 w/o molybdenum, 5.4 w/o fissium, and 10.8 w/o fissium (the Pu^{239} fissium composition was used, for which see Table 2). The test conditions and dimensional changes are summarized in Table 3. The growth and volume changes were so small for the three sets of specimens as to make the differences between them meaningless. Typical of the results is the uranium–20 w/o plutonium–10.8 w/o fissium alloy pin shown in Figure 2, which was irradiated to 0.9 per cent burnup of all the atoms. The original smooth machined surfaces were preserved, and, except for slight volume increases, which can be attributed to the new elements formed by fission, the dimensions were remarkably unchanged. The uranium–20 w/o plutonium–5 w/o molybdenum alloy pins, which also showed little effect, were not irradiated to as high a burnup as the 10.8 w/o fissium alloy pins.

Irradiation of uranium-plutonium-fissium alloys at temperatures above 450° C showed severe damage from swelling (12), as shown in Table 4. Density decreases in all specimens were of the order of 40 per cent. The specimens are shown in Figure 3.

TABLE 4

EFFECTS OF IRRADIATION AT HIGH TEMPERATURES ON CAST U–20 W/O PU–10 W/O FS ALLOYS

Spec. No.	Est. Burnup (a/o)	Est. Temp. (° C)		Length Change (Per Cent)	Diameter Change (Per Cent)	Weight Change (Gm)	Density Decrease (Per Cent)	% Den. Dec./a/o Burnup	Remarks
		Surface	Center						
EA-1...	2.4	410	500	....	20.1–25.7	−1.3777	39.41	16.4	Porous material on upper end of specimen exploded before measurements and photographs could be made
EA-2...	2.2	620	730	− 2.64	18.5–55.6	....	....	....	Specimen crumbled to powder before weight and density measurements could be made
EA-3...	1.8	370	460	....	6.8–16.6	−1.8854	36.18	20.1	Specimen was found in two pieces; the smaller piece, which was distorted from swelling, ignited and burned before measurements and photographs could be made
EA-4...	1.6	500	600	17.9–22.8	9.58–22.8	0.1046	39.05	24.4	Specimen was found in two pieces
EA-5...	2.2	470	570	22.1	20.5–27.6	0.0616	41.30	18.8	Upper end of specimen distorted from swelling
EA-6...	2.8	560	670	14.8	13.5–46.8	0.1027	38.68	13.8	Upper end of specimen distorted from having grown into restricted free space in top of irradiation capsule
EA-7...	3.5	620	730	....	98.6	....	....	....	Specimen had swelled to inside diameter of irradiation capsule and could not be removed

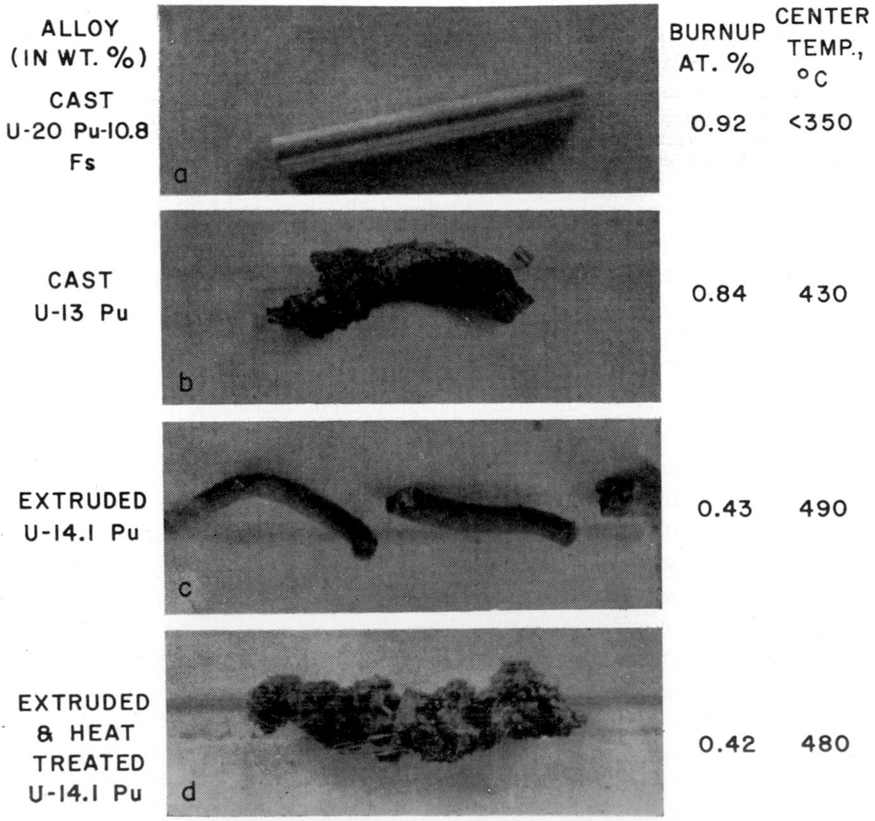

FIG. 2.—Typical irradiation results obtained with uranium-plutonium base alloys

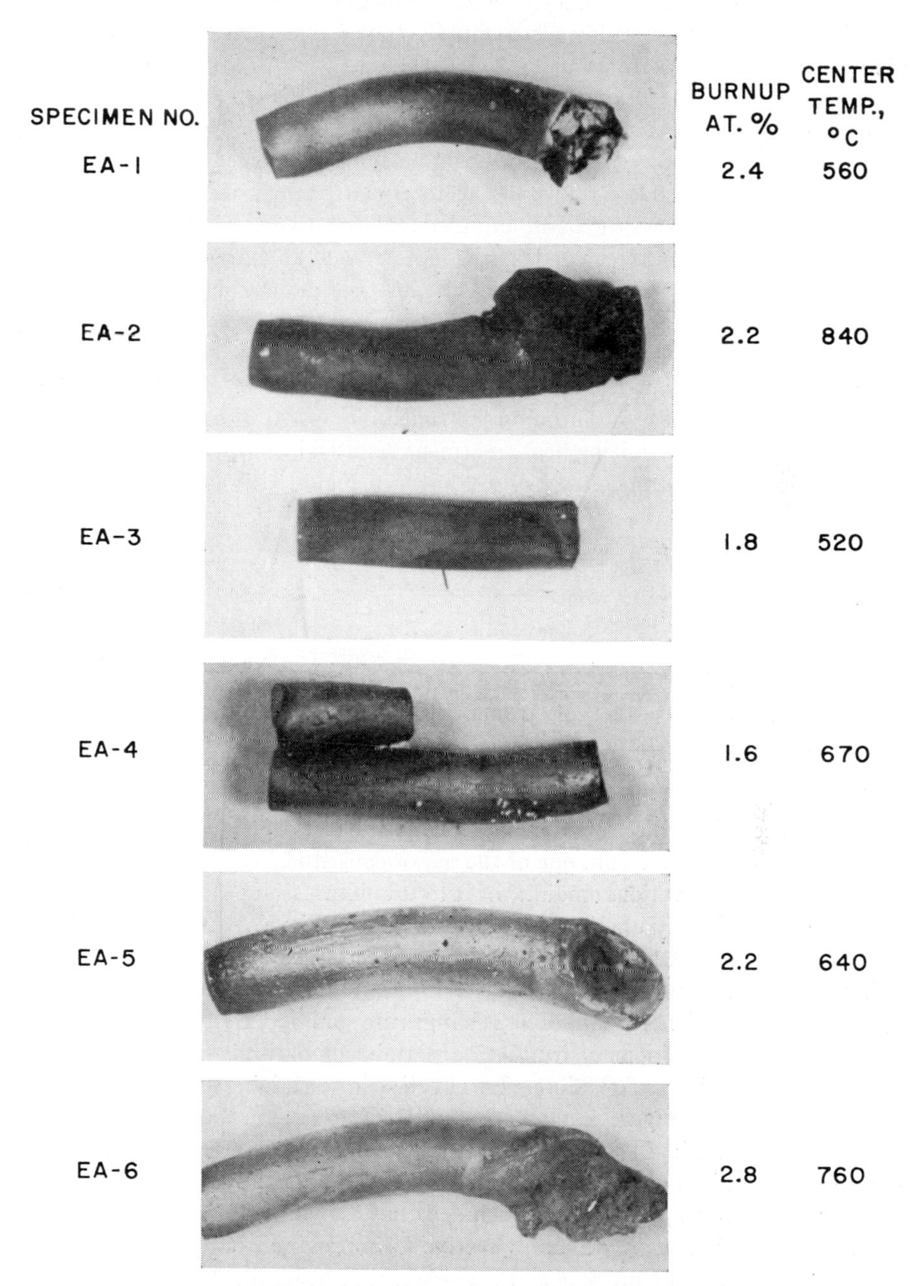

FIG. 3.—The effects of irradiation at high temperatures on cast uranium–20 wt. per cent plutonium–10 wt. per cent fissium alloys.

By X-ray diffraction, dilatometric, and metallographic studies, we have shown that the 10.8 w/o fissium alloy when rapidly cooled is essentially body-centered cubic gamma uranium with an a_0 = 3.42 A. The 5.4 w/o fissium alloy shows evidence of an appreciable amount of a second phase.

Compared with the uranium-plutonium-fissium alloys, we found the binary uranium-plutonium alloys to be more unstable to irradiation (13). Castings containing 3.7, 6.6, and 13.0 w/o plutonium were irradiated to as high as 0.84 per cent burnup. Figure 2, *b*, shows the badly bumped and distorted results that were typical of this group. The surface condition is indicative of an extremely coarse grain structure and made meaningless the measurements of dimensional changes.

Binary uranium-plutonium alloys containing 9.5 and 14.1 w/o plutonium developed fiber textures on extrusion, as evidenced by dilatometric studies and also by elongation when irradiated. Figure 2, *c*, shows the effect of irradiating to 0.4 burnup a uranium–14.1 w/o plutonium extruded pin at 490° C central metal temperature. A 645° C randomizing heat treatment resulted in excessive grain coarsening, which led to the surface roughening shown in Figure 2, *d*.

TABLE 5

COMPOSITIONS OF U-FS-TYPE ALLOYS THAT WERE IRRADIATED

Composition (w/o)	No. of Specimens	Composition (w/o)	No. of Specimens	Composition (w/o)	No. of Specimens
U–2.5Mo.........	16	U–5Fs..........	24	U–5Fs–7.5Mo....	15
U–3.17Fs.........	27	U–5Fs–2.5Mo....	18		

The alloy to be used first in EBR-II, uranium–5 w/o fissium, and some related alloys have been the subject of numerous irradiation tests by Smith (14) and Kittel (15). The role of molybdenum, one of the major constituents of fissium alloys, was examined by adding various amounts of it to uranium-fissium alloy and by including in the tests a binary uranium-molybdenum alloy. The alloys studied are listed in Table 5. Both cast and wrought specimens were included, and the effects of various heat treatments were studied.

Results of dimensional and hardness changes are summarized in Table 6 for tests in which similar conditions of irradiation permit comparison. Both the 2.5 w/o molybdenum and the 5 w/o fissium were effective in restraining length and volume changes, as shown in Figure 4, *a*, whereas the 3.2 w/o fissium was not so effective. However, all these alloys had somewhat rougher surfaces as the result of irradiation than did the uranium-plutonium-fissium alloys, even though they were not irradiated to as high burnup. It can be concluded that alloying to stabilize the gamma structure of uranium is, in general, effective in improving irradiation stability.

From Table 6 and Figure 4, *a–e*, it is evident that water quenching from 850° C is not a desirable heat treatment, whereas furnace cooling does no harm to castings and effectively disorients the wrought structures. Work done at ANL (14) and at Battelle Memorial Institute (16) helps to explain some of the results of these heat treatments, but clarifying irradiation tests are still needed. As can be seen in Table

6, water quenching of the cast and the wrought uranium–5 w/o fissium alloy from 850° C results in a relatively soft structure, presumably retained gamma uranium, whereas neither 3.2 w/o fissium nor 2.5 w/o molybdenum had this effect. The soft gamma structure was not retained in the uranium–5 w/o fissium alloy by furnace cooling from 850° C. Irradiation did not appreciably change the hardness of the as-cast or the furnace-cooled specimens but did harden the water-quenched uranium–5 w/o fissium specimens. The BMI studies (16) of the heat-treating and transformation characteristics of the uranium–5 w/o fissium composition substantially verify and expand these findings by dilatometric, metallographic, X-ray diffraction, and hardness evidence. Their results indicate a transformation at about 625° C, and they retained a single-phase gamma-uranium structure by water

TABLE 6

EFFECTS OF IRRADIATION ON URANIUM-FISSIUM ALLOYS

Alloy	Fabrication and Heat Treat.*	Range of Burnup (a/o)	G_i† Av.	V_i‡ Av.	Hardness (R_A)		No. of Specimens
					Before Irradiation	After Irradiation	
U-3.2Fs.......	As cast	0.33–0.39	11.1	11.4	72	76	3
U-5Fs........	As cast	.32– .64	1.8	7.7	66	64	4
U-2.5Mo......	As cast	.40– .53	2.0	3.9	73	74	8
U-3.2Fs.......	Cast (W.Q.)	.19– .21	24.4	18.4	66	75	3
U-3.2Fs.......	Cast (F.C.)	.19– .20	3.1	17.8	66	73	3
U-5Fs........	Cast (W.Q.)	.19– .21	8.1	28.9	46	73	3
U-5Fs........	Cast (F.C.)	.19– .21	4.7	13.3	72	73	3
U-3.2Fs.......	Wrought (W.Q.)	.32– .69	53.8	27.5	67	69	4
U-3.2Fs.......	Wrought (F.C.)	.33– .56	1.2	8.2	67	74	3
U-5Fs........	Wrought (W.Q.)	.50– .67	0.6	11.0	45	66	3
U-5Fs........	Wrought (F.C.)	.25– .56	2.7	6.7	72	69	3
U-2.5Mo......	Wrought (W.Q.)	.42– .54	32.7	39.5	66	70	4
U-2.5Mo......	Wrought (F.C.)	0.42–0.56	3.4	4.1	64	71	4

* All heat treatments were for 1 hour at 850° C, followed by water quenching (W.Q.) or furnace cooling (F.C.). The maximum irradiation temperature is estimated not to have exceeded 400° C.

† G_i = Per cent length change/a/o burnup.

‡ V_i = Per cent volume change/a/o burnup.

quenching from 800° C. Subsequent heat treatment of this metastable structure at the temperatures that were attained in the above-described irradiation tests results in substantial precipitation hardening. Overaging just below the transformation temperature softens the alloy.

Irradiation studies have also been made on full-size EBR-II fuel elements of uranium–5 w/o fissium alloy by Sowa and co-workers (17) with very encouraging results. In one test the upper half of the fuel pin was irradiated in the MTR to at least 1.5 per cent burnup at a maximum central metal temperature of about 425° C. This resulted in a 2.8 per cent increase in length (G_i = approximately 2), with only slight surface roughening and a density decrease of about 3 per cent.

In another series of tests by Sowa and co-workers (17) on three full-size fuel elements in CP-5, irradiations were carried to as high as 0.5 a/o burnup and a central metal temperature of about 400° C. The fuel pins elongated less than 2 per cent (G_i = less than 3), with some surface roughening and a density decrease of 1–2.6 per cent. Special attention was given in these tests to measuring the volume of fission products released from the fuel core. The ratio of the fission products in the bond to that in the fuel was $10^{-3}:10^{-5}$, and the fission gases that accumulated above the sodium were too small in amount to detect.

Thermal-Cycling Stability

Thermal-cycling tests in and out of the reactor indicate that the EBR-II fuel alloys have extremely good thermal stability. We have cycled them to reactor

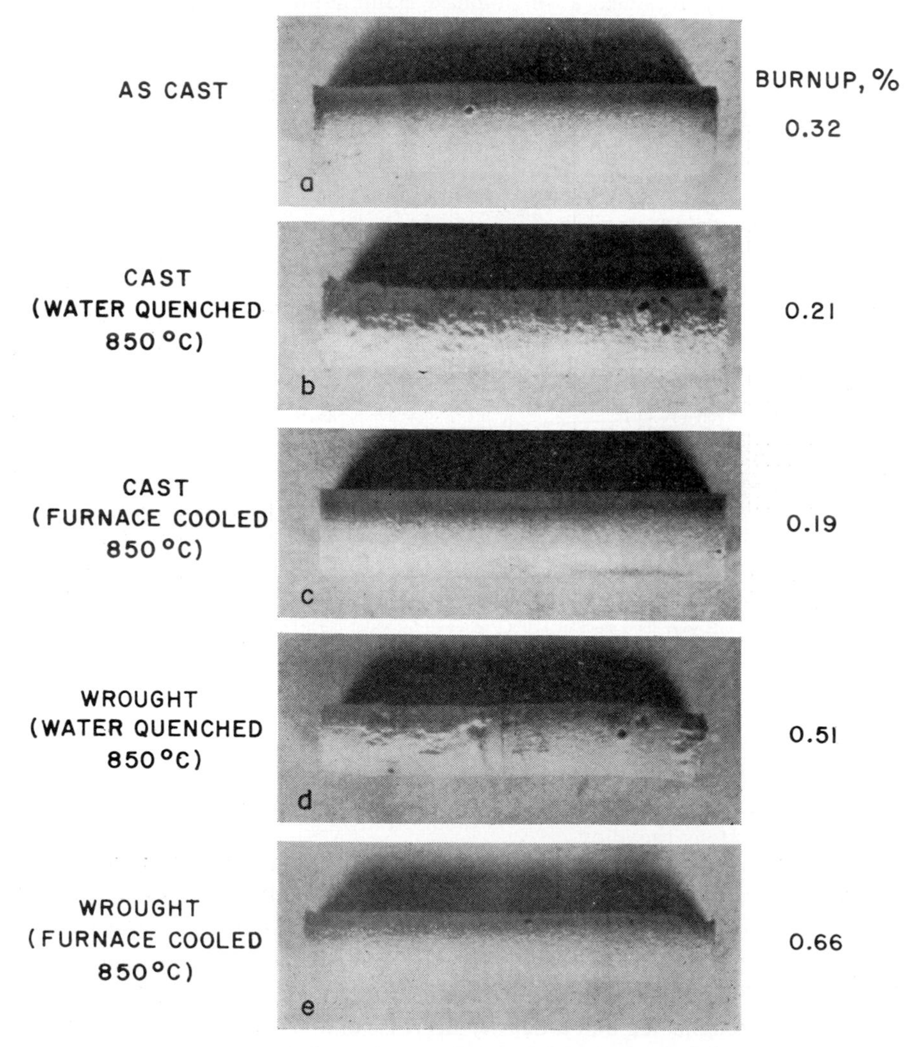

Fig. 4.—The effects of irradiation at moderate temperatures on uranium–5 wt. per cent fissium alloys.

temperatures more than 200 times, which far exceeds the number of such cycles that the fuel pins will experience in the reactor.

Cast, as well as extruded, uranium–20 w/o plutonium–fissium alloy pins (5.4 and 10.8 w/o fissium) were practically unchanged as the result of 200 cycles between 25° and 500° C. In contrast, both cast and extruded binary uranium-plutonium alloy pins spalled and cracked badly in less than 200 similar cycles.

Full-size EBR-II fuel elements containing injection-cast uranium–5 w/o fissium alloy pins have been thermally cycled by Sowa and co-workers (17) 200 times to 550° C without significant dimensional changes. Also, cast specimens and gamma heat-treated wrought specimens of this alloy have been cycled 250 times from 66° to 620° C, with negligible growth or surface roughening (4). Hot-rolled uranium–5 w/o fissium specimens elongated 3–6 per cent when cycled from 150° to 750° C, as reported by Saller *et al.* (16). They suggest that differences in the fabrication history of this alloy may have a considerable effect on the tendency to deform on cycling.

CORROSION

The EBR-II fuel pin must be compatible with the environment in the reprocessing cell, as well as with the liquid sodium bond within the jacket. Even though the reprocessing cell will contain a relatively pure atmosphere of argon, the problem of reaction of the uranium-plutonium base alloys with the working environment was not solved until the fission elements were added. Our glove-box facility in which these alloys were studied at ANL (18) has a much purer helium atmosphere than the reprocessing cell will have. Yet the binary uranium-plutonium alloys containing over 15 w/o plutonium tend to disintegrate to a powder on standing in this atmosphere. Binary alloys containing 20–30 w/o plutonium disintegrate more rapidly in our helium atmosphere and are pyrophoric when exposed to air. This undesirable property has not been explained, although several theories have been proposed involving various combinations of the presence of the zeta phase, microcracks in the castings, internal stresses, and hydride and oxide formation. Studies now in progress at ANL on arc-melted binary alloys may help to clarify this problem.

Fortunately, the addition of the fissium elements made in preparing the alloys that are favored for the EBR-II fuel has solved the pyrophoricity problem. Also, these alloys have been stored for long periods in our glove-box atmosphere without manifesting the slightest tendency to disintegrate.

With regard to the compatibility of the fuel alloy with the thermal bond, unalloyed uranium has such a small tendency to dissolve in, or alloy with, clean sodium or NaK that it is hardly measurable. Addition of plutonium and the fission-product elements to uranium does not appreciably change this picture. Numerous tests under a wide variety of conditions, including long-time pile irradiations at elevated temperatures, bear this out. (Practically all ANL's irradiation tests use NaK as the thermal bond between the specimen and the container.) However, as in the case of unalloyed uranium, the sodium or NaK must be clean with respect to oxygen and probably other impurities. When present in small amounts, oxygen results in weight changes, and, when present in large amounts, spalling and cracking result.

Fabrication

In the EBR-II reprocessing cycle, the pin fabrication technique demands very little of the fuel alloy in the way of mechanical properties. The alloy must be castable and must possess sufficient strength and ductility to allow stripping the mold and subsequent handling operations. All the pertinent alloys—uranium-plutonium, uranium–5 w/o fissium, uranium–20 w/o plutonium–10 w/o fissium—can be cast into molds. The binary uranium-plutonium alloys containing greater than 15 w/o plutonium tend to crack, however, even before they are removed from the mold. Uranium-plutonium alloys, when melted in MgO crucibles and cast, were not fabricable. Although this brittleness has been attributed to the zeta phase in this system, recent studies at ANL by Biver (19) on arc-melted buttons show that all compositions from 5 to 50 w/o can be hot-pressed and that they have much less tendency to crack or powder on standing in the glove-box helium atmosphere. The fissium alloys, both uranium-fissium and uranium-plutonium-fissium, can be readily gravity-cast into molds or injection-cast into Vycor tubes. Saller *et al.* (16) consider the fissium-containing alloys to have low fluidity, and the work of Yaggee *et al.* on injection casting shows these alloys to have less fluidity than uranium (20). The melting range of the uranium–5 w/o fissium composition is between 1,002° and 1,081° C, as reported by Saller (16). We have found that the solidus of uranium–10 w/o fissium is 1,000° C and the solidus of uranium–20 w/o plutonium–10 w/o fissium is 930° C, which is about the same as the solidus of uranium–20 w/o plutonium.

The uranium–5 w/o fissium alloy can be hot-rolled. The alloy was forged at 870° C, as reported by Saller *et al.* (16). Yaggee rolled the cast material at 800° C (20). Attempts to hot-roll the uranium-plutonium and uranium-plutonium-fissium alloys failed even at 800° C. We have extruded the plutonium alloys (by an inverse extrusion method) at temperatures between 450° and 500° C. The billets were extruded bare with a 7:1 reduction at a pressure of 100,000 psi.

These alloys are somewhat harder than uranium. For compositions from 0 to 20 w/o plutonium the hardness of the binary uranium-plutonium alloys increases from 59 R_A to 69 R_A. The hardness of the uranium-fissium and uranium-plutonium-fissium alloys is in a range of 70–75 R_A. As may be expected, the higher hardness of the fissium alloys makes machining more difficult. Carbide-tipped tools and slow machining speeds are essential for any reasonable production. The presence of plutonium (to 15 w/o), however, in uranium-plutonium alloys makes machining somewhat easier than with unalloyed uranium, even though the hardness is greater.

Thermal Expansion and Density

For reactor safety reasons it is desirable that the thermal expansion of the EBR-II fuel alloy be positive and reliable on heating. The cast uranium-plutonium-fissium and uranium-plutonium alloys behave dilatometrically about the same as cast unalloyed uranium. The thermal expansions of the cast uranium–20 w/o plutonium–10.8 w/o fissium and uranium–5 w/o fissium alloys are given in Figure 5. The thermal-expansion coefficients and room-temperature densities of alloys pertinent to the EBR-II fuel-alloy development are summarized in Table 7.

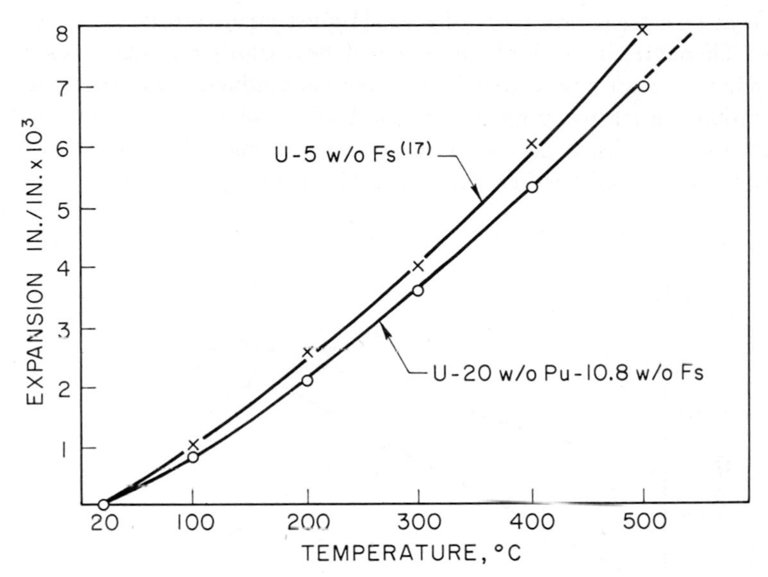

FIG. 5.—The thermal expansions of uranium–5 wt. per cent fissium and uranium–20 wt. per cent plutonium–10.8 wt. per cent fissium alloys.

TABLE 7

THERMAL-EXPANSION COEFFICIENTS AND DENSITIES OF CASTINGS

ALLOY (w/o)	THERMAL EXPANSION		DENSITY (R.T.) (GM/CM³)
	Temp. Range (° C)	Coef. (10^{-6}/° C)	
U–20Pu–5.4Fs	25–438	17.5	17.6
U–20Pu–10.8Fs	{20–100 20–500	12.4 15.2}	16.6
U–20Pu–5Mo	25–530	20	17.7
U–15Pu	25–300	13	18.8
U–5Fs*	{20–100 20–500	13.7 16}	17.95
U†	{20–100 20–500	14.5 18.5}	19.1

* See Ref. 16.
† See Ref. 4.

THERMAL CONDUCTIVITY

The heat-transfer performance of EBR-II must far exceed normal engineering practice. Therefore, in spite of the excellent heat-transfer characteristics of the liquid sodium coolant, the relatively poor thermal conductivity of the alloys under consideration is a matter of great concern. Addition of 10 w/o plutonium to uranium lowers the thermal conductivity of uranium, and additions of fissium are expected to lower it still further. Saller *et al.* (16) have shown that the thermal con-

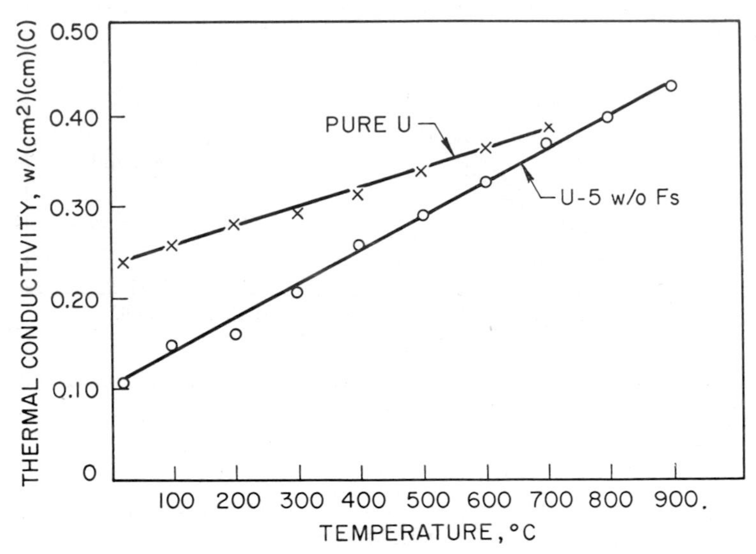

FIG. 6.—The thermal conductivities of unalloyed uranium and uranium–5 wt. per cent fissium alloys.

TABLE 8

COMPARISON OF THERMAL CONDUCTIVITIES OF URANIUM–5 W/O FISSIUM ALLOY AND UNALLOYED URANIUM [Cal/(Cm²)(Sec)(C)]

TEMP. (° C)	U–5 w/o FISSIUM		URANIUM†
	BMI*	ANL	
20	0.026	...	0.066
100	.034	...	.068
200	.044	0.040	.069
300	.053	.048	.074
400	.062	.056	.081
500	.071	.063	.088
600	.080	0.071	.096
700	.088	...	.105
800	.097	...	.115
900	0.125	...	0.127

* See Ref. 16. † See Ref. 4.

ductivity of uranium–5 w/o fissium at room temperature is about half that of unalloyed uranium. However, as shown in Figure 6 and Table 8, the conductivity of the uranium–5 w/o fissium alloy increases more rapidly with temperature than does that of uranium, so that the difference is small at pile-operating temperatures.

SUMMARY

Pu^{239} is the favored fuel for fast-neutron power-breeder reactors. However, its unattractive properties make it extremely unlikely that the unalloyed metal will find use as a solid fuel. Heat-transfer considerations and the limitations of normal fabrication methods suggest that the plutonium be diluted by alloying, to make it usable in fast breeder reactors. The degree of dilution necessary makes plutonium the minor element, and therefore its undesirable properties are no longer manifested. However, the precautions necessary for safe handling of intense alpha-emitters still apply.

EBR-II, a prototype fast power-breeder reactor, has a fuel element that is designed for high thermal performance and high burnup. Long, thin fuel pins fit loosely into a thin-walled tube with liquid sodium as the thermal bond. The simple design lends itself to a simple remote reprocessing cycle. EBR-II has been designed to receive plutonium fuel at the earliest possible loading. A promising plutonium alloy is being developed and appears to satisfy EBR-II fuel-element requirements. Because this element is still being tested, EBR-II will be fueled at first with U^{235}. The alloys that have been developed for EBR-II fuel elements result from a combination of reactor physics, reactor engineering, reprocessing, and metallurgical requirements. Diluting the fissionable material helps to solve the heat-transfer problem, and, by using U^{238} as the diluent, considerable internal breeding is achieved. The metallurgical refining process that is integral with the EBR-II does not remove certain fission-product elements (referred to as "fissium" or "Fs"), and this determines to a considerable degree the fuel-alloy composition. Some of these fission-product elements tend to stabilize the gamma phase of uranium and are therefore desirable alloying elements. The Pu^{239} alloy consists of uranium–20 w/o plutonium–10 w/o fissium, with about 20 w/o U^{235} included because of the relatively small reactor core size; a full-size power reactor would require less than 20 w/o plutonium and no U^{235}. The non-plutonium first loading consists of a uranium–5 w/o fissium alloy with 49 per cent enrichment.

The uranium-plutonium-fissium alloys that have been studied showed practically no irradiation-damage effects at moderate irradiation temperatures but swelled badly at elevated temperatures. The uranium–5 w/o fissium alloy that will be used at first in EBR-II behaves similarly under irradiation. In general, the studies indicate that alloying to stabilize the gamma structure of uranium is effective in improving irradiation stability. These alloys also possess excellent thermal cycling stability; they can be remotely injection-cast into long, small-diameter pins; they resist attack by the liquid sodium thermal bond and are compatible with the reprocessing cell environment; and they have reasonable thermal-expansion characteristics. A limitation is that they are relatively poor thermal conductors.

REFERENCES

1. WALTON, G. N. (ed.). *Glove Boxes and Shielded Cells.* New York: Academic Press, Inc., 1958.
2. LICHTENBERGER, H. V., THALGOTT, F. W., KATO, W. Y., and NOVICK, M. "Operating Design of EBR-II and Experimental Results Obtained from an NaK-cooled Fast Reactor," *Proceedings of the International Conference on the Peaceful Uses of Atomic Energy*, **3**:345–60. New York: United Nations, 1956.
3. BARNES, A. H., KOCH, L. J., MONSON, H. O., and SMITH, F. A. "The Engineering Design of EBR-II, a Prototype Fast Neutron Reactor Power Plant," *Proceedings of the International Conference on the Peaceful Uses of Atomic Energy*, **3**:330–44. New York: United Nations, 1956.
4. KOCH, L. J. "Engineering Report on EBR-II," USAEC Rept. ANL-5788 being prepared for publication.
5. OKRENT, D., AVERY, R., and HUMMEL, H. H. "A Survey of the Theoretical and Experimental Aspects of Fast Reactor Physics," *Proceedings of the International Conference on the Peaceful Uses of Atomic Energy*, **5**:347–63. New York: United Nations, 1956.
6. KITTEL, J. H., and TEDESCHI, P. *A Capsule Design for Experimental High-Flux Irradiations of Fuel Materials.* (USAEC Rept. ANL-4900, October, 1952.)
7. CHISWIK, H. H., and KELMAN, L. R. "Thermal Cycling Effects in Uranium," *Proceedings of the International Conference on the Peaceful Uses of Atomic Energy*, **9**:147–58. New York: United Nations, 1956.
8. DWIGHT, A. E. Private communication.
9. SMITH, K. F., ZEGLER, S. T., and MAYFIELD, R. M. *Mechanical Stabilization of Uranium Fuel Elements.* (USAEC Rept. ANL-5377, November, 1956.)
10. HOWE, J. P. "Problems in Materials for Nuclear Power," in *Nuclear Metallurgy: A Symposium Conducted by the Institute of Metals Division of the American Institute of Mining and Metallurgical Engineers*, pp. 9–27. ("IMD Special Report Series," No. 1.) New York: Metallurgical Society of the American Society of Mining and Metallurgical Enginners, October, 1955.
11. SMITH, K. F., and KELMAN, L. R. *Irradiation of Cast Uranium-Plutonium Base Alloys.* (USAEC Rept. ANL-5677, May, 1957.)
12. KITTEL, J. H. Unpublished information.
13. KITTEL, J. H., and KELMAN, L. R. *Effects of Irradiation on Some Uranium-Plutonium Alloys.* (USAEC Rept. ANL-5706, June, 1958.)
14. SMITH, K. F. *Irradiation of Uranium-Fissium Alloys and Related Compositions.* (USAEC Rept. ANL-5736, September, 1957.)
15. KITTEL, J. H. Unpublished information.
16. SALLER, H. A., DICKERSON, R. F., BAUER, A. A., and DANIEL, N. E. *Properties of a Fissium-Type Alloy.* (USAEC Rept. BMI-1123, August, 1956.)
17. SOWA, E. S. Private communication.
18. KELMAN, L. R., WILKINSON, W. D., SHUCK, A. B., and GOERTZ, R. C. *The Safe Handling of Radioactive-pyrophoric Materials.* (USAEC Rept. ANL-5509, December, 1955.)
19. BIVER, C. J., JR. Private communication.
20. YAGGEE, F. L., AYER, J. E., and JELINEK, H. F. "The Injection Casting of Uranium-Fissium Alloy Pins," in *Nuclear Metallurgy*, **4**:51–62. ("Institute of Metals Division Special Report Series.") New York: Metallurgical Society of the American Institute of Mining, Metallurgical and Petroleum Engineers, 1957.

CHAPTER XXXI

PLUTONIUM FUELS FOR POWER REACTORS

S. F. Pugh

INTRODUCTION

At Harwell one interest in plutonium-bearing fuel elements relates to the possibility of recycling plutonium in combination with natural uranium as the fuel in thermal reactors. Although the U^{235} in the natural uranium would constitute a part of the fissionable material, Pu^{239} would be bred from the fertile U^{238} in the natural uranium. Pu^{239} would then be extracted from the spent fuel elements and used to enrich the natural uranium in new fuel elements. This could be done either by uniformly distributing the plutonium throughout the reactor core in identical fuel elements containing both plutonium and uranium in a single alloy or by introducing a number of plutonium-alloy "spike" elements distributed among the natural uranium elements. In the present chapter the advantages and disadvantages of these alternative procedures are discussed, as well as the type and composition of plutonium alloy required for each. Because good resistance to irradiation damage is of paramount importance in a solid-fuel alloy, a number of plutonium alloys of each type have been irradiation-tested at Harwell, and the results of these tests are reported.

PLUTONIUM RECYCLING

A great advantage of plutonium recycling in thermal reactors using uranium fuel is that greater utilization of the natural uranium feed can be obtained by extracting the plutonium from the spent fuel and recycling this together with some of the depleted uranium from the fuel and a small addition of natural uranium. In this way it is calculated that about 2 per cent of all atoms of the uranium feed can be burned, as opposed to the 0.3–0.6 per cent which can be burned in a once-through process.

The second advantage of plutonium recycling is that an enriched core can be obtained without the need of a plant for separating U^{235} from natural uranium. For example, it is possible to break into the thorium-U^{233} cycle by using plutonium for the initial enrichment and possibly for makeup to the system and thus avoid the

S. F. Pugh is at the Atomic Energy Research Establishment, Harwell, Berkshire, United Kingdom.

need for separated U^{235}. The main advantage of enriched cores is that a greater power density per unit volume in the core can be obtained, thus reducing the capital cost per unit of installed power. The increased power rating of the core may lead to lower costs per kilowatt-hour of electrical output even if the fuel costs are increased.

The processing and refabrication of plutonium fuel are more difficult than those of uranium fuel because of the radioactivity of the plutonium, as well as its criticality hazard. For this reason the fuel element should be as simple to fabricate as possible. Fuel elements consisting of thorium bars containing PuO_2 have recently been fabricated from powder at Harwell by a hydrostatic pressing technique followed by sintering. The process was found to be eminently suitable for handling active materials. Fully automatic fabrication of fuels may be desirable for producing very large quantities of a single type of fuel element. Since, however, the evolution of fuel elements is usually fairly rapid and minor modifications in design are frequent, it is doubtful whether automatic remote fabrication will be profitable in the early stages of the use of plutonium fuels.

UNIFORM AND SPIKED CORES

Plutonium can be recycled through a solid-fueled reactor either dispersed uniformly through the fertile material or in the form of separate spike fuel elements. In the latter scheme the plutonium is diluted with a non-fissile metal preferably of low neutron-capture cross-section. Choice of recycling schemes would depend on the relative costs of fabricating, operating, and processing the fuel charge. No extra diluent for the plutonium is required in a uniform core, so that extra-parasitic absorption of neutrons is avoided and operation of the reactor is simpler, in that only one kind of fuel element is handled. The disadvantage of the uniform core is that the freshly formed Pu^{239} is mixed with degraded plutonium and there is thus no easy way of rejecting degraded plutonium without rejecting some of the newly formed Pu^{239}. ("Degraded" plutonium contains significant amounts of isotopes higher than Pu^{239} not all of which are fissionable by thermal neutrons but which may, nevertheless, act as neutron absorbers.) This disadvantage is not very great and could be avoided completely, if desired, by making fuel elements consisting of two parts which could be physically or chemically separated after irradiation. An important feature of the uniform core is that the whole of the fuel charge must be processed remotely.

The main disadvantage of a uniform core in a reactor based on uranium metal fuel is that uranium does not provide the most satisfactory type of matrix for the plutonium for operating temperatures up to 400° C, since, in this temperature range, alpha uranium undergoes severe distortion because of the phenomena of growth and wrinkling peculiar to it. Thorium does not have this disadvantage and would be preferable in this respect for low-temperature operation. In its resistance to distortion at temperatures above 500° C and below the temperature of transformation to beta uranium, uranium is as good as most other diluents for certain ranges of enrichment of the core. The choice of diluent for the plutonium may depend on resistance to irradiation damage of the fuel. To reduce the cost of chemical processing, it is desirable that the burnup of fissile material be limited only by the nuclear properties of the charge and not by irradiation damage in the fuel. The problem of obtaining satisfactory burnup is greater in the uniform core because there is less scope for choice of the composition of the plutonium-bearing fuel. In a spiked core, if burnup is limited by irradiation damage, the burnup of fissile material per cycle may be greater than that in a uniform core, other things being equal, since the fissile material is more dilute in the non-spiked elements. It is possible, however, that the burnup per unit cost of fuel-element fabrication and processing will not be increased under these conditions. In a spiked core, independent of the choice of fuel material for the main charge, it might be desirable to use ceramic or cermet fuel for the spike-fuel element, in order to obtain high burnup of the plutonium in a single reactor cycle.

Spiking gives greater flexibility in reactivity adjustment but poses a number of operational problems due to change in heat rating in the spikes during burnup. There are, of course, other schemes for using plutonium as a fuel, in addition to the two described above.

COMPOSITION OF PLUTONIUM FUELS

In thermal reactors the plutonium content of the uranium fuel in a uniform core with complete recycling would eventually be about 1 per cent. If, however, the plutonium is introduced in a separate spiked-fuel element, the composition of the spike would probably be in the range of 1–10 at. per cent plutonium. In large fast reactors the plutonium content of the fuel would be about 2 gm of plutonium per cm^3, or, if the plutonium were separate from the fertile material, the content would be about 5–6 gm of plutonium per cm^3 in the spike.

Choice of materials for plutonium fuels is governed by the factors also operating in the choice of uranium fuels, but some of these factors have extra emphasis. The factors to be considered are as follows:

1. Type of equipment required for remote fabrication. Use of simple fuel shapes will, to some extent, avoid fabrication processes that require complex equipment.

2. Chemical processing. Some materials are difficult to dissolve or require a specialized plant for chemical reprocessing. It is an advantage if the new fuel material can be handled in an existing chemical plant.

3. Dimensional stability of the fuel and its expected life in the reactor. The life of a particular fuel element can be determined only by fairly extensive trials under the conditions it will be required to withstand during burnup in the reactor. A good idea of the behavior of a new fuel material can be obtained by irradiating specimens in capsules or loops, followed by careful measurement of changes in structure and dimensions.

4. Retention of both fission products and plutonium in the event of a fuel element's bursting. Quantitative measurements can be made of emission of fission products from burst fuels in loops containing the coolant to be used in the final reactor.

5. Compatibility of the fuel with the canning material and with the coolant. The fuel material might be either metal, ceramic, cermet, or liquid.

Unalloyed plutonium metal has very poor properties for use as a fuel material. It undergoes a number of phase changes between room temperature and the melting point of 640° C, which result in a total increase in volume over this temperature range of about 20 per cent; also, some of the phases are brittle, and others are very soft. The thickness of the fuel in a flat-plate fuel element of undiluted plutonium is limited by the reactor heat ratings to about 0.01 cm. Control of the thickness of such a thin fuel sandwich would be very difficult, and the extra thickness required in a diluted fuel would therefore lead to greater ease of fabrication. (At this thickness, there would be no problem with neutron blackness of the fuel, and considerations of neutron blackness therefore do not arise, once the fuel-rating parameters have been fixed.) Hence the use of plutonium in the unalloyed form is unlikely to be contemplated as a solid-fuel material. The choice of alloying additions is made on grounds of neutron economy, and materials are chosen that will, when alloyed with plutonium or chemically combined with it, produce a fuel material that resists irradiation damage and preferably is compatible with the coolant to be used. For reasons of neutron economy and ease of operation, the most desirable addition to the plutonium is that of a fertile material, such as uranium or thorium, provided that these resulting alloys do not have undesirable properties or high fabrication costs. For this reason, the irradiation properties of uranium-plutonium and thorium-plutonium alloys, with possible additions of third elements to improve the properties, are being investigated, and preliminary results are described below. If such alloys are not suitable for a specific application, the next choice would be plutonium alloyed with a metal of low neutron-capture cross-section, such as aluminum, magnesium, zirconium, or beryllium. The behavior of zirconium-plutonium alloys has been studied under irradiation. The aluminum-plutonium and magnesium-plutonium alloys have melting points too low to be considered for use as solid fuel in advanced gas-cooled or sodium-cooled reactors, and beryllium-plutonium is expected to behave in much the same way as beryllium-uranium, since in both cases the fissile material forms with the beryllium a hard brittle compound dispersed in a matrix of beryllium.

The chief disadvantage of metallic fuels other than those based on beryllium is their reaction with coolants such as CO_2 and water. For this reason the use of ceramic fuels has been proposed, in particular PuO_2 diluted with UO_2 or ThO_2.

The physical and chemical properties of PuO_2 and UO_2 show such strong similarities that the irradiation behavior of fuels containing PuO_2 is unlikely to be very different from that of UO_2 itself. There are, however, some differences in physical and chemical properties that may be important in certain cases; for example, PuO_2 does not form higher oxides and therefore will be more stable in oxidizing environments than UO_2. Changes in ceramic fuels during irradiation are caused by the very high temperatures developed at the center of such fuels due to their very low thermal conductivity. Melting and volatilization may occur. The presence of PuO_2 may alter such processes quantitatively by causing changes in melting point and vapor pressure.

The properties of UO_2 and ThO_2 are not greatly different from those of other metallic oxides, and there is therefore not much to be gained by using inert diluents such as BeO, Al_2O_3, MgO, and ZrO_2. In some applications, however, such diluents have been considered, and the important properties required in a ceramic fuel are fairly good thermal conductivity, a high melting point, and resistance to cracking due to thermal stress or thermal shock. Data relevant to the choice of these materials are shown in Tables 1 and 2. At 1,000° C beryllia appears to have a thermal conductivity next best to that of graphite. There are no very great differences in thermal stress and thermal shock resistance among the various oxides when considered over the whole range of temperatures in which they are expected to operate.

TABLE 1

THERMAL-CONDUCTIVITY VALUES FOR CERAMIC MATERIALS
(Cal Sec^{-1} ° C^{-1} Cm^{-2} Cm)

MATERIAL	TEMPERATURE (° C)		
	100	400	1,000
Graphite	0.426	0.268	0.149
BeO	.525	.222	.049
MgO	.086	.039	.017
Al_2O_3	.072	.031	.015
Zircon	(.016)	.012	.016
ThO_2	.025	.014	.008
UO_2	.023	.014	.008
ZrO_2 stabilized	.0047	.0049	.0055
Fused SiO_2	.0038	.0045	(.0048)
SiC, self-bonded	....	.181	.106
SiC, nitride-bonded	....	0.0525	0.033
Metallic monocarbides	0.05	....	....

TABLE 2

THERMAL STRESS RESISTANCE FACTORS USED TO ASSESS BEHAVIOR DURING RAPID AND SLOW COOLING

MATERIAL	R (° C)* (FAST COOLING)			R' (CAL SEC^{-1} CM^{-1}) (SLOW COOLING)		
	100	400	1,000	100	400	1,000
Al_2O_3	37	36	40	2.7	1.1	0.60
BeO	33	31	14	17.3	6.9	.69
MgO	22	24	26	1.9	0.94	.45
ThO_2	51	47	37	1.3	0.66	.30
Zircon	137	115	73	(2.2)	1.4	.73
ZrO_2 stabilized	66	62	53	0.31	0.30	0.29
Fused SiO_2	2,500	2,400	....	9.5	10.1	....
Graphite	700	700	700	910	197	....
Tungsten	2,300	2,300	2,300	1,100	....	....

* R is defined by $R = S_t(1 - \mu)/E\alpha$ for fast-quenching conditions; R' is defined by $R' = KS_t(1 - \mu)/E\alpha$ for slow-cooling conditions, where E is Young's modulus in dyne cm^{-2}; S_t is the breaking stress in dyne cm^{-2}; α is the coefficient of linear expansion, ° C^{-1}; μ is Poisson's ratio; and K is thermal conductivity in cal sec^{-1} ° C^{-1} cm^{-2} cm.

Dilution of the fuel material with moderator is proposed in certain reactor systems, and in these systems the irradiation behavior of the fuel is unlikely to be very different when the fissile material is changed from uranium to plutonium, since the content of such material in the moderator is only about one part in a thousand. Irradiation behavior is therefore not likely to be a factor affecting the behavior of such fuels. It must be remembered, however, that plutonium forms a lower oxide, Pu_2O_3; thus PuO_2 is not so chemically stable as UO_2 in the presence of reducing agents such as graphite at temperatures above 1,000° C.

RESULTS OF IRRADIATION TESTS OF METALLIC PLUTONIUM-BEARING FUELS

Experimental Method

The plutonium-bearing alloys were examined before and after irradiation to burnups of between 0.2 and 0.8 per cent of all atoms. The alloys were chosen as being likely fuel materials for use in cores enriched by recycling plutonium. The alloys can be divided into (1) those based on uranium; (2) those based on uranium plus 28 at. per cent molybdenum; (3) thorium-based alloys; and, finally, (4) a zirconium-plutonium alloy which is a possible material for a spike-fuel element.

Specimens containing plutonium were arc-melted and cast into cylinders, and, with one exception, the specimens were machined on all surfaces to a right cylindrical shape. The zirconium-based specimen was extruded into a cylindrical shape. The specimens containing U^{235} as fissile content were in the form of arc-melted buttons. All specimens were irradiated immersed in sodium and supported in small zirconium cups. Temperatures of 500° and 700° C were maintained by furnaces surrounding the sodium capsules. In a separate series of tests the specimens were heated in sodium to insure that the changes observed during irradiation were not due to reaction with sodium. It was shown that all materials were compatible with sodium in the absence of irradiation. The irradiation time for the specimens containing plutonium was about 240 days in a neutron flux of about 5×10^{12} n/cm²/sec, and during this period specimens were cooled from the irradiation temperature about 40 times to a temperature 100° C below the irradiation temperature and about 4 times to 100° C because of shutdown of the reactor.

The principal damage caused by burnup in the specimens irradiated in this series of experiments was swelling in the ductile alloys, due to the formation of the inert fission-product gases xenon and krypton, and cracking, due to a similar cause in those alloys which were brittle at the irradiation temperature. Density was measured by weighing in air and in water before and after irradiation, since, if distortion is severe, only a rough estimate of density can be obtained from measurement of changes in dimensions. Where a fairly uniform increase in volume occurred, the distortion due to irradiation was expressed in terms of the R value, calculated by dividing the percentage decrease in density by the percentage burnup of all atoms in the specimen. It was assumed at this stage that the R value was independent of burnup in the range 0 to 0.6 per cent of all atoms.

The results of irradiation of the series of specimens containing plutonium are summarized in Table 3, which includes, for comparison, results of the irradiation of similar specimens in which the fissile material was U^{235} irradiated under the same conditions and to about the same burnup.

TABLE 3

BEHAVIOR OF NUCLEAR FUELS CONTAINING PLUTONIUM DURING BURNUP, COMPARED WITH URANIUM FUELS

Composition	Temperature (° C)	Burnup of All Atoms	Volume Increase (Per Cent)	*R* Value*	Remarks on Appearance after Irradiation
U unalloyed	500	0.26	10.2†	34†	Good
U unalloyed	700	.27	11.5	33	Good
U+7½ at. per cent Pu	500	.33	9.7	17	Good
U+7½ at. per cent Pu	700	.32	14	28	Small crack
U+10 at. per cent Pu	400	.29	18.3	53‡	Distortion
U+10 at. per cent Pu	700§	.29	25.4‡	70‡	Large crack
U+28 at. per cent Mo	500	.50	3.5	6	Very good
	700	.47	1.6	4	Very good
U+28 at. per cent Mo+6½ at. per cent Pu	500	.22	11.2‡	45‡	Cracked
	700	.22	2.7	12	Good
U+28 at. per cent Mo+18 at. per cent Pu	500	.44	....	....	Bulge on side
	700	.42	....	....	Cracked and distorted
Th+47 at. per cent U	500	.62	9.2	13.6	Good
Th+15 at. per cent Pu	500	.54	14.0	22	Good
Th+47 at. per cent U arc melt+cast	800‖	.56	20.4	30.0	Good
Zr+20 at. per cent U	500	.20	1.8	8.5	Very good
Zr+40 at. per cent Pu	500	0.83	5.4	6	Good

* *R* value = Per cent decrease in density/Per cent burnup of all atoms.
† Mean of three specimens.
‡ Estimated from change in dimensions.
§ For 40 days, then 150 days, at 315° C.
‖ After 0.3 per cent burnup the furnace failed, the specimen then ran at 450° C.

PREVIOUS WORK

In studying the dimensional stability of uranium and its alloys under irradiation, two main phenomena have been discovered: (1) growth and wrinkling, which are confined to an alpha-uranium structure at temperatures up to 500° C, and (2) swelling due to the accumulation of fission-product gases, which occurs at temperatures of about 450° C and above. In a specimen in which there is a steep temperature gradient, the temperature of the surface layer is rather more important than the temperature at the center, since the surface layers will constrain any distortion at the center of the specimen. While the mechanism of growth under irradiation is not known, the main variables affecting growth and wrinkling can be prevented by alloying in such a way as to avoid the anisotropic alpha-uranium structure. In specimens having the alpha-uranium structure, growth and wrinkling effects are mitigated by a fine grain size and the absence of preferred orientation. In this connection the only phenomenon likely to be obtained at low temperatures in plutonium-rich specimens is that of growth in some of the anisotropic structures found in plutonium and its alloys. The mechanism of swelling is now fairly clear, namely, the diffusion of fission-product gases into bubbles and the increase in volume of the bubbles due to the pressure in them, which is resisted more or less by the uranium matrix. Although the mechanism is fairly well known, the relative importance of such factors as the creep strength of the matrix, the effect of phase transformations, and effects due to growth or thermal cycling of anisotropic phases have not yet been determined. Investigations so far using uranium as the fissile material have indicated that strengthening the matrix at the temperature of irradiation by, for example, alloying uranium with molybdenum will reduce considerably the amount of swelling; but the possibility remains that, in uranium itself and in dilute alloys irradiated at temperatures above the transformation temperature, swelling may have been increased by cycling through the alpha-beta-gamma phase changes.

Uranium-Plutonium Alloys

The behavior of metallic fuel under irradiation is structure-sensitive and therefore can be altered by heat treatment or alloying. In particular, the amount of

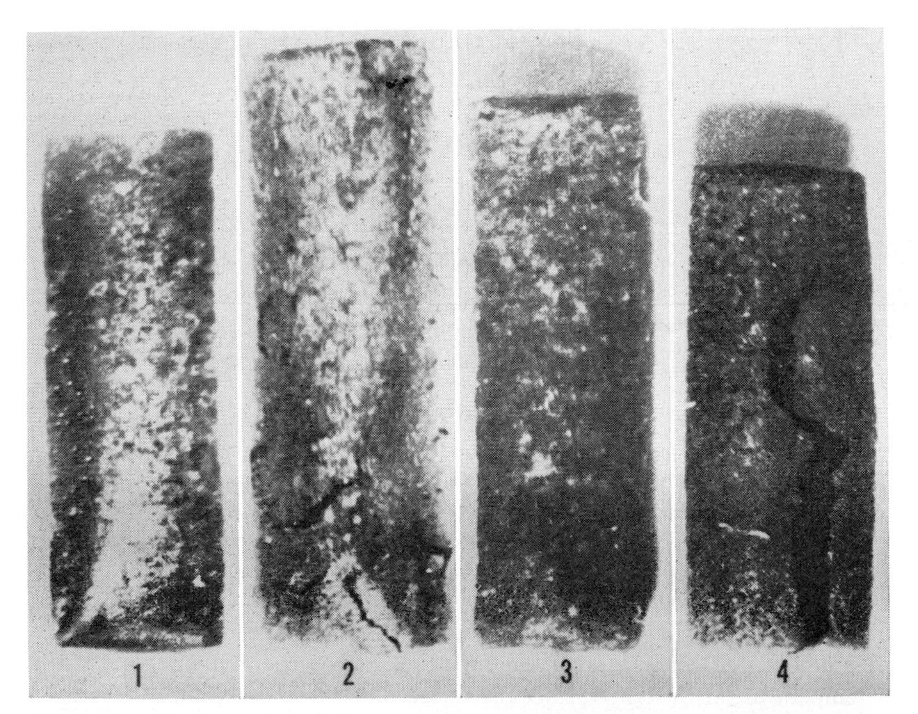

Figs. 1–4.—Fig. 1: Uranium–7.5 at. per cent plutonium alloy irradiated at 500° C to a burnup 0.33 per cent of all atoms. Increase in volume, 9.7 per cent; *R* value 17. Uniform increase in volume. Fig. 2: Uranium–7.5 at. per cent plutonium alloy irradiated at 700° C to a burnup of 0.32 per cent of all atoms. Increase in volume, 14 per cent; *R* value 28. Small cracks mainly at one end. Fig. 3: Uranium–10 at. per cent plutonium alloy irradiated at 315°–350° C to a burnup of 0.29 per cent of all atoms. Increase in volume, 18.3 per cent, estimated from change in dimensions; *R* value 53. Specimen shows distortion. Fig. 4: Uranium–10 at. per cent plutonium alloy irradiated at 350°–400° C to a burnup of 0.29 per cent of all atoms. Increase in volume, 25.4 per cent, estimated from change in dimensions; *R* value 70. A large crack has appeared.

swelling or cracking caused by the formation of fission-product gases might be affected by the change in mechanical properties of uranium or its alloys caused by the addition of plutonium.

The first batch of specimens irradiated consisted of a pair of unalloyed uranium specimens, a pair composed of uranium alloy containing $7\frac{1}{2}$ at. per cent plutonium, and a pair composed of uranium alloy containing 10 at. per cent plutonium. It was intended that one of each pair should be irradiated at 500° and the other at 700° C, but in one case the lower irradiation temperature was 400° C, and one specimen ran for only half the time at 700° C because of failure of furnaces in the pile. In both the unalloyed and the alloyed specimens the structure of the specimens irradiated at 500° C was that of alpha uranium, and the structure of specimens irradiated at 700° C was that of beta uranium. Plutonium is completely soluble in the amounts added in both these phases. The distortion under irradiation of specimens containing $7\frac{1}{2}$ per cent plutonium was somewhat less than that found in unalloyed uranium. After irradiation at 500° C the R value of the $7\frac{1}{2}$ per cent plutonium alloy was 17, while that of unalloyed uranium was 34. The distortion consisted of a uniform increase in volume, with no sign of roughening or cracking (Fig. 1). Unalloyed uranium and uranium containing $7\frac{1}{2}$ per cent plutonium, after irradiation at 700° C, showed a rather greater increase in volume, and the $7\frac{1}{2}$ per cent plutonium alloy contained small cracks (Fig. 2). The greater distortion in these specimens is due partly to the rather poorer mechanical properties of beta uranium, which is noted for its brittleness, and due partly to cycling through the phase changes. The specimens containing 10 per cent plutonium showed a large amount of cracking and distortion after irradiation and large R values, presumably because of the presence of cracks in the material (Figs. 3 and 4). It is suspected that segregation of the plutonium in these alloys may partly account for their poor behavior under irradiation. The appearance of uranium-plutonium specimens after irradiation is illustrated in Figures 1–4.

Uranium-Molybdenum-Plutonium Alloys

The next batch of fuels was based on uranium alloy containing 28 at. per cent molybdenum, which, without the addition of plutonium, has proved to be dimensionally very stable during burnup at temperatures up to 800° C and to have good oxidation resistance, as compared with unalloyed uranium. In the absence of plutonium, this alloy has a body-centered cubic structure above 565° C or if quenched to room temperature, but it becomes slightly tetragonal because of ordering on soaking in the range 565°–300° C.

When $6\frac{1}{2}$ at. per cent plutonium is substituted for uranium in this U-Mo alloy, the structure at 500° C is still single-phase ordered gamma, but some alpha uranium is precipitated out at lower temperatures, as in the alloy free from plutonium. After irradiation at 500° C the alloy with $6\frac{1}{2}$ at. per cent plutonium was found to be severely cracked (Fig. 5), and the R value based on measurements of external dimensions was rather high. However, after irradiation at 700° C, the R value was only 12, and the specimen was free from distortion or cracking (Fig. 6). This material therefore gives the peculiar result of showing a better irradiation behavior at a

FIGS. 5–8.—Fig. 5: Uranium–28 at. per cent molybdenum–6.5 at. per cent plutonium alloy irradiated at 500° C to a burnup of 0.50 per cent of all atoms. Increase in volume estimated from change in dimensions, 11.2 per cent; *R* value 45. Specimen is badly cracked. Fig. 6: Uranium–28 at. per cent molybdenum–6.5 at. per cent plutonium alloy irradiated at 700° C to a burnup of 0.22 per cent of all atoms. Increase in volume, 2.7 per cent; *R* value 12. Appearance is good. Fig. 7: Uranium–28 at. per cent molybdenum–18 at. per cent plutonium alloy irradiated at 500° C to a burnup of 0.44 per cent of all atoms. Specimen is badly distorted. Fig. 8: Uranium–28 at. per cent molybdenum–18 at. per cent plutonium alloy irradiated at 700° C to a burnup of 0.42 per cent of all atoms. Surface is cracked and distorted; one end has erupted.

FIGS. 9–10.—Fig. 9: Thorium–15 at. per cent plutonium alloy irradiated at 500° C to a burnup of 0.54 per cent of all atoms. Increase in volume, 14 per cent; *R* value 22. Appearance is good. Fig. 10: Zirconium–40 at. per cent plutonium alloy irradiated at 500° C to a burnup of 0.83 per cent of all atoms. Increase in volume, 5.4 per cent; *R* value 8.5. Appearance is good.

higher irradiation temperature than at a lower temperature, probably because of the decrease in brittleness of the alloy as the temperature is raised. This alloy was probably single-phase body-centered cubic during irradiation. The uranium-molybdenum alloy containing 18 at. per cent plutonium showed gross distortion and cracking after irradiation (Figs. 7 and 8), and the dimensional changes were so non-uniform that segregation of the plutonium or molybdenum within the specimen was suspected. The structure of these alloys containing 18 per cent plutonium is complex at 500° C, consisting of three phases. At 700° C, in the absence of segregation, the structure is single-phase gamma, but only slight segregation is required to cause the appearance of free molybdenum. It would appear that an improvement in these ternary alloys could be obtained by reducing either the plutonium or the molybdenum content to obtain less brittle materials. The appearance of irradiated specimens of the uranium-molybdenum-plutonium alloys is shown in Figures 5–8.

Thorium Alloys

Of a pair of specimens based on thorium, one was enriched with 15 at. per cent plutonium and the other with 47 at. per cent U^{235}. Both specimens were irradiated at 500° C to a burnup of greater than 0.5 per cent of all atoms. R values of the Th-U and Th-Pu specimens irradiated at 500° C were 13.6 and 22, respectively; the increase in volume was uniform and not accompanied by wrinkling or cracking. Figure 9 shows the Th-Pu alloy after irradiation. The R values were thus somewhat lower than that of unalloyed uranium also irradiated at 500° C. The change in irradiation behavior in going from uranium-based to thorium-based fuel, or in changing the fissile content from uranium to plutonium, appears to be rather small for irradiation at 500° C. The structures of the thorium-plutonium and thorium-uranium alloys are quite different, the thorium-plutonium alloy consists of a single-phase face-centered cubic structure having no phase changes during cooling from the melting point. The thorium-uranium alloy consists of two phases, based on uranium and thorium, respectively, and forming an interpenetrating network.

Zirconium Alloys

The final pair of specimens was based on zirconium, one containing 40 at. per cent plutonium and the other 20 at. per cent uranium, irradiated at 500° C. The R values obtained for both specimens were very low, and the appearance after irradiation was good (Fig. 10). Zirconium would therefore be an excellent, though expensive, diluent for plutonium, since it also has a low neutron-capture cross-section.

I wish to thank members of the Irradiation Section, Plutonium Section, and Chemical Metallurgy Group of the Metallurgy Division, A.E.R.E., Harwell, for permission to refer to work as yet unpublished, and to thank J. Williams, J. Syrett, and M. B. Waldron for useful discussions.

CHAPTER XXXII

THE HANFORD PLUTONIUM-RECYCLE TEST-REACTOR DESIGN CONCEPT

R. M. Fryar

The General Electric Company, at the request of the Atomic Energy Commission, has undertaken a program to develop technology for the use of uranium-recycle–plutonium fuel cycles in thermal heterogeneous power reactors. The program embraces the optimization of technical bases for plutonium-recycle operation, the minimization of fuel-cycle costs, and the actual demonstration of uranium-plutonium fuel cycles, self-sustaining in respect to plutonium requirements. An essential tool in the conduct of this program is a test reactor. The criteria for this reactor and the reactor design scope that have been developed will be the subject of this chapter. I should first like to emphasize that the selection of this reactor design is not meant to imply that this reactor type is necessarily optimum for the uranium-plutonium cycle. This type was selected because we believe that it has certain outstanding advantages as a facility for the experimental and testing program.

As a research tool, the prime prerequisite of the facility is that it provide the maximum amount of flexibility in as many parameters as possible, consistent with safety and available funds. Operation of the Plutonium-Recycle Test Reactor (PRTR) is to provide data on nuclear physics, fuel-element design and performance, and engineering technology which are essential to the plutonium-recycle concept, including, but not necessarily restricted to, the self-sustaining cycle. Physics tests will yield information on critical start-up conditions, reactor-control characteristics, and reactivity effects of various lattice parameters and conditions of operation. Such information as long-exposure characteristics, isotope composition, and material balance under both equilibrium and non-equilibrium conditions is necessary to reduce the cycle concept to practice. Testing of a number of fuel-element designs will be required in developing feasible and economical high-exposure uranium and plutonium elements. Plutonium-bearing fuel elements have received comparatively little study and are particularly vital to the economics of plutonium recycle. Finally, the test reactor must provide the development-scale quantities of irradiated fuel elements required for conduct of chemical reprocessing and refabricating studies.

R. M. Fryar is with the General Electric Company, Hanford Atomic Products Operation, Richland, Washington.

The PRTR is a vertical-pressure-tube type of reactor, is heavy-water moderated and cooled, and has a thermal power rating of 70 megawatts. A perspective view of the reactor and building is shown in Figure 1. Fuel elements are charged into and discharged from the 85 Zircaloy-2 process tubes from the top face. For the initial loading, about 35 per cent of the fuel elements will be spike-enrichment plutonium elements, and the remainder will be natural UO_2 elements, although subsequent loadings may be uniformly enriched (see chap. xxxi). Goal exposure for the uranium oxide fuel elements will be about 5,000 megawatt days per ton.

The reactor proper, including the process equipment, will be housed in an all-welded steel cylindrical containment vessel, 80 feet in diameter by 122 feet high, extending about 75 feet above grade. The containment vessel will be designed and

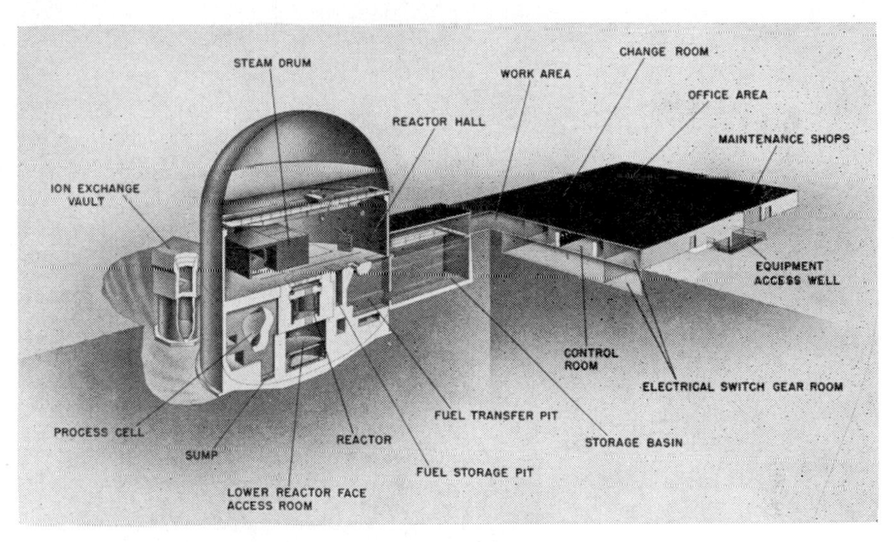

FIG. 1.—The PRTR

constructed in accordance with the pressure-vessel codes, at a design pressure of 15 psig. This design pressure is based upon the pressure (12.5 psi) that would be generated inside the tank in case of the worst credible incident, which is defined as the most serious possible incident, even though with a negligibly small probability of occurrence. In the case of the PRTR, this incident relates to failure of the coolant headers and would occur if ordinary water were used instead of heavy water as the fuel-element coolant. If, as planned, heavy water is used as the fuel-element coolant, the resulting pressure within the containment shell would be only about 3 psi.

Next to the containment shell and attached thereto is a single-story service building, 100 feet long by 80 feet wide. This wing, of conventional construction, is to provide for necessary services, such as a control room, office area, reception and conference room, and non-process work area.

In the reactor proper the heavy-water moderator, which is unpressurized, is contained in an aluminum tank called the "calandria," as shown in Figure 2. This tank is 84 inches in inside diameter and 115 inches in height. Passing vertically through the calandria are 85 fuel channels, 18 shim-rod control channels, and 13 flux-monitoring channels. The 85 process tubes are arranged on an 8-inch equilateral triangular lattice. To insulate the low-temperature moderator from the high-temperature primary coolant, a double-tube arrangement is used. The outer or shroud tube, which is part of the calandria, is made of aluminum. A quarter-inch gap is provided between the shroud tubes and the vertical process tubes. This quarter-inch helium thermal insulating gap is sufficient to restrict the flow of heat from the process tubes to the calandria to less than 1 megawatt for all process tubes.

Process tubes will be fabricated of Zircaloy-2 and will be 3.250 ± 0.010 inches in inside diameter, with a 0.154 ± 0.008-inch wall thickness in the reactor core. The

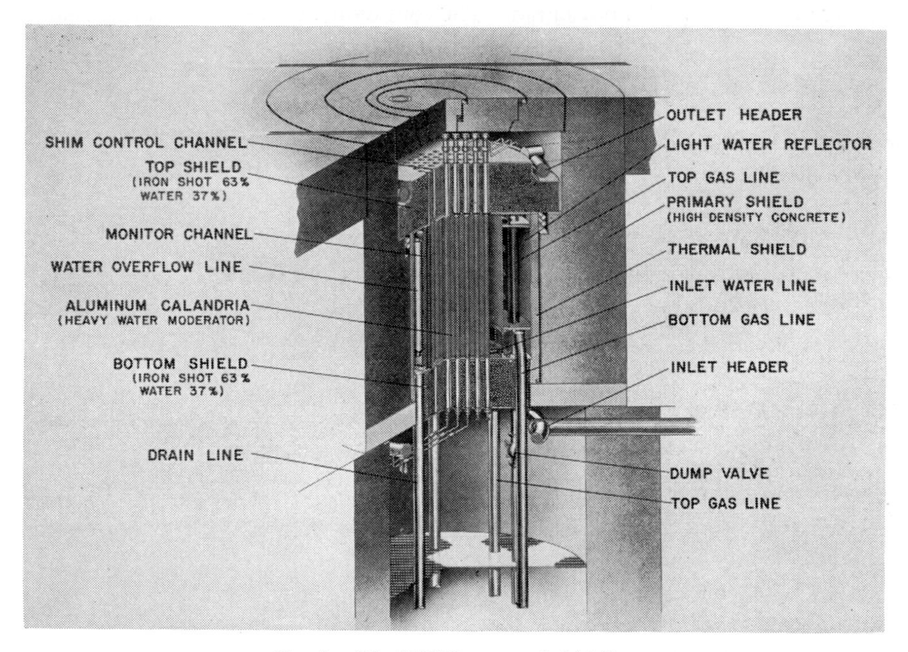

FIG. 2.—The PRTR core and shielding

lower ends of the tubes will be tapered to a smaller diameter (2 inches O.D.), with a greater wall thickness for ease of assembly at the lower face piping.

Attached to the lower end of the calandria is an annular dump chamber which is connected to the calandria through a weir. In addition to being a collecting header for part of the moderator effluent which is always spilling over the weir, the dump chamber is designed to serve as a temporary storage volume for a significant fraction of the moderator during the reactor scram. By having this volume here, it is possible to obtain a very rapid initial drop in moderator level upon receipt of a scram signal. This allows a very rapid decrease in reactivity in the first fraction of a second after an emergency condition has developed.

Near the top of the calandria is an annular drain header connected to the moderator tank by orifices. About 65 per cent of the moderator circulation of effluent flows through this drain header, the remainder flowing over the dump-chamber weir. The moderator flow rate is approximately 1,100 gallons per minute, accompanied by a temperature change from 137° to 160° F.

Outside the calandria tank proper is a water reflector, which is contained in an annular cylinder 11 feet in outside diameter, 7 feet in inside diameter, and 6 feet, 10 inches in height. This tank is integral with three common walls. The reflector, which may be either light or heavy water, is cooled by the recirculation of approximately 300 gallons per minute at the same inlet and outlet temperatures as the moderator.

The calandria and reflector are surrounded on all sides by biological shielding. The side biological shield is made of high-density concrete, 71 inches thick and about 21 feet deep. The inside surface is faced with a steel liner ½ inch thick, which also functions as the gas seal for the reactor atmosphere of helium. Supported inside this steel liner is the thermal shield, consisting of a series of 12 iron slabs, 6 inches thick. The purpose of this auxiliary cast-iron shield is to remove about 90 per cent of the energy escaping from the reactor core before it reaches the concrete biological shield.

The calandria and core components of the reactor rest on the bottom shield, which is supported by the lower end of the concrete biological shield. The bottom shield and the top shield of the calandria consist of cylindrical steel tanks, 40 inches thick, pierced by process tube, access, and monitoring channels. These shielding tanks are filled with a mixture of iron pellets and water, 63 per cent iron and 37 per cent water by volume, giving a mixture density of approximately 285 lb/ft^3. The top shield is similar to the bottom shield and supports the process piping assembly and the reactor fuel elements.

Directly above the top shield is the rotating-shield assembly, which consists of two rotating shielding disks. The center disk is eccentrically located and supported on ball bearings, so that the eccentrically located access hole in this disk can be positioned over any process tube. The rotating-shield components are 27 inches thick and consist of steel cylinders filled with a mixture of iron punching and limonite grout having an average density of about 320 lb/ft^3. All components of the reactor core, including process tubes, piping, calandria, and top and bottom shields, are designed in such a manner that they can be readily removed at some future date to permit replacement of the components.

Since a major objective of the plutonium-recycle program is the development of suitable fuel elements for use with the recycle concept, it is impossible to predict what the final fuel elements will look like; however, the developments are sufficiently far along that the likely candidates for the initial loading can be selected. A leading candidate for the uranium loading is the concentric-cylinder fuel-element assembly shown in Figure 3. The feed-fuel elements will be made of sintered uranium dioxide. It is expected that the uranium dioxide will be of PWR grade, extruded, or pressed and sintered, to about 90 per cent of theoretical density. These fuel elements will be jacketed in Zircaloy-2 or perhaps Zircaloy-3.

For the initial loading, it is anticipated that the plutonium enrichment (spike) elements will have plutonium-alloy cores, which will contain plutonium in aluminum or aluminum-silicon alloy. These elements may be in the same form as the feed-fuel elements previously shown or may be in the form of a 19-rod fuel-element cluster, as shown in Figure 4.

Heat generated in the fuel elements is removed by circulating heavy water through the process tubes and a boiler-type heat exchanger. Steam is generated in the exchanger and dissipated in the river. Approximately 66 megawatts of reactor heat are removed by the coolant experiencing a temperature change from 480° to 530° F. The 8,400 gallons per minute of primary coolant flow to the steam-generator heat exchanger is cooled by boiling about 206,000 pounds of water into 425 psia saturated steam. Most of the steam is piped to the Columbia River, where it is condensed in a jet condenser.

The primary control system of the reactor is based on the principle of variation

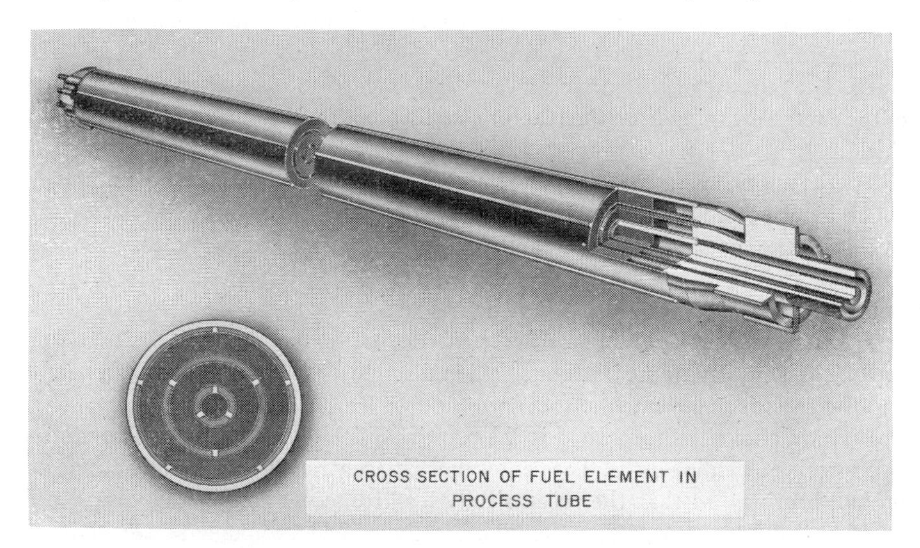

FIG. 3.—The nested tubular fuel element

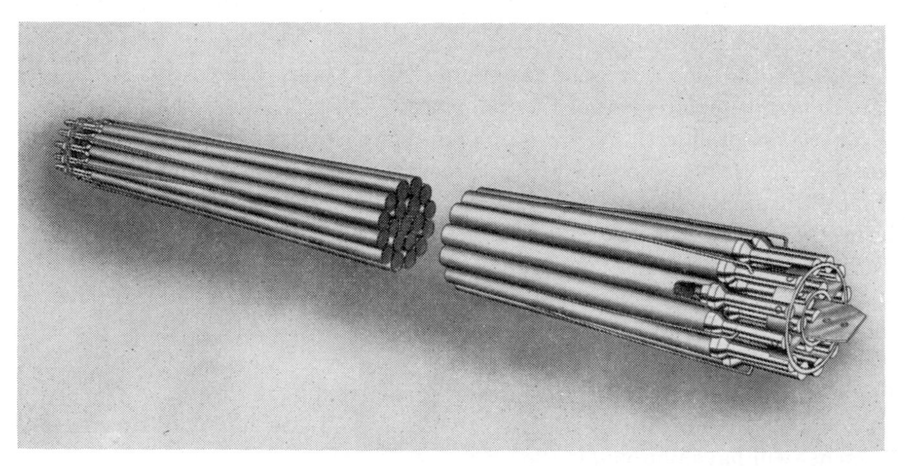

FIG. 4.—The 19-rod cluster fuel element

of reactivity by varying the level of the moderator in the calandria. The moderator level in the calandria will be maintained and varied by a helium gas-pressure balance system. Figure 5 is the schematic diagram of this system. The moderator level is controlled by applying the differential helium gas pressure to the surface of the moderator in the calandria and to its surface at the annular weir in the dump chamber. When the pressure differential is established, flow over the weir will be suppressed, and the moderator will seek a level at which the liquid head exactly balances the applied pressure differential. The differential pressure will be established and maintained by a helium compressor and associated control valves in a dynamic-type control system. In general, the maximum rate of increase in reactivity,

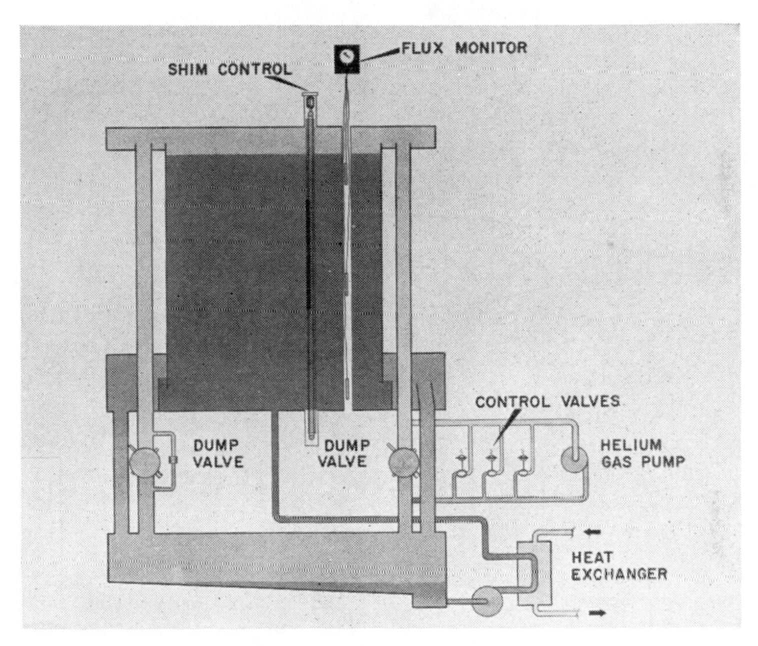

FIG. 5.—The PRTR control system

governed by increases in moderator level, will be limited by the capacity of the helium system. This helium system is sized sufficiently small to maintain the rate of reactivity change at a value less than the response time of the control system. One advantage of moderator-level adjustment for reactivity control is that the system is particularly adaptable to automatic control. In the PRTR, adjustments to the moderator level will be made by means of an automatic controller which takes its signal from the reactor neutron-flux level and/or the reactor period.

In addition to the primary control furnished by the moderator level, a shim control system is provided for close adjustment of reactivity. The shim control rods are shown diagramatically in Figure 6. The primary purposes of these shim controls are (1) maintenance of the moderator level within the normal operating range; (2) compensation for fuel burnout; (3) provision for xenon override and flattening of the neutron flux; and (4), if desired for experimental purposes, depression of the flux in certain portions of the reactors. Operation of the shim control is entirely manual, by means of switches located in the control room. The shim control is not intended as a safety device. Each shim unit consists of two half-rods about 3 feet long. The position of each rod is individually adjustable by a miniature stainless-steel stud chain driven by a geared miniature electric motor. Indication of rod position is transmitted to the control room by synchro transmitter linkages. The rods are exposed directly to the moderator, eliminating the need for a separate cooling system. In the withdrawn position, the rods are contained within the bottom bio-

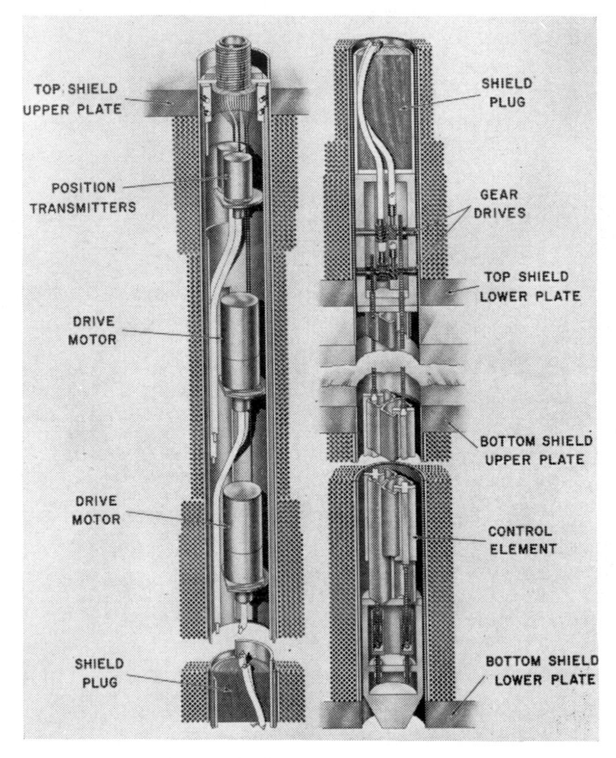

FIG. 6.—The PRTR shim control

logical shield in tubes extending downward from the calandria proper. Each of the 18 shim rods is a self-contained unit and is easily replaceable by means of the normal charge-discharge vehicle.

A reactor scram or emergency shutdown is quickly accomplished by a rapid drainage of the moderator from the calandria. A scram is initiated automatically whenever a condition exists which threatens the integrity of the reactor or its auxiliaries, or it may be initiated manually by the operator. The scram is accomplished by equalizing the gas pressure between the top gas plenum and the moderator storage tank. This results in extremely rapid flow of the moderator out of the calandria, at an average rate of about 20,000 gallons per minute, so that within 0.7 second after receiving the scram signal the moderator level will have fallen 2 feet.

The charge-discharge operation is based on the use of a large self-propelled fueling vehicle, which carries an unshielded tube for charging operations of unirradiated fuel elements and a heavily shielded cask for both the charging and the discharging of irradiated fuel elements. The schematic arrangement of the charge-discharge operation is shown in Figure 7. The charge-discharge machine proper is about 25 feet high and weighs 50 tons. Much of the load is concentrated in the 12.5 inches of lead required as shielding to protect personnel during the discharge operation. Control of the vehicle will be by the operator riding the vehicle. New non-irradiated elements are removed from shipping crates and stored in the pit under the floor of the reactor hall. From this location the fueling vehicle using the charging tool can withdraw an element, carry it to the reactor, and insert it into one of

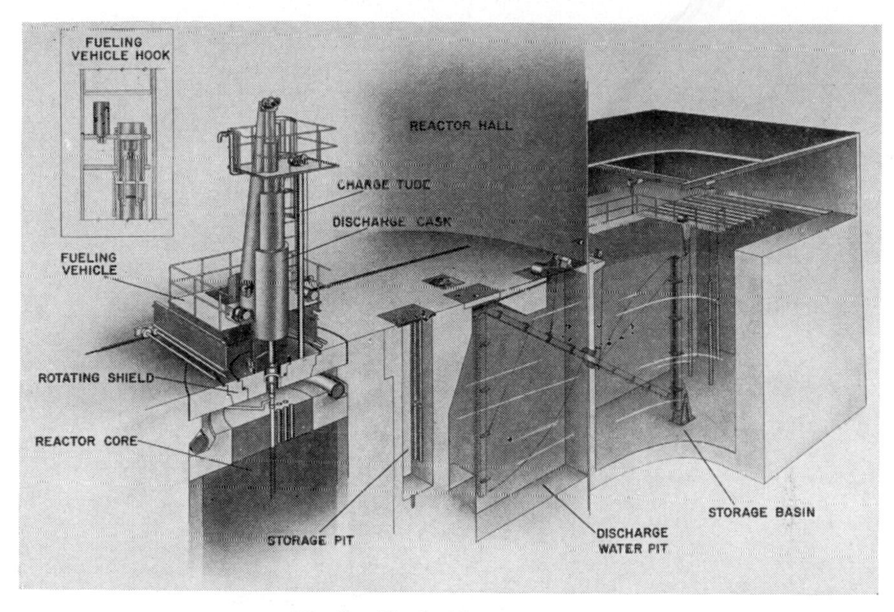

FIG. 7.—The fuel-handling system

the process tubes. Similarly, when discharging, the vehicle and discharge cask can be positioned over the proper process tube and the hook lowered into the tube, where it may be attached to the fuel element. The fuel element is then withdrawn into the cask and transported to an underwater transfer carrier. This carrier, which is part of the transfer conveyor, receives the element from the cask, tips it up into an inclined position, and, by gravity, allows it to roll out of the reactor containment vessel endwise onto a similar set of conveyor tracks in the storage basin, where the whole assembly is again set upright. The basin is equipped with a bridge crane for moving the fuel elements from the carrier to the desired storage position.

The plutonium-recycle program being carried on by the General Electric Company is scoped to give broad coverage to pertinent problems in the field of nuclear physics, uranium and plutonium fuels technology, reactor engineering, and chemical reprocessing of fuel elements. As an essential tool for the program, the test reactor is being designed not only to serve as a facility for the collection of experimental data but also to demonstrate the fuel cycle under various conditions. We who are associated with the design of the PRTR feel that we have met the design objectives by providing a facility that is characterized by a high degree of flexibility, by low neutron temperature, by independent accessibility of the fuel elements, and by comprehensive instrumentation for experimental purposes.

CHAPTER XXXIII

A PLUTONIUM LIQUID-METAL-FUEL REACTOR

David H. Gurinsky

For the past several years personnel of the Nuclear Engineering Department at Brookhaven National Laboratory have been working on the components of a liquid-metal-fuel reactor (LMFR) because such a reactor shows promise as a cheap fission heat source for the generation of electrical power. The principal advantages and disadvantages of such a system have been given in a number of papers (1, 2). A few of the expected advantages which have particular bearing on the metallurgi-

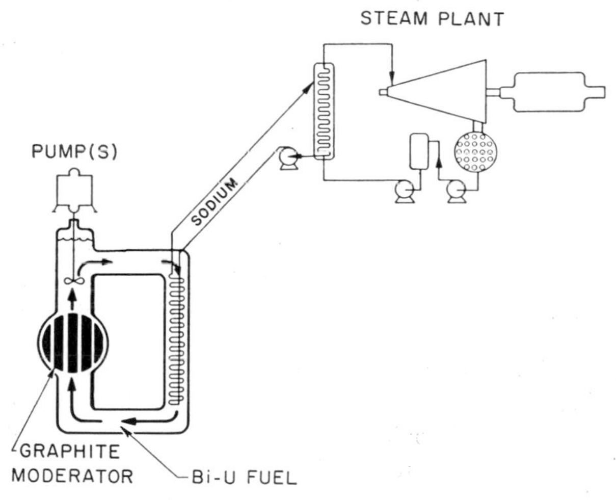

FIG. 1.—The liquid-metal-fuel reactor. A schematic diagram showing the reactor and the steam plant.

cal problems are (1) high-temperature operation and resulting thermodynamic efficiency; (2) simple fuel makeup; (3) lack of radiation damage to the liquid-metal fuel; and (4) unlimited burnup, if continuous chemical processing is incorporated.

A schematic diagram of the reactor system that is receiving the most attention at BNL is given in Figure 1. In this reactor the fuel is circulated through the moderator or core of the reactor and then to the heat exchanger. The core is of such a

David H. Gurinsky is at the Brookhaven National Laboratory, Upton, New York.

shape and size that a chain reaction can occur. The heat generated in the fuel is carried to the heat exchanger and is transferred first to sodium and then to water.

The fuel which is currently under investigation at BNL is a solution of uranium in bismuth. This fuel was selected because it was shown that sufficient uranium could be dissolved in bismuth for a fission chain reaction if a highly enriched uranium was used. Uranium was initially selected for study because it was hoped that ultimately a thermal breeder could be designed using the U^{233}-thorium cycle.

If breeding is not the prime consideration, it is apparent that any fissile material which promises to give economic power can be employed in this reactor scheme. The reactor schematically shown in Figure 1 has certain features, i.e., high temperature and good specific power, which make it attractive as a "burner" of fissile

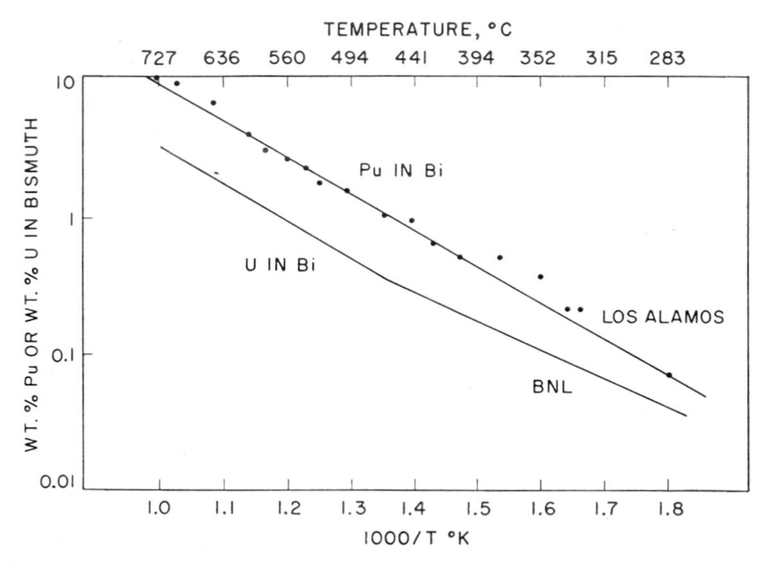

FIG. 2.—The solubilities of plutonium and uranium in bismuth

material if the technical problems can be solved and if the price of the fissile material is low enough. In the future it may be desirable to design a reactor that can efficiently burn the recycled plutonium of fast reactors and the plutonium produced in slightly enriched thermal reactors. This is true because most of the reactors currently under design employ solid-fuel elements. The costs of these elements is high at present, and these costs are expected to go higher when the fuel elements have to be refabricated remotely from fissile material that cannot be decontaminated. This will have the effect of lowering the value of such radioactively contaminated fissile material. Since a fluid fuel lends itself to remote makeup (fabrication), decontamination of the fissile material prior to dissolution in the fluidizing medium will be governed primarily by the physics requirements, i.e., neutron losses in the contaminants.

In Figure 2 the solubilities of plutonium and uranium in bismuth are given. It is seen that the solubility of the former is roughly twice that of uranium. Further, since the fission cross-section of plutonium is roughly twice that of uranium, the amount of plutonium required for a fuel system utilizing plutonium is half that for uranium. F. T. Miles (3) of BNL, using the data of Jaffey (4) and P. R. Fields *et al.* (5) has made some calculations on the concentrations required for a reactor in which plutonium has come to equilibrium with its higher isotopes, Pu^{240}, Pu^{241}, and Pu^{242}. These build up as a result of recycling plutonium in a reactor or when natural or slightly enriched uranium fuel is taken to a high burnup. Miles finds that the concentrations required are about twice that for Pu^{239} in one of the designs he investigated. Because of the higher fission cross-section of plutonium, the solubility advantage of plutonium-bismuth over a uranium-bismuth fuel remains even after buildup of the higher isotopes of plutonium.

A metallurgical research and development program will have to be carried out before a plutonium fuel system can be used in the LMFR. Based on the experience gained in the study of the uranium-bismuth fuel system, a minimum program would have to include the following: (1) a determination of the effect of uranium, fission products, corrosion inhibitors, deoxidants, and dissolved constituents of materials of construction on the solubility of plutonium in bismuth; (2) a study of the stability of the fuel system with respect to a moderator and materials of construction; (3) an evaluation of the corrosiveness of the fuel system on materials of containment; and (4) the testing of construction materials and moderating materials in the presence of radiation.

This program will now be discussed briefly. It has been shown in the studies of the uranium-bismuth fuel that uranium is not quantitatively recovered as dissolved uranium when it is added to pure hydrogen-treated bismuth. Apparently, some of it reacts with impurities present in the bismuth and the test crucible. To obtain a quantitative recovery of the uranium, magnesium or other strong deoxidants such as thorium or calcium must be added to the bismuth. A practical fuel must therefor econtain deoxidants.

Because it is economically desirable to use standard materials for piping, heat exchangers, and vessels and because these materials are corroded by bismuth, it has been necessary to add corrosion inhibitors to the uranium-bismuth fuel. Zirconium and magnesium have been found useful for this purpose. The low chrome steels are the standard materials of construction that meet the requirements of strength, air-oxidation resistance at elevated temperature, and economic considerations. These steels dissolve to a limited extent in uranium-bismuth fuel. Further, when fission takes place, fission products are produced and dissolve in the fuel. Thus it is seen that a real fuel, as a result of the above considerations, is a complex alloy which contains deoxidants (magnesium), inhibitors (zirconium), fission products (strontium, rare earths, etc.), and construction material constituents (iron, chromium, nickel, etc.).

Figures 3 and 4 show the effects of magnesium, nickel, and zirconium on the solubility of uranium in bismuth. It is seen that all these metals lower the solubility if they are present in sufficient quantities. Recent experiments at BNL suggest that if the fission products are allowed to build up, they also depress the solubility. Since a knowledge of the solubility effects is so basic to the design of this type of reactor, the first work to be done on the plutonium LMFR is a determination of the effects of uranium, magnesium, and zirconium on the solubility of plutonium in bismuth.

A few exploratory experiments (6) on the compatibility of the plutonium-bismuth fuel with respect to moderators and materials for containment have been performed. The intermetallic compounds Pu_6Fe and $PuFe_2$ were added to molten bismuth and held at 800° C for 200 hours. Both compounds reacted to form $PuBi_2$ with the rejection of free iron. This result is similar to one obtained at BNL when U_6Fe and UFe_2 were contacted with bismuth. Also, the examination of a graphite crucible in which a bismuth alloy containing 10 a/o plutonium was held for 2 days at 800° C showed very little evidence of any reaction.

Although the initial experiments with iron and graphite indicate that these may be useful as crucible materials for the solubility and phase-relationship studies, much more work will have to be done before these can be considered for construction and moderator materials. For graphite, it will have to be shown that plutonium does not react with it to form carbides at expected reactor temperatures. With the

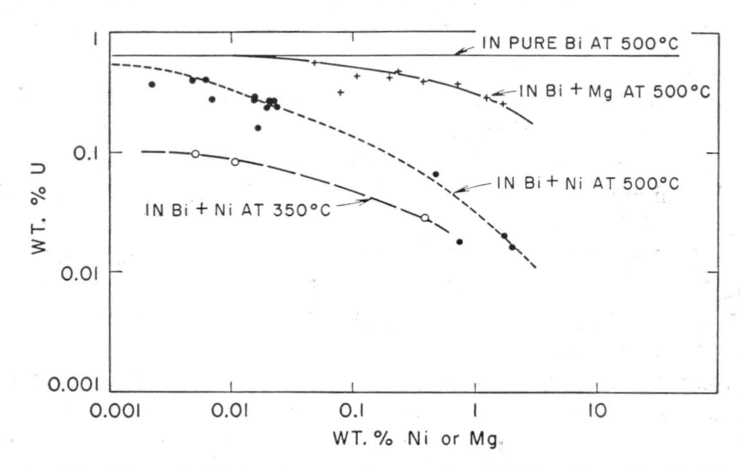

FIG. 3.—The effects of nickel and magnesium on the solubility of uranium in bismuth

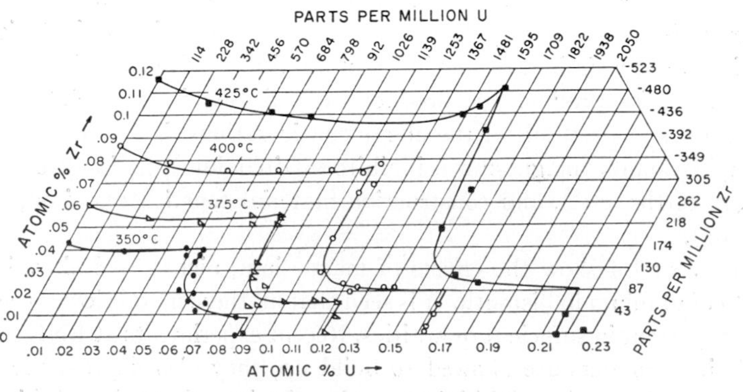

FIG. 4.—Liquidus isotherms in U-Zr-Bi system

large surface area available for reaction even in the so-called impervious dense graphites now under development, considerable fissile material can be tied up. To prevent the reaction of impurities on the graphite with the uranium in the uranium-bismuth alloy, it has been found necessary to pretreat the graphite by contacting it with bismuth containing magnesium and/or zirconium prior to dissolving the uranium in the bismuth. Figure 5 shows the effect of adding zirconium to a lead-bearing uranium-bismuth alloy contained in a graphite crucible. Zirconium reacts with the graphite to form ZrC, which is more stable thermodynamically and forms in preference to UC. From the available thermodynamic data on PuC (7) and ZrC, it is expected that zirconium will react with graphite—a possible moderator material—and thereby prevent the tieup of plutonium on the graphite surfaces. However, although the thermodynamics may appear favorable, the relative activities of plutonium and zirconium in bismuth and the kinetics of the reaction will determine whether zirconium will prevent the formation of plutonium carbide. Since beryllium is a possible moderator in a plutonium LMFR, it must also be tested in plutonium-bismuth solutions to determine whether plutonium reacts with it.

In the evaluation of the feasibility of steels as materials for construction it will be necessary to determine (1) whether plutonium reacts with any of the constituents of the steels, and (2) whether the plutonium-bismuth fuel system plus additives corrodes or dissolves the steels. In the work on the uranium-bismuth system it has been shown that, in loops with a temperature differential (thermally circulated or mechanically pumped), the constituents (iron, chromium, manganese, silicon) of the low chrome steels dissolve in the fuel stream in the hot section and precipitate from the fuel in the colder sections. To minimize this mass transfer dependent on temperature differential, zirconium was added to the uranium-bismuth alloy.

The effectiveness of the zirconium (8) is apparently due to the formation of zirconium nitride and/or carbide. These compounds are formed on the surface of the steel by a reaction of the carbon and/or nitrogen in the steel with the zirconium in solution. The corrosion/mass-transfer problem has not been adequately solved for the uranium-bismuth fuel. It will pose an even more difficult problem in the plutonium-bismuth fuel if the published thermodynamic data on PuN (7) are correct, since PuN is stated to be more stable than ZrN. It will, therefore, be necessary to run static and dynamic tests to determine whether zirconium inhibits mass transfer in a plutonium-bismuth fuel. Such a program is at present in the planning stages. As soon as facilities become available, convectively circulated loops will be set up to assess this problem.

With respect to item 4 of the minimum program listed above, it will be necessary to repeat some of the out-of-pile loop tests in a reactor to determine whether radiation neutrons and fission-product bombardment destroy the protective films formed or cause other unexpected reactions between the fuel and the moderator and construction materials. The most serious problem in any fluid-fuel reactor is corrosion. Whereas, in a solid-fueled reactor, radiation effects are limited to the core, in a fluid reactor the whole system is subjected to radiation (delayed neutrons and fission-product bombardment). The corrosion problem is made more difficult by the de-

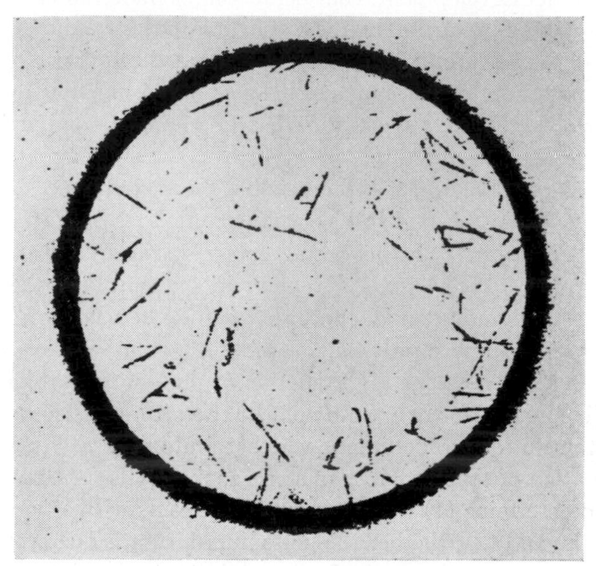

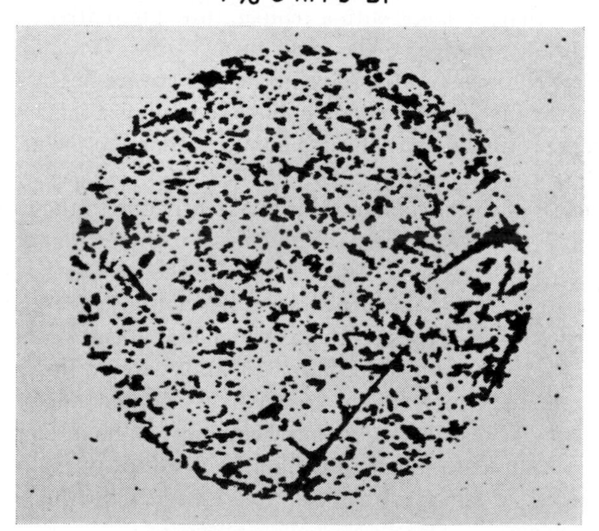

FIG. 5.—Autoradiographs of cross-sections of graphite crucibles and their alloy charges after 100 hours at 1,000° C. In the case of the U-Zr-Pb-Bi alloy, all the uranium remained in the charge. In the absence of zirconium in the alloy, much of the uranium in the charge reacted to form a thick coating of UC on the inside of the crucible.

sirability of high temperature for high thermodynamic efficiency. Although there are many problems that must be solved before liquid-metal reactors are feasible, the economic advantage which may be gained by the use of a liquid-metal fuel offers sufficient incentive for the study of these systems.

REFERENCES

1. *The Reactor Handbook*, **2**:741–97. (Rept AECD-3646, 1955.) Also the same pages in *Selected Reference Material on Atomic Energy*, Vol. **3.** Geneva: U.S. Atomic Energy Program, August, 1955. Or in *Reactor Handbook: Engineering*. New York: McGraw-Hill Book Co., Inc., 1955.
2. GURINSKY, DAVID H., and DIENES, G. J. *Nuclear Fuels*, pp. 296–334. Princeton, N.J.: D. Van Nostrand Co , Inc., 1956.
3. MILES, F. T. Personal communication.
4. JAFFEY, A. H. "Long Term Variation in Composition and Neutron Yield in Pile Plutonium," *Nuclear Science and Engineering*, **1**:204–15, 1956.
5. FIELDS, P. R., *et al.* "Pile Neutron Cross Sections of the Heavier Plutonium Isotopes," *Nuclear Science and Engineering*, **1**:62–67, 1956.
6. COFFINBERRY, A. S. Personal communication (July 19, 1957).
7. SEABORG, G. T., KATZ, J. J., and MANNING, W. M. *The Transuranium Elements*, p. 863 ("National Nuclear Energy Series, Div. IV, Vol. **14B.**) New York: McGraw-Hill Book Co., Inc., 1949.
8. KAMMERER, O. F., *et al.* "Zirconium and Titanium Inhibit Corrosion and Mass Transfer of Steels by Heavy Metals," *Transactions of the American Institute of Mining, Metallurgical and Petroleum Engineers*, **212**:20–25, 1958.

CHAPTER XXXIV

A FLUID-FUELED FAST-POWER-REACTOR CONCEPT

Russell E. L. Stanford

Atomic Power Development Associates, Inc., is investigating the potential of sodium-cooled, fast breeder reactors for central-station power applications. To develop their full potential, fast breeders must burn and breed plutonium. Therefore, the efforts of research groups, both here and abroad, aimed at the development of plutonium technology are of interest to us.

We are particularly interested in reactor designs that use plutonium in a molten, or at least mobile, form. At present, we are considering two basic approaches. One is called the "paste project." This involves the use of fissionable and fertile materials in the form of small spherical particles settled in a liquid metal carrier, presumably sodium. The other concept, based on the LAMPRE work, involves the use of a molten, or intermittently molten, alloy of plutonium in the core, with probably a paste blanket. I shall limit my remarks to the paste reactor. Our approach to the design of a new reactor is one of developing a very preliminary design concept—only to the point where it can be used in defining the critical problems involved—then seeking workable solutions to these problems.

The APDA paste is a mixture of fine particles of a metal alloy or compound at their settled density, approximately 60 v/o solids, in sodium. The particles are approximately spherical and near uniform size. The size can be from 25 to 250 μ, the limits set by flow considerations. This is not a toothpaste-like or putty-like material but is more on the order of the moving beds used in the chemical and petroleum industries. It is illustrated in Figure 1.

The first block represents a true solution, that is, a single-phase system. The second and third blocks represent two-phase systems of solid particles in a liquid carrier. Slurries hold solids in suspension. The particles in a paste are in the settled or near-settled state and thus not subject to further concentration.

We have found that little is known about the fluid mechanics of settled systems. To develop information in this field, a research project was established at the University of Michigan. In this project, work is done with simulated systems such as sand, glass beads, and various types of metal shot in water and other liquids. A

Russell E. L. Stanford is with the Atomic Power Development Associates, Inc., Detroit, Michigan.

companion project, using U-Na and UO_2-Na pastes, was set up at the Southern Research Institute. Results to date indicate that dense down-flow through a column can be obtained, provided that a constriction is placed in the outlet. The system is illustrated in Figure 2.

The paste flows slowly downward through the narrow channels in the core at a rate influenced by gravity and drag of the liquid on the particles. At the bottom it

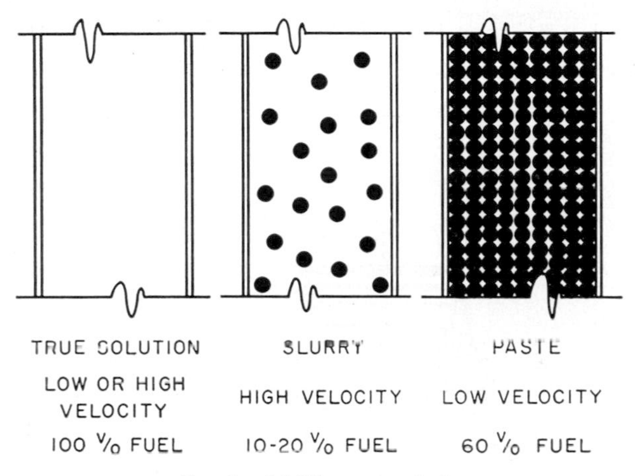

FIG. 1.—Mobile reactor fuels

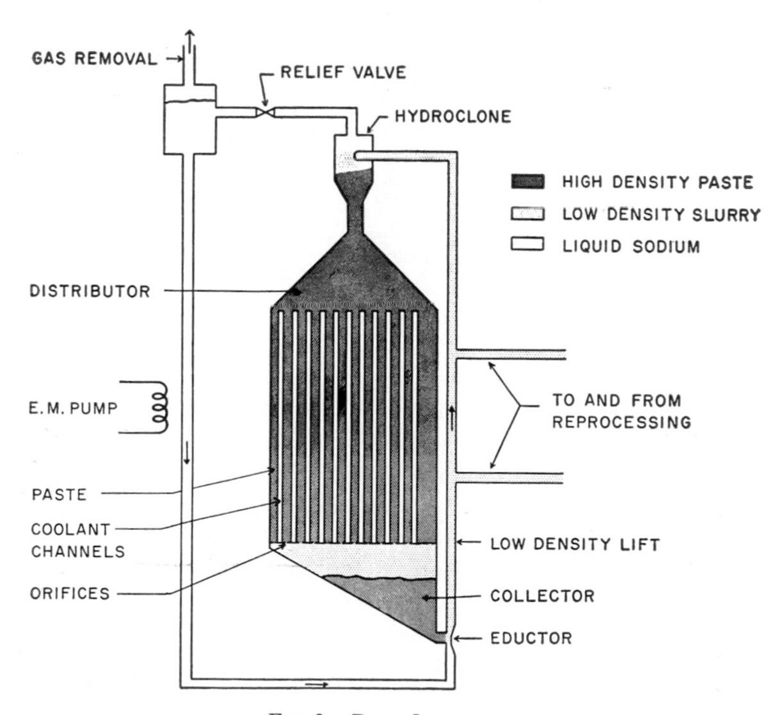

FIG. 2.—Paste flow system

is collected and picked up through an eductor as a suspension in a stream of higher-velocity carrier liquid. This stream returns the particles to a settling unit, possibly a hydroclone, at the top of the reactor. The settling unit discharges the settled paste to the distributor for recirculation through the reactor. The bulk of the carrier stream is returned by a pump to the bottom of the reactor. Side streams are drawn off for gas removal and reprocessing.

The irradiation stability of paste systems is unknown. It has been calculated that small particles, less than 10 μ, will lose most of their fission products by recoil to the liquid medium and be relatively stable. Large particles, greater than 100 μ,

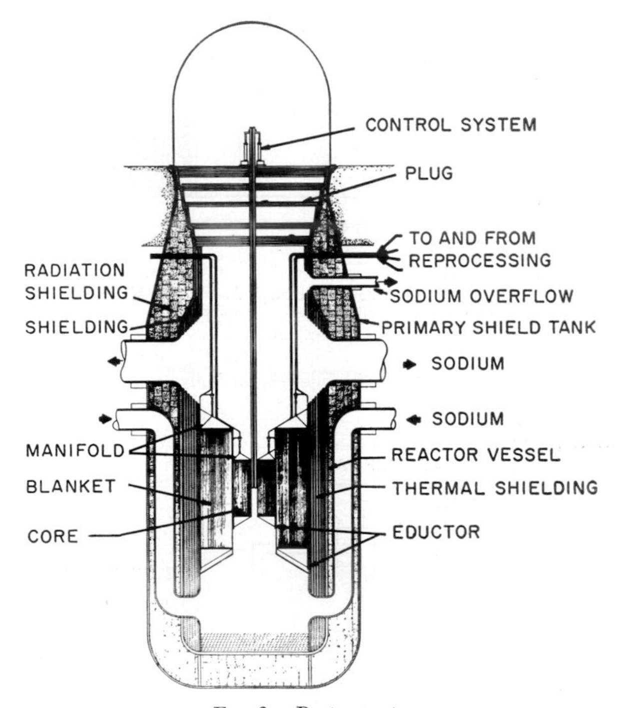

FIG. 3.—Paste reactor

may retain enough of their gaseous fission products to fracture because of internal pressures. These are only two of many possibilities. To investigate what does happen, an irradiation program is being planned at Battelle Memorial Institute.

An idea of the principal features of a 600-MW (thermal) paste reactor is shown in Figure 3. The over-all height is 35–40 feet. The outer tanklike vessel is about 14 feet in diameter. The inner or reactor vessel, about 8 feet in diameter, would be designed to withstand high thermal stresses but only nominal pressures. The space between the outer and inner vessels would be packed with neutron-absorbing material, such as borated graphite.

Within the reactor vessel would be a layer of thermal and gamma shielding, about 6–8 inches thick, probably built up of steel plates. Concentrically located within this built-up structure is the reactor itself. It consists of an outer blanket region about 2 feet thick, surrounding a very compact core which is about 36 inches in diameter.

Also shown are the fuel and blanket paste circulation systems, a provision for some type of control device, a plug through which reactor components could be removed if necessary, and the relative sizes of primary coolant piping required, using sodium in four loops having 16-inch-diameter inlet lines and 30-inch-diameter outlet lines.

Our preliminary paste-reactor design calculations were based on the assumptions shown in Table 1. Of particular significance are the resulting calculated fig-

TABLE 1

PRELIMINARY DESIGN CALCULATIONS FOR A PASTE REACTOR

Total power	600 MW_{th}
Core power	540 MW_{th}
Blanket power	60 MW_{th}
Fuel composition	
Pu-U-X alloy particles	60 v/o
Sodium	40 v/o
Effective thermal conductivity of paste fuel	22 Btu/hr-ft²-° F
Coolant temperature rise	400° F
Coolant inlet temperature	600° F
Coolant outlet temperature	1,000° F
Maximum coolant velocity	40 ft/sec
Core geometry—right hexagonal prism	L/D 1
Allowable heat flux—maximum	1.5×10^6 Btu/ft²-hr
Core composition	
Pu-U-X alloy	32.5 v/o
Stainless steel	12.5 v/o
Sodium (in paste and coolant)	55.0 v/o
Average fuel channel thickness	0.14 inch
Coolant channel thickness	0.09 inch
Plate wall thickness	0.015 inch
Maximum fuel temperature, with 1.5 hot-channel factor	1,540° F
Coolant pressure drop through core	50 psi
Paste fuel flow rate	2 gpm
Paste fuel flow velocity	6 ft/hr
Core dimensions	
Across flats	31 inch
Across corners	36 inch
Height	33 inch
Volume	16 ft³
Heat transfer surface	1,500 ft²
Average heat flux	1.24×10^6 Btu/ft²-hr
Maximum to average heat flux	1.21
Power density	33,800 kw/ft³
Coolant flow rate through core	15×10^6 lb/hr
Coolant flow rate through core	35,000 gpm
Coolant flow area in core	2 ft²

ures of 1,540° F for the maximum fuel temperature, allowing a 1.5 hot-channel factor, and the required 1,500 square feet of heat-transfer surface.

A maximum fuel temperature of 1,540° F is only 80° F below the boiling point of sodium at atmospheric pressure. As yet, we do not know the effect of boiling the carrier liquid in the pastes, but we suspect that boiling will be undesirable. In a fast power reactor, provision must be made for power excursions 2 to 3 times the design power level. This indicates an undesirably low maximum temperature limit for sodium pastes and the possible need for a carrier liquid having a higher boiling point. Likewise, a stainless-steel container is incompatible with the use of a U-Pu base alloy if power excursions result in temperatures above U-Pu-Fe melting temperatures. This may mean that an oxide particle will be required. Problems of this type have not yet been studied because, so far, we have concentrated on the flow characteristics of pastes and the heat-removal aspects of systems using this type of fuel.

Elaborating on the core design, 1,500 ft^2 of coolant surface must be built into a structure that is approximately a 3-foot right cylinder. This structure has requirements of very close tolerances, a high degree of integrity against leaks, and volumetric change limitations unique to fluid-fuel systems. Recognizing that many design features of this core structure are yet far from solved, a design which seems to hold promise and on which our current calculations are based is illustrated in Figure 4.

Alternate channels in a radial direction are for fuel, those in between for coolant. There is also the possibility of radial flux flattening by varying the fuel channel thickness. This structure would be fabricated of 15-mil stock, with the plates about 0.1 inch apart, braced every 0.25 inch.

In the design of this core, provisions must be made to compensate for effects of the following temperature and power coefficients of reactivity, which are unique to internally cooled, fluid-fueled reactors: (1) a relatively large negative temperature coefficient due to fuel expansion; (2) a positive temperature coefficient due to expansion of the core structure, which increases the amount of fuel contained in the core; (3) a negative power coefficient due to the increased temperature drop through the fuel-channel container walls, which reduces the amount of fuel contained in the core; and (4) a power coefficient caused by varying pressure differences between fuel and coolant. The resulting wall deflection affects the amount of fuel contained in the core. The kinetics of these effects are very important.

The description of a reactor design concept has stressed the application of pastes as fuel. Paste systems are also applicable to breeder blankets for other types of fast reactors. Use as a blanket may actually be the first application of pastes, since the thermal and irradiation requirements should be much less severe.

Pastes have some apparent advantages over molten alloys, for several reasons.

1. Pastes have a wide latitude, composition-wise. This allows the use of either U^{235} or plutonium, depending on their relative availability and cost. Pastes make use of the heavy metal in the solid state, and there is a wider range of compositions available in the solid than in the liquid state for desired operating temperatures.

2. The corrosion problems and their imposed restrictions on core structural ma-

terials appear to be less severe for at least certain specific types of pastes than are the problems encountered in molten plutonium and uranium alloy fuel systems.

3. The bulk of fission products are expected to remain in the particle of their origin. This indicates that the deposition of gases on heat-transfer surfaces and other features of the fission-product gas problem may be less serious.

However, in addition to the upper temperature limitations of certain types of pastes—U-Na, for example—the ultimate feasibility of paste systems depends to a

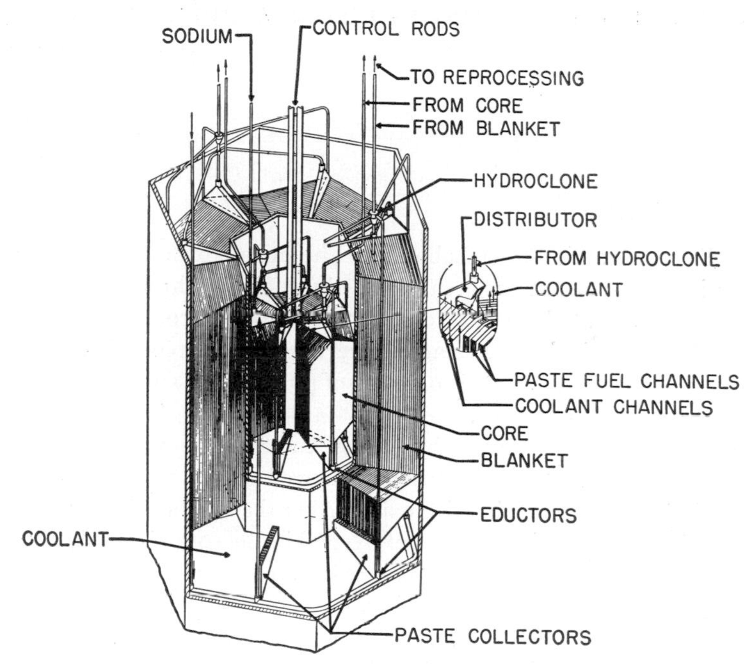

FIG. 4.—Core and blanket of paste reactor

large extent on their irradiation stability, particularly the gross effects of irradiation on their flow characteristics. Until at least preliminary results are obtained, we have little basis for predictions in this regard.

The foregoing description is a very brief treatment of the paste reactor concept. Although not all the answers are known, it is believed that pastes will be applicable to fast-breeder power reactors. For this reason, we should like to see more development work on the fabrication of various types of particles usable in paste systems, particularly those containing plutonium.

CHAPTER XXXV

A MIXED-OXIDE CONCEPT FOR A PLUTONIUM-FUELED POWER REACTOR

W. M. Cashin

INTRODUCTION

At the Knolls Atomic Power Laboratory the mixed-oxide concept for utilizing plutonium in nuclear power has been extended principally through a study of a fuel cycle applicable to a fast-breeder reactor. The mixed-oxide concept involves utilization, as the fuel material in a fast-reactor core, a mixture of plutonium and uranium oxides, generally considered as a 1 to 5 ratio of plutonium to uranium, the uranium being either natural or depleted.

Preliminary physics calculations by Luebke and Sampson were concerned with a fast-reactor concept (1). Their calculations showed that in a breeder reactor some breeding gain is lost (0.2 loss) in using oxide rather than metal, since, in general, the fuel density is lower and the presence of oxygen in the core gives a lower neutron-energy spectrum. However, the loss incurred thereby is believed to be more than compensated for by the economies realized in the fuel reprocessing. Later work (2) on reactor physics has confirmed and extended the original work.

For purposes of discussion, therefore, let us grant a feasible fast-reactor design for oxide and concern ourselves with the fuel cycle. This essentially has been the KAPL emphasis. The status of the over-all program developing the reactor and fuel-cycle concept is summarized in a recent KAPL report (3).

First, what do we hope to achieve by mixed oxide? Economy—economy in fuel preparation, irradiation, and reprocessing. These economic requirements have been of major consideration in all phases—the fuel-element design, its fabrication, its extent of depletion in the reactor, its reprocessing. Where we have had a choice—to do or not to do things in a process, such as to swage or not to swage our final element—the attitude adopted was that of doing the least possible necessary to obtain a successful fuel element. Each step in a process costs money, and each additional step costs more money.

THE FUEL CYCLE (4)

Let us review the fuel cycle in brief form and then devote some time to those details which appear to be of major importance. The cycle is as follows. Mixed plu-

W. M. Cashin is at the Knolls Atomic Power Laboratory, General Electric Company, Schenectady, New York.

tonium-uranium oxide is irradiated in stainless-steel tubing to 50 per cent depletion of the original plutonium. During the irradiation some of the U^{238} in the core is converted to plutonium, so that at the maximum burnup the plutonium content is still more than half that originally present. For this depletion, the reactor residence time is near 1 year.

For a reactor core power of 680 megawatts, the daily through-put of plutonium is near 1.1 kg, and, in terms of conventional nomenclature, exposures are 620 MWD/kg plutonium. Upon removal from the reactor and following a 30-day cooling period, the fuel-containing tubes are sheared into short lengths, and the irradiated oxide is dissolved in nitric acid. This solution is processed by a modification of a currently employed aqueous reprocessing method (Purex process) with the elimination of the plutonium-uranium partition step. Essentially, following the separation of fission products, one has an aqueous nitrate solution of plutonium and uranium to which additional plutonium may be added to bring the ratio up to 1 to 5. From such a solution, the fuel is precipitated with ammonia, yielding ammonium diuranate and hydrated plutonium oxide. This mixture is decomposed and reduced in hydrogen to give a solid solution of plutonium and uranium dioxides. The obtaining of a solid solution, rather than a mechanical mixture of the two oxides, is of essential import to the cycle, for, when in solid solution, the plutonium is soluble in nitric acid (5). Otherwise, PuO_2 is soluble only with great difficulty. The solid solubility holds over the entire composition range, as shown by the work of Mulford and Ellinger (6). The solubility in HNO_3 has been observed at least up to a 1:4 ratio of PuO_2 to UO_2. Beyond this we have made no determination.

The mixed oxide is then fabricated into oxide compacts, which are loaded into stainless-steel tubing with relatively loose fit, a 1–2-mil clearance within a 0.1875-inch diameter; and this fuel element is then ready for reloading into the reactor. Originally, the oxide compacts were formed by dry-pressing the powder into pellets of the right diameter and loading these unsintered into stainless tubing to make the final element. Subsequently, it appeared that the pelletizing of dry powder is a more complex operation than the following: (1) making a paste of the oxide, (2) extruding this paste, and (3) drying and sintering the oxide to give the final compact. With this process, which includes the sintering, we place a requirement on the quality of the oxide produced from the chemical separation, and the requirement is not that it be readily sinterable but that it be consistent, either good or bad in sintering characteristics, but consistent. Fortunately, the process gives very sinterable oxide, partially because of the presence of PuO_2 and partially because of the low temperature of diuranate decomposition.

We use 65–70 per cent dense oxide because it is not difficult to obtain; high-density oxide too often requires special oxide preparation steps, and they cost money. We put a loose tolerance on the fuel-compact diameter; note the 1–2-mil clearance permitted in the fit into the tubing. Such annular voids raise the oxide operating temperatures, but we have not convinced ourselves that such temperatures must be avoided. As a matter of fact, in the fuel-element design we have chosen to place the structural responsibility for the fuel element on the cladding. The principal concern about the effects of high temperature on the oxide is that the exposed fuel undergo no gross sudden displacement during reactor operation. The reactor design assumes complete release of fission gases from the oxide and the fuel cladding in a pressure vessel strong enough to contain it. In the light of results obtained in experiments on UO_2, the assumption of 100 per cent release is a very conservative approach.

Finally, the proof of such a concept is, as in all reactors, the performance of the fuel. I should like to describe some of the experiments which give us knowledge of how this fuel performs under the thermal and irradiation environment of a reactor. Also mentioned are the possible clouds—the problems which need more work.

THERMAL EFFECTS

Before irradiation testing, several heating experiments were performed to get a feeling for what might occur upon irradiation. Heat calculations by Fritz and

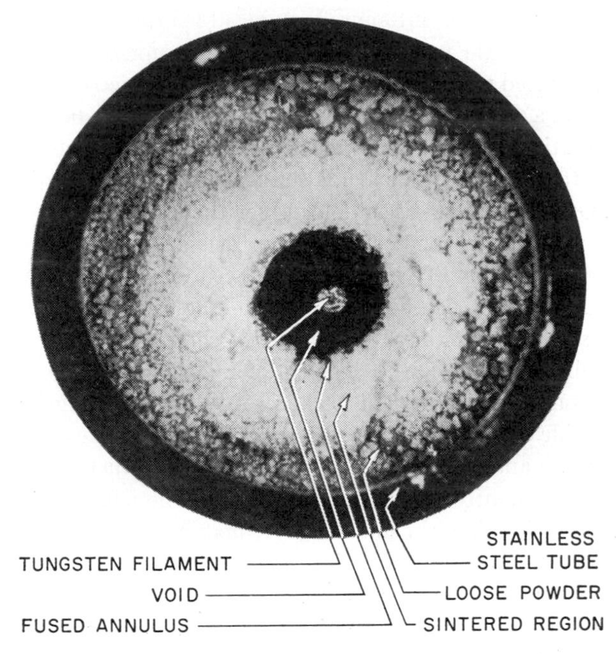

FIG. 1.—Cross-section of a model fuel element after thermal testing

others at KAPL showed that internal temperatures of the oxide might exceed its melting point, and at the Chalk River NRX reactor early in 1951 one test of a 100-mil homogeneous UO_2 fuel pin did show void formation.

The performance of oxide under a temperature gradient was investigated in the laboratory by heating an aluminum oxide powder by means of a central tungsten filament. Figure 1 is a photograph of the cross-section of this laboratory-model fuel element. Note that the periphery has remained a loose powder; there are a sintered region, a fused glassy region, and a central void. The void is formed by radial migration of the oxide. In such a fuel element the curve of temperature versus r/r_0 (distance from the central axis of the fuel element divided by its radius) will be a logarithmic, rather than a hyperbolic, function, but the desired steep radial temperature gradient is achieved. Figure 2 compares the thermal curve of this experiment with that expected in a fuel.

Similar behavior is to be expected from a fuel powder in which the central operating temperature exceeds the sintering temperature of the oxide. Under the thermal gradient existing during reactor operation, the oxide would be expected to attain the sintering temperature during first reactor startup. This would cause radial migration of the fuel, formation of a central void, and a marked increase in

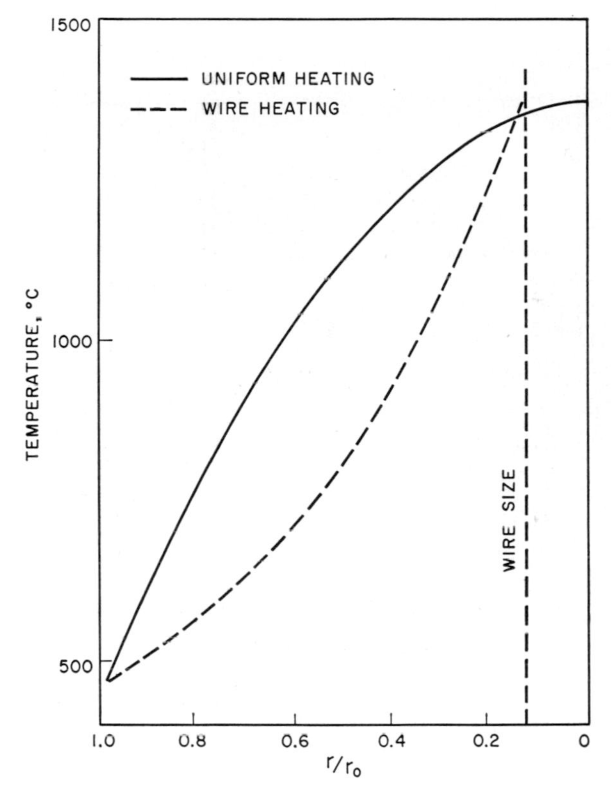

FIG. 2.—Comparison of the temperature gradients in an actual fuel element (uniform heating) and in the model fuel element of Fig. 1 (wire heating).

thermal conductivity. This increase in thermal conductivity would lower the central temperature, and no further sintering would be expected. The additional effect of PuO_2 is to increase the sinterability of the oxide, and this effect is most evident within the concentration range that is of interest.

IRRADIATION TESTS (7)

Samples of $1PuO_2$ to $5UO_2$ mixed oxide were prepared, dry-pressed into 0.1-inch-high pills of 0.185 inch diameter, and loaded into a stainless 0.25-inch O.D. tube having an I.D. of 0.1875 inch. The final capsule contained 0.5 inch of MgO as a thermal barrier at the bottom, 1.0 inch of fuel pills containing 0.42 gm of fuel at the center, and another 0.5 inch of MgO at the top. Finally, a cap was welded on.

The capsules were irradiated in a flux near 0.7×10^{14} thermal. The heat output was 470 watts/cm. It may be noted from the earlier figure that if the heat generation exceeded 420 watts/cm without a change in thermal conductivity, melting could occur.

The first capsule (KAPL-26-1) in the series was irradiated for one MTR cycle to an estimated 5 per cent depletion of the original plutonium. The coolant temperature was near 650°, with a maximum fluctuation to 725° C. After 30 days, the element was examined in the KAPL Radioactive Materials Laboratory. In this and all other tests no deformation of the steel tube occurred, nor was there any expected. The specimen was sectioned and photographed. Such a photograph is shown in Figure 3.

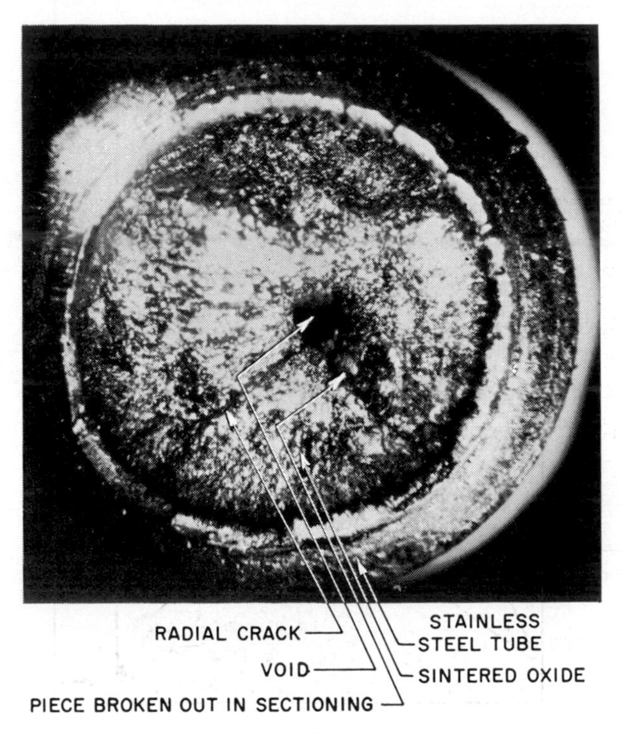

FIG. 3.—Cross-section of irradiated fuel capsule KAPL 26-1

Note that a central void was formed. The oxide is somewhat sintered. There is no evidence for melting. Similar experience was obtained with mixed oxide irradiated to 35 per cent depletion. In both cases the irradiated oxide was readily soluble in nitric acid. In no experiment was there any distortion of the fuel-element cladding; however, this is not at all surprising, since the tubing had a heavier wall (30 mil) than that to be used in the final reactor design. The experiments were not designed as cladding tests.

The occurrence of a central void in this and other irradiations had led to concern as to whether the fuel would spall off and settle in the void. Two experiments were done to provide an answer to this question.

The first of these (KAPL-26-3) was designed to eliminate the void. In it, a lead thermal bond was placed between core and cladding. In detail, a $\frac{3}{8}$-inch I.D. tube (30-mil wall) was filled with lead and a 0.1875-inch hold drilled therein, leaving a lead annulus. The pellets of oxide were then loaded as before. In the MTR the lead melted, giving a good thermal bond from oxide to stainless steel. The sample was irradiated for one MTR cycle, and the element subsequently sectioned.

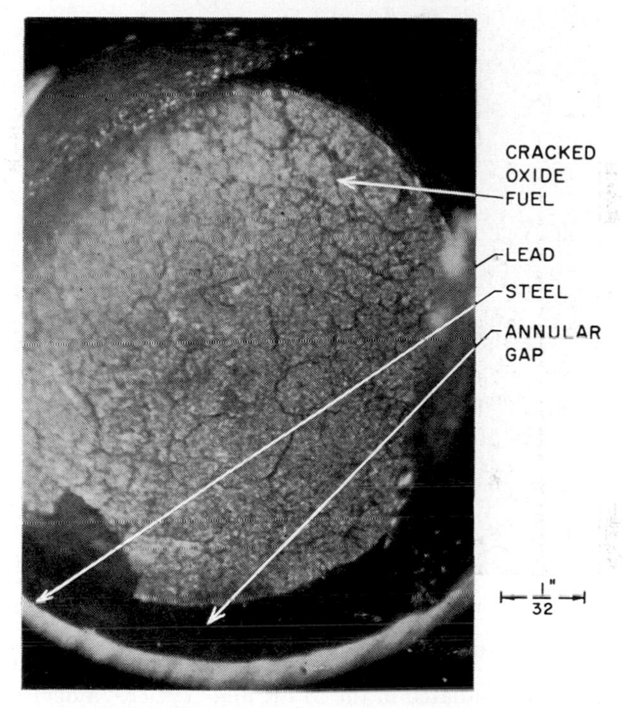

FIG. 4.—Cross-section near the top of irradiated fuel capsule KAPL 26-3

It was apparent that the fuel had floated on the lead because there were void spaces near the top. Figure 4 shows a section at the top of the fuel. The cracking is not typical of the sintered irradiated fuel observed previously. The oxide was quite friable, as checked with a probe, which indicated that little, if any, sintering occurred. A section through the mid-plane is shown in Figure 5. At this point, the maximum temperature was attained, and there was still no evidence of sintering.

Another liquid metal such as sodium should work as well, since it, too, does not react with the oxide. Though the addition of liquid metal does prevent central-void formation, at the present time, because it is not certain that such coring must be avoided, the addition of lead or sodium is an additional operation and an additional expense to be avoided unless its need is clearly shown.

The second experiment suggested by the central-void problem was specifically designed to determine what, if any, settling does occur in a representative type fuel under conditions at least as severe as those to be expected in the Fast Oxide Breeder design. For this an element (KAPL-39-1) containing 20 per cent enriched UO_2 at 70 per cent of theoretical density was assembled. As before, the element was clad in stainless steel, with MgO end plugs. The oxide was prepared by extrusion of oxide with binder at 40 per cent of theoretical density, followed by drying and sintering

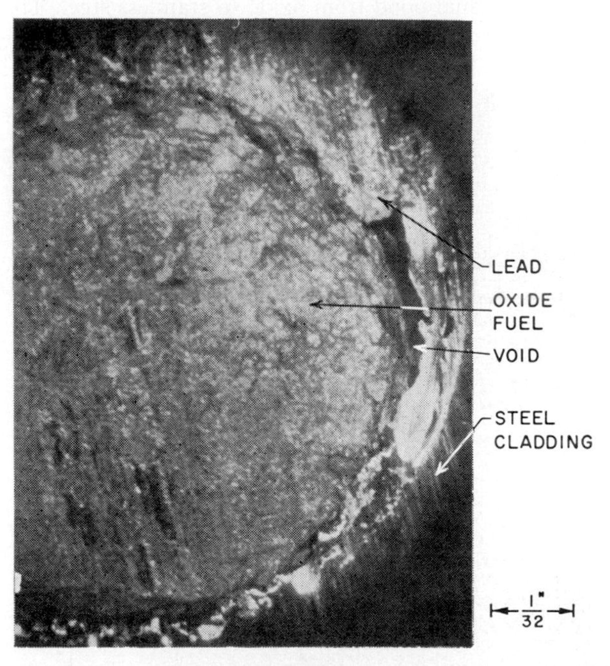

FIG. 5.—Cross-section through the center of irradiated fuel capsule KAPL 26-3

at 1,600° C in hydrogen. After assembly, the balance point of the element was determined. It was then irradiated in the MTR in a "cycler," a device which moves the element in and out of a reflector position (flux = 10^{14}). Depletion of U^{235} in this element was less than 5 percent, but heat fluxes were near 400 watts/cm^2, the power density being 350 watts/cm^3. The element was irradiated for 3 weeks in the MTR and cycled 50 times. The balance point for the element was unchanged by the power cycling. Figure 6 is a photograph of three sections taken along the element. A central void indeed formed, but no settling occurred.

KINETIC EXPERIMENTS (8)

In discussions of fast reactors the question frequently arises as to what sort of contribution to a negative temperature coefficient one may expect during a reactor transient because of expansion of the fuel. With solid metallic fuel, the evidence is strong that one will get linear expansion of the fuel and a corresponding reactivity decrease when a 200° C/sec temperature rise occurs. Uranium oxide rods, both with and without a central void, were heated at such a rate, and in those cases studied, the expansion, which occurred without time lag, was just that to be expected from the equilibrium coefficient of thermal expansion. Such "cold" experiments do not, of course, answer the question as to what will occur with highly irradiated and cracked rods.

We have concluded from our work that the mixed-oxide concept is indeed feasible. No step in the process appears now to be difficult to achieve. The major effort in the fuel study should be irradiation, settling, and transient experiments on full-length (1 meter) fuel assemblies.

Regarding costs for the process, which have not been discussed at all, rough but conservative estimates give a fuel-cycle cost, with plutonium credit at \$12/gm, of

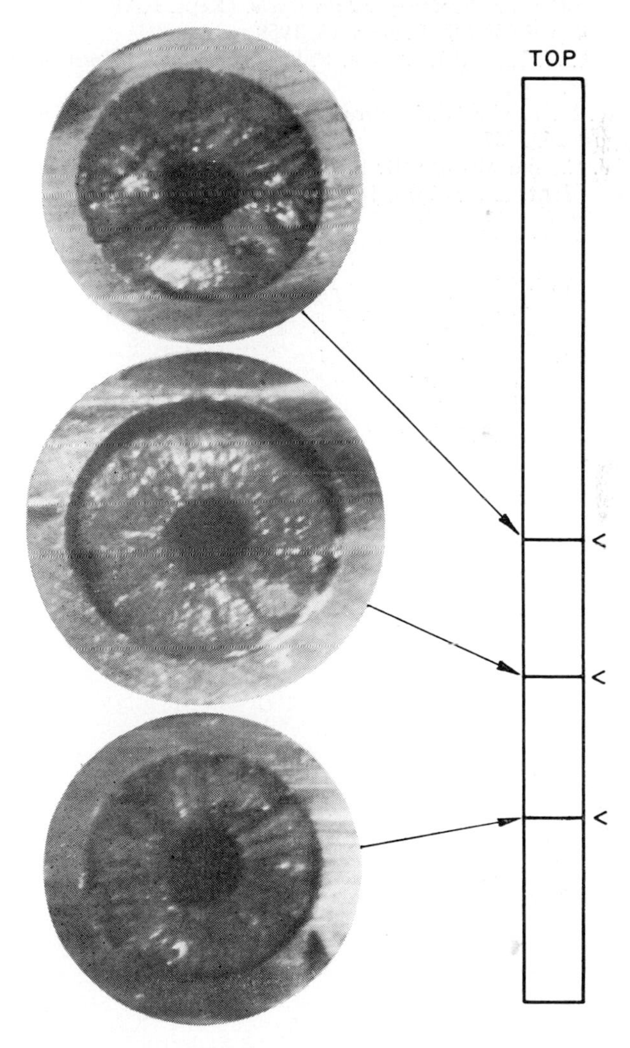

FIG. 6.—Three cross-sections of irradiated fuel capsule KAPL 39-1

1.5 mils/kwh. This cost is based on a $3,000,000 core-processing plant at the reactor and includes the cost of reprocessing the blanket fuel at a low-enrichment reprocessing plant (4).

REFERENCES

1. LUEBKE, E. A., and SAMPSON, J. B. *Plutonium Power Reactor with Oxide Fuel Elements.* (Rept. TID-2017, April, 1955.)
2. MOLINO, D. F., and ASTRA. *The Fast Oxide Breeder—Physics and Reactor Calculations.* (Rept. KAPL-1756, Parts I and II, June 30, 1957.)
3. DAVIDSON, J. K. *The Fast Oxide Breeder—a Summary.* (Rept. KAPL-1701, July 1, 1957.)
4. DAVIDSON, J. K., HAAS, W. O., JR., MEWHERTER, J. L., MILLER, R. S., and SMITH, D. J. *The Fast Oxide Breeder—the Fuel Cycle.* (Rept. KAPL-1757, July 1, 1957.)
5. U.S. Patent No. 2,868,707, January 13, 1959.
6. MULFORD, R. N. R., and ELLINGER, F. H. *Journal of the American Chemical Society,* **80**:2023, 1958.
7. CASHIN, W. M. *The Fast Oxide Breeder—Fuel Irradiation Experiments.* (Rept. KAPL-1784, August 26, 1957.)
8. COHEN, W. C., and MOLINO, D. F. *Behavior of UO_2 Bodies Subjected to Temperature Transients.* (Rept. KAPL-1873, July 1, 1957.)

INDEX

INDEX

PRINTED IN U.S.A.